The Water Waves Problem

Mathematical Analysis and Asymptotics

Mathematical
Surveys
and
Monographs

Volume 188

The Water Waves Problem

Mathematical Analysis and Asymptotics

David Lannes

American Mathematical Society
Providence, Rhode Island

2010 *Mathematics Subject Classification.* Primary 76B15, 35Q53, 35Q55, 35J05, 35J25.

For additional information and updates on this book, visit
www.ams.org/bookpages/surv-188

Library of Congress Cataloging-in-Publication Data

Lannes, David, 1973–
The water waves problem : mathematical analysis and asymptotics / David Lannes.
pages cm. – (Mathematical surveys and monographs ; volume 188)
Includes bibliographical references and index.
ISBN 978-0-8218-9470-5 (alk. paper)
1.Water waves–Mathematical models. 2. Hydrodynamics–Mathematical models. I. Title.
TC172.L36 2013
551.46′301515–dc23

2012046540

∞ The paper used in this book is acid-free and falls within the guidelines
established to ensure permanence and durability.
Visit the AMS home page at http://www.ams.org/

10 9 8 7 6 5 4 3 2 1 18 17 16 15 14 13

Contents

Preface

There is a vast literature devoted to the study of water waves ranging from coastal engineering preoccupations to a very theoretical mathematical analysis of the equations. Since all the scientific communities involved have their own approach and terminology, it is sometimes quite a challenge to check that they actually speak about the same things. The rationale of this book is to propose a simple and robust framework allowing one to address some important issues raised by the water waves equations. Of course, such a global approach is sometimes not compatible with sharpness; it has been a deliberate choice to sacrifice the latter when a choice was necessary (on minimal regularity assumptions, for instance). Hopefully, experts on the well-posedness of the water waves equations or on the mathematical properties of some asymptotic system, or on any other aspect, may, however, discover here other related subjects of interest and some open problems.

Since this book is addressed to various audiences, there has been an effort to make each chapter as thematically focused as possible. For instance, physical aspects are not often present in the chapters devoted to the well-posedness of the equations, and, vice versa, the chapter devoted to the derivation of the asymptotic models is rather oriented by physical considerations and requires only basic mathematical tools. More precisely,

- Chapter 1 is a general and basic introduction. It also includes a review of various approaches developed in recent years for the mathematical analysis of the water waves equations. Some extensions raising several new problems are also presented (such as moving bottoms and rough topography), and the physical assumptions made to derive the water waves equations are discussed.
- Chapters 2, 3, and 4 are devoted to the well-posedness of the water waves equations. They are addressed to mathematicians looking for a basic introduction to the Cauchy problem for water waves, as well as to researchers more familiar with these equations but not necessarily with their behavior in shallow water. Some aspects of the proof (e.g., the study of the Laplace equation, properties of the Dirichlet-Neumann operator) can also be of interest by themselves and we therefore gave sharper results than those actually needed to prove the well-posedness of the water waves equations.
- In Chapter 5, we derive many shallow water asymptotic models used in coastal oceanography. This chapter is addressed to oceanographers and people working on such models and those who want to know precisely what their range of validity is. This chapter uses only basic mathematical tools and should be readable independently of Chapters 2, 3, and 4. Note that the derivation of various models presented here (the so-called models with

improved frequency dispersion, for instance) is dictated by experimental considerations rather than mathematical ones (they are just likely to improve a multiplicative constant in some error estimates). Some numerical computations compared with experimental data show their relevance.

- Chapter 6 is a more mathematically oriented complement to Chapter 5. It gathers all the more technical steps necessary to provide a full justification of the models derived in Chapter 5 (i.e., error estimates between the approximation they furnish and the exact solution of the water waves equations constructed in Chapter 4). Some mathematical properties of these systems are also given here.
- Chapter 7 deals with approximations of water waves by *scalar* models. Many of these equations (KdV/BBM, KP, Camassa-Holm) have played a central role in mathematical physics in various contexts; they are derived and justified here for water waves.
- In Chapter 8 we address the deep water approximations and modulation equations. The latter are used to model the propagation of wave packets. These modulation equations appear in many physical situations and therefore play an important role in mathematical physics, for instance, the nonlinear Schrödinger, Davey-Stewartson, or Benney-Roskes equations. The full justification of these models still raises many open problems. We also derive several new variants of these equations, such as "full dispersion" versions that are expected to behave better than the standard models in some situations where the latter furnish a poor approximation.
- Chapter 9 analyzes the influence of surface tension. Since this physical effect is generally not relevant for applications to coastal oceanography, we did not include it in the previous chapters. However, capillary gravity waves raise many mathematical problems of interest and the results gathered here (e.g., well-posedness theory for capillary gravity waves, influence of the capillary effects on asymptotic models, etc.) may prove useful to those working in this area.
- In Appendix A, we provide additional results on the Dirichlet-Neumann operator (for instance, its analyticity properties, self adjointness, and invertibility). These results are not needed for our main purposes but are often needed in related topics. We tried to provide new proofs and/or improvements of the versions of these results that can be found in the literature.
- In Appendix B, we gather the product and commutator estimates used throughout this book so that the uninterested reader can use them as a black box.
- Finally, Appendix C presents in an extremely condensed form the models derived in Chapters 5, 7, and 9. It is aimed at helping the hurried reader to locate any asymptotic model among the dozens that are used to describe waves in several physical regimes (shallow water, modulation equations, etc.).

The material in this book can essentially be found in several papers by the author, such as [**12, 13, 223, 36, 107, 96, 225**]; therefore, this book owes a lot to the coauthors of these papers. However, most of the proofs of the results corresponding to these references have been changed; they have been simplified and/or improved using recent advances by different authors, in particular T. Iguchi [**181**], F. Rousset and N. Tzvetkov [**277**], and T. Alazard, N. Burq, and C. Zuily [**4, 6**].

New material has also been added on several theoretical aspects (in particular on the Dirichlet-Neumann operator); though these new results are not needed for the applications presented in this book, they are of general interest, and we hope that they will be useful to people working on this subject. We also derived several new models that are expected to improve existing versions. Most of them remain to be analyzed mathematically, and the quality of the approximation they furnish is yet to be investigated theoretically, numerically, and experimentally.

Many of the improvements and most of the new material gathered in this book were motivated by conversations with T. Alazard, J. Bona, P. Bonneton, A. Constantin, J.-C. Saut, and E. Wahlen, to whom I address my warmest thanks. I also want to express my gratitude to A. Castro, F. Chazel, V. Duchêne, and M. Tissier, who helped me with some illustrations and numerical simulations.

David Lannes

Index of notations

General notations

- d refers to the surface dimension (and thus, $d = 1$ or 2).
- $|\alpha| = \sum_{j=1}^{d} \alpha_j$ for all multi-index $\alpha \in \mathbb{N}^d$.
- Cst is a generic notation for a constant whose exact value is of no importance.
- $A \lesssim B$ means $A \leq \mathrm{Cst}\, B$.
- $\lceil s \rceil$ ($s \in \mathbb{R}$) denotes the smallest integer larger or equal to s.
- $A + \langle B \rangle_{s>r}$ is equal to A if $s \leq r$ and equal to $A + B$ if $s > r$.
- $A \vee B$ stands for $\max\{A, B\}$.
- c.c. means "complex conjugate".
- Ω denotes the fluid domain.
- $\mathcal{S}$ is the flat strip $\mathcal{S} = \{(X, z) \in \mathbb{R}^d \times \mathbb{R}, -1 < z < 0\}$.
- Γ is the boundary of Ω. We sometimes write $\Gamma = (\zeta, b)$, where ζ and b are the surface and bottom parametrizations.
- $\mathbf{\Gamma} \subset H^{t_0+2}(\mathbb{R}^d)^2$ is the set of all ζ and b such that $\inf_{\mathbb{R}^d}(1 + \varepsilon\zeta - \beta b) > 0$.
- $\psi^\dagger$ is an explicit smoothing extension of ψ on $\mathcal{S}$ (see Notation 2.28).
- $\psi^\mathfrak{h}$ is the "harmonic extension" of ψ (see Notation 2.33).

Matrices and vectors

- $\mathbf{e}_z$ denotes the unit (upward) vector in the vertical direction.
- $\mathbf{n}$ always denotes the unit upward normal vector evaluated at the boundary of the fluid domain.
- I always stands for the identity matrix (its size is usually obvious from the context).
- $I_{\mu,\gamma}$ is the $(d+1) \times (d+1)$ diagonal matrix with diagonal $(\sqrt{\mu}, \gamma\sqrt{\mu}, 1)$ when $d = 2$ and $(\sqrt{\mu}, 1)$ when $d = 1$.
- $P_{\mu,\gamma}(\Sigma)$: see Equation (2.13).
- $|\mathbf{v}|$, where $\mathbf{v}$ is a vector of $\mathbb{R}^{d+1}$ stands for the euclidean norm of $\mathbf{v}$.
- $|M|$, where M is a $(d+1) \times (d+1)$ matrix is the matricial norm subordinated to the euclidean norm.

Variables and standard operators

- $X \in \mathbb{R}^d$ always refers to the horizontal variables; we sometimes write $X = (x, y)$ or $X = (x_1, x_2)$ when $d = 2$ and $X = x$ or $X = x_1$ when $d = 1$.
- z always refers to the vertical variable (also sometimes denoted x_{d+1})
- $\nabla = (\partial_x, \partial_y)^T$ when $d = 2$ and $\nabla = \partial_x$ when $d = 1$.
- $\Delta = \partial_x^2 + \partial_y^2$ when $d = 2$ and $\Delta = \partial_x^2$ when $d = 1$.

- $\nabla_{X,z}$ denotes the $(d+1)$-dimensional gradient with respect to the variables X and z.
- We sometimes write ∂_1 and ∂_2 rather than ∂_x and ∂_y.
- div denotes the $(d+1)$-dimensional divergence operator with respect to X and z.
- curl denotes the $(d+1)$-dimensional rotational operator with respect to X and z.
- ∂^α when $\alpha \in \mathbb{N}^d$: $\partial^\alpha = \partial_x^{\alpha_1}\partial_y^{\alpha_2}$ if $d=2$ and $\partial^\alpha = \partial_x^\alpha$ if $d=1$.
- ∂^α when $\alpha \in \mathbb{N}^{d+1}$: $\partial^\alpha = \partial_x^{\alpha_1}\partial_y^{\alpha_2}\partial_t^{\alpha_3}$ if $d=2$, a straightforward adaptation if $d=1$.
- $\partial_{\mathbf{n}} u$ denotes the upward *conormal* derivative of u (see Notation 1.12).
- $\mathcal{F}\cdot$ or $\widehat{\cdot}$ stands for the Fourier transform.
- $\mathcal{F}^{-1}\cdot$ or $\check{\cdot}$ denote the inverse Fourier transform.
- $D = \frac{1}{i}\nabla$ and $f(D)$ stand for the Fourier multiplier.

$$\forall \xi \in \mathbb{R}^d, \qquad \widehat{f(D)u}(\xi) = f(\xi)\widehat{u}(\xi).$$

- $\Lambda = (1-\Delta)^{1/2} = (1+|D|^2)^{1/2}$ is the fractional derivative.
- ∂_x^{-1}: the Fourier-multiplier of symbol $-i/\xi_1$.

Parameter depending quantities

- μ, ε, γ, β, ν, ϵ: see §1.3.1.
- α is the roughness coefficient: see §1.7.2.
- $\nabla^\gamma = (\partial_x, \gamma\partial_y)^T$ if $d=2$ and $\nabla^\gamma = \partial_x$ if $d=1$.
- $\mathrm{div}_\gamma = (\nabla^\gamma)^T$.
- $\Delta^\gamma = \mathrm{div}_\gamma \nabla^\gamma = \partial_x^2 + \gamma^2\partial_y^2$ $(d=2)$ and $\Delta^\gamma = \partial_x^2$ $(d=1)$.
- $D^\gamma = \frac{1}{i}\nabla^\gamma$.
- $\xi^\gamma = (\xi_1, \gamma\xi_2)^T$ if $d=2$ and $\xi^\gamma = \xi$ if $d=1$.
- $I_{\mu,\gamma}$: see (2.1).
- $\nabla^{\mu,\gamma}$: see (2.2).
- $\Delta^{\mu,\gamma}$: see (2.3).
- $\mathfrak{P} = \dfrac{|D^\gamma|}{(1+\sqrt{\mu}|D^\gamma|)^{1/2}}$.
- $\mathcal{G} = \mathcal{G}_{\mu,\gamma}[\varepsilon\zeta, \beta\gamma]$: nondimensionalized Dirichlet-Neumann operator: see Chapter 3.
- $\mathcal{w}[\varepsilon\zeta,\beta b]$ and $\mathcal{V}[\varepsilon\zeta,\beta b]$: see (4.12) and (4.13).
- $\underline{w} = \mathcal{w}[\varepsilon\zeta,\beta b]\psi$ and $\underline{V} = \mathcal{V}[\varepsilon\zeta,\beta b]\psi$ are, respectively, the vertical and horizontal components of the velocity evaluated at the surface.
- $\zeta_{(\alpha)}$, $\psi_{(\alpha)}$: see (4.9).
- M_0, M, $M(s)$: see (2.10).
- $\mathcal{E}^N(U)$: see (4.8).
- $\mathfrak{m}^N(U)$: see (4.11).
- $M(s)$ and $N(s)$: see (4.5) and (5.6).
- $\mathcal{T}$ and $\mathcal{T}_b$: see (3.32) or (5.12) and (5.24).
- $\mathcal{Q}$, $\mathcal{Q}_b$ and $\mathcal{Q}_1$: see (5.13), (5.14), and (5.16).

Functional spaces

Functional spaces on $\mathbb{R}^d$.

- $\mathcal{D}(\mathbb{R}^d)$ denotes the space of C^∞ functions on $\mathbb{R}^d$ with compact support.
- $\mathcal{D}'(\mathbb{R}^d)$ is the space of distributions on $\mathbb{R}^d$.
- $\mathfrak{S}(\mathbb{R}^d)$ denotes the Schwartz space of rapidly decaying smooth functions on $\mathbb{R}^d$.
- $\mathfrak{S}'(\mathbb{R}^d)$ stands for the space of all tempered distribution on $\mathbb{R}^d$.
- $C(\mathbb{R}^d)$ denotes the space of real valued continuous functions on $\mathbb{R}^d$.
- $C^k(\mathbb{R}^d)$ denotes the space of real valued continuous functions on $\mathbb{R}^d$ with continuous derivatives up to order k.
- $C_b^k(\mathbb{R}^d)$ denotes the subspace of $C^k(\mathbb{R}^d)$ of functions that are bounded together with all their derivatives of order less or equal to k.
- $L^p(\mathbb{R}^d)$ $(1 \leq p \leq \infty)$ denotes the standard Lebesgue space on $\mathbb{R}^d$ with associated norm $|f|_p = (\int_{\mathbb{R}^d} |f|^p)^{1/p}$ when $p < \infty$ and $|f|_\infty = \text{supess}_{\mathbb{R}^d}|f|$.
- $W^{k,\infty}(\mathbb{R}^d)$ is defined for all $k \in \mathbb{N}$ as
$$W^{k,\infty}(\mathbb{R}^d) = \{f \in L^\infty, |f|_{W^{k,\infty}} < \infty\},$$
where $|f|_{W^{k,\infty}} = \sum_{\alpha\in\mathbb{N}^d, |\alpha|\leq k} |\partial^\alpha f|_\infty$.
- $C_*^r(\mathbb{R}^d)$ is the Zygmund space of order r as defined in (B.8).
- $H^s(\mathbb{R}^d)$ $(s \in \mathbb{R})$ denotes the usual Sobolev space
$$H^s(\mathbb{R}^d) = \{f \in \mathfrak{S}', |f|_{H^s} < \infty\} \quad \text{where} \quad |f|_{H^s} = |\Lambda^s f|_2.$$
- $H^s_{ul}(\mathbb{R}^d)$ $(s \in \mathbb{R})$ denotes the uniformly local Sobolev space: see Definition B.20.
- $\dot{H}^s(\mathbb{R}^d)$: Beppo-Levi spaces on $\mathbb{R}^d$: see Definition 2.2.
- $\overset{\circ}{H}{}^{s+1/2}(\mathbb{R}^d) = \{f \in \mathcal{S}'(\mathbb{R}^d), \quad |D|^{1/2}\psi \in H^s(\mathbb{R}^d)\}$.
- $H^s_{(2k)}(\mathbb{R})$: the set of $f \in H^{s+2k}(\mathbb{R})$ such that for all $j = 0, \dots, n$, $x^j f \in H^{s+2(k-j)}$: see (7.23).
- $\partial_x H^s(\mathbb{R}^2)$ (resp. $\partial_x^2 H^s(\mathbb{R}^2)$): space of $H^{s-1}(\mathbb{R}^2)$ (resp. $H^{s-2}(\mathbb{R}^2)$) functions that are the first (resp. second) x-derivative of an $H^s(\mathbb{R}^2)$ function: see Notation 7.15.
- E_T^N: see (4.10).
- $X^s = \{V \in H^s(\mathbb{R}^d)^d, |V|_{X^s} < \infty\}$ with $|V|_{X^s}^2 = |V|_{H^s}^2 + \mu|\nabla^\gamma \cdot V|_{H^s}^2$.
- $Y^s = H^s(\mathbb{R}^d) \times X^s$.

Functional spaces on a domain $\Omega \subset \mathbb{R}^{d+1}$.

- $\mathcal{D}(\Omega)$ $(\Omega \subset \mathbb{R}^{d+1})$ is the space of all smooth functions with compact support in Ω.
- $\mathcal{D}'(\Omega)$ is the space of distributions on Ω.
- $C(\Omega)$ (resp. $C(\overline{\Omega})$) denotes the space of real valued continuous functions on Ω (resp. on its adherence $\overline{\Omega}$).
- $L^p(\Omega)$ $(1 \leq p \leq \infty$ and $\Omega = \Omega$ or $\mathcal{S})$ are the standard Lebesgue spaces on Ω, with norm $\|u\|_p = (\int_\Omega |u|^p)^{1/p}$ when $p < \infty$ and $\|u\|_\infty = \text{supess}_\Omega|u|$. The notation for the norm does not make the dependence on Ω explicit, but this is always obvious from the context.

- $H^k(\Omega)$ ($k \in \mathbb{N}$) denotes the Sobolev space given by

$$H^k(\Omega) = \{u \in L^2(\Omega), \|u\|_{H^k} < \infty\}, \qquad \text{with} \qquad \|u\|_{H^k} = \sum_{\alpha \in \mathbb{N}^d |\alpha| \leq k} \|\partial^\alpha u\|_2.$$

- $H^1_{0,surf}(\Omega)$: see Definition 2.6.
- $L^\infty H^s$: see Definition 2.11.
- $H^{s,k}$: see Definition 2.11.
- $\dot{H}^{k+1}(\Omega)$: homogeneous Sobolev space on Ω: see Definition 2.2.

CHAPTER 1

The Water Waves Problem and Its Asymptotic Regimes

We derive here various equivalent mathematical formulations of the water waves problem (and some extensions to the two-fluids problem). We then propose a dimensionless version of these equations that is well adapted to the qualitative description of the solutions. The way we nondimensionalize the water waves equations relies on a rough analysis of their linearization around the rest state and shows the relevance of various dimensionless parameters, namely, the *amplitude parameter* ε, the *shallowness parameter* μ, the *topography parameter* β, and the *transversality parameter* γ. The linear analysis of the equations is also used to introduce the concept of wave packets and modulation equations.

With the relevant physical dimensionless parameters introduced, we then identify *asymptotic regimes* (the shallow water regime for instance) as conditions on these dimensionless parameters (e.g., $\mu \ll 1$ for the shallow water regime). Finally, we present two natural extensions of the problems addressed in this book: the case of moving bottoms and of rough topographies. A discussion of the main physical assumptions (e.g., homogeneity, inviscidity, incompressibility, etc.) and some comments on possible extensions (such as taking into account Coriolis effects, or a nonconstant external pressure) are then briefly addressed in the last section.

As everywhere throughout this book, $d = 1, 2$ denotes the spacial dimension of the surface of the fluid. The spatial variable is written $X \in \mathbb{R}^d$ and the vertical variable is denoted by z. We also write $X = (x, y)$ when $d = 2$ and $X = x$ when $d = 1$.

1.1. Mathematical formulation

1.1.1. Basic assumptions. The water waves problem consists in describing the motion, under the influence of gravity, of a fluid occupying a domain delimited below by a fixed bottom and above by a free surface that separates it from vacuum (that is, from a fluid whose density is considered negligible, such as for the air-water interface). The following assumptions are made on the fluid and the flow:

(H1) The fluid is homogeneous and inviscid.
(H2) The fluid is incompressible.
(H3) The flow is irrotational.
(H4) The surface and the bottom can be parametrized as graphs above the still water level.
(H5) The fluid particles do not cross the bottom.
(H6) The fluid particles do not cross the surface.

(H7) There is no surface tension[1] and the external pressure is constant.
(H8) The fluid is at rest at infinity.
(H9) The water depth is always bounded from below by a nonnegative constant.

These assumptions are discussed with more detail in §1.8.1 below. Assumptions (H1) and (H2) imply that the fluid motion is governed by the incompressible Euler equation inside the fluid domain.

The irrotationality assumption (H3) is useful but not necessary (see §1.8.1) but is commonly made in coastal oceanography since for most applications rotational effects are negligible up to the breaking point of the waves.

The assumption (H4) excludes overhanging waves; this is of course not a restriction for our present purpose of describing asymptotic dynamics of interest in coastal oceanography. Describing the free surface with a parametrized hypersurface (see §1.8), it is, however, possible to take overhanging waves into account.

Assumptions (H5) and (H6) provide boundary conditions to the Euler equations: (H5) implies that the normal component of the velocity must vanish at the bottom while (H6) provides a (nonlinear) kinematic boundary condition at the surface.

Neglecting the surface tension as in (H7) is completely reasonable in coastal oceanography since the typical scale for which surface tension occurs is 1.6 cm (see Chapter 9 for more details). Dealing with a nonconstant external pressure does not raise any particular difficulty (see §1.8.1).

Assumption (H8) is quite natural as long as one considers infinite domains that satisfy (H9). The latter condition excludes beaches (seen as vanishing shorelines); this is obviously a serious restriction for applications to coastal oceanography, but removing this assumption remains to this day an open mathematical problem.

1.1.2. The free surface Euler equations. The free surface Euler equations are just a mathematical restatement of assumptions (H1)–(H9). Let us first introduce the following notation:

- The domain occupied by the fluid at time t is denoted $\Omega_t \subset \mathbb{R}^{d+1}$.
- The velocity of the fluid particle located at $(X, z) \in \Omega_t$ at time t is written $\mathbf{U}(t, X, z) \in \mathbb{R}^{d+1}$. Its horizontal and vertical components are denoted by $V(t, X, z) \in \mathbb{R}^d$ and $w(t, X, z) \in \mathbb{R}$, respectively.
- We write $P(t, X, z) \in \mathbb{R}$ for the pressure at time t at the point $(X, z) \in \Omega_t$.
- The (constant) acceleration of gravity is denoted by $-g\mathbf{e}_z$, where $g > 0$ and $\mathbf{e}_z$ is the unit (upward) vector in the vertical direction.
- The (constant) density of the fluid is written ρ.

The motion of a homogeneous, inviscid, incompressible, and irrotational fluid (assumptions (H1)–(H3)) is governed by the Euler equation, with constraints on the divergence and curl of the velocity field:

$$\begin{array}{lll} (\mathrm{H1})' & \partial_t \mathbf{U} + (\mathbf{U} \cdot \nabla_{X,z})\mathbf{U} = -\frac{1}{\rho}\nabla_{X,z} P - g\mathbf{e_z} & \text{in} \quad \Omega_t, \\ (\mathrm{H2})' & \operatorname{div} \mathbf{U} = 0 & \text{in} \quad \Omega_t, \\ (\mathrm{H3})' & \operatorname{curl} \mathbf{U} = 0 & \text{in} \quad \Omega_t. \end{array}$$

There also exist two functions $b : \mathbb{R}^d \to \mathbb{R}$ and $\zeta : [0, T) \times \mathbb{R}^d \to \mathbb{R}$ ($T > 0$) such that

$$(\mathrm{H4})' \qquad \forall t \in [0, T), \qquad \Omega_t = \{(X, z) \in \mathbb{R}^{d+1}, -H_0 + b(X) < z < \zeta(t, X)\},$$

[1] We remove this assumption in Chapter 9.

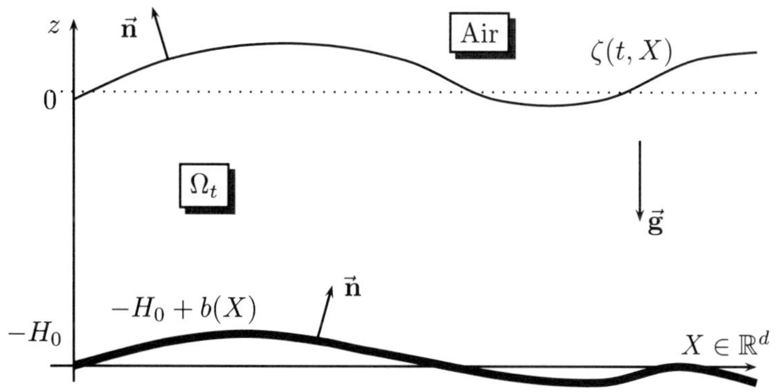

FIGURE 1.1. Main notation.

where $H_0 > 0$ is a constant reference depth introduced for later convenience; note that $z = 0$ corresponds to the still water level.

Denoting by $\mathbf{n}$ the unit normal vector to the fluid domain pointing upwards, we can reformulate[2] (H5) and (H6) as

$$\begin{array}{lll} (\mathrm{H5})' & \mathbf{U}\cdot\mathbf{n} = 0 & \text{on} \quad \{z = -H_0 + b(X)\}, \\ (\mathrm{H6})' & \partial_t\zeta - \sqrt{1+|\nabla\zeta|^2}\,\mathbf{U}\cdot\mathbf{n} = 0 & \text{on} \quad \{z = \zeta(t,X)\}. \end{array}$$

Denoting by P_{atm} the (constant) atmospheric pressure, assumption (H7) can be restated as

$$(\mathrm{H7})' \qquad P = P_{atm} \quad \text{on} \quad \{z = \zeta(t,X)\},$$

while (H8) and the nonvanishing shoreline assumption (H9) can be written, respectively, as

$$(\mathrm{H8})' \qquad \forall t \in [0,T), \qquad \lim_{(X,z)\in\Omega_t, |(X,z)|\to\infty} |\zeta(t,X)| + |\mathbf{U}(t,X,z)| = 0$$

and

$$(\mathrm{H9})' \qquad \exists H_{min} > 0, \quad \forall (t,X) \in [0,T)\times\mathbb{R}^d, \qquad H_0 + \zeta(t,X) - b(X) \geq H_{min}.$$

Equations (H1)′–(H8)′ are called *free surface Euler equations.*

1.1.3. The free surface Bernoulli equations. The free surface Bernoulli equations are another formulation of the free surface Euler equations based on the representation of the velocity field in terms of avelocity potential. More precisely, there exists a mapping $\Phi : [0,T)\times\mathbb{R}^{d+1} \to \mathbb{R}$ such that

$$\begin{array}{lll} (\mathrm{H3})'' & \mathbf{U} = \nabla_{X,z}\Phi & \text{in} \quad \Omega_t, \\ (\mathrm{H2})'' & \Delta_{X,z}\Phi = 0 & \text{in} \quad \Omega_t, \\ (\mathrm{H1})'' & \partial_t\Phi + \frac{1}{2}|\nabla_{X,z}\Phi|^2 + gz = -\frac{1}{\rho}(P - P_{atm}) & \text{in} \quad \Omega_t. \end{array}$$

[2]Let Γ_t be a hypersurface given implicitly by an equation $\gamma(t,X,z) = 0$, and denote by $M(t) = (X(t), z(t))$ the position of a fluid particle at time t. It is on the hypersurface Γ_t if and only if $\gamma(t, M(t)) = 0$ and stays on Γ_t for all times if $\frac{d}{dt}\gamma(t,M(t)) = 0$, or equivalently $\partial_t\gamma + \frac{d}{dt}M\cdot\nabla_{X,z}\gamma = 0$. Since by definition $\frac{d}{dt}M = \mathbf{U}$, we get $\partial_t\gamma + \mathbf{U}\cdot\nabla_{X,z}\gamma = 0$. The conditions (H5)′ and (H6)′ are therefore deduced from (H5) and (H6) by taking $\gamma(t,X,z) = z - H_0 + b(X)$ and $\gamma(t,X,z) = z - \zeta(t,X)$, respectively.

Note that (H1)$''$ is obtained upon integrating (H1)$'$ with respect to the spatial variables and is thus defined up to a time-dependent function. Changing Φ if necessary, it is possible to assume, as we did, that this function identically vanishes.

Recalling that $\partial_{\mathbf{n}}$ always stands for the upwards normal derivative, assumptions (H5)$'$ and (H6)$'$ can be recast in terms of the velocity potential Φ:

$$\begin{array}{lll} (\mathrm{H5})'' & \partial_{\mathbf{n}}\Phi = 0 & \text{on} \quad \{z = -H_0 + b(X)\}, \\ (\mathrm{H6})'' & \partial_t\zeta - \sqrt{1+|\nabla\zeta^2|}\partial_{\mathbf{n}}\Phi = 0 & \text{on} \quad \{z = \zeta(t,X)\}. \end{array}$$

Equations (H1)$''$–(H6)$''$, complemented with (H7)$'$, are called *free surface Bernoulli equations.*

1.1.4. The Zakharov/Craig-Sulem formulation. In his paper [**333**], Zakharov remarked that the knowledge of the free surface elevation ζ and the trace of the velocity potential at the surface $\psi = \Phi_{|_{z=\zeta}}$ fully define the flow, and Craig, Sulem, and Sulem [**110, 111**] gave an elegant formulation of the equations involving the Dirichlet-Neumann operator. The latter is the mathematical formulation of the water waves problem that will be used throughout these notes because it is well adapted to the study of the asymptotic dynamics of the water waves.

Let us now proceed to the derivation of these equations. As mentioned above, the first step consists in noting that ζ and $\psi = \Phi_{|_{z=\zeta}}$ fully determine the flow. Indeed, the velocity potential $\Phi(t,\cdot,\cdot)$ is recovered with (H2)$''$ and (H5)$''$ by solving

$$\left\{ \begin{array}{ll} \Delta_{X,z}\Phi = 0 & \text{in } \Omega_t, \\ \Phi_{|_{z=\zeta}} = \psi, & \partial_{\mathbf{n}}\Phi_{|_{z=-H_0+b}} = 0. \end{array} \right. \tag{1.1}$$

Note that the resolution of the Laplace equation with Neumann (at the bottom) and Dirichlet (at the surface) boundary conditions is possible under reasonable regularity assumptions on ζ and ψ if the nonvanishing shoreline assumption (H9)$'$ is satisfied and if the flow is at rest at infinity as in (H8)$'$; we refer to Chapter 2 for more details.

Now, knowing Φ, one gets the velocity field $\mathbf{U}$ through (H3)$''$ and the pressure P through (H1)". We are thus led to find a set of two equations that determines ζ and ψ (and thus all the physical quantities relevant to the water waves problem). To this end, it is convenient to introduce the Dirichlet-Neumann operator:

$$G[\zeta,b] : \psi \mapsto \sqrt{1+|\nabla\zeta|^2}\partial_{\mathbf{n}}\Phi_{|_{z=\zeta}}, \tag{1.2}$$

where Φ solves (1.1). This operator is linear with respect to ψ but highly nonlinear with respect to the surface and bottom parametrizations ζ and b, which play a role through the fluid domain Ω_t on which (1.1) is solved. We refer to Chapter 3 for more details on the construction of the Dirichlet-Neumann operator (1.2). A straightforward application of the chain rule then yields the relations

$$\begin{array}{rcl} (\partial_t\Phi)_{|_{z=\zeta}} & = & \partial_t\psi - (\partial_z\Phi)_{|_{z=\zeta}}\partial_t\zeta, \\ (\nabla\Phi)_{|_{z=\zeta}} & = & \nabla\psi - (\partial_z\Phi)_{|_{z=\zeta}}\nabla\zeta, \\ (\partial_z\Phi)_{|_{z=\zeta}} & = & \dfrac{G[\zeta,b]\psi + \nabla\zeta\cdot\nabla\psi}{1+|\nabla\zeta|^2}. \end{array}$$

Note now that, owing to (H7)$'$, the r.h.s. of (H1)$''$ vanishes at the free surface, so that using the above relations provides us with an evolution equation on ψ in terms of ζ and ψ (and b) only. Since (H6)$''$ provides an evolution equation on ζ in terms of the same quantities, the water waves problem can be reduced to the following system of two scalar evolution equations:

$$(1.3)\qquad \begin{cases} \partial_t\zeta - G[\zeta,b]\psi = 0, \\ \partial_t\psi + g\zeta + \frac{1}{2}|\nabla\psi|^2 - \frac{(G[\zeta,b]\psi + \nabla\zeta\cdot\nabla\psi)^2}{2(1+|\nabla\zeta|^2)} = 0. \end{cases}$$

As noted by Zakharov [**333**], this system has a Hamiltonian structure in the canonical variables (ζ,ψ). Indeed, (1.3) takes the form

$$\partial_t\begin{pmatrix}\zeta\\ \psi\end{pmatrix} = \begin{pmatrix}0 & I\\ -I & 0\end{pmatrix}\begin{pmatrix}\partial_\zeta H\\ \partial_\psi H\end{pmatrix},$$

with the Hamiltonian H given by

$$(1.4)\qquad H = K + E,$$

where the kinetic and potential energies K and E are defined as

$$K = \frac{1}{2}\int_{\mathbb{R}^d}\int_{-H_0+b(X)}^{\zeta(X)} |\nabla_{X,z}\Phi(X,z)|^2 dz dX = \frac{1}{2}\int_{\mathbb{R}^d}\psi G[\zeta,b]\psi$$

(the second formula following from Green's identity) and

$$E = \frac{1}{2}\int_{\mathbb{R}^d} g\zeta^2.$$

REMARK 1.1. It follows from the analysis above that the Hamiltonian H is a conserved quantity. This is not the only one. For instance, in the case of a flat bottom ($b = 0$), the following quantities are conserved:

$$Q_x = \int_{\mathbb{R}^d}\zeta\partial_x\psi, \qquad Q_y = \int_{\mathbb{R}^d}\zeta\partial_y\psi, \quad m = \int_{\mathbb{R}^d}\zeta;$$

the first two are related to the horizontal momentum and the third one is the excess mass. We refer to [**24**] for more details on conserved quantities.

1.2. Other formulations of the water waves problem

Throughout these notes, we use the Zakharov-Craig-Sulem formulation of the water waves equations (1.3), which seems to be the one most adapted for the derivation of shallow water asymptotic models. However, many other formulations of the water waves problem exist and may be adapted in other contexts (singularity formation for instance). We briefly review some of these alternative formulations here. Since most of them have been derived in the context of infinite depth, we consider in this section that $H_0 = \infty$.

1.2.1. Lagrangian parametrizations of the free surface. The free surface Euler equations (H1)$'$–(H7)$'$ can be reformulated in Lagrangian variables.

Let $M(t,X) \in \mathbb{R}^{d+1}$ be the Lagrangian representation of the free surface defined as

$$(1.5)\qquad \begin{cases} \partial_t M(t,\alpha,\beta) = \mathbf{U}\big(t, M(t,\alpha,\beta)\big), \\ M(0,\alpha,\beta) = (\alpha,\beta,\zeta^0(\alpha,\beta)), \end{cases}$$

for all $\alpha,\beta\in\mathbb{R}$ (with straightforward adaptation if $d=1$). We also denote by Γ_t the free surface at time t,

$$\Gamma_t = \{M(t,\alpha,\beta), \alpha,\beta\in\mathbb{R}\},$$

with here again straightforward adaptation if $d = 1$.

REMARK 1.2. The surface is here represented at time t as a parametrized hypersurface, which is not necessarily a graph. We could also have chosen a more general initial condition

$$M(0,\alpha,\beta) = \big((m_1^0(\alpha,\beta), m_2^0(\alpha,\beta), m_2^0(\alpha,\beta)\big)$$

allowing, for instance, overhanging waves.

Time differentiation of (1.5) yields

$$\begin{aligned} \partial_t^2 M &= \partial_t \mathbf{U} + \partial_t M \cdot \nabla_{X,z}\mathbf{U} \\ &= \partial_t \mathbf{U} + \mathbf{U}\cdot\nabla_{X,z}\mathbf{U} \\ &= -g\mathbf{e}_z - \frac{1}{\rho}\nabla_{X,z}P(M), \end{aligned}$$

where we used the Euler equation (H1)$'$ to derive the last identity. We also deduce from (H7)$'$ that tangential derivatives of the pressure vanish at the surface, so that $\nabla_{X,z}P(M) = (\partial_{\mathbf{n}}P)\mathbf{n}$. We deduce therefore that

$$\partial_t^2 M + g\mathbf{e}_z = \frac{1}{\rho}(-\partial_{\mathbf{n}}P)\mathbf{n}, \tag{1.6}$$

where $\mathbf{n}$ is the unit normal pointing out of the fluid domain.

1.2.1.1. *Nalimov's formulation in dimension $d = 1$.* Let us identify canonically $\mathbb{R}^{d+1}$ ($d = 1$) with the complex plane $\mathbb{C}$: $(x,z) \in \mathbb{R}^d \rightsquigarrow x + iz \in \mathbb{C}$. This identification allows one to see the incompressibility and irrotationality assumptions (H2)$'$–(H3)$'$ as the Cauchy-Riemann relations for the conjugate of the velocity field (seen as a complex-valued vector field), denoted $\overline{\mathbf{U}} = u - iw$. It follows that $\overline{\mathbf{U}}$ is holomorphic in the fluid domain Ω_t, and there exists a singular integral operator on the surface recovering boundary values of w from boundary values of u — denoted, respectively, by $\underline{w}(t,\alpha) = w(t,M(t,\alpha))$ and $\underline{u}(t,\alpha) = u(t,M(t,\alpha))$,

$$\underline{w} = K[\Gamma_t]\underline{u} \quad \text{ or, owing to (1.5),} \quad \partial_t M_2 = K[\Gamma_t]\partial_t M_1,$$

where we used the notation $M(t) = (M_1(t), M_2(t))$.

REMARK 1.3. The operator $K[\Gamma_t]$ is related to the Hilbert transform associated to the fluid domain. Actually, it is often convenient [**328, 315**] to replace the relation $\underline{w} = K[\Gamma_t]\underline{u}$ by

$$\underline{\overline{\mathbf{U}}} = \mathfrak{H}[\Gamma_t]\underline{\overline{\mathbf{U}}} \qquad (\text{with } \underline{\overline{\mathbf{U}}} = \underline{u} - i\underline{v}).$$

Here, $\mathfrak{H}[\Gamma_t]$ is the Hilbert transform on Γ_t associated with the parametrization $M(t,\cdot)$,

$$\mathfrak{H}[\Gamma_t]f(t,\alpha) = \frac{1}{\pi i}\text{p.v.}\int \frac{f(t,\alpha')\partial_\alpha M(t,\alpha')}{M(t,\alpha) - M(t,\alpha')}d\alpha' \tag{1.7}$$

where p.v. stands for "principal value" and $\mathbb{R}^2$ has been identified with $\mathbb{C}$ (we also recall that a function f is holomorphic in a domain Ω with boundary $\partial\Omega$ if and only if $\mathfrak{H}[\partial\Omega]f_{|\partial\Omega} = f_{|\partial\Omega}$).

Multiplying (1.6) by the tangent vector $(\partial_\alpha M_1, \partial_\alpha M_2)^T$, we get Nalimov's formulation of the one-dimensional water waves problem,

$$\begin{cases} \partial_\alpha M_1 \partial_t^2 M_1 + (g + \partial_t^2 M_2)\partial_\alpha M_2 = 0, \\ \partial_t M_2 = K[\Gamma_t]\partial_t M_1. \end{cases} \tag{1.8}$$

Using this formulation, Nalimov [**265**] was the first to prove a local well-posedness result for the water waves equation with Sobolev (small) initial data (other results deal with analytic data [**300, 198**]). Nalimov dealt with the case of infinite depth, but Yosihara extended his work to finite depth [**330**], while Craig used it [**101**] to provide the first justification of the KdV approximation (see Chapter 7).

1.2.1.2. *Wu's formulation.* The main limitation of Nalimov's result is that it requires a smallness condition on the initial data. In [**326**], S. Wu managed to remove this smallness condition using a modification of Nalimov's formulation. One of the key points of [**326**] is the observation that $(-\partial_{\mathbf{n}}P) > 0$ in the infinite depth case considered in [**326**] (see Proposition 4.32 in Chapter 4), so that one gets from (1.6) that $N := (-\partial_\alpha M_2, \partial_\alpha M_1) = |\partial_\alpha M|\mathbf{n}$ is given by

$$N = \frac{\partial_t^2 M + g\mathbf{e}_z}{|\partial_t^2 M + g\mathbf{e}_z|}|\partial_\alpha M|.$$

Denoting by $\mathfrak{a}$ the Rayleigh-Taylor coefficient,

$$\mathfrak{a} = \frac{1}{|\partial_\alpha M|}\frac{1}{\rho}\big(- \partial_{\mathbf{n}}P(M)\big), \tag{1.9}$$

and differentiating (1.6) with respect to time, we get

$$\partial_t^3 M - \mathfrak{a}\partial_t N = \partial_t \mathfrak{a} N;$$

using the expression derived above for N and recalling that $\partial_t M = \underline{\mathbf{U}}(t,M)$, we get, with the notation $\underline{\mathbf{U}}(t,\alpha,\beta) = \mathbf{U}(t, M(t,\alpha,\beta))$,

$$\partial_t^2 \underline{\mathbf{U}} - \mathfrak{a}\partial_t N = \frac{\partial_t \underline{\mathbf{U}} + g\mathbf{e}_z}{|\partial_t \underline{\mathbf{U}} + g\mathbf{e}_z|}|\partial_\alpha M|\partial_t \mathfrak{a}.$$

Moreover, since $\partial_t N = (-\partial^2_{\alpha t} M_2, \partial^2_{\alpha t} M_1)$, we get after differentiating (1.5) with respect to α that

$$\partial_t N = \big(- \partial_\alpha M \cdot \nabla_{X,z} w(M), \partial_\alpha M \cdot \nabla_{X,z} u(M)\big)^T.$$

From the irrotationality and incompressibility conditions, we also have $\partial_x w = \partial_z u$ and $\partial_z w = -\partial_x u$, so that

$$\begin{aligned} -\partial_\alpha M \cdot \nabla_{X,z} w &= -\partial_\alpha M_1 \partial_x w - \partial_\alpha M_2 \partial_z w \\ &= -\partial_\alpha M_1 \partial_z u + \partial_\alpha M_2 \partial_x u \\ &= -|\partial_\alpha M|\partial_{\mathbf{n}} u, \end{aligned}$$

and similarly $\partial_\alpha M \cdot \nabla_{X,z} u = -|\partial_\alpha M|\partial_{\mathbf{n}} w$. It is consistent[3] with the definition (1.2) of the Dirichlet-Neumann operator to use the notation

$$G[\Gamma_t]\underline{\mathbf{U}} = \big(G[\Gamma_t]\underline{u}, G[\Gamma_t]\underline{w}\big)^T := |\partial_\alpha M|(\partial_{\mathbf{n}} u, \partial_{\mathbf{n}} v)^T.$$

In the one-dimensional case $d = 1$, Wu's formulation of the water waves equation can then be put under the form[4]

$$\begin{cases} \partial_t^2 \underline{\mathbf{U}} + \mathfrak{a} G[\Gamma_t]\underline{\mathbf{U}} = \dfrac{\partial_t \underline{\mathbf{U}} + g\mathbf{e}_z}{|\partial_t \underline{\mathbf{U}} + g\mathbf{e}_z|}|\partial_\alpha M|\partial_t \mathfrak{a}, \\ \underline{\mathbf{U}} = \mathfrak{H}[\Gamma_t]\underline{\mathbf{U}}, \end{cases} \tag{1.10}$$

[3]If Γ_t is a graph parametrized by $(x, \zeta(x))$, then $G[\Gamma_t]\psi = G[\zeta]\psi$ where $G[\zeta]$ is defined as in (1.1)–(1.2) with the homogeneous Neumann condition at the bottom in (1.1) replaced by $\nabla_{X,z}\Phi \to 0$ as $z \to -\infty$.

[4]This formulation is actually not the one used by S. Wu in [**326**] but rather the adaptation to the one-dimensional case of the formulation Wu used in [**327**] for the case $d = 2$.

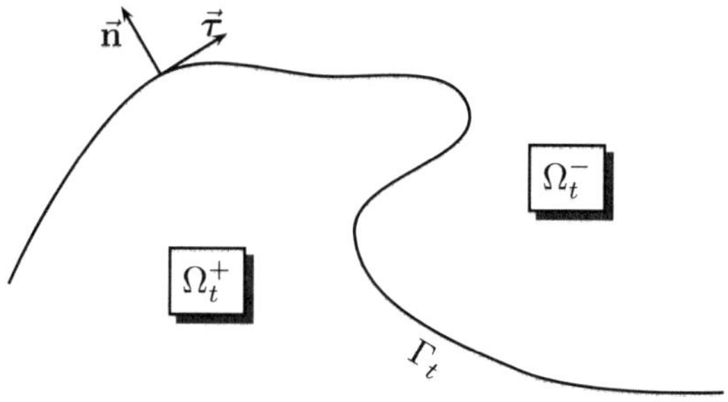

FIGURE 1.2. A generalization of the water waves problem: two-fluids interfaces.

where $\mathfrak{a} = \frac{1}{|\partial_\alpha M|}|\partial_t \underline{\mathbf{U}} + g\mathbf{e}_2|$ and $\mathfrak{H}(\Gamma_t)$ is the Hilbert transform introduced in Remark 1.3. S. Wu showed [**326**] that the right-hand side of the first equation consists of lower order terms in $\underline{\mathbf{U}}$, so that (1.10) is a quasilinear weakly hyperbolic problem, for which Wu constructed solutions by an iterative scheme without any smallness assumption on the initial data. This formulation is also the starting point for [**328**] where an almost global existence result is proved, for [**337**] where nonzero vorticity is allowed and for [**315**] where the nonlinear Schrödinger approximation is justified.

Wu's formulation (1.10) also holds in dimension $d = 2$. The derivation of the first equation of (1.10) is absolutely similar to the derivation sketched above in the case $d = 1$. The second equation of (1.10) also holds true when $d = 2$; this only requires an extension of the formula (1.7) for the Hilbert transform to the two-dimensional case. This extension is derived in [**327**] using Clifford analysis instead of complex analysis, but more standard tools could of course be used. In [**329**], global well-posedness in dimension $d = 2$ is proved using this formulation of the water waves equations.

1.2.2. Other interface parametrizations and extension to two-fluids interfaces. We have considered in §1.2.1 the equations obtained when using a Lagrangian parametrization of the surface Γ_t. However, other parametrizations are possible. In order to underline the importance of the choice of the parametrization of the interface, we consider here a generalization of the water waves problem, namely, the case of two-fluids interfaces. For the sake of simplicity, we consider the one dimensional case $d = 1$.

The interface Γ_t separates two fluids of densities $\rho_+ \geq \rho_-$, the heavier fluid being below the interface (see Figure 1.2). The water waves problem corresponds therefore to the endpoint case $\rho_+ = \rho$, $\rho_- = 0$. The discussion below is inspired by [**18**], devoted to a more general review on interface evolution.

Let $r(t, \cdot)$ be a parametrization of the interface,

$$\Gamma_t = \{r(t,\alpha), \alpha \in \mathbb{R}\};$$

we assume that the flow is incompressible on the whole space $\mathbb{R}^{d+1}$ and irrotational above and below the interface. We therefore have

$$\nabla_{X,z} \cdot \mathbf{U} = 0, \qquad \nabla_{X,z} \times \mathbf{U} = \tilde{\omega} \otimes \delta_{\Gamma_t}, \tag{1.11}$$

where δ_{Γ_t} is the Dirac measure associated to the interface, and $\tilde{\omega}$ is the *vorticity density* associated to the parametrization $r(t,\alpha)$,

$$\tilde{\omega}(t,\alpha) = |\partial_\alpha r|(\mathbf{U}_+^\tau - \mathbf{U}_-^\tau)_{|(t,r(t,\alpha))},$$

with $\mathbf{U}_\pm^\tau$ the tangential components of $\mathbf{U}_\pm$ evaluated at the interface. We therefore deduce from (1.11) the following jump relations on the normal and tangential components of the velocity at the interface,

$$[\mathbf{U}_\pm^n] = 0, \qquad [\mathbf{U}_\pm^\tau] = \frac{1}{|\partial_\alpha r|}\tilde{\omega}.$$

Let us now consider the conservation equations of mass and momentum,

$$\begin{aligned} \partial_t(\rho\mathbf{U}) + \nabla_{X,z}\cdot(\rho\mathbf{U}\otimes\mathbf{U}) + \nabla_{X,z}P &= -\rho g\mathbf{e}_z, \\ \partial_t\rho + \nabla_{X,z}\cdot(\rho\mathbf{U}) &= 0, \end{aligned}$$

written in the sense of distributions in the whole space $\mathbb{R}^{d+1}$, with $\rho = \rho_+$, $\mathbf{U} = \mathbf{U}_+$ below Γ_t and $\rho = \rho_-$, $\mathbf{U} = \mathbf{U}_-$ above. We deduce from the weak formulation of these equations the following two evolution equations (see [**306**], and [**97**] for a slightly different but equivalent formulation),

$$(1.12) \qquad (\partial_t r - v)\cdot\vec{n} = 0$$

$$(1.13) \qquad \begin{aligned} \partial_t\Big(\frac{1}{2}\tilde{\omega} - a|\partial_\alpha r|v^\tau\Big) &+ \partial_\alpha\Big(\frac{1}{|\partial_\alpha r|}\big(\frac{1}{2}\tilde{\omega} - a|\partial_\alpha r|v^\tau\big)(v-\partial_t r)\cdot\vec{\tau}\Big) \\ &- a\partial_\alpha\Big(\frac{1}{8}\frac{\tilde{\omega}^2}{|\partial_\alpha r|^2} - \frac{1}{2}|v|^2\Big) = -a|\partial_\alpha r|g\mathbf{e}_z\cdot\vec{\tau}, \end{aligned}$$

where v is the average velocity at the interface, and a the Atwood number,

$$v = \frac{1}{2}(\mathbf{U}_+ + \mathbf{U}_-)_{|\Gamma_t}, \qquad a = \frac{\rho_- - \rho_+}{\rho_- + \rho_+}.$$

Observe that the above system does not fully determine the evolution of the vector $r(\lambda,t)$. Equation (1.12) concerns only the normal component of its time derivative. This corresponds to the fact that it is only the global evolution of the interface and not its parametrization that matters. Depending on the situation, different choices can be made for the parametrization $\alpha \mapsto r(t,\alpha)$ of Γ_t:

- When $a = 0$ (this corresponds to $\rho_- = \rho_+$ i.e., to the so-called Kelvin-Helmholtz problem), it is convenient to choose a Lagrangian parametrization,

$$r_t(t,\alpha) = v(t,r(t,\alpha)) = \frac{\mathbf{U}_+ + \mathbf{U}_-}{2}(t,r(t,\alpha));$$

  equations (1.12) and (1.13) then reduce to

$$\partial_t\tilde{\omega} = 0 \quad \text{and thus} \quad \tilde{\omega}(t,\alpha) = \tilde{\omega}(0,\alpha).$$

  Moreover, since we can recover v in terms of $\tilde{\omega}$ through Biot-Savart's law

$$v(t,r(t,\alpha)) = \text{p.v.}\int_{\mathbb{R}} \nabla^\perp G(r(t,\alpha),r(t,\alpha'))\tilde{\omega}(t,\alpha')d\alpha',$$

  where G is the Green function of the Laplacian on $\mathbb{R}^{1+1}$, we get

$$\partial_t r(t,\alpha) = \frac{1}{2\pi}R_{\frac{\pi}{2}}\text{p.v.}\int \frac{r(t,\alpha)-r(t,\alpha')}{|r(t,\alpha)-r(t,\alpha')|^2}\tilde{\omega}(0,\alpha')d\alpha',$$

where $R_{\frac{\pi}{2}}$ is the rotation of angle $\pi/2$.
Therefore, it is natural to reparametrize the interface by the (time independent) arc length *of its initial position*, $\Gamma_t = \{r(t, s_0^{-1}(s))\}$, where $s_0(\alpha) = \int_0^\alpha \partial_\alpha r(0,\alpha')d\alpha'$. One can further identify the fluid domain with the complex plane, i.e. $z(t,s) \sim r(t, s_0^{-1}(s))$, to obtain:

$$\partial_t \overline{z}(t,s) = \frac{1}{2\pi i}\text{p.v.}\int \frac{\gamma(0,s')}{z(t,s)-z(t,s')}ds',$$

where $\gamma(0,s)$ is the *vortex strength*[5] at $t=0$.
Moreover, if $\gamma(s,0)$ has *a distinguished sign* (which is the case when the initial vorticity is a Radon measure), then we can use the mapping

$$\alpha_0(s) = \int_0^s \gamma(0,s')ds'$$

to perform another (time independent) reparametrization of the interface, namely, $\Gamma_t = \{z(t, \alpha_0^{-1}(\alpha))\}$. One thus obtains the *Birkhoff-Rott equation* for the variable $\mathbf{z}(t,\alpha) = z(t, \alpha_0^{-1}(\alpha))$,

$$\partial_t \overline{\mathbf{z}}(t,\alpha) = \frac{1}{2\pi i}\text{p.v.}\int \frac{d\alpha'}{\mathbf{z}(t,\alpha)-\mathbf{z}(t,\alpha')}. \tag{1.14}$$

The Birkhoff-Rott equation corresponds to a parametrization by the circulation; indeed, one readily checks that $|\partial_\alpha \mathbf{z}(t,\alpha_0(s))| = \frac{1}{\gamma(0,s)}$.

- When $a=-1$ (this corresponds to $\rho_-=0$, and thus to the water waves problem), several choices are possible. One can, for instance, as in [**306**], choose a Lagrangian parametrization with the velocity given by the trace of the velocity field in Ω_t^+ at the surface; this situation has been described in §1.2.1.
  Another choice, adopted in [**14**] (and [**15**] for a generalization to the two-dimensional case $d=2$), consists of parametrizing the surface by its (renormalized) arc-length. It is also possible to use the particular case $a=-1$ of the following point when the surface is a graph.
- When the interface is a graph, we can also choose the canonical parametrization by x, $\Gamma_t = \{(x, \zeta(t,x))\}$. This is the choice made in the derivation of (1.3) and it will be used throughout these notes.
- The fact that one can add any tangential term to the velocity field at the interface without modifying the geometric evolution of the curve has been used by several authors for the water waves problem or two-fluids interfaces [**175, 97**] but also for interface evolutions in other contexts such as the Muskat problem [**66, 98**].

1.2.3. Variational formulations.

Variational formulations for Euler's equations go back at least to [**171, 143**]. We briefly sketch here two different variational approaches. The first approach follows the geometric path of Arnold [**16**], who showed that Euler's equations can be seen as a geodesic flow on the group of volume-preserving homeomorphisms. The second approach (see Luke [**243**]), is more formal but has been historically used to derive several important asymptotic models.

---

[5] The vortex strength is defined as the vorticity density $\tilde{\omega}$ associated to the arc-length parametrization.

1.2.3.1. *The geometric approach.* It was observed in [**16**] that Euler equations for the motion of an inviscid incompressible fluid can be viewed as the geodesic flow on the infinite-dimensional manifold of volume-preserving diffeomorphisms. This approach has been used by several authors, including [**53, 142, 301**]. More recently, Beyer and Günther [**29, 30**] used this geometric approach to study the free surface for capillary water waves (surface tension but no gravity) in a bounded domain. In a series of papers [**297, 298, 299**], Shatah and Zeng dealt with the rotational case and addressed several related problems (such as two-fluid interfaces). Here we sketch their approach.

Let us denote by Σ the Lagrangian parametrization of the fluid domain associated to the velocity field $\mathbf{U}$,

$$\begin{cases} \partial_t\Sigma(t,X,z) = \mathbf{U}\big(t,\Sigma(t,X,z)\big), \\ \Sigma(0,X,z) = (X,z), \end{cases} \tag{1.15}$$

for all $(X,z) \in \Omega_0$ (the fluid domain at time $t=0$). Since $\mathbf{U}$ is divergence free, one has $\Sigma \in \mathcal{H}$, where $\mathcal{H}$ is the set of all volume-preserving homeomorphisms,

$$\begin{aligned} \mathcal{H} \;=\; & \{\Sigma : \Omega_0 \to \mathbb{R}^{d+1}, \\ & \Sigma \text{ is a volume-preserving homeomorphism from } \Omega_0 \text{ onto } \Sigma(\Omega_0)\}. \end{aligned}$$

Let us now write the conserved energy H introduced in (1.4) under the form

$$\begin{aligned} H &= \frac{1}{2}\int_{\Omega_t} |\mathbf{U}|^2 + \frac{g}{2}\int_{\mathbb{R}^d} \zeta^2 \\ &= \frac{1}{2}\int_{\Omega_0} |\mathbf{U}\circ\Sigma|^2 + gG(\Sigma), \end{aligned}$$

where $G(\Sigma)$ is the volume delimited by the free surface and the asymptotic (rest) state. Owing to (1.15), we have

$$H = L(\Sigma,\partial_t\Sigma) := \frac{1}{2}\int_{\Omega_0} |\partial_t\Sigma|^2 + gG(\Sigma).$$

In order to derive the Euler-Lagrange equations associated with the Lagrangian action L, we can consider $\mathcal{H}$ as a manifold with tangent space

$$T_\Sigma\mathcal{H} = \{\Sigma' : \Omega_0 \to \mathbb{R}^{d+1}, \nabla_{X,z}\cdot(\Sigma'\circ\Sigma^{-1}) = 0\}.$$

The conservation of energy suggests that $T\mathcal{H}$ can be endowed with the L^2 metric, and one can show [**16, 297**] that the water waves equations coincide with the geodesic equation on $\mathcal{H}$, namely,

$$\partial_t^2\Sigma + gG'(\Sigma) = \text{Lagrange multiplier},$$

where $G'(\Sigma)$ denotes the tangential gradient of $G(\Sigma)$; the covariant derivative $\mathcal{D}_t\partial_t\Sigma$ (or equivalently the tangential component of $\partial_t^2\Sigma$) therefore satisfies

$$\mathcal{D}_t\partial_t\Sigma + gG'(\Sigma) = 0. \tag{1.16}$$

Shatah and Zeng studied the linearization of (1.16) to obtain local well-posedness results under the assumption that the Rayleigh-Taylor criterion is satisfied (i.e., that the Rayleigh-Taylor coefficient (1.9) is strictly positive).

1.2.3.2. *Luke's variational formulation.* The water waves equation can also be derived formally from Hamilton's principle using the Lagrangian derived by Luke [**243**]. Though we do not use this formulation in these notes, we briefly mention it for the sake of completeness. The Lagrangian $\widetilde{l}$ is given by the vertical integration of the pressure,

$$\widetilde{l} = \int_{-H_0+b(X)}^{\zeta(X)} (P - P_{atm})(t, X, z)dz.$$

Using the Bernoulli equation (H1)″, we can define a new Lagrangian $l = l(\zeta, \Phi)$ (deduced from $\widetilde{L}$ by dropping constant terms that obviously have no incidence on the minimization principle),

$$l = -\rho\big[g\frac{\zeta^2}{2} + \int_{-H_0+b}^{\zeta} (\partial_t \Phi + \frac{1}{2}|\nabla_{X,z}\Phi|^2)dz\big].$$

In [**243**] (which we refer to for the details of the computations) Luke observed that the water waves equations could be recovered from the variational principle,

$$\delta J(\zeta, \Phi) = 0 \quad \text{where} \quad J(\zeta, \Phi) = \int\int l(\zeta, \Phi)dXdt.$$

As shown by Miles and Salmon [**261**], adding some constraints to this minimization principle allows one to recover some of the asymptotic models derived here (e.g., the Green-Naghdi equation over flat bottom). See also [**82**] for a similar approach.

1.2.4. Free surface Euler equations in Lagrangian formulation. Written in the Lagrangian coordinates (1.15), and denoting

$$\begin{aligned} \tilde{\mathbf{U}}(t, X, z) &= \mathbf{U}(t, \Sigma(t, X, z)), \\ \tilde{P}(t, X, z) &= P(t, \Sigma(t, X, z)), \\ A(t, X, z) &= [\nabla_{X,z}\Sigma(t, X, z)]^{-1}, \end{aligned}$$

the free surface Euler equations become (without the irrotationality assumption (H3)′)

$$\begin{aligned} \Sigma = \mathrm{Id} + \int_0^t \tilde{\mathbf{U}} \qquad & \text{in } \Omega_0,\ t \geq 0, \\ \partial_t \tilde{\mathbf{U}} + A\nabla_{X,z}\tilde{P} + g\mathbf{e}_z = 0 \qquad & \text{in } \Omega_0,\ t \geq 0, \\ \mathrm{Tr}(A\nabla_{X,z}\tilde{\mathbf{U}}) = 0 \qquad & \text{in } \Omega_0,\ t \geq 0, \\ \tilde{P} = P_{atm} \qquad & \text{on } \Gamma_0,\ t \geq 0. \end{aligned}$$

The advantage of this formulation is that it is cast on a fixed domain Ω_0, which eases the construction of a solution by an iterative scheme. This has been done in [**78, 236, 237**] using refined geometric estimates and in [**99**] using a clever smoothing of the equations (see also [**75**] for an extension of this method to two-fluids interfaces).

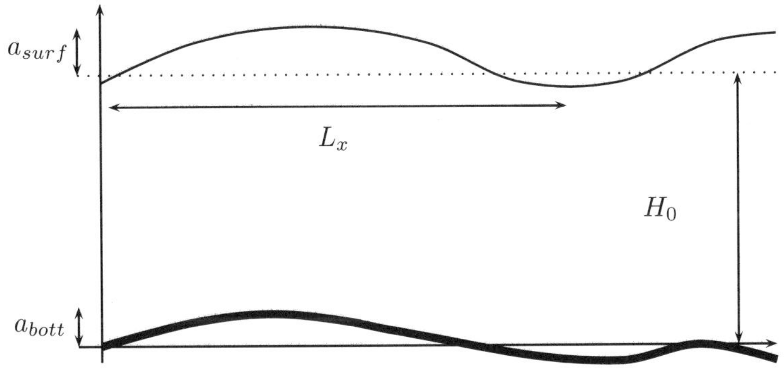

FIGURE 1.3. Length scales involved in the wave motion.

1.3. The nondimensionalized equations

The water waves equations (1.3) have a particularly rich structure and it is possible to exhibit solutions with dramatically different properties, depending on the physical characteristic of the flow. For instance, dispersive effects are more important in deep water than in shallow water, and nonlinear effects become more important when the amplitude of the waves grows larger, etc.

It is easier to comment on the qualitative properties of the solutions to (1.3) if we nondimensionalize these equations using the physical characteristics of the flow. This is the goal of this section.

1.3.1. Dimensionless parameters. We introduce here dimensionless parameters based on the characteristic scales of the wave motion. Four (when $d = 1$) or five (when $d = 2$) main length scales are involved in this problem:

(1) The characteristic water depth H_0
(2) The characteristic horizontal scale L_x in the longitudinal direction
(3) The characteristic horizontal scale L_y in the transverse direction (when $d = 2$)
(4) The order of the free surface amplitude a_{surf}
(5) The order of bottom topography variation a_{bott}

Four independent dimensionless parameters can be formed from these five scales. We choose:

$$\frac{a_{surf}}{H_0} = \varepsilon, \quad \frac{H_0^2}{L_x^2} = \mu, \quad \frac{a_{bott}}{H_0} = \beta, \quad \frac{L_x}{L_y} = \gamma, \tag{1.17}$$

where ε is often called the *nonlinearity* parameter, while μ is the *shallowness* parameter. We will also refer to β and γ as the *topography* and *transversality* parameters.

REMARK 1.4.

i- It is implicitly assumed here that the characteristic horizontal scales for the variations of the bottom are the same as for the free surface. This is of course not the case when a wave propagates over a "rough" bottom; in such configurations, the characteristic length is much smaller for the bottom than for the surface. See §1.7 for more comments on this point.

ii- Quite obviously, the set of dimensionless parameters (1.17) is not adapted to study the propagation of waves over infinite depth. So in this case, one uses

$$\epsilon = \frac{a_{surf}}{L_x},$$

where ϵ is the *steepness* of the wave. However, since $\epsilon = \varepsilon\sqrt{\mu}$ one can stick to (1.17) even in very deep water, provided that the dependence on ε and μ appears through $\varepsilon\sqrt{\mu}$ only.

EXAMPLE 1.5 (Tsunami). For the 2004 Indian ocean tsunami, the characteristic length L_x is given by the size of the fault slip, between 160 and 240 km. The depth H_0 of the ocean ranges between 1 km and 4 km, and satellite observations suggest an amplitude a_{surf} of the surface wave of 60 cm. The nonlinearity and shallowness parameters are thus estimated by

$$1.5 \times 10^{-4} < \varepsilon < 6 \times 10^{-4} \quad \text{and} \quad 1.7 \times 10^{-5} < \mu < 6.2 \times 10^{-4}.$$

A tsunami is therefore an example of wave propagating in a shallow-water regime (even though the depth H_0 is very large). We refer to [**115, 138**] and references therein for details on tsunami modeling and simulation.

EXAMPLE 1.6 (Coastal oceanography). For a swell of wavelength $L_x \sim 100m$ and amplitude $a_{surf} \sim 1m$ over a continental shelf of depth $H_0 \sim 10m$, one has

$$\varepsilon \sim 10^{-1} \quad \text{and} \quad \mu \sim 10^{-2}.$$

Though the shallow water assumption $\mu \ll 1$ is satisfied, this is far less obvious than for the tsunami mentioned in Example 1.5—even though the depth is 400 times smaller here! It is therefore important to keep in mind that *shallowness always concerns the ratio* H_0/L_x *and not* H_0 *itself.* Another important consideration is that since μ is not extremely small here, it is important to find asymptotic models with a wide range of validity, i.e., that remain correct for not so small values of μ and not only as $\mu \to 0$ (see, for instance, §5.2).

1.3.2. Linear wave theory. The easiest way to get some information on the order of magnitude of wave motion variables is to investigate the behavior of the water waves equations around the rest state. Linearizing (1.3) around $(\zeta, \psi) = (0, 0)$ (and considering flat bottoms, $b = 0$), one gets the following system

$$\begin{cases} \partial_t \zeta - G[0,0]\psi = 0, \\ \partial_t \psi + g\zeta = 0. \end{cases} \tag{1.18}$$

In particular, the surface elevation ζ solves a second-order evolution equation,

$$\partial_t^2 \zeta + gG[0,0]\zeta = 0. \tag{1.19}$$

Using the definition of the Dirichlet-Neumann operator (1.2), we have

$$G[0,0]\psi = \partial_z \Phi_{|z=0},$$

where

$$\begin{cases} \Delta_{X,z}\Phi = 0 \qquad \text{for } -H_0 < z < 0, \\ \Phi_{|z=0} = \psi, \quad \partial_z \Phi_{|z=-H_0} = 0. \end{cases}$$

Taking the Fourier transform of this equation with respect to the horizontal variables, one is led to solve a second-order ODE in z with boundary conditions at

$z = 0$ and $z = -H_0$; one readily checks that there exists a unique solution given by

$$\forall(\xi, z) \in \mathbb{R}^{d+1}, \qquad \widehat{\Phi}(\xi, z) = \frac{\cosh\big((z + H_0)|\xi|\big)}{\cosh(H_0|\xi|)}\widehat{\psi}(\xi),$$

where $\widehat{\cdot}$ denotes the Fourier transform. Therefore, we have

$$\widehat{\partial_z\Phi}(\xi, 0) = \partial_z\widehat{\Phi}(\xi, 0) = |\xi|\tanh(H_0|\xi|)\widehat{\psi}(\xi)$$

and the Dirichlet-Neumann operator admits the following explicit expression,

$$G[0,0]\psi = |D|\tanh(H_0|D|)\psi,$$

where we used the Fourier multiplier notation:

NOTATION 1.7 (Fourier multipliers). Let $f \in L^\infty(\mathbb{R}^d)$; the Fourier multiplier $f(D) : L^2(\mathbb{R}^d) \to L^2(\mathbb{R}^d)$ is then defined by

$$\forall u \in L^2(\mathbb{R}^d), \quad \forall\xi \in \mathbb{R}^d, \qquad \widehat{f(D)u}(\xi) = f(\xi)\widehat{u}(\xi).$$

This definition can be generalized to other functional settings. (For instance, it is easy to check that $G[0,0] : H^1(\mathbb{R}^d) \to L^2(\mathbb{R}^d)$ is well defined and bounded.)

The second-order evolution equation (1.19) can therefore be recast as

$$\partial_t^2\zeta + g|D|\tanh(H_0|D|)\zeta = 0. \tag{1.20}$$

Taking initial conditions for (1.20) of the form

$$\zeta_{|_{t=0}} = \zeta_0 \quad \text{and} \quad \partial_t\zeta_{|_{t=0}} = |D|\tanh(H_0|D|)\psi_0$$

(which corresponds to initial condition $(\zeta, \psi)_{|_{t=0}} = (\zeta_0, \psi_0)$ to (1.18)), the general form of the solutions of (1.20) is given by

$$\begin{aligned}\zeta(t, X) = &\frac{1}{2}\int_{\mathbb{R}^d} e^{i(\xi\cdot X - \omega(\xi)t)}\big(\widehat{\zeta_0}(\xi) + i\frac{\omega(\xi)}{g}\widehat{\psi_0}(\xi)\big)d\xi \\ &+ \frac{1}{2}\int_{\mathbb{R}^d} e^{i(\xi\cdot X + \omega(\xi)t)}\big(\widehat{\zeta_0}(\xi) - i\frac{\omega(\xi)}{g}\widehat{\psi_0}(\xi)\big)d\xi,\end{aligned} \tag{1.21}$$

where

$$\omega(\xi) = \big(g|\xi|\tanh(H_0|\xi|)\big)^{1/2} \tag{1.22}$$

is the *dispersion relation* of the linearized water waves equations; the *wave celerity* at the wave number ξ is defined as

$$c(|\xi|) = \frac{\omega(\xi)}{|\xi|} = \big(g\frac{\tanh(H_0|\xi|)}{|\xi|}\big)^{1/2}. \tag{1.23}$$

These relations can be used to determine some typical order of magnitude for various characteristics of the flow in terms of the physical quantities introduced in the previous section. Let us focus on the one-dimensional case $d = 1$ for the sake of simplicity.
If the typical wavelength of the wave under consideration is L_x, as assumed in §1.3.1, then in Fourier space, the wave is concentrated around the wave number $\xi_0 = \frac{2\pi}{L_x}$. The typical wave celerity $c_0 = \omega(\xi_0)/|\xi_0|$ is then given by

$$c_0 = \big(gL_x\frac{1}{2\pi}\tanh(2\pi\frac{H_0}{L_x})\big)^{1/2} = \sqrt{gH_0\nu}, \tag{1.24}$$

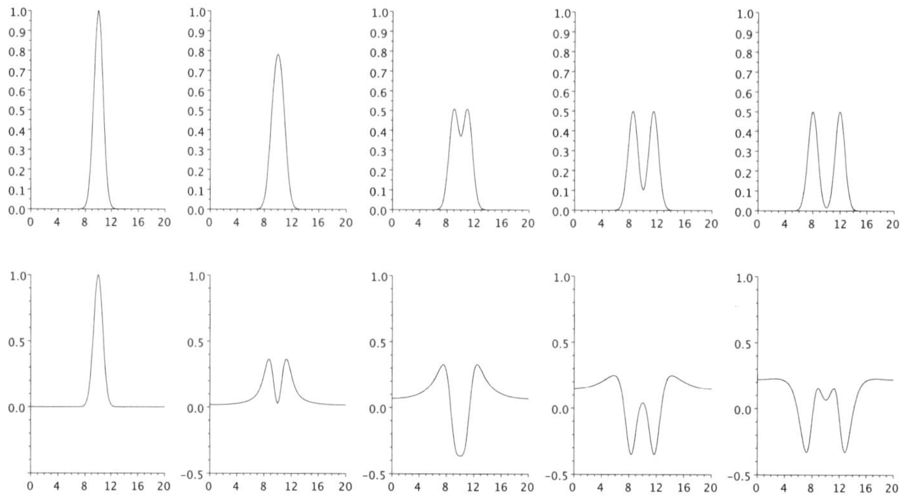

FIGURE 1.4. Time evolution of the same initial data in shallow (up) and deep (down) water for the linear equation (1.20).

where ν is a dimensionless parameter defined as

$$\nu = \frac{\tanh(2\pi\sqrt{\mu})}{2\pi\sqrt{\mu}}. \tag{1.25}$$

Similarly, it follows from (1.21) that if the typical order of magnitude of ζ (and ζ_0) is a_{surf}, as assumed in §1.3.1, then $\frac{\omega(\xi_0)}{g}\psi_0$ must also be roughly of size a_{surf}. A typical order of magnitude for the velocity potential ψ is thus given by

$$\Phi_0 = 2\pi g\frac{a_{surf}}{\omega(\xi_0)} = \frac{a_{surf}}{H_0}L_x\sqrt{\frac{gH_0}{\nu}} \tag{1.26}$$

(the factor 2π is used here to simplify computations to be done later).

REMARK 1.8. In shallow water ($\mu \ll 1$), one has $\nu \sim 1$ while in deep water ($\mu \gg 1$), one has $\nu \sim (2\pi\sqrt{\mu})^{-1}$. It follows that the typical wave celerity is $\sqrt{gH_0}$ in shallow water and $\sqrt{\frac{gL_x}{2\pi}}$ in deep water. In particular, water waves are *dispersive* in deep water (their celerity depends on the wavelength) but *nondispersive* in shallow water. In view of this fundamental qualitative difference, it is not surprising that different nondimensionalizations are used to describe shallow and deep water waves propagation (see §1.3.3 below). See also Figures 1.4 and 1.5 for examples of the very different behavior of shallow and deep water waves.

1.3.3. Nondimensionalization of the variables and unknowns. Some variables and unknowns can be nondimensionalized in a very simple way using the typical lengths introduced in §1.3.1. Namely, let us introduce

$$x' = \frac{x}{L_x}, \qquad y' = \frac{y}{L_y}, \qquad \zeta' = \frac{\zeta}{a_{surf}}, \qquad b' = \frac{b}{a_{bott}}.$$

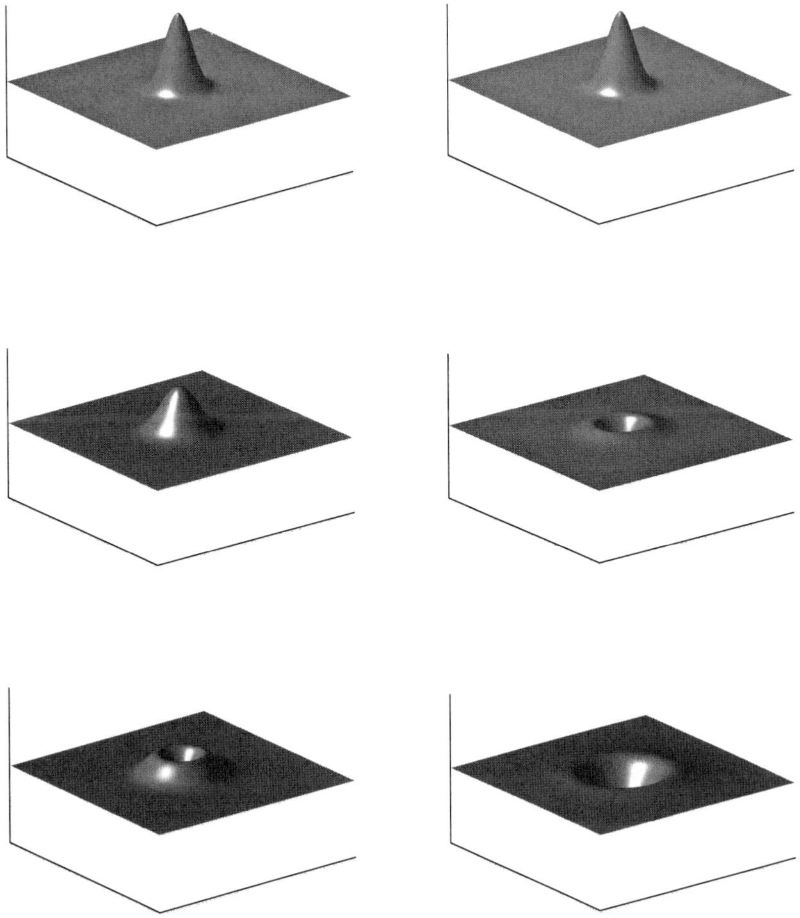

FIGURE 1.5. Same as Figure 1.4 in dimension $d = 2$ (left: shallow water; right: deep water).

In other words, we take L_x and L_y as unit lengths in the x and y direction respectively, and the water and bottom parametrizations have amplitude variations roughly equal to 1.

The nondimensionalization of the variable t and of the velocity potential Φ (and thus of ψ) are less intuitive, and we appeal to the linear analysis performed in §1.3.2. It is quite natural to scale the time variable on L_x/c_0, where c_0 is the wave celerity. Owing to (1.24), this yields

$$t' = \frac{t}{t_0} \quad \text{with} \quad t_0 = \frac{L_x}{\sqrt{gH_0\nu}}.$$

We have already seen that the typical order of magnitude for the velocity potential is given by (1.26). Consequently, we set

$$\Phi' = \frac{1}{\Phi_0}\Phi, \quad \text{with} \quad \Phi_0 = \frac{a_{surf}}{H_0} L_x \sqrt{\frac{gH_0}{\nu}}.$$

REMARK 1.9. Two distinct nondimensionalizations are used in oceanography (e.g., [**125**]) depending on the value of μ: shallow water scaling ($\mu \ll 1$) and Stokes wave scaling for intermediate to deep water ($\mu \not\ll 1$). They correspond to the specializations of the present nondimensionalization in the cases $\nu = 1$ and $\nu = 2\pi\mu^{-1/2}$ respectively, that is, to the asymptotic value of ν as $\mu \to 0$ and $\mu \to \infty$.

Even though the vertical variable z does not appear in (1.3), it is implicitly involved in the definition (1.2) of the Dirichlet-Neumann operator. We therefore need to nondimensionalize it also, but the choice of the scaling length is technically rather than physically motivated. For the convenience of working with a reference depth equal to 1, we set

$$z' = \frac{z}{H_0}.$$

But any other choice (such as $z' = z/L_x$ in infinite depth) would be equivalent. See Remark 1.13.

REMARK 1.10. As a wave propagates towards the beach, its wavelength shortens, but its period[6], denoted T_0, remains constant. It is sometimes convenient to use T_0 rather than L_x to nondimensionalize the equations. The typical wavelength L_x is therefore deduced from the constitutive relation $c_0 = L_x/T_0$, by solving the following equation for L_x,

$$\frac{L_x}{\sqrt{gH_0}}\Big(\frac{L_x}{H_0}\frac{1}{2\pi}\tanh(2\pi\frac{H_0}{L_x})\Big)^{-1/2} = T_0.$$

For instance, a swell of period T_0 will have a wavelength $L_0 \sim \frac{1}{2\pi}gT_0^2$ offshore (infinite depth limit in the above formula), while this wavelength will shorten as $L_0 \sim \sqrt{gH_0}T_0$ as the waves will approach the shore (take $\tanh(2\pi H_0/L_x) \sim 2\pi H_0/L_x$ in the above formula).

1.3.4. Nondimensionalization of the equations. Before addressing the nondimensionalization of (1.3), let us introduce a few notations for parameter depending operators.

NOTATION 1.11. We define "twisted" horizontal gradient and $\mathbb{R}^{d+1}$-dimensional Laplace operator as follows:

$$\begin{array}{llll} \nabla^\gamma = (\partial_x, \gamma\partial_y)^T & \text{if } d=2 \quad \text{and} & \nabla^\gamma = \partial_x & \text{if } d=1, \\ \Delta^{\mu,\gamma} = \mu\partial_x^2 + \gamma^2\mu\partial_y^2 + \partial_z^2 & \text{if } d=2 \quad \text{and} & \Delta^{\mu,\gamma} = \mu\partial_x^2 + \partial_z^2 & \text{if } d=1. \end{array}$$

In order to write the water waves equations (1.3) in nondimensionalized form let us first focus our attention on the Dirichlet-Neumann operator. One gets readily from (1.2) that

$$\begin{aligned} G[\zeta, b]\psi &= (\partial_z\Phi)_{|z=\zeta} - (\nabla\Phi)_{|z=\zeta}\cdot\nabla\zeta \\ &= \frac{\Phi_0}{H_0}\big(\partial_{z'}\Phi' - \mu\nabla'^\gamma(\varepsilon\zeta')\cdot\nabla'^\gamma\Phi'\big)_{|z'=\varepsilon\zeta'}, \end{aligned}$$

and therefore

$$G[\zeta, b]\psi = \frac{\Phi_0}{H_0}\mathcal{G}_{\mu,\gamma}[\varepsilon\zeta', \beta b']\psi', \tag{1.27}$$

[6]Roughly defined as the time interval between two consecutive "crests".

where

$$\mathcal{G}_{\mu,\gamma}[\varepsilon\zeta', \beta b']\psi' = \sqrt{1+|\nabla'(\varepsilon\zeta')|^2}\partial'_{\mathbf{n}}\Phi'_{|_{z'=\varepsilon\zeta'}}$$

and where Φ' solves the nondimensionalized version of (1.1), namely,

$$\begin{cases} \Delta'^{\mu,\gamma}\Phi' = 0 \quad \text{for} \quad -1+\beta b' \leq z' \leq \varepsilon\zeta', \\ \Phi'_{|_{z'=\varepsilon\zeta'}} = \psi', \qquad \partial'_{\mathbf{n}}\Phi'_{|_{z'=-1+\beta b'}} = 0; \end{cases}$$

here, $\partial'_{\mathbf{n}}\Phi'_{|_{z'=-1+\beta b'}} = \big(\partial_{z'}\Phi' - \mu\nabla'^{\gamma}(\beta b')\cdot\nabla'^{\gamma}\Phi'\big)_{|_{z'=-1+\beta b'}}$, according to the notation below.

NOTATION 1.12. As always in these notes, $\partial_{\mathbf{n}}$ refers to the upward *conormal* derivative associated to an elliptic operator $\nabla_{X,z}\cdot P\nabla_{X,z}$ (where P is a $(d+1)\times(d+1)$ coercive matrix). It is defined as

$$\partial_{\mathbf{n}} = \mathbf{n}\cdot P\nabla_{X,z},$$

where $\mathbf{n}$ is the upward normal vector. When $P = I$, the elliptic operator is the standard Laplace operator and the conormal derivative coincides with the normal derivative.

REMARK 1.13. As noted in §1.3.3, the choice of a scaling length for the vertical variable is purely technical. Let us for instance consider what happens if the vertical variable z is scaled by L_x (which is usually done in deep or infinite depth), i.e.,

$$z' = \frac{z}{L_x}.$$

Instead of (1.27), one would then obtain

$$G[\zeta, b]\psi = \frac{\Phi_0}{L_x}\tilde{\mathcal{G}}_{\mu,\gamma}[\epsilon\zeta', \beta\sqrt{\mu}b']\psi',$$

where

$$\tilde{\mathcal{G}}_{\mu,\gamma}[\epsilon\zeta', \beta\sqrt{\mu}b']\psi' = \sqrt{1+|\nabla'(\epsilon\zeta')|^2}\partial'_{\mathbf{n}}\Phi'_{|_{z'=\epsilon\zeta'}} = \big(\partial'_z - \epsilon\nabla'^{\gamma}\zeta'\cdot\nabla'^{\gamma}\big)\Phi'_{|_{z'=\epsilon\zeta'}}$$

and where Φ' solves the nondimensionalized version of (1.1), namely,

$$\begin{cases} \Delta^{\gamma}_{X',z'}\Phi' = 0 \quad \text{for} \quad -\sqrt{\mu}+\beta\sqrt{\mu}b' \leq z' \leq \epsilon\zeta', \\ \Phi'_{|_{z'=\epsilon\zeta'}} = \psi', \qquad \partial'_{\mathbf{n}}\Phi'_{|_{z'=-\sqrt{\mu}+\beta\sqrt{\mu}b'}} = 0, \end{cases}$$

where $\Delta^{\gamma}_{X',z'} = \partial_z^2 + \gamma^2\partial_y^2 + \partial_z^2$ (if $d = 2$). Because of the change in the scaling of z, the dependence on μ no longer appears in the Laplace operator, but in the bottom parametrization. Both formulations are of course completely equivalent in the sense that they give the same expression for $G[\zeta, b]\psi$. While the first scaling is more convenient to study shallow water regimes, the second scaling is the relevant one to handle the case of infinite depth.

Using (1.27) together with the nondimensionalizations described in §1.3.3, it is easy to deduce the dimensionless version of the water waves equations[7] (1.3), namely (omitting the primes for the sake of clarity),

$$\begin{cases} \partial_t\zeta - \dfrac{1}{\mu\nu}\mathcal{G}_{\mu,\gamma}[\varepsilon\zeta, \beta b]\psi = 0, \\ \partial_t\psi + \zeta + \dfrac{\varepsilon}{2\nu}|\nabla^{\gamma}\psi|^2 - \dfrac{\varepsilon\mu}{\nu}\dfrac{\big(\frac{1}{\mu}\mathcal{G}_{\mu,\gamma}[\varepsilon\zeta, \beta b]\psi + \nabla^{\gamma}(\varepsilon\zeta)\cdot\nabla^{\gamma}\psi\big)^2}{2(1+\varepsilon^2\mu|\nabla^{\gamma}\zeta|^2)} = 0, \end{cases} \tag{1.28}$$

[7]See also (4.65) for the case of infinite depth.

where

$$\begin{aligned}\mathcal{G}_{\mu,\gamma}[\varepsilon\zeta,\beta b]\psi &= \sqrt{1+\varepsilon^2|\nabla\zeta|^2}\partial_{\mathbf{n}}\Phi_{|_{z=\varepsilon\zeta}}\\ &= -\mu\nabla^\gamma(\varepsilon\zeta)\cdot\nabla^\gamma\Phi_{|_{z=\varepsilon\zeta}}+\partial_z\Phi_{|_{z=\varepsilon\zeta}},\end{aligned}\tag{1.29}$$

and with

$$\begin{cases}\Delta^{\mu,\gamma}\Phi=0, & -1+\beta b\le z\le\varepsilon\zeta,\\ \Phi_{|_{z=\varepsilon\zeta}}=\psi, & \partial_{\mathbf{n}}\Phi_{|_{z=-1+\beta b}}=0.\end{cases}\tag{1.30}$$

(Recall that ∇^γ and $\Delta^{\mu,\gamma}$ are defined in Notation 1.11 and that $\partial_{\mathbf{n}}$ stands for the upwards conormal derivative. See Notation 1.12.)

REMARK 1.14. It is of course possible to perform the nondimensionalization directly on the Euler equations (H1)′–(H9)′. For instance, (H1)′ reads in dimensionless form,

$$\begin{cases}\partial_t V+\frac{\varepsilon}{\nu}(V\cdot\nabla^\gamma+\frac{1}{\mu}w\partial_z)V=-\frac{1}{\varepsilon}\nabla^\gamma P,\\ \partial_t w+\frac{\varepsilon}{\nu}(V\cdot\nabla^\gamma+\frac{1}{\mu}w\partial_z)w=-\frac{1}{\varepsilon}(\partial_z P+1),\end{cases}$$

where, for the sake of clarity, we have split the equation into its horizontal and vertical components, and where $P_0=\rho g H_0$ (the typical order of the hydrostatic pressure) has been used to nondimensionalize the pressure, while V and w have been naturally[8] scaled with $V_0=\Phi_0/L$ and $w=\Phi_0/H$.
Similarly, the dimensionless version of (H6)′ is given by

$$\partial_t\zeta+\frac{\varepsilon}{\nu}\underline{V}\cdot\nabla^\gamma\zeta-\frac{1}{\mu\nu}\underline{w}=0,$$

where $\underline{\mathbf{U}}=(\underline{V},\underline{w})$ stands for the velocity evaluated at the surface $\underline{\mathbf{U}}=\mathbf{U}_{|_{z=\varepsilon\zeta}}$.

1.4. Plane waves, waves packets, and modulation equations

The goal of this section is to give a brief and formal introduction to the study of wave packets and modulation equations that will be addressed in Chapter 8.

One easily deduces from the analysis performed in §1.3.2 that the linear water waves equations (1.18) possess *plane waves* solution of the form

$$\begin{pmatrix}\zeta(t,X)\\ \psi(t,X)\end{pmatrix}=\begin{pmatrix}\zeta_{01}\\ \psi_{01}\end{pmatrix}e^{i(\mathbf{K}\cdot X-\Omega t)}+\text{c.c.}$$

[8] This nondimensionalization comes from the fact that the velocity can be expressed as the gradient of a potential, $V=\nabla^\gamma\Phi$, $w=\partial_z\Phi$. If one is not interested in working with the potential—for instance, to handle the irrotational case—it is possible to use a different nondimensionalization. Typically, in the shallow water regime ($\mu\ll 1$, and we thus set $\nu=1$), we know that with the above nondimensionalization, one has $w'=O(\mu)$ (see §5.6.2.3). We can therefore choose to scale the vertical velocity by $\mu w_0=\frac{a}{L_x}\sqrt{gH}$, so that $w'=O(1)$. The equations are then (with $\nu=1$)

$$\begin{cases}\partial_t V+\varepsilon(V\cdot\nabla^\gamma+w\partial_z)V=-\frac{1}{\varepsilon}\nabla^\gamma P,\\ \partial_t w+\varepsilon(V\cdot\nabla^\gamma+w\partial_z)w=-\frac{1}{\varepsilon\mu}(\partial_z P+1).\end{cases}$$

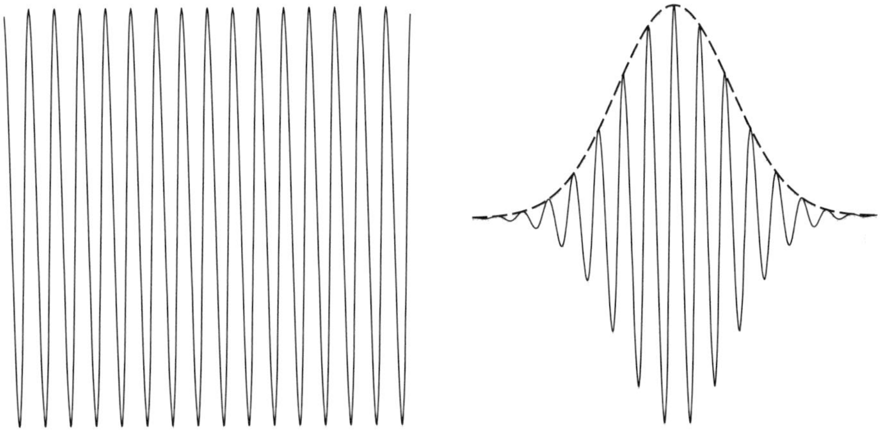

FIGURE 1.6. A plane wave and a wave packet (in dashed line, the envelope of the wave packet).

(c.c. stands for "complex conjugate"), that is, oscillations of constant amplitude ζ_{01} and ψ_{01}, wave number $\mathbf{K}$ and pulsation Ω, provided that

$$\Omega = \omega(\mathbf{K}) \quad \text{and} \quad \zeta_{01} = i\frac{\omega(\mathbf{K})}{g}\psi_{01},$$

where $\omega(\cdot)$ is as in (1.22).
Oscillating waves observed at the surface of an ocean are of course not of infinite extent, and their amplitude is slowly modulated (i.e., it varies with a typical scale much larger than the period of the oscillations). These slowly modulated oscillations are called *wave packets* (see Figure 1.6), and they are of the form

$$\begin{pmatrix} \zeta(t,X) \\ \psi(t,X) \end{pmatrix} = \begin{pmatrix} \zeta_{01}(\epsilon t,\epsilon X) \\ \psi_{01}(\epsilon t,\epsilon X) \end{pmatrix} e^{i(\mathbf{K}\cdot X - \Omega t)} + \text{c.c.},$$

where ϵ is some dimensionless parameter assumed to be small[9]. The difference between wave packets and plane waves is that the amplitudes ζ_{01} and ψ_{01} are not constant anymore but depend on space and time through $X' = \epsilon X$ and $t' = \epsilon t$. (The functions ζ_{01} and ψ_{01} are often called *envelopes* of the oscillations.)

Contrary to what happened with plane waves, one cannot find *exact* solutions of the linear water waves equations under the form of wave packets. We (formally) show below that one can, however, find approximate solutions under the form of wave packets. Let us therefore consider the linear water waves equations (1.18) with initial conditions

$$\begin{pmatrix} \zeta(0,X) \\ \psi(0,X) \end{pmatrix} = \begin{pmatrix} \zeta_{01}^0(\epsilon X) \\ \psi_{01}^0(\epsilon X) \end{pmatrix} e^{i\mathbf{K}\cdot X} + \text{c.c.},$$

and let us assume for simplicity that the Fourier transforms of ζ_{01} and ψ_{01} are compactly supported. From our brief analysis of the plane wave solutions, we naturally impose that

$$\zeta_{01}^0 = i\frac{\omega(\mathbf{K})}{g}\psi_{01}^0.$$

[9]In Chapter 8 where the study of wave packets is addressed, this small parameter is the *steepness* of the wave (see §1.5 below for a definition of this parameter).

Remarking that

$$\mathcal{F}(\zeta_{01}^0(\epsilon X)e^{i\mathbf{K}\cdot X})(\xi) = \frac{1}{\epsilon^d}\widehat{\zeta_{01}^0}(\frac{\xi-\mathbf{K}}{\epsilon})$$

we get from the representation formula (1.21) that

$$\begin{aligned}\zeta(t,X) &= \frac{i}{2}\int_{\mathbb{R}^d} e^{i(\xi\cdot X-\omega(\xi)t)}\frac{\omega(\mathbf{K})+\omega(\xi)}{g}\frac{1}{\epsilon^d}\widehat{\psi_{01}^0}(\frac{\xi-\mathbf{K}}{\epsilon})d\xi\\ &+ \frac{i}{2}\int_{\mathbb{R}^d} e^{i(\xi\cdot X+\omega(\xi)t)}\frac{\omega(\mathbf{K})-\omega(\xi)}{g}\frac{1}{\epsilon^d}\widehat{\psi_{01}^0}(\frac{\xi-\mathbf{K}}{\epsilon})d\xi\end{aligned}$$

and therefore, with a simple change of the integration variable,

$$\begin{aligned}\zeta(t,X) &= e^{i(\mathbf{K}\cdot X-\Omega t)}\frac{i}{2}\int_{\mathbb{R}^d} e^{i(\epsilon\xi\cdot X+(\omega(\mathbf{K})-\omega(\mathbf{K}+\epsilon\xi)t)}\frac{\omega(\mathbf{K})+\omega(\mathbf{K}+\epsilon\xi)}{g}\widehat{\psi_{01}^0}(\xi)d\xi\\ &+ \frac{i}{2}\int_{\mathbb{R}^d} e^{i(\epsilon\xi\cdot X+\omega(\mathbf{K}+\epsilon\xi)t)}\frac{\omega(\mathbf{K})-\omega(\mathbf{K}+\epsilon\xi)}{g}\widehat{\psi_{01}^0}(\xi)d\xi + \text{c.c.},\end{aligned}$$

with $\Omega = \omega(\mathbf{K})$.

Since we assumed that $\widehat{\psi_{01}}$ is compactly supported, a first-order Taylor expansion of $\omega(\mathbf{K}+\epsilon\xi)$ shows that the second term in the above identity is of order $O(\epsilon)$; up to another error term of size $O(\epsilon)$, we can also replace $\omega(\mathbf{K}+\epsilon X)$ by $\Omega=\omega(\mathbf{K})$ in the first term of the right-hand-side. We can therefore write

$$\zeta(t,X) = \zeta_{01}(\epsilon t,\epsilon X)e^{i(\mathbf{K}\cdot X-\Omega t)} + \text{c.c.} + O(\epsilon); \tag{1.31}$$

denoting by $(t',X')=(\epsilon t,\epsilon X)$ the slow time and space variables, and recalling that $\zeta_{01}^0 = i\frac{\Omega}{g}\psi_{01}^0$, the envelope $\zeta_{01}(t',X')$ is explicitly given by

$$\zeta_{01}(t',X') = \int_{\mathbb{R}^d} e^{i(\xi\cdot X' + \frac{\omega(\mathbf{K})-\omega(\mathbf{K}+\epsilon\xi)}{\epsilon}t)}\widehat{\zeta_{01}^0}(\xi)d\xi.$$

Equivalently, ζ_{01} can be defined as the solution of the *linear full-dispersion equation*

$$\partial_t'\zeta_{01} + i\frac{\omega(\mathbf{K}+\epsilon D')-\omega(\mathbf{K})}{\epsilon}\zeta_{01} = 0, \qquad \zeta_{01}\,_{|_{t'=0}} = \zeta_{01}^0, \tag{1.32}$$

with $D'=-i\nabla'$ and where we used Notation 1.7 for Fourier multipliers.

The linear full-dispersion equation (1.32) is a nonlocal equation that might be delicate to handle in some contexts. It is therefore of interest to approximate it by a differential equation. This approximation depends on the time scale of interest:

- Time scale $t=O(1/\epsilon)$ (equivalently, $t'=O(1)$). Up to another error term of size $O(\epsilon)$ in (1.31), one can replace $\omega(\mathbf{K}+\epsilon D')$ by its *first-order* Taylor expansion in (1.32). This leads us to change the definition of ζ_{01} in (1.31) into the solution of the following *transport equation at the group velocity*

$$\partial_t'\zeta_{01} + \nabla\omega(\mathbf{K})\cdot\nabla'\zeta_{01} = 0, \qquad \zeta_{01}\,_{|_{t'=0}} = \zeta_{01}^0. \tag{1.33}$$

- Time scale $t=O(1/\epsilon^2)$ (equivalently, $t'=O(1/\epsilon)$). The approximation error made when replacing (1.32) by (1.33) add up to finally be of order $O(1)$ when $t'=O(1/\epsilon)$. This is why a more accurate approximation of (1.32) is necessary, and $\omega(\mathbf{K}+\epsilon D')$ is therefore replaced by its *second-order* Taylor expansion. Consequently, ζ_{01} in (1.31) is now defined as the solution to the *linear Schrödinger equation*,

$$\partial_t'\zeta_{01} + \nabla\omega(\mathbf{K})\cdot\nabla'\zeta_{01} - \epsilon\frac{i}{2}\nabla'\cdot\mathcal{H}_\omega(\mathbf{K})\nabla'\zeta_{01} = 0, \qquad \zeta_{01}\,_{|_{t'=0}} = \zeta_{01}^0. \tag{1.34}$$

By analogy with optics, we say that the time scale $t' = O(1)$ corresponds to geometric optics for which wave packets travel at the group velocity, while the time scale $t' = O(1/\epsilon)$ corresponds to diffractive optics where dispersive diffractive effects significantly affect the envelope of the wave packet.

The equations (1.32), (1.33), and (1.34) do not directly describe the solution to the (linear) water waves equation, but of the envelope of the wave packet; an approximate solution to the water waves equation is then recovered through (1.31). Such approximations are called *modulation approximations*, and (1.32), (1.33), and (1.34) are *modulation equations.*

This book is mainly focused on the approximation of shallow water models (see §1.5 below) that describe the transformation of waves as they approach the shore, but we nevertheless consider modulation approximations in Chapter 8. Dealing with the full nonlinear water waves equations makes the analysis much more technical than in the above presentation. An important nonlinear effect is the possible creation of a nonoscillating term by quadratic interaction of oscillating wave packets (the equivalent phenomenon in optics is called *optical rectification*). Such nonoscillating terms may interact with the modulation equations governing the evolution of wave packets and therefore there exist many models (such as the nonlinear Schrödinger, the Benney-Roskes, or the Davey-Stewartson equations) that describe the very rich dynamics of wave packets. Most of these models are not fully justified yet.

1.5. Asymptotic regimes

The significance of introducing the dimensionless parameters ε, μ, β and γ in §1.3.3 is that it is often possible to deduce from their values some insight on the behavior of the flow. More precisely, it is possible to derive from (1.30) some (much simpler) asymptotic models more amenable to numerical simulations and whose properties are more transparent. Various asymptotic models, with different properties, can be derived from (1.30); each of them corresponds to a specific *asymptotic regime*, that is, to a specific range of the values of ε, μ, β and γ. Before commenting on these asymptotic regimes, let us briefly discuss the typical values of the dimensionless parameters mentioned earlier.

- The amplitude parameter $\varepsilon = a_{surf}/H_0$. Since the amplitude of the wave is at most of the same order as the depth, it is perfectly reasonable to assume that $0 \leq \varepsilon \leq 1$. (See Figure 1.7 for experimental data near the shore and Figure 1.8 for typical data about the ocean.)
- The topography parameter $\beta = a_{bott}/H_0$. Since the order of the bottom variations does not exceed the typical depth, we also have $0 \leq \beta \leq 1$.
- The shallowness parameter $\mu = H_0^2/L_x^2$. This ratio can be either small or very large. We thus take $\mu \geq 0$.
- The transversality parameter $\gamma = L_x/L_y$. Up to a permutation of the horizontal axis, it is possible to assume that $0 \leq \gamma \leq 1$.
- The steepness of the wave $\epsilon = a_{surf}/L_x = \varepsilon\sqrt{\mu}$. Physical observations show that the steepness ϵ never exceeds a critical value because of the occurrence of wave breaking. We thus assume that $0 \leq \epsilon \leq \epsilon_{max}$ (for some $\epsilon_{max} > 0$).
- The steepness of the bottom variations $\epsilon_{bott} = a_{bott}/L_x = \beta\sqrt{\mu}$. As for the steepness of the surface variations, we impose that $0 \leq \epsilon_{bott} \leq \epsilon_{max}$.

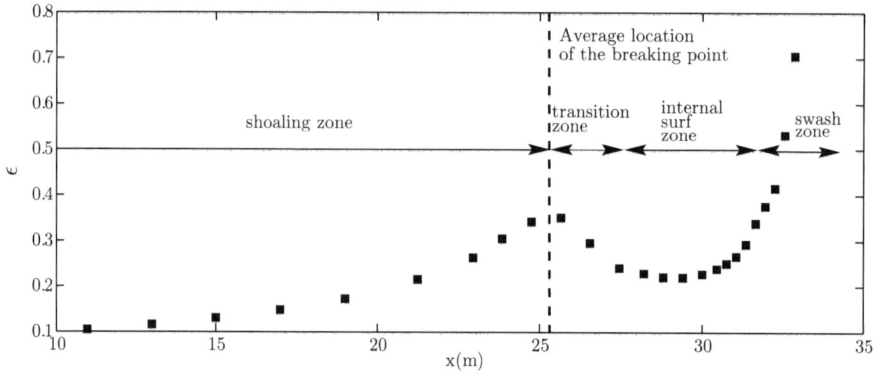

FIGURE 1.7. Variation of the nonlinear parameter ε averaged over several wave groups in terms of the position. Computed by Tissier [**313**] using laboratory experiments by van Dongeren et al. [**127**] on bichromatic waves breaking on a plane beach.

This limitation is essentially due to technical reasons, but is completely harmless for the applications.

To sum up the above discussion, we do not lose much generality if we assume that the parameters ε, μ, β and γ are subject to the following constraints

$$0 \leq \varepsilon, \beta, \gamma \leq 1, \quad 0 \leq \mu, \quad 0 \leq \varepsilon\sqrt{\mu} \leq \epsilon_{max}, \quad 0 \leq \beta\sqrt{\mu} \leq \epsilon_{max}. \tag{1.35}$$

It is possible to work in this framework, as in [**12**]. However, we can simplify the work if we assume further that μ remains bounded (but not necessarily small). This assumption excludes very deep water configurations, but we are mostly concerned here with applications to coastal oceanography where such configurations are not relevant. (For the sake of completeness, several infinite depth models are derived in Chapter 8.) We refer to [**12**] for a general approach that allows for very large values of μ and assume throughout these notes that

$$0 \leq \varepsilon, \beta, \gamma \leq 1, \quad 0 \leq \mu \leq \mu_{max}, \tag{1.36}$$

for some $\mu_{max} > 0$ (not necessarily small). Consequently, we can take $\nu = 1$ in (1.28).

What we call *asymptotic regimes* correspond therefore to some subdomains of the set of all the parameters satisfying (1.36). These asymptotic regimes will be studied in full detail in Chapters 5–8 (see also Appendix C for a reader's digest of all the asymptotic models derived in these notes), but it is worth introducing some terminology at this point:

- Small/large amplitude regimes. It is said that the flow under consideration is in a *small amplitude* regime if $\varepsilon \ll 1$. If no smallness assumption is made on ε (i.e. if $\varepsilon = O(1)$), then the flow is said to be in a *large amplitude* regime.
- Shallow/deep water regime. The *shallow water* regime corresponds to $\mu \ll 1$. If this condition is not true, then we are in *deep water* (the situation $\mu \sim 1$ is sometimes referred to as *intermediate depth*).
- Small/large bottom variations. The bottom variations are of *small amplitude* if $\beta \ll 1$ and large amplitude if $\beta = O(1)$.

Wind Speed (Kts)	Sea State	Significant Wave (Ft)	Significant Range of Periods (Sec)	Average Period (Sec)	Average Length of Waves (FT)
3	0	<.5	<.5 - 1	0.5	1.5
4	0	<.5	.5 - 1	1	2
5	1	0.5	1 - 2.5	1.5	9.5
7	1	1	1 - 3.5	2	13
8	1	1	1 - 4	2	16
9	2	1.5	1.5 - 4	2.5	20
10	2	2	1.5 - 5	3	26
11	2.5	2.5	1.5 - 5.5	3	33
13	2.5	3	2 - 6	3.5	39.5
14	3	3.5	2 - 6.5	3.5	46
15	3	4	2 - 7	4	52.5
16	3.5	4.5	2.5 - 7	4	59
17	3.5	5	2.5 - 7.5	4.5	65.5
18	4	6	2.5 - 8.5	5	79
19	4	7	3 - 9	5	92
20	4	7.5	3 - 9.5	5.5	99
21	5	8	3 - 10	5.5	105
22	5	9	3.5 - 10.5	6	118
23	5	10	3.5 - 11	6	131.5
25	5	12	4 - 12	7	157.5
27	6	14	4 - 13	7.5	184
29	6	16	4.5 - 13.5	8	210
31	6	18	4.5 - 14.5	8.5	236.5
33	6	20	5 - 15.5	9	262.5
37	7	25	5.5 - 17	10	328.5
40	7	30	6 - 19	11	394
43	7	35	6.5 - 21	12	460
46	7	40	7 - 22	12.5	525.5
49	8	45	7.5 - 23	13	591
52	8	50	7.5 - 24	14	655
54	8	55	8 - 25.5	14.5	722.5
57	8	60	8.5 - 26.5	15	788
61	9	70	9 - 28.5	16.5	920
65	9	80	10 - 30.5	17.5	1099
69	9	90	10.5 - 32.5	18.5	1182
73	9	100	11 - 34.5	19.5	1313.5

FIGURE 1.8. The Pierson-Moskowitz sea-state table. Table from Resolute Weather.

- Weakly transverse regime. A flow is said to be *weakly transverse* if $\gamma \ll 1$ (the $1D$ case corresponds formally to $\gamma = 0$).

1.6. Extension to moving bottoms

As discussed in §1.1.1, we consider in most of these notes the motion of a fluid domain delimited below by a fixed bottom and above by a free surface. It is possible to extend the study to the case of a moving bottom. This has been done for instance in [**4**] for the local well-posedness theory and in [**182**] in the shallow water setting. In this case, the bottom parametrization b is now time dependent (but, contrary to the free surface, it is known). Therefore, assumption (H4)′ in §1.1.2 must be replaced by

$$(\mathrm{H4})' \qquad \forall t \in [0,T), \qquad \Omega_t = \{(X,z) \in \mathbb{R}^{d+1}, -H_0 + b(t,X) < z < \zeta(t,X)\},$$

while the boundary condition at the bottom (H5)′ is now a kinematic condition similar to the one imposed at the surface,

$$(\mathrm{H5})' \qquad \sqrt{1+|\nabla b|^2}\mathbf{U}\cdot\mathbf{n} = \partial_t b \quad \text{on} \quad \{z = -H_0 + b(t,X)\},$$

or, equivalently, in terms of the velocity potential Φ,

$$\text{(H5)}'' \qquad \sqrt{1+|\nabla b|^2}\partial_{\mathbf{n}}\Phi = \partial_t b \quad \text{on} \quad \{z = -H_0 + b(X)\}.$$

The first consequence of these modifications is that we recover the velocity potential from its trace at the surface by solving a boundary problem with a *nonhomogeneous* Neumann condition at the bottom. More precisely, (1.1) must be replaced by

$$\text{(1.37)} \qquad \begin{cases} \Delta_{X,z}\Phi = 0 & \text{in } \Omega_t, \\ \Phi_{|_{z=\zeta}} = \psi, & \sqrt{1+|\nabla b|^2}\partial_{\mathbf{n}}\Phi_{|_{z=-H_0+b}} = \partial_t b, \end{cases}$$

so that we can decompose Φ into its "fix bottom" and "moving bottom" components,

$$\Phi = \Phi_{fb} + \Phi_{mb}$$

with

$$\begin{cases} \Delta_{X,z}\Phi_{fb} = 0 & \text{in } \Omega_t, \\ \Phi_{fb\,|_{z=\zeta}} = \psi, & \sqrt{1+|\nabla b|^2}\partial_{\mathbf{n}}\Phi_{fb\,|_{z=-H_0+b}} = 0, \end{cases}$$

and

$$\begin{cases} \Delta_{X,z}\Phi_{mb} = 0 & \text{in } \Omega_t, \\ \Phi_{mb\,|_{z=\zeta}} = 0, & \sqrt{1+|\nabla b|^2}\partial_{\mathbf{n}}\Phi_{mb\,|_{z=-H_0+b}} = \partial_t b. \end{cases}$$

We therefore have

$$\sqrt{1+|\nabla\zeta|^2}\partial_{\mathbf{n}}\Phi_{|_{z=\zeta}} = G[\zeta,b]\psi + G^{NN}[\zeta,b]\partial_t b,$$

where the "Neumann-Neumann" operator $G^{NN}[\zeta,b]$ is defined as

$$G^{NN}[\zeta,b] : b_t \mapsto \sqrt{1+|\nabla\zeta|^2}\partial_{\mathbf{n}}\Phi_{mb\,|_{z=\zeta}}.$$

The Zakharov-Craig-Sulem formulation (1.3) can therefore be extended to the case of moving bottoms in the following way,

$$\text{(1.38)} \qquad \begin{cases} \partial_t\zeta - G[\zeta,b]\psi = G^{NN}[\zeta,b]\partial_t b, \\ \partial_t\psi + g\zeta + \frac{1}{2}|\nabla\psi|^2 - \frac{(G[\zeta,b]\psi + G^{NN}[\zeta,b]\partial_t b + \nabla\zeta\cdot\nabla\psi)^2}{2(1+|\nabla\zeta|^2)} = 0. \end{cases}$$

REMARK 1.15. We can generalize the linear analysis of Section 1.3.2 to the case of moving bottoms. With the same kind of Fourier analysis used to compute $G[0,0] = |D|\tanh(H_0|D|)$, we obtain $G^{NN}[0,0] = \frac{1}{\cosh(H_0|D|)}\partial_t b$, so that (1.19) must be replaced by

$$\text{(1.39)} \qquad \partial_t^2\zeta + gG[0,0]\zeta = G^{NN}[0,0]\partial_t^2 b.$$

We will now give a dimensionless version of (1.38) that generalizes (1.28) to the case of moving bottoms. We need to consider the time scale t_{mb} of the bottom (temporal) variations, which is not necessarily of the same order as the time scale $t_0 = \frac{L_x}{gH_0\nu}$ deduced from the linear wave theory and used in §1.3.3 to get the dimensionless version of the water waves equations. We therefore introduce a new dimensionless parameter δ

$$\delta = \frac{t_{mb}}{t_0} = \frac{t_{mb}}{L_x/gH_0\nu},$$

so that in dimensionless variables, the bottom parametrization is given by $-1 + \beta b(t/\delta, X)$, and its time derivative $\frac{\beta}{\delta}\partial_\tau b$, with $\tau = t/\delta$. The dimensionless form of the potential equation (1.37) generalizing (1.30) is therefore given by

$$(1.40)\qquad \begin{cases} \Delta^{\mu,\gamma}\Phi = 0, & -1+\beta b \le z \le \varepsilon\zeta, \\ \Phi_{\vert z=\varepsilon\zeta} = \psi, & \sqrt{1+\beta^2\vert\nabla b\vert^2}\partial_{\mathbf{n}}\Phi_{\vert z=-1+\beta b} = \mu\nu\frac{\beta}{\varepsilon\delta}\partial_\tau b \end{cases}$$

and the equations (1.38) that generalize (1.28) can be written in the form

$$(1.41)\qquad \begin{cases} \partial_t\zeta - \frac{1}{\mu\nu}\mathcal{G}\psi = \frac{\beta}{\delta\varepsilon}\mathcal{G}^{NN}\partial_\tau b, \\ \partial_t\psi + \zeta + \frac{\varepsilon}{2\nu}\vert\nabla^\gamma\psi\vert^2 - \frac{\varepsilon\mu}{\nu}\frac{\left(\frac{1}{\mu}\mathcal{G}\psi + \frac{\beta\nu}{\delta\varepsilon}\mathcal{G}^{NN}\partial_\tau b + \nabla^\gamma(\varepsilon\zeta)\cdot\nabla^\gamma\psi\right)^2}{2(1+\varepsilon^2\mu\vert\nabla^\gamma\zeta\vert^2)} = 0, \end{cases}$$

where $\mathcal{G} := \mathcal{G}_{\mu,\gamma}[\varepsilon\zeta,\beta b]$ is as defined in (1.29)–(1.30) while the dimensionless Neumann-Neumann operator $\mathcal{G}^{NN} := \mathcal{G}^{NN}_{\mu,\gamma}[\varepsilon\zeta,\beta b]$ is given by

$$(1.42)\qquad \begin{aligned} \mathcal{G}^{NN}_{\mu,\gamma}[\varepsilon\zeta,\beta b]\partial_\tau b &= \sqrt{1+\varepsilon^2\vert\nabla\zeta\vert^2}\partial_{\mathbf{n}}\Phi_{mb\,\vert z=\varepsilon\zeta} \\ &= -\mu\nabla^\gamma(\varepsilon\zeta)\cdot\nabla^\gamma\Phi_{mb\,\vert z=\varepsilon\zeta} + \partial_z\Phi_{mb\,\vert z=\varepsilon\zeta}, \end{aligned}$$

and with

$$(1.43)\qquad \begin{cases} \Delta^{\mu,\gamma}\Phi_{mb} = 0, & -1+\beta b < z < \varepsilon\zeta, \\ \Phi_{mb\,\vert z=\varepsilon\zeta} = 0, & \sqrt{1+\beta^2\vert\nabla b\vert^2}\partial_{\mathbf{n}}\Phi_{mb\,\vert z=-1+\beta b} = \partial_\tau b. \end{cases}$$

(Recall that ∇^γ and $\Delta^{\mu,\gamma}$ are defined in Notation 1.11 and that $\partial_{\mathbf{n}}$ stands for the upwards conormal derivative. See Notation 1.12.)

T. Iguchi studied the system (1.41) in [**182**] with the goal of investigating tsunamis created by submarine earthquakes. He considered a shallow water regime ($\mu \ll 1$, $\nu = 1$) with large surface and bottom deformations ($\varepsilon = \beta = 1$) and fast bottom variations ($\delta \ll 1$). A model frequently used to describe tsunami propagation is the standard Nonlinear Shallow Water system (or Saint-Venant; see Chapter 5) with zero initial velocity and initial surface elevation equal to the permanent shift of the seabed. Iguchi deduced from his study of (1.41) that this is a correct approximation if $\mu = o(\delta)$ but that when $\mu \sim \delta$, the initial velocity field cannot be taken equal to zero.

There are many other asymptotic regimes where (1.41) is of interest; see for instance §5.4 of Chapter 5.

1.7. Extension to rough bottoms

We have already commented in Remark 1.4 that it is implicitly assumed here that we do not consider "rough" bottoms. There can actually be several kinds of rough bottoms, for instance:

(1) The bottom parametrization is not regular (see Figure 1.9).
(2) The bottom parametrization is regular, but its typical scale of variation l_b is much smaller than the typical horizontal length L_x (see Figure 1.10) for the surface wave.

1.7.1. Nonsmooth topographies.

1.7.1. Nonsmooth topographies. Nonsmooth topographies raise two kinds of problems. The first one is the issue of the local well-posedness of the water waves equations (1.3). Indeed the well-posedness theorem provided in Chapter 4 requires a lot of smoothness for the bottom parametrization b. Using the fact that the bottom contribution to the surface evolution is of lower order (it is analytic by

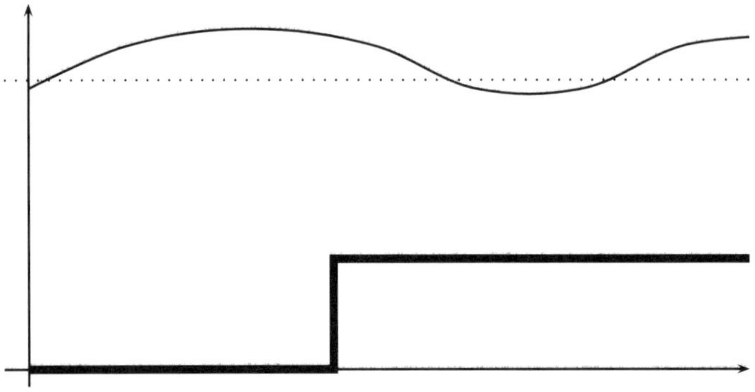

FIGURE 1.9. Rough bottom (1): the bottom parametrization is not smooth.

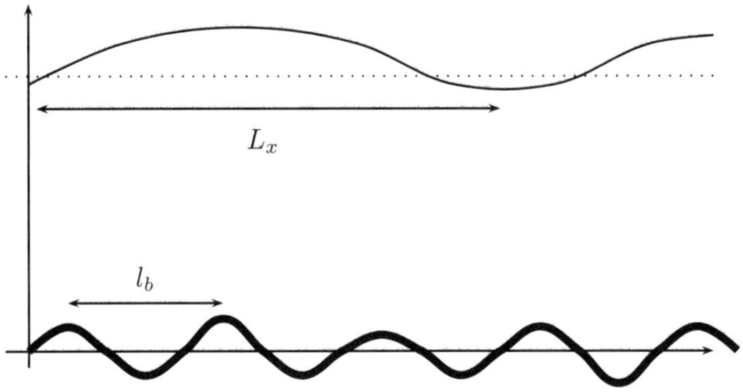

FIGURE 1.10. Rough bottom (2): the typical scale l_b for the bottom variations is much smaller than the typical scale L_x for the surface waves (e.g., distance from crest to crest).

ellipticity; see §A.4 for more comments on this point), Alazard, Burq, and Zuily showed in [**4**] that the water waves equations are locally well posed without any assumption on the bottom regularity. However, the existence time thus obtained shrinks to zero in the shallow water limit $\mu \to 0$ (see §A.4). There is still no result ensuring that the water waves equations are well posed on a time scale compatible with the shallow water limit.

This limitation raises the second issue related to nonsmooth topographies: since no result ensures that the solution to the water waves equations exists on the relevant time scale, no shallow water type model has been rigorously justified when the bottom parametrization is not smooth. For practical purposes, the shallow water models derived in Chapter 5 for smooth topographies are often used with nonsmooth topographies. For instance, the Nonlinear Shallow Water (or Saint-Venant) equations

$$\begin{cases} \partial_t \zeta + \nabla^\gamma \cdot \big((1 + \varepsilon\zeta - \beta b)\overline{V}\big) = 0, \\ \partial_t \overline{V} + \nabla^\gamma \zeta + \varepsilon(\overline{V} \cdot \nabla^\gamma)\overline{V} = 0, \end{cases} \tag{1.44}$$

are typically used when the bottom is a step (see Figure 1.9). Here,

$$\overline{V} = \frac{1}{1+\varepsilon\zeta-\beta b}\int_{-1+\beta b}^{\varepsilon\zeta} \nabla^{\gamma}\Phi(X,z)dz$$

is the vertically averaged horizontal component of the velocity. While we know from Chapter 6 that the Nonlinear Shallow Water (NSW) equations are fully justified for smooth topographies, there is no reason to believe that such a result holds in the case of a step, for instance. Indeed, while the contribution of the bottom to the surface evolution in (1.44) is singular (it is the divergence of a discontinuous term), we know that it must be infinitely smooth for the full water waves equations. In [**167**] (see also [**67**]), it is argued that in the case of a step, the topography term in (1.44) must indeed be modified if one wants to keep the same $O(\mu)$ precision associated to the NSW model with smooth topographies (see Theorem 6.10). However, the full justification of this new model remains to be done.

1.7.2. Rapidly varying topographies. Rapidly varying topographies do not raise any local well-posedness issues when the bottom parametrization is smooth. The issue is instead about the scale of the existence time. More precisely, if we denote by α the relative size of the bottom variations with respect to the typical horizontal scale L_x,

$$\alpha = \frac{l_b}{L_x},$$

the existence time furnished by standard local existence theorems shrinks to zero as $\alpha \to 0$. Getting an existence time independent of α is an homogenization issue that has not been addressed so far.

In order to fully justify shallow water models with rapidly varying topographies, the existence time should be uniform with respect to α *and* μ, and all the asymptotic models proposed in the literature are, in the best case, only partially justified. The issue is that the shallow water limit $\mu \to 0$ and the homogenization limit $\alpha \to 0$ *do not necessarily commute.*

In their seminal paper [**276**], Rosales and Papanicolaou use standard shallow water models (as those given in Chapter 5) without modifying them when the bottom is rapidly varying ($\alpha \ll 1$). When the bottom variations are periodic, or more generally when they are given by a stationary ergodic process, the techniques of homogenization theory are used to obtain effective shallow water models. The two most important examples are of periodic bottom topography and of topography given by a stationary random process. The approach of [**276**] has been followed in several works, unveiling interesting homogenization mechanisms (such as apparent diffusion). See [**17, 264, 152, 153**].

In the paper by Craig et al. [**105**], another approach is used. The homogenization approximation is performed on the Hamiltonian H given by (1.4), and (Hamiltonian) asymptotic models are deduced from the homogenized Hamiltonian. The periodic case is addressed in [**105**], while random topographies are treated in [**43**].

None of these two approaches brings a complete justification of the homogenized models they propose: The shallow water models used as a starting point in the first approach are *a priori* not valid for rapidly varying topographies, and the homogenization limit performed on the Hamiltonian in the second approach is not fully justified. This observation was the motivation for [**107**], where a shallow water model for rapidly varying periodic bottoms is obtained by handling *simultaneously* the shallow water and homogenization limits ($\mu \to 0$, $\alpha \to 0$). More precisely, the following regime is investigated in [**107**],

$$\beta = \sqrt{\mu} = 1 \ll 1, \tag{1.45}$$

and it is shown that at leading order, the solution solves the Nonlinear Shallow Water equations (1.44) with flat bottom, and that a rapidly oscillating corrector term must be added. It is also pointed out that these corrector terms may grow to destabilize the leading order term when a resonance occurs between the mean (or homogenized) flow and the bottom topography; this resonance can be understood as a nonlinear generalization of the Bragg resonance [**256**]. This approximation is shown to be consistent with the full water waves equations, but the full justification (see §C.1) remains an open problem. Similarly, handling random topography and/or other asymptotic regimes than (1.45) remains to be done.

1.8. Supplementary remarks

1.8.1. Discussion on the basic assumptions. We will first discuss further the relevance of various of the assumptions made in this chapter, namely, (H1)–(H9) in §1.1.1.

- *The fluid is homogeneous and inviscid (H1).*

(i) Homogeneity. The density ρ of seawater is given by an equation of state depending on the pressure P (in bars), the temperature θ (in degree C), and practical salinity S. "The equation of state defined by the Joint Panel on Oceanographic Tables and Standards (UNESCO, 1981) fits available measurements with a standard error of 3.5 ppm for pressure up to 1000 bars, for temperature between freezing and $40C$, and for salinities between 0 and 42" [**160**]. We are mainly interested here in models describing the transformation of the wave at the scale of a beach, for which this equation of state does not exhibit significant variations. A typical value for ρ at the surface is 1025 kg m^{-3}. At larger scales (e.g., the scale of the ocean), density variations of up to 2% are observed. They are mainly due to temperature variations; the region of large temperature gradient is called *thermoclyne* and is generally located at depth smaller than 1500m (a particular case is the *seasonal thermoclyne* observed near the surface in summer and autumn). The corresponding change of density is called *pycnocline.* Density variations are also sometimes due to a large gradient of salinity called *halocline.* Such density variations are responsible for important physical phenomena such as *internal waves.* We refer to [**160**] for a general physical description of these phenomena, to [**170**] for a review on internal waves, and to [**37, 131, 132**] for a mathematical and asymptotic description of these phenomena related to the approach developed in this book for surface waves.

(ii) Inviscidity. Throughout these notes, energy dissipations mechanisms are neglected. In order to take them into account, one should replace the Euler equation (H1)$'$ by the *Navier-Stokes equation*

$$\partial_t \mathbf{U} + (\mathbf{U}\cdot\nabla_{X,z})\mathbf{U} = -\frac{1}{\rho}\nabla_{X,z}P - g\mathbf{e_z} + \frac{v}{\rho}\Delta\mathbf{U} \quad \text{in} \quad \Omega_t, \tag{H1$'_v$}$$

where v is the *viscosity* [10]. One must also replace the boundary condition (H5)′ at the bottom by the *no-slip condition*,

$$(\mathrm{H5})'_v \qquad \mathbf{U} = 0 \quad \text{on} \quad \{z = -H_0 + b(X)\},$$

and the continuity of the pressure at the surface (H7)′ by the *continuity of the stress tensor*,

$$(\mathrm{H7})'_v \qquad (P - P_{atm})\mathbf{n} = 2v(S\mathbf{U})\mathbf{n},$$

where $S\mathbf{U} = \frac{1}{2}(\nabla_{X,z}\mathbf{U} + \nabla_{X,z}\mathbf{U}^T)$ is the symmetric part of the gradient. The dissipation induced by these changes can be roughly decomposed into three categories:

(1) Bottom friction. Tangential motion at the bottom is allowed by the boundary condition (H5)' but not in the viscous case $(\mathrm{H5})'_v$. It is well known that this induces a bottom boundary layer whose size is proportional to $\sqrt{v}$ and whose profile is governed by the Prandtl equation. This raises many difficulties: the Prandtl equation can be ill-posed for nonanalytic data [**154**] and the approximate solution can be unstable [**164, 166**]. This is why the only existing justification of the inviscid limit of the Navier-Stokes equation with a fixed boundary has been done in an analytic framework [**283**]. Handling the bottom boundary layer in the context of water waves therefore remains a widely open problem. If the dissipation effects due to this boundary layer are not significant for large depth (because the fluid is essentially at rest in deep water), they can be important in some situations in a shallow water regime (for which significant horizontal motion exists[11]). Coastal engineers usually model these effects by adding a friction term to the asymptotic models derived in the inviscid case (see §5.6.6). Of course, a complete mathematical justification of this friction term remains at the moment out of reach.

(2) Internal dissipation. This corresponds to the energy dissipated by viscous stresses in the fluid. In shallow water, it is much smaller than the energy dissipated by bottom friction. In deep water, it is the main cause of energy dissipation, but its effects can be considered as negligible. It is indeed shown by qualitative arguments in [**235**] that "the time needed for the energy in waves of length 1m and 10m to be reduced by a factor of e is 8000 periods and 250000 periods respectively".

(3) Surface dissipation. The presence of viscosity also induces a boundary layer near the surface. However, and contrary to what happens at the bottom, the boundary conditions on the velocity are not of Dirichlet type. The boundary conditions (H6)′–$(\mathrm{H7})'_v$ are of Navier type and the amplitude of the surface boundary layer profile is therefore smaller than for the bottom; in particular, it is natural to expect a uniform (with respect to the viscosity) control of the velocity in Lipschitz norm near the surface [**251**]. This fact has recently been used in [**252**] to justify the inviscid limit of the free surface Navier-Stokes equations towards the water waves equation in infinite depth. For practical applications, surface dissipation is almost

[10]The quantity v/ρ is often called *kinematic viscosity* (roughly 10^{-6} m^2 s^{-1} for water).

[11]Approximate particle trajectories can be deduced from the expression of the velocity field provided in the Appendix to Chapter 5. We refer to [**91**] and references therein for more refined considerations on particle trajectories.

always neglected (except in certain situations such as water covered with a thin film or surface contaminant [**235**]).

Due to the difficulty of handling rigorous viscous effects on water waves, various simple models have been proposed from the early works of Lamb [**214**] and Boussinesq [**50**]. For a recent reference, see [**122**].
- *Incompressibility assumption (H2).* The sound speed in water is approximatively $c_s = 1400$ m s^{-1}. Since the typical wave celerity is given by $c_0 = \sqrt{gH_0\nu}$, with $\nu \leq 1$, we have

$$\frac{c_0}{c_s} \leq \frac{\sqrt{gH_0}}{1400} \sim 2.2\,10^{-3}\sqrt{H_0}.$$

We therefore have $c_0 \ll c_s$ (and therefore incompressibility) provided that $H_0 \ll (2.2\,10^{-3})^{-2}$m $\sim$ 200km, which is obviously satisfied on Earth.

- *Irrotationality assumption (H3).* Irrotationality allows one to reduce the problem to equations cast on the interface. If this assumption is not made, an equation on the whole fluid domain must be kept for the vorticity. This makes the construction of an iterative scheme for the construction of solutions more delicate. Among the approaches described in §1.2 the variational approach of Shatah-Zeng [**297**] (see §1.2.3.1), and the Lagrangian approaches of Lindblad [**237**] and Coutand-Shkoller [**99**] (see §1.2.4) do not require the irrotationality assumption (H3). Moreover, Wu's formulation (see §1.2.1.2) has been extended to the rotational case in [**337, 338**], and Zakharov's Hamiltonian formulation (1.3) of the water waves equations can be generalized when $d = 1$ to the case of nonzero, constant vorticity (see also [**244, 90**] for variants in Lagrangian coordinates). When studying steady water waves with vorticity in horizontal dimension $d = 1$, a stream-function formulation or the related Dubreil-Jacotin formulation can be very useful (see for instance [**91**]).
For the applications we have in mind here (coastal oceanography), rotational effects are generally not very important and we therefore consider the irrotational case only.

- *Nonoverhanging waves (and bottoms) assumption (H4).* Our intention here is to describe the wave transformation as it approaches the shore, and it is therefore natural to assume that it is parametrized by a graph (when it ceases to be a graph, the wave enters the breaking zone that is not described here). However, it would be possible to describe the wave through a parametrized curve ($d = 1$) or surface ($d = 2$), thus allowing overhanging of waves; see [**326, 327, 14, 15, 97**]. In particular, all the methods described in §1.2 do not require the surface to be a graph. This is necessary, for instance, to study the formation of "splash" singularities as in [**65, 66, 100**] (see Figure 1.11).

The assumption that the bottom is a graph is not necessary either. In fact, it can be shown that the bottom only contributes to lower order to the dynamics of the surface and does not affect the well-posedness theory. In [**4**], the authors show that it is possible to take very wild bottoms (as far as the depth does not vanish). However, in the shallow water regime, the contribution of the bottom is very important (even if it remains of lower order in terms of regularity), and it is necessary to make stronger assumptions on the bottom in order to describe correctly the dynamics of the waves in this very important (for the applications) regime. The assumption that the bottom is a graph is standard in coastal oceanography [**125**].

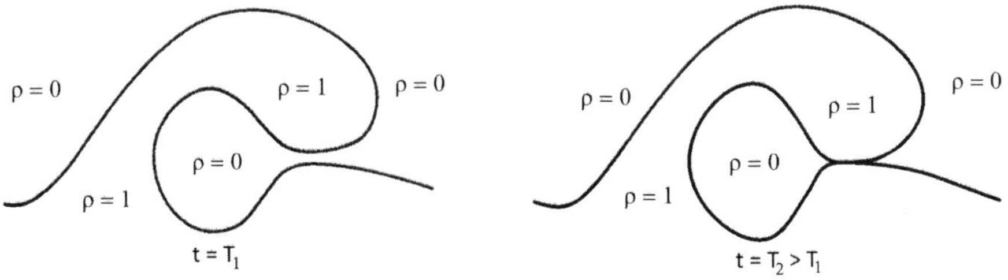

FIGURE 1.11. A splash singularity at time T_2: the interface between water ($\rho = 1$) and vacuum ($\rho = 0$) remains smooth but intersects itself.

- *No surface tension and constant external pressure (H7).* As discussed in §1.1.1, surface tension is in general irrelevant in coastal oceanography and we do not take it into account in the central part of these notes. For the sake of completeness, and because surface tension is relevant in other situations (such as ripples), we show in Chapter 9 how to take it into account, both at the level of the well-posedness of the water waves equations and for their description by asymptotic models.
In order to deal with a nonconstant external pressure, we could replace (H7)′ by

$$(\mathrm{H7})'_{ext} \qquad P = P_{atm} + P_{ext}(t,X) \quad \text{on} \quad \{z = \zeta(t,X)\},$$

for some nonconstant external pressure $P_{ext}(t,X)$; the equations (1.3) should then be replaced by

$$(1.46) \qquad \begin{cases} \partial_t\zeta - G[\zeta,b]\psi = 0, \\ \partial_t\psi + g\zeta + \frac{1}{2}|\nabla\psi|^2 - \dfrac{(G[\zeta,b]\psi + \nabla\zeta\cdot\nabla\psi)^2}{2(1+|\nabla\zeta|^2)} = -\dfrac{1}{\rho}P_{ext}. \end{cases}$$

- *The fluid is at rest at infinity (H8).* This assumption is very natural but could be removed without major difficulty (for wave-current interactions, see for instance [**91**]).

- *Nonvanishing water depth (H9).* This is a serious limitation for applications, but it remains a completely open mathematical problem. Many of the shallow water models derived in this book (and in particular the nonlinear shallow water equations and the Green-Naghdi equations) are used for applications up to the shoreline where the water depth vanishes (see for instance [**41**] and the review [**40**]). The use of these asymptotic models in such configurations is of course far from being rigorously justified.

1.8.2. Related frameworks. We are interested here in applications to coastal oceanography where the typical scale of interest is the size of the wave or the size of the beach. Shallow water models are also derived in oceanography for much larger scales for which *rotation effects* should be included. If we identify the vertical axis with the direction of gravity, we must consider that the equations are written in a rotating frame, and therefore add a Coriolis force. In full generality (i.e., including viscous effects), the Euler equation (H1)′ should then be replaced by

$$(\mathrm{H1})'_{\upsilon,\Omega} \qquad \partial_t\mathbf{U} + (\mathbf{U}\cdot\nabla_{X,z})\mathbf{U} + 2\mathbf{\Omega}\times\mathbf{U} = -\frac{1}{\rho}\nabla_{X,z}P - \nabla_{X,z}\Phi_\Omega - g\mathbf{e_z} + \frac{\upsilon}{\rho}\Delta\mathbf{U},$$

where $\mathbf{\Omega}$ is the rotation vector of the Earth (assumed to be constant for the scales under consideration) and $\Phi_\Omega = -\frac{1}{2}|\mathbf{\Omega} \times \mathbf{r}|^2$ ($\mathbf{r}$ being the position vector) is the centrifugal potential. This latter term does not modify the analysis since it can be included in the pressure term,[12] and it is the Coriolis term $2\mathbf{\Omega} \times \mathbf{U}$ that induces the effects of rotation on the fluid motion.

- *Effects of the rotation.* The effects of the Earth's rotation can be estimated by comparing the typical frequency of the phenomenon under study to the Coriolis frequency (also called the Coriolis parameter) $f_{cor} := 2\Omega \sin\phi$, where $\Omega = |\mathbf{\Omega}| = 7.3\,10^{-5}\mathrm{s}^{-1}$ is the Earth's angular velocity, and ϕ is the latitude.
  - At mid-latitudes, one has $f_{cor} \sim 10^{-4}$ Hz.
  - The typical frequency of the wave motion under consideration is given by $f = V_0/L_x$, where V_0 is the typical horizontal velocity.

In order to compare these two quantities, it is convenient to introduce the *Rossby number* defined as

$$\mathrm{Ro} = \frac{f}{f_{cor}} = \frac{V_0}{f_{cor} L_x}.$$

Coriolis effects will have a significant role if $\mathrm{Ro} < 1$, that is, if $L_x > \frac{V_0}{f_{cor}}$. For a typical[13] horizontal velocity $V_0 \sim 1\mathrm{m/s}$, the horizontal scale L_x must therefore be larger than 10km in order to notice Coriolis effects. For slower motions $V_0 \sim 0.1\mathrm{m/s}$, Coriolis effects become relevant at a shorter scale $L_x > 1\mathrm{km}$. These scales are much larger than the typical scale of a beach; this is the reason why we neglected Coriolis effects in this book.

- *Eddy viscosity.* The same reasoning used as in §1.8.1 can be applied to the effect of the viscosity. However, at large scales, an artificial viscosity, called *eddy viscosity*, is often introduced to model the dissipation caused by various phenomena such as white caps, local wave breaking, small-scale dissipation mechanisms, etc. Since these dissipation processes are much more important, this artificial eddy viscosity is anisotropic. Typically, vertical eddy viscosity is 10^3 the molecular value, while horizontal ones can be 10^{10} or 10^{11} times this value [**160**]. Introducing eddy viscosity is of course only a rough approximation of the dissipation processes involved at large scales, and much effort is made in oceanography to improve the modeling of these phenomena.

The large-scale models obtained after taking the above effects into consideration share many common points with the shallow water models derived in this book (see for instance [**155, 249, 54, 250**]). The pioneering work of Laplace[14] on tides [**226**],

---

[12]We should consequently modify the boundary condition $(\mathrm{H7})'$ on the pressure. However, the centrifugal potential is generally considered as constant at the surface and since the pressure is defined up to a constant, this does not affect the analysis.

[13]As previously seen in Remark 1.14, a typical scale for the horizontal velocity is $V_0 = \Phi_0/L_x = \frac{a_{surf}}{H_0}\sqrt{\frac{gH_0}{\nu}}$ and therefore $V_0 \sim \frac{a_{surf}}{H_0}\sqrt{gH_0}$ in shallow water and $V_0 \sim \frac{a_{surf}}{L_x}\sqrt{gL_x}$ in deep water. For a wave of amplitude $1m$, with a depth $H_0 \sim 10m$, and of wave length $L_x \gg 1$, therefore one has $V_0 \sim 1\mathrm{m/s}$.

[14]In this work, Laplace not only derived the Coriolis force due to the Earth's rotation more than fifty years before Coriolis; he also took into account the Earth's curvature, an effect that should be added to $(\mathrm{H1})'_{v,\Omega}$ at very large scales. For such scales, the variations of the rotation vector $\mathbf{\Omega}$ cannot be neglected as in $(\mathrm{H1})'_{v,\Omega}$ and this has important consequences, such as, for instance, the specific dynamics observed in tropical zones (see for instance [**207, 149, 150**]).

has motivated many mathematicians to study the effects of the Earth's rotation. (See [**72, 137, 149, 150, 238, 247, 312**]).

Shallow water models have also been derived in other contexts, such as the motion of a thin layer of viscous down an inclined plane, avalanches, etc. (see for instance [**318, 57, 43, 44, 45, 51**] and the references in the handbook by D. Bresch [**54**]).

CHAPTER 2

The Laplace Equation

As discussed in Chapter 1, an essential step in the formulation of the water waves equations adopted here (see Equations (1.28)–(1.30)) consists of recovering the velocity potential Φ from the knowledge of its trace at the surface, provided of course that the fluid domain is known.

We first construct a variational solution to the Laplace equation in the fluid domain. We then show the link between this variational solution and the variational solution of the elliptic boundary value problem obtained by transforming the fluid domain into a flat strip, using a diffeomorphism. Some emphasis is put on the choice of this diffeomorphism; in particular, we introduce so-called *regularizing diffeomorphisms* that are used to derive optimal regularity estimates on the solution. An important attempt is made to control precisely the dependence on the various dimensionless parameters in these estimates (hence the introduction of the operator $\mathfrak{P}$ that appears in the regularity estimates of Proposition 2.36). These regularity estimates are then used to construct strong solutions to the original Laplace equation in the fluid domain.

Some extensions of these results are addressed in §2.5, such as the case of nonasymptotically flat bottom and surface parametrizations and infinite depth. We also discuss the case of rough bottoms. Other aspects, such as the analytic dependence of the solution to the (transformed) Laplace equation on the surface parametrization and the case of nonhomogeneous Neumann conditions at the bottom, are addressed in Appendix A.

Since the time variable does not play any role in the problem addressed in this chapter, all the functions considered will be time independent. It is also convenient to introduce some notations. We define the $(d+1)\times(d+1)$ matrix $I_{\mu,\gamma}$ by

$$(2.1)\quad I_{\mu,\gamma} = \begin{pmatrix} \sqrt{\mu} & 0 & 0 \\ 0 & \gamma\sqrt{\mu} & 0 \\ 0 & 0 & 1 \end{pmatrix} \text{ if } d=2 \quad \text{and} \quad I_{\mu,\gamma} = \begin{pmatrix} \sqrt{\mu} & 0 \\ 0 & 1 \end{pmatrix} \text{ if } d=1$$

and define parameter depending $(d+1)$-dimensional gradient and Laplace operators by

$$(2.2)\qquad \nabla^{\mu,\gamma} = I_{\mu,\gamma}\nabla_{X,z} = \begin{cases} (\sqrt{\mu}\partial_x, \partial_z)^T & \text{when} \quad d=1, \\ (\sqrt{\mu}\partial_x, \gamma\sqrt{\mu}\partial_y, \partial_z)^T & \text{when} \quad d=2, \end{cases}$$

and

$$(2.3)\qquad \Delta^{\mu,\gamma} = \nabla^{\mu,\gamma}\cdot\nabla^{\mu,\gamma} = \begin{cases} \mu\partial_x^2 + \partial_z^2 & \text{when} \quad d=1, \\ \mu\partial_x^2 + \gamma^2\mu\partial_y^2 + \partial_z^2 & \text{when} \quad d=2. \end{cases}$$

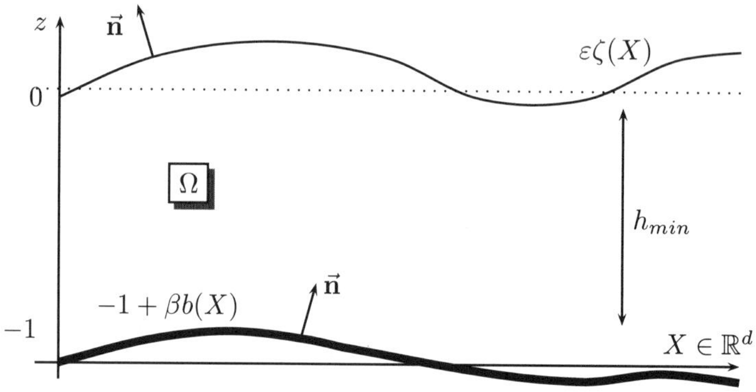

FIGURE 2.1. Main notations.

We also assume throughout this chapter that the dimensionless parameters ε, μ, β, γ satisfy the constraint

$$0 \leq \varepsilon, \beta, \gamma \leq 1, \quad 0 \leq \mu \leq \mu_{max}, \tag{2.4}$$

for some $\mu_{max} > 0$ (not necessarily small).

2.1. The Laplace equation in the fluid domain

2.1.1. The equation. The fluid domain Ω is given by

$$\Omega = \{(X, z) \in \mathbb{R}^d \times \mathbb{R}, -1 + \beta b(X) < z < \varepsilon\zeta(X)\}, \tag{2.5}$$

where ε and β are two parameters (see §1.3.1 for their physical meaning), while ζ and b are two continuous real valued functions defined over $\mathbb{R}^d$ and such that

$$\exists h_{min} > 0, \quad \forall X \in \mathbb{R}^d, \quad 1 + \varepsilon\zeta(X) - \beta b(X) \geq h_{min}, \tag{2.6}$$

that is, the water height is always bounded from below (this follows from Assumption (H9)′ of §1.1.1, with $h_{min} = H_{min}/H_0$).

As discussed in Chapter 1, we want to recover the velocity potential Φ from the knowledge of its trace ψ at the surface by solving the (nondimensionalized) Laplace equation

$$\begin{cases} \Delta^{\mu,\gamma}\Phi = 0 & \text{in } \Omega, \\ \Phi_{|z=\varepsilon\zeta} = \psi, & \partial_{\mathbf{n}}\Phi_{|z=-1+\beta b} = 0, \end{cases} \tag{2.7}$$

where, as always, $\partial_{\mathbf{n}}\Phi_{|z=-1+\beta b}$ stands for the (upwards) *conormal* derivative,

$$\partial_{\mathbf{n}}\Phi_{|z=-1+\beta b} = I_{\mu,\gamma}\mathbf{n} \cdot \nabla^{\mu,\gamma}\Phi_{|z=-1+\beta b}, \qquad \mathbf{n} = \frac{1}{\sqrt{1+\beta^2|\nabla b|^2}} \begin{pmatrix} -\beta\nabla b \\ 1 \end{pmatrix}.$$

REMARK 2.1. We refer to §A.5.1 of Appendix A for a generalization of (2.7) with nonhomogeneous Neumann boundary conditions at the bottom (we have seen in the introduction to §1.6 that this situation arises when handling moving bottoms).

2.1.2. Functional setting and variational solutions. A technical difficulty arises from the fact that in general the physical assumption (H8) (see §1.1.1) that the velocity vanishes at infinity does not imply that this is also the case for the velocity potential. Looking for solutions to the water waves equations (1.28)–(1.30) with ψ in some Sobolev space would therefore be too restrictive. It is, however, physically relevant to assume that the derivatives of ψ belong to some Sobolev space; this observation motivates the introduction of *Beppo-Levi spaces* [**121**].

DEFINITION 2.2. Let $s \in \mathbb{R}$ and $k \in \mathbb{N}$. Then

(1) If $\Omega \subset \mathbb{R}^{d+1}$ is as in (2.5)–(2.6), we denote by $\dot{H}^{k+1}(\Omega)$ the topological vector space

$$\dot{H}^{k+1}(\Omega) = \{f \in L^2_{loc}(\Omega), \quad \nabla_{X,z} f \in H^k(\Omega)^{d+1}\}$$

endowed with the (semi) norm $\|f\|_{\dot{H}^{k+1}(\Omega)} = \|\nabla_{X,z} f\|_{H^k(\Omega)}$.

(2) We denote by $\dot{H}^{s+1}(\mathbb{R}^d)$ the topological vector space

$$\dot{H}^{s+1}(\mathbb{R}^d) = \{f \in L^2_{loc}(\mathbb{R}^d), \quad \nabla f \in H^s(\mathbb{R}^d)^d\}$$

endowed with the (semi) norm $|f|_{\dot{H}^{s+1}(\mathbb{R}^d)} = |\nabla f|_{H^s(\mathbb{R}^d)}$.

The spaces $\dot{H}^{k+1}(\Omega)$ and $\dot{H}^{s+1}(\mathbb{R}^d)$ are not separated; since the adherence of 0 in these spaces is given by the constants, the spaces $\dot{H}^{k+1}(\Omega)/\mathbb{R}$ and $\dot{H}^{s+1}(\mathbb{R}^d)/\mathbb{R}$ are separated. Important properties of these spaces are given in the proposition below. We use the notation Ω_{bott} for the union of Ω and its lower boundary,

$$\Omega_{bott} = \Omega \cup \{z = -1 + \beta b\}.$$

PROPOSITION 2.3. *Let $s \in \mathbb{R}$, $k \in \mathbb{N}$ and $\Omega \subset \mathbb{R}^{d+1}$ be as in (2.5)–(2.6). Then*

(1) *The spaces $\dot{H}^{k+1}(\Omega)/\mathbb{R}$ and $\dot{H}^{s+1}(\mathbb{R}^d)/\mathbb{R}$ are Banach spaces.*

(2) *The space $\dot{H}^1(\Omega)/\mathbb{R}$ (resp. $\dot{H}^1(\mathbb{R}^d)/\mathbb{R}$) is a Hilbert space for the scalar product $\langle f, g\rangle_{\dot{H}^1} = (\nabla_{X,z} f, \nabla_{X,z} g)_{L^2}$ (resp. $\langle f, g\rangle_{\dot{H}^1} = (\nabla f, \nabla g)_{L^2}$).*

(3) *The completion of $\mathcal{D}(\Omega_{bott})$ for the norm $\|\cdot\|_{\dot{H}^1}$ coincides with the closure of $\mathcal{D}(\Omega_{bott})$ in $\dot{H}^1(\Omega)$.*

(4) *The space $C^\infty(\Omega) \cap \dot{H}^1(\Omega)$ is dense in $\dot{H}^1(\Omega)$.*

(5) *The space $H^s(\mathbb{R}^d)$ is dense in $\dot{H}^s(\mathbb{R}^d)$.*

REMARK 2.4. The fact that Ω is bounded in the vertical direction is crucial for the third point. In general, such a property is false for unbounded domains. For instance, the completion of $\mathcal{D}(\mathbb{R}^d)$ for $|\cdot|_{\dot{H}^1}$ is not a distribution space when $d = 1$ or 2. (It is quite easy to construct a sequence that goes to zero for the topology of $\dot{H}^1(\mathbb{R}^d)$, $d = 1, 2$, but that does not converge to zero in the distribution sense. See, for instance, Remarque 4.1 of [**121**].)

REMARK 2.5. Instead of working with the Beppo-Levi spaces $\dot{H}^s(\mathbb{R}^d)$, one can handle functions that do not vanish at infinity by working in the uniformly local Sobolev spaces introduced by Kato [**200**]; they have the same local regularity property as the standard Sobolev spaces, but they contain functions that do not vanish at infinity (such as smooth step functions, or periodic functions). This framework has been adopted in [**7**]. The definition and some basic properties of these spaces are given in §B.4 of Appendix B so that the interested reader may adapt the proofs of this chapter to this framework.

PROOF. The following proof is adapted from [**121, 268**]. Let us first prove that $\dot{H}^{k+1}(\Omega)/\mathbb{R}$ is a Banach space (the proof for $\dot{H}^{s+1}(\mathbb{R}^d)/\mathbb{R}$ is absolutely similar). Therefore let $(f_n)_{n\in\mathbb{N}}$ be a Cauchy sequence in $\dot{H}^{k+1}(\Omega)/\mathbb{R}$. Then $(\nabla_{X,z} f_n)_{n\in\mathbb{N}}$ is a Cauchy sequence in $H^k(\Omega)^{d+1}$ and therefore there exists $G \in H^k(\Omega)^{d+1}$ such that $\nabla_{X,z} f_n \to G$ in $H^k(\Omega)^{d+1}$ as $n \to \infty$. By de Rham's theorem (or Theorem VI of Chapter II in [**294**] in the case of $\dot{H}^{s+1}(\mathbb{R}^d)$), there exists $f \in \mathcal{D}'(\Omega)$ such that $\nabla_{X,z} f = G$. In order to conclude, we still have to prove that $f \in L^2_{loc}(\Omega)$. We refer to Corollary 2.1 of [**121**] for the proof of this rather technical result.
The second point is a direct consequence of the first one. We will prove the third claim. From the Poincaré inequality

$$\forall u \in \overline{\mathcal{D}(\Omega_{bott})}^{H^1(\Omega)}, \qquad \|u\|_2 \leq \sup_{X\in\mathbb{R}^d}(1+\varepsilon\zeta(X)-\beta b(X))\|\partial_z u\|_2, \tag{2.8}$$

the norms $\|\cdot\|_{\dot{H}^1}$ and $\|\cdot\|_{H^1}$ are equivalent and the result follows immediately. We are thus left to prove (2.8). For all $u \in \mathcal{D}(\Omega_{bott})$, one has

$$\begin{aligned}
|u(X,z)|^2 &= \Big|\int_z^{\varepsilon\zeta(X)} \partial_z u(X,z')dz'\Big|^2 \\
&\leq \sup_{\mathbb{R}^d}(1+\varepsilon\zeta(X)-\beta b(X)) \int_{-1+\beta b(X)}^{\varepsilon\zeta(X)} |\partial_z u(X,z')|^2 dz'
\end{aligned}$$

from which we deduce (2.8) by integrating both sides in X and z and by a density argument if $u \notin \mathcal{D}(\Omega_{bott})$.
We now prove the fourth point of the proposition. Let $f \in \dot{H}^{k+1}(\Omega)$ and $\epsilon > 0$. We want to construct $g \in C^\infty(\Omega) \cap \dot{H}^{k+1}(\Omega)$ such that $\|\nabla_{X,z} f - \nabla_{X,z} g\|_{H^k} < \epsilon$. Let $(\omega_j)_{j\in\mathbb{N}}$ be a partition of unity on Ω such that $\operatorname{Supp}\omega_j \cap \operatorname{Supp}\omega_l = \emptyset$ if $|j-l| > 1$. We then write $f = \sum_{j\in\mathbb{N}} f_j$ with $f_j = \omega_j f$ (the series converges in $\mathcal{D}'(\Omega)$). We now consider an approximation of unity $(\rho_j)_{j\in\mathbb{N}}$ with ρ_j supported in a small enough compact set to have $\operatorname{Supp}(\rho_j * f_j) \cap \operatorname{Supp}(\rho_l * f_l) = \emptyset$ if $|j-l| > 1$. The series $g = \sum_{l\in\mathbb{N}} \rho_l * f_l$ therefore converges in $C^\infty(\Omega)$ and one has

$$\nabla_{X,z} f - \nabla_{X,z} g = \sum_{l\in\mathbb{N}} (\rho_l - \delta) * \nabla_{X,z} f_l.$$

Since $\nabla_{X,z} f_l \in H^k(\Omega)^{d+1}$, we can choose $\epsilon_l > 0$ and ρ_l such that $\sum_{l\in\mathbb{N}} \epsilon_l < \epsilon$ and $\|(\rho_l - \delta) * \nabla_{X,z} f_l\|_{H^k} < \epsilon_l$, and the result follows.
For the fifth point of the proposition, let $f \in \dot{H}^s(\mathbb{R}^d)$ and χ be a smooth, even, compactly supported function equal to one in a neighborhood of the origin. The idea is to show that f_ι, defined as $f_\iota = (1-\chi(\frac{1}{\iota}|D|)f$, is in $H^s(\mathbb{R}^d)$ for all $\iota > 0$ and converges to f in $\dot{H}^s$ as $\iota \to 0$. The problem here is that since f is in $\dot{H}^s$, which is not a distribution space, the Fourier multiplier $\chi(\frac{1}{\iota}|D|)f$ does not necessarily make sense. However, noting that

$$\begin{aligned}
(1-\chi(\frac{1}{\iota}|\xi|)) &= -\int_0^1 \frac{d}{dz}\chi(\frac{z}{\iota}|\xi|)dz \\
&= -\int_0^1 \frac{|\xi|}{\iota}\chi'(\frac{1}{\iota}z|\xi|)dz \\
&= \int_0^1 \frac{1}{\iota}\chi'(\frac{1}{\iota}z|\xi|)\frac{i\xi}{|\xi|}\cdot i\xi dz,
\end{aligned}$$

we can use the following alternative definition for f_ι,

$$f_\iota = \int_0^1 \frac{1}{\iota}\chi'(\frac{1}{\iota}z|D|)\frac{\nabla}{|D|}\cdot\nabla f dz,$$

which makes sense since $\nabla f \in H^{s-1}(\mathbb{R}^d)^d$. It follows from this expression that $f_\iota \in H^{s-1}(\mathbb{R}^d)$. One observes that $\nabla f_\iota = (1-\chi(\frac{1}{\iota}|D|))\nabla f \in H^{s-1}(\mathbb{R}^d)^d$, so that we also have $f_\iota \in H^s(\mathbb{R}^d)$. We also deduce from this last expression and Lebesgue's dominated convergence theorem that $\nabla f_\iota \to \nabla f$ in $H^{s-1}(\mathbb{R}^d)^d$ as $\iota \to 0$, or equivalently, that $f_\iota \to f$ in $\dot{H}^s(\mathbb{R}^d)$. □

Motivated by the third point of Proposition 2.3, we introduce the following space.

DEFINITION 2.6. We denote by $H^1_{0,surf}(\Omega)$ the completion of $\mathcal{D}(\Omega_{bott})$ in $H^1(\Omega)$.

This space plays a central role in the variational formulation of the boundary value problem (2.7) as defined below.

DEFINITION 2.7. Let $\psi \in \dot{H}^1(\mathbb{R}^d)$. We say that Φ is a variational solution to (2.7) if there exists $\widetilde{\Phi} \in H^1_{0,surf}(\Omega)$ such that $\Phi = \psi + \widetilde{\Phi}$ and

$$\int_\Omega \nabla^{\mu,\gamma}\widetilde{\Phi}\cdot\nabla^{\mu,\gamma}\varphi = -\int_\Omega \nabla^{\mu,\gamma}\psi\cdot\nabla^{\mu,\gamma}\varphi$$

for all $\varphi \in H^1_{0,surf}(\Omega)$.

REMARK 2.8. Since we have, for all $\varphi \in H^1_{0,surf}(\Omega)$,

$$-\int_\Omega \Delta^{\mu,\gamma}\psi\varphi = \int_\Omega \nabla^{\mu,\gamma}\psi\cdot\nabla^{\mu,\gamma}\varphi + \int_{\{z=-1+\beta b(X)\}} \partial_{\mathbf{n}}\psi\varphi,$$

an equivalent way to formulate the above definition is to say that $\widetilde{\Phi} = \Phi - \psi$ is the unique variational solution in $H^1_{0,surf}(\Omega)$ to the boundary value problem (with homogeneous Dirichlet condition at the surface)

$$\begin{cases} \Delta^{\mu,\gamma}\widetilde{\Phi} = -\Delta^{\mu,\gamma}\psi & \text{in } \Omega,\\ \widetilde{\Phi}_{|_{z=\varepsilon\zeta}} = 0, \quad \partial_{\mathbf{n}}\widetilde{\Phi}_{|_{z=-1+\beta b}} = -\partial_{\mathbf{n}}\psi_{|_{z=-1+\beta b}}. \end{cases}$$

2.1.3. Existence and uniqueness of a variational solution. The next proposition shows that (2.7) is well posed in the variational sense (it will later be proved in Corollary 2.44 that it is in fact well posed in the strong sense).

PROPOSITION 2.9. *Let $\varepsilon, \beta > 0$ and $\zeta, b \in W^{1,\infty}(\mathbb{R}^d)$ be such that (2.6) is satisfied. Then for all $\psi \in \dot{H}^1(\mathbb{R}^d)$, there exists a unique variational solution (in the sense of Definition 2.7) to (2.7).*

PROOF. Since ψ does not depend on z, one has $\nabla^{\mu,\gamma}\psi = (\sqrt{\mu}\partial_x\psi, \gamma\sqrt{\mu}\partial_y\psi, 0)^T$ and the assumption $\psi \in \dot{H}^1(\mathbb{R}^d)$ therefore states that $\nabla^{\mu,\gamma}\psi \in L^2(\Omega)^{d+1}$. The linear form $\varphi \in H^1_{0,surf}(\Omega) \mapsto -\int_\Omega \nabla^{\mu,\gamma}\psi\cdot\nabla^{\mu,\gamma}\varphi$ is therefore continuous; the bilinear form $(\varphi_1,\varphi_2) \mapsto \int_\Omega \nabla^{\mu,\gamma}\varphi_1\cdot\nabla^{\mu,\gamma}\varphi_2$ is also coercive on $H^1_{0,surf}$ by Poincaré's inequality (2.8). The result is a direct consequence of Lax-Milgram's theorem. □

2.2. The transformed Laplace equation

In this section, we transform the constant coefficients elliptic problem (2.7) on the nonflat domain Ω into a variable coefficients elliptic problem on the flat strip $\mathcal{S}$ defined as

$$\mathcal{S} = \mathbb{R}^d \times (-1, 0). \tag{2.9}$$

This transformation will allow us to give some precise regularity estimates.

2.2.1. Notations and new functional spaces.

Let us start with a standard notation.

NOTATION 2.10. For all $a, b \in \mathbb{R}$, we write $a \vee b = \max\{a, b\}$.

It is convenient for what follows to denote by M_0, M, and $M(s)$ constants of the form

$$\begin{aligned} M_0 &= C(\frac{1}{h_{min}}, \mu_{max}, |\zeta|_{H^{t_0+1}}, |b|_{H^{t_0+1}}), \\ M &= C(\frac{1}{h_{min}}, \mu_{max}, |\zeta|_{H^{t_0+2}}, |b|_{H^{t_0+2}}), \\ M(s) &= C(\frac{1}{h_{min}}, \mu_{max}, |\zeta|_{H^{s\vee(t_0+1)}}, |b|_{H^{s\vee(t_0+1)}}), \end{aligned} \tag{2.10}$$

with a nondecreasing dependence on each argument, and where h_{min} is as in (2.6), $t_0 > d/2$ and where μ_{max} serves as an upper bound for the *shallowness* parameter μ as assumed in (2.4).

To study the regularity properties of the solution of the transformed problem, we also introduce the following functional spaces on the flat strip $\mathcal{S}$. In particular, the spaces $H^{s,k}$ are introduced to control functions that are differentiated s times in X and z, with at most k derivatives in z. We recall that $\Lambda = (1-\Delta)^{1/2} = (1+|D|^2)^{1/2}$, so that $|f|_{H^s} = |\Lambda^s f|_2$.

DEFINITION 2.11. Let $s \in \mathbb{R}$ and $k \in \mathbb{N}$.

(1) We write $L^\infty H^s = L^\infty((-1,0); H^s(\mathbb{R}^d))$ endowed with the canonical norm $\|u\|_{L^\infty H^s} = \operatorname{supess}_{z\in(-1,0)}|u(\cdot,z)|_{H^s}$.

(2) The Banach space $(H^{s,k}, \|\cdot\|_{H^{s,k}})$ is defined by

$$H^{s,k} = \bigcap_{j=0}^{k} H^j((-1,0); H^{s-j}(\mathbb{R}^d)), \qquad \|u\|_{H^{s,k}} = \sum_{j=0}^{k} \|\Lambda^{s-j}\partial_z^j u\|_2.$$

Basic properties of these spaces are given in the following proposition.

PROPOSITION 2.12. *For all $s \in \mathbb{R}$,*

(1) *The mapping $u \mapsto u_{|_{z=0}}$ extends continuously from $H^{s+1,1}$ to $H^{s+1/2}(\mathbb{R}^d)$.*

(2) *The space $H^{s+1/2,1}$ is continuously embedded in $L^\infty H^s$.*

PROOF. Let $u \in \mathcal{D}(\{(X,z), -1 < z \le 0\})$ and denote by $\xi \mapsto \widehat{u}(\xi, z)$ the Fourier transform of $u(\cdot, z)$. One has

$$|\widehat{u}(\xi,0)|^2 \le 2\int_{-1}^{0} |\partial_z \widehat{u}(\xi,z)\widehat{u}(\xi,z)|dz,$$

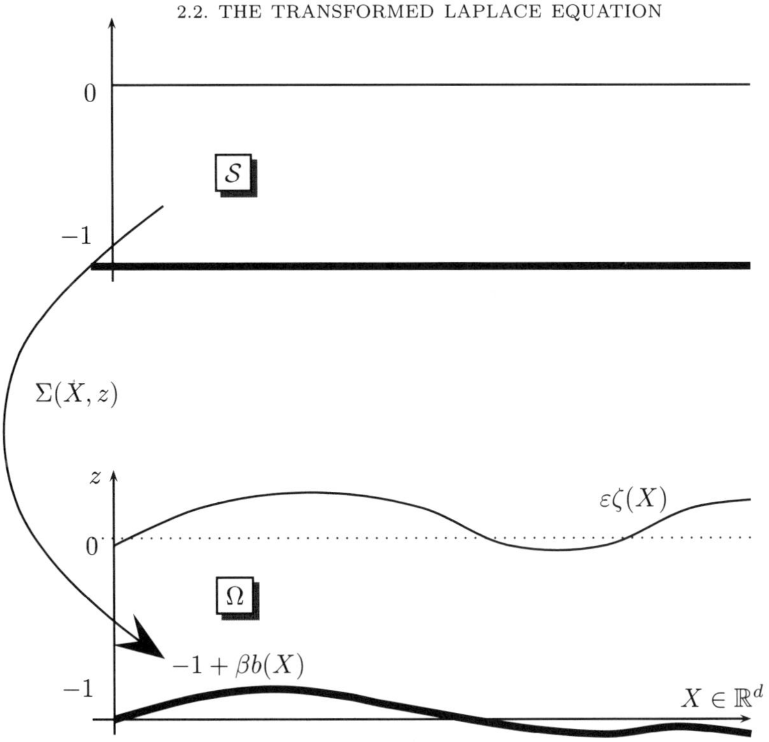

FIGURE 2.2. Straightening the fluid domain with the diffeomorphism Σ.

and therefore

$$\langle\xi\rangle^{2s+1}|\widehat{u}(\xi,0)|^2 \leq \int_{-1}^0 \langle\xi\rangle^{2s}|\partial_z\widehat{u}(\xi,z)|^2dz + \int_{-1}^0 \langle\xi\rangle^{2s+2}|\widehat{u}(\xi,z)|^2dz,$$

which yields, after integration with respect to ξ,

$$|u(\cdot,0)|_{H^{s+1/2}} \leq \|u\|_{H^{s+1,1}}.$$

The first point of the lemma follows by a density argument. The proof of the second point is very similar and we omit it. □

2.2.2. Choice of a diffeomorphism. Let $\zeta, b \in H^{t_0+1}(\mathbb{R}^d)$ $(t_0 > d/2)$ be such that (2.6) is satisfied[1], and let Σ be a diffeomorphism mapping $\mathcal{S}$ onto Ω,

$$\Sigma : \begin{array}{ccc} \mathcal{S} & \to & \Omega \\ (X,z) & \mapsto & \Sigma(X,z). \end{array}$$

For most of what follows, an explicit choice of Σ is not required, provided it is an admissible diffeomorphism in the following sense:

DEFINITION 2.13 (Admissible diffeomorphisms). Let $t_0 > d/2$ and $\zeta, b \in H^{t_0+1}(\mathbb{R}^d)$ be such that (2.6) is satisfied. A diffeomorphism $\Sigma : \mathcal{S} \to \Omega$ is *admissible* if

[1]For much of what follows, the assumption that ζ and b belong to $H^{t_0+1}(\mathbb{R}^d)$ is stronger than necessary and could be replaced by $\zeta, b \in W^{1,\infty}(\mathbb{R}^d)$. See §2.5.2 below.

(1) It can be extended as a continuous mapping $\overline{\mathcal{S}} \to \overline{\Omega}$ and

$$\forall X \in \mathbb{R}^d, \qquad \Sigma(X,0) = (X, \varepsilon\zeta(X)) \quad \text{and} \quad \Sigma(X,-1) = (X, -1+\beta b(X)).$$

(2) The coefficients of the Jacobian matrix $J_\Sigma = d_{X,z}\Sigma$ all belong to $L^\infty(\mathcal{S})$ and

$$\|J_\Sigma\|_{L^\infty} \le M_0.$$

(3) The determinant of the Jacobian matrix J_Σ is uniformly bounded from below on $\mathcal{S}$ by a nonnegative constant c_0 such that

$$\frac{1}{c_0} \le M_0.$$

We give below two examples of diffeomorphisms.

EXAMPLE 2.14. The most simple example one can think of is to take

$$\Sigma(X,z) = \big(X, (1+\varepsilon\zeta(X) - \beta b(X))z + \varepsilon\zeta(X)\big).$$

EXAMPLE 2.15. Another useful example can be constructed as follows. Let $\chi : \mathbb{R} \to \mathbb{R}$ be a positive, compactly supported, smooth, even function equal to 1 in a neighborhood of the origin. For any $\delta > 0$, we define

$$\zeta^{(\delta)}(\cdot,z) = \chi(\delta z|D|)\zeta, \quad \text{and} \quad b_{(\delta)}(\cdot,z) = \chi(\delta(z+1)|D|)b,$$

where we used the Fourier multiplier notation (see Notation 1.7), and

$$\Sigma(X,z) = \big(X, (1+\varepsilon\zeta^{(\delta)}(X,z) - \beta b_{(\delta)}(X,z))z + \varepsilon\zeta^{(\delta)}(X,z)\big).$$

The following proposition shows that these two examples provide admissible diffeomorphisms.

PROPOSITION 2.16. *Let $t_0 > d/2$ and $\zeta, b \in H^{t_0+1}(\mathbb{R}^d)$ be such that (2.6) is satisfied. Then*
(1) *Example 2.14 provides an admissible diffeomorphism with $c_0 = h_{min}$ as lower bound for $|\det J_\Sigma|$.*
(2) *Moreover, if $\delta > 0$ satisfies*

$$\delta < \frac{h_{min}}{C(\chi)(\varepsilon|\zeta|_{H^{t_0+1}} + \beta|b|_{H^{t_0+1}})} \qquad \textit{with } C(\chi) = |\chi'|_\infty\Big(\int_{\mathbb{R}^d} \frac{1}{(1+|\xi|^2)^{t_0}} d\xi\Big)^{1/2} < \infty,$$

then Example 2.15 provides an admissible diffeomorphism and one can take $c_0 = h_{min} - \delta C(\chi)(\varepsilon|\zeta|_{H^{t_0+1}} + \beta|b|_{H^{t_0+1}})$.

PROOF. Only the second point of the proposition deserves a short proof. First note that when Σ is given by Example 2.15, one has

$$\begin{aligned}\det J_\Sigma &= (1+\varepsilon\zeta^{(\delta)} - \beta b_{(\delta)}) + \varepsilon(z+1)\partial_z\zeta^{(\delta)} - \beta z\partial_z b_{(\delta)} \\ &= (1+\varepsilon\zeta - \beta b) + \varepsilon\int_0^z \partial_z\zeta^{(\delta)} - \beta\int_{-1}^z \partial_z b_{(\delta)} + \varepsilon(z+1)\partial_z\zeta^{(\delta)} - \beta z\partial_z b_{(\delta)},\end{aligned}$$

where the second identity comes from the observation that $\zeta^{(\delta)}_{|_{z=0}} = \zeta$ and $b_{(\delta)}\,_{|_{z=-1}} = b$ (which also ensures that condition (1) in the definition of an admissible diffeomorphism is satisfied). It follows that

$$|\det J_\Sigma| \ge h_{min} - \varepsilon\|\partial_z\zeta^{(\delta)}\|_\infty - \beta\|\partial_z b_{(\delta)}\|_\infty. \tag{2.11}$$

Note that $\partial_z\zeta^{(\delta)} = \delta\chi'(z|D|)|D|\zeta$, so that for all $z \in [-1,0]$,

$$\begin{aligned} |\partial_z\zeta^{(\delta)}(\cdot,z)|_\infty &= \delta\big|\chi'(z|D|)|D|\zeta\big|_\infty \\ &\leq \delta\big|\mathcal{F}(\chi'(z|D|)|D|\zeta)\big|_1. \end{aligned}$$

Since $\chi'(z|\xi|)|\xi|\widehat{\zeta}(\xi) = \dfrac{\chi'(z|\xi|)}{(1+|\xi|^2)^{t_0/2}}(1+|\xi|^2)^{t_0/2}|\xi|\widehat{\zeta}(\xi)$, a Cauchy-Schwarz inequality shows that

$$|\partial_z\zeta^{(\delta)}(\cdot,z)|_\infty \leq \delta C(\chi)|\zeta|_{H^{t_0+1}},$$

with $C(\chi)$ as in the statement of the proposition. It follows directly that $\|\partial_z\zeta^{(\delta)}\|_\infty \leq \delta C(\chi)|\zeta|_{H^{t_0+1}}$ and one can prove with the same arguments that $\|\partial_z b_{(\delta)}\|_\infty \leq \delta C(\chi)|b|_{H^{t_0+1}}$. Therefore, the result follows from (2.11). $\square$

Finally, we introduce a particular class of admissible diffeomorphisms called *regularizing diffeomorphisms* (see [**218**]). Their introduction is motivated by a classical issue in elliptic theory where it is important to keep track of the regularity of the boundary of the domain. See [**29**] for a related context.

DEFINITION 2.17 (Regularizing diffeomorphisms). Let $t_0 > d/2$, $s \geq 0$ and $\zeta, b \in H^{s+1/2} \cap H^{t_0+1}(\mathbb{R}^d)$ be such that (2.6) is satisfied. An admissible diffeomorphism $\Sigma : \mathcal{S} \to \Omega$ is *regularizing* if

(1) There exists a mapping $\sigma : \mathcal{S} \to \mathbb{R}$ such that

$$\forall (X,z) \in \mathcal{S}, \qquad \Sigma(X,z) = (X, z + \sigma(X,z)).$$

(2) The following estimate holds

$$\|\nabla^{\mu,\gamma}\sigma\|_{H^{s,0}} \leq M_0(|\varepsilon\zeta|_{H^{s+1/2}} + |\beta b|_{H^{s+1/2}}),$$

with $H^{s,0}$ as in Definition 2.11.

The following proposition shows that the diffeomorphism given by Example 2.15 is regularizing.

PROPOSITION 2.18. *Let $t_0 > d/2$, $s \geq 0$ and $\zeta, b \in H^{s+1/2} \cap H^{t_0+1}(\mathbb{R}^d)$ be such that (2.6) is satisfied. If $\delta > 0$ satisfies the smallness condition given in Proposition 2.16, then the diffeomorphism provided by Example 2.15 is regularizing.*

REMARK 2.19. The diffeomorphism provided by Example 2.14 is not regularizing since one gets only $\|\nabla^{\mu,\gamma}\sigma\|_{H^{s,0}} \leq M_0(|\varepsilon\zeta|_{H^{s+1}} + |\beta b|_{H^{s+1}})$, while $(|\varepsilon\zeta|_{H^{s+1/2}} + |\beta b|_{H^{s+1/2}})$ is needed for regularizing diffeomorphisms. See Figure 2.3 for an illustration of the different behavior of standard and regularizing diffeomorphisms.

PROOF. The first condition of Definition 2.17 is obviously satisfied. To check the second one, let us prove the following lemma (which, for later use, gives a more general result than is needed here). We recall that $D^\lambda = \frac{1}{i}\nabla^\lambda = \frac{1}{i}(\partial_x, \lambda\partial_y)^T$ if $d = 2$, and $D^\lambda = D = \frac{1}{i}\partial_x$ if $d = 1$.

LEMMA 2.20. *Let $s \in \mathbb{R}$ and $\lambda_1, \lambda_2 > 0$. For all real valued, compactly supported function f, one has*

$$\|\Lambda^s f(\sqrt{\lambda_1}z|D^{\lambda_2}|)u\|_2^2 \leq C(f)\Big|\frac{1}{(1+\sqrt{\lambda_1}|D^{\lambda_2}|)^{1/2}}u\Big|_{H^s}^2,$$

where $C(f)$ is a constant that depends only on f.

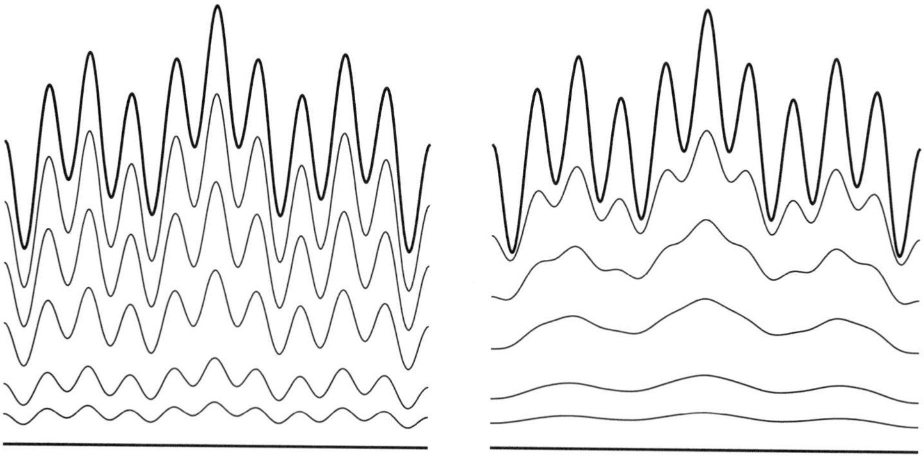

FIGURE 2.3. Level lines for the standard diffeomorphism of Example 2.14 (left) and for the regularizing diffeomorphism of Example 2.15 (right). The bottom is flat here.

PROOF OF THE LEMMA. Remark that

$$\begin{aligned}\|\Lambda^s f(\sqrt{\lambda_1}z|D^{\lambda_2}|)u\|_2^2 &= \int_{\mathbb{R}^d}\int_{-1}^0 \langle\xi\rangle^{2s} f(\sqrt{\lambda_1}z|\xi^{\lambda_2}|)^2|\widehat{u}(\xi)|^2 d\xi \\ &= \int_{\mathbb{R}^d}\langle\xi\rangle^{2s}\frac{F(0)-F(-\sqrt{\lambda_1}|\xi^{\lambda_2}|)}{\sqrt{\lambda_1}|\xi^{\lambda_2}|}|\widehat{u}(\xi)|^2 d\xi,\end{aligned}$$

where $F(\cdot)$ is any primitive of f^2. Moreover, since

$$\Big|\frac{F(0)-F(-\sqrt{\lambda_1}|\xi^{\lambda_2}|)}{\sqrt{\lambda_1}|\xi^{\lambda_2}|}\Big| \leq C(f)\frac{1}{1+\sqrt{\lambda_1}|\xi^{\lambda_2}|},$$

we deduce the result. (Note that F is bounded since f is compactly supported.) □

For the sake of simplicity, we assume for the rest of the proof that $b=0$ (flat bottom). It is straightforward to extend it to the general case.
Note first that

$$\begin{aligned}\nabla^{\mu,\gamma}\sigma &= (\varepsilon\sqrt{\mu}(z+1)(\nabla^\gamma\zeta^{(\delta)})^T, \varepsilon\zeta^{(\delta)}+\varepsilon(z+1)\partial_z\zeta^{(\delta)})^T, \\ \partial_z\nabla^{\mu,\gamma}\sigma &= (\varepsilon\sqrt{\mu}(\nabla^\gamma\zeta^{(\delta)})^T+\varepsilon\sqrt{\mu}(z+1)(\partial_z\nabla^\gamma\zeta^{(\delta)})^T, 2\varepsilon\partial_z\zeta^{(\delta)}+\varepsilon(z+1)\partial_z^2\zeta^{(\delta)})^T\end{aligned}$$

so that (recall that $\gamma\leq 1$)

$$\|\nabla^{\mu,\gamma}\sigma\|_{H^{s,1}} \leq C(\varepsilon,\mu_{max})\|\zeta^{(\delta)}\|_{H^{s+1,2}}.$$

Since Lemma 2.20 (with $\lambda_1=\delta^2$, $\lambda_2=1$) ensures that $\|\zeta^{(\delta)}\|_{H^{s+1,2}} \leq C(\frac{1}{\delta})|\zeta|_{H^{s+1/2}}$, the result is proved. □

2.2.3. Transformed equation. We show here that any admissible diffeomorphism can be used to transform the boundary value problem (2.7) satisfied by Φ in the fluid domain Ω into the following boundary value problem satisfied by $\phi=\Phi\circ\Sigma$ on the flat strip $\mathcal{S}$:

$$\begin{cases}\nabla^{\mu,\gamma}\cdot P(\Sigma)\nabla^{\mu,\gamma}\phi = 0 \quad \text{in} \quad \mathcal{S}, \\ \phi_{|z=0}=\psi, \qquad \partial_{\mathbf{n}}\phi_{|z=-1}=0,\end{cases} \tag{2.12}$$

where, as always, $\partial_{\mathbf{n}}\phi_{|_{z=-1}} = \mathbf{e_z}\cdot P(\Sigma)\nabla^{\mu,\gamma}\phi_{|_{z=-1}}$ is the upwardconormal derivative and where

$$P(\Sigma) = |\det J_\Sigma| I_{\mu,\gamma}^{-1} J_\Sigma^{-1} I_{\mu,\gamma}^2 (J_\Sigma^{-1})^T I_{\mu,\gamma}^{-1}, \tag{2.13}$$

where $I_{\mu,\gamma}$ is defined in (2.1) and J_Σ is the Jacobian matrix of Σ.

EXAMPLE 2.21. When the diffeomorphism Σ is such that, for some function $\sigma : \mathcal{S} \to \mathbb{R}$,

$$\forall (X,z) \in \mathcal{S}, \qquad \Sigma(X,z) = (X, z + \sigma(X,z)),$$

which is the case for both Examples 2.14 and 2.15, one can check that

$$P(\Sigma) = I + Q(\Sigma) \quad \text{with} \quad Q(\Sigma) = \begin{pmatrix} \partial_z \sigma I_{d\times d} & -\sqrt{\mu}\nabla^\gamma \sigma \\ -\sqrt{\mu}\nabla^\gamma \sigma^T & \dfrac{-\partial_z\sigma + \mu|\nabla^\gamma\sigma|^2}{1+\partial_z\sigma} \end{pmatrix}.$$

Before stating the main proposition of this section, we must define the notion of variational solutions for (2.12) with Dirichlet data $\psi \in \dot{H}^1(\mathbb{R}^d)$. (This definition coincides of course with Definition 2.7 when the surface and bottom are flat and Σ is the identity mapping.)

DEFINITION 2.22. Let $\psi \in \dot{H}^1(\mathbb{R}^d)$. We say that ϕ is a variational solution of (2.12) if there exists $\widetilde{\phi} \in H^1_{0,surf}(\mathcal{S})$ such that $\phi = \psi + \widetilde{\phi}$ and

$$\int_{\mathcal{S}} \nabla^{\mu,\gamma}\widetilde{\phi}\cdot P(\Sigma)\nabla^{\mu,\gamma}\varphi = -\int_{\mathcal{S}} \nabla^{\mu,\gamma}\psi\cdot P(\Sigma)\nabla^{\mu,\gamma}\varphi$$

for all $\varphi \in H^1_{0,surf}(\mathcal{S})$.

REMARK 2.23. Let $\psi_1 = \psi + u$, with $u \in H^1_{0,surf}(\mathcal{S})$. Then ϕ is a variational solution to (2.12) if and only $\phi = \psi_1 + \widetilde{\phi}_1$, where $\widetilde{\phi}_1$ belongs to $H^1_{0,surf}(\mathcal{S})$ and satisfies

$$\int_{\mathcal{S}} \nabla^{\mu,\gamma}\widetilde{\phi}_1\cdot P(\Sigma)\nabla^{\mu,\gamma}\varphi = -\int_{\mathcal{S}} \nabla^{\mu,\gamma}\psi_1\cdot P(\Sigma)\nabla^{\mu,\gamma}\varphi$$

for all $\varphi \in H^1_{0,surf}(\mathcal{S})$ (just take $\widetilde{\phi}_1 = \widetilde{\phi} - u$ to recover the variational equation of the definition).

REMARK 2.24. The above definition says that $\widetilde{\phi} = \phi - \psi$ must be the unique variational solution in $H^1_{0,surf}(\mathcal{S})$ of the boundary value problem

$$\begin{cases} \nabla^{\mu,\gamma}\cdot P(\Sigma)\nabla^{\mu,\gamma}\widetilde{\phi} = -\nabla^{\mu,\gamma}\cdot P(\Sigma)\nabla^{\mu,\gamma}\psi \quad \text{in} \quad \mathcal{S}, \\ \widetilde{\phi}_{|_{z=0}} = 0, \qquad \partial_{\mathbf{n}}\widetilde{\phi}_{|_{z=-1}} = -\partial_{\mathbf{n}}\psi_{\,|_{z=-1}}. \end{cases}$$

We can now state the main result of this section.

PROPOSITION 2.25. *Let $t_0 > d/2$ and $\zeta, b \in H^{t_0+1}(\mathbb{R}^d)$ be such that (2.6) is satisfied. Also let $\psi \in \dot{H}^1(\mathbb{R}^d)$ and Φ be the variational solution to (2.7) provided by Proposition 2.9.*
If Σ is an admissible diffeomorphism, then there exists a unique variational solution ϕ to (2.12) in the sense of Definition 2.22, and it is given by $\phi = \Phi\circ\Sigma$.

PROOF. Let us first prove the coercivity of $P(\Sigma)$, which ensures the uniqueness of the variational solution.

LEMMA 2.26. *If Σ is an admissible diffeomorphism, then $P(\Sigma)$ is coercive and*

$$\forall \Theta \in \mathbb{R}^{d+1}, \quad \forall (X,z) \in \mathcal{S}, \qquad P(\Sigma)(X,z)\Theta \cdot \Theta \geq k(\Sigma)|\Theta|^2,$$

with

$$k(\Sigma) = \frac{c_0}{\sup_{(X,z)\in\mathcal{S}} \| I_{\mu,\gamma}^{-1} J_\Sigma(X,z) I_{\mu,\gamma} \|^2_{\mathcal{L}(\mathbb{R}^{d+1},\mathbb{R}^{d+1})}} > 0.$$

PROOF. From the definition (2.13) of $P(\Sigma)$, one deduces that for all $\Theta \in \mathbb{R}^{d+1}$ and $(X,z) \in \mathcal{S}$,

$$\begin{aligned} P(\Sigma)(X,z)\Theta \cdot \Theta &= |\det J_\Sigma| |I_{\mu,\gamma}(J_\Sigma^{-1}(X,z))^T I_{\mu,\gamma}^{-1}\Theta|^2 \\ &\geq |\det J_\Sigma| \frac{|\Theta|^2}{\| I_{\mu,\gamma}^{-1} J_\Sigma(X,z) I_{\mu,\gamma}\|^2_{\mathcal{L}(\mathbb{R}^{d+1},\mathbb{R}^{d+1})}}, \end{aligned}$$

and the result follows easily. □

We will now prove that $\phi = \Phi \circ \Sigma$ is a (and therefore the) variational solution to (2.12). Since Φ is a variational solution of (2.7), we know that $\widetilde{\Phi} = \Phi - \psi$ satisfies

$$\int_\Omega \nabla^{\mu,\gamma}\widetilde{\Phi} \cdot \nabla^{\mu,\gamma}\varphi = - \int_\Omega \nabla^{\mu,\gamma}\psi \cdot \nabla^{\mu,\gamma}\varphi,$$

for all $\varphi \in \mathcal{D}(\Omega_{bott}) = \mathcal{D}(\Omega \cup \{z = -1 + \beta b\})$. Using the diffeomorphism Σ to perform a change of variable in both integrals, we obtain

$$\int_\mathcal{S} \nabla^{\mu,\gamma}\widetilde{\Phi} \circ \Sigma \cdot \nabla^{\mu,\gamma}\varphi \circ \Sigma |\det J_\Sigma| = - \int_\mathcal{S} \nabla^{\mu,\gamma}\psi \circ \Sigma \cdot \nabla^{\mu,\gamma}\varphi \circ \Sigma |\det J_\Sigma|.$$

Defining $\widetilde{\phi} = \widetilde{\Phi} \circ \Sigma$, one can remark that $\nabla^{\mu,\gamma}\widetilde{\Phi} \circ \Sigma = I_{\mu,\gamma}(J_\Sigma^{-1})^T I_{\mu,\gamma}^{-1} \nabla^{\mu,\gamma}\widetilde{\phi}$, and similar formulas can be derived for the other components of the integrands; it follows that

$$\int_\mathcal{S} P(\Sigma)\nabla^{\mu,\gamma}\widetilde{\phi} \cdot \nabla^{\mu,\gamma}(\varphi \circ \Sigma) = - \int_\mathcal{S} P(\Sigma)\nabla^{\mu,\gamma}(\psi \circ \Sigma) \cdot \nabla^{\mu,\gamma}(\varphi \circ \Sigma).$$

Since $\{\varphi \circ \Sigma, \varphi \in \mathcal{D}(\Omega \cup \{z = -1 + \beta b\})\}$ is dense in $H^1_{0,surf}(\mathcal{S})$ and $\psi \circ \Sigma - \psi \in H^1_{0,surf}(\mathcal{S})$, we can conclude (see Remark 2.23) that ϕ is the variational solution to (2.12). □

REMARK 2.27. One readily observes from Lemma 2.26 that if Σ is given by Examples 2.14 or 2.15, then

$$\frac{1}{k(\Sigma)} \leq M_0,$$

with M_0 given by (2.10).

2.2.4. Variational solutions for data in $\dot{H}^{1/2}(\mathbb{R}^d)$. We show here how to extend the definition of a variational solution to Dirichlet data less regular than in Definition 2.22 (namely, $\dot{H}^{1/2}(\mathbb{R}^d)$ rather than $\dot{H}^1(\mathbb{R}^d)$). Let us first introduce a particular extension $\psi^\dagger$ of ψ:

NOTATION 2.28. Let $\chi : \mathbb{R} \to \mathbb{R}$ be a smooth, even, compactly supported function that is constant equal to 1 in a neighborhood of the origin. For all $\psi \in \mathfrak{S}'(\mathbb{R}^d)$, we write

$$\psi^\dagger(\cdot, z) = \chi(\sqrt{\mu} z |D^\gamma|)\psi.$$

We keep this notation if $\psi \in \dot{H}^{s+1/2}(\mathbb{R}^2)$, $s \geq 0$ (and therefore not necessarily in $\mathfrak{S}'(\mathbb{R}^d)$); this makes sense because we can write

$$\psi^\dagger(\cdot, z) = \psi + \int_0^1 \chi'(s\sqrt{\mu}z|D^\gamma|)ds\sqrt{\mu}z|D^\gamma|\psi,$$

with $|D^\gamma|\psi$ well defined in $H^{s-1/2}(\mathbb{R}^d)$.

We can then extend the definition of a variational solution to Dirichlet data in $\dot{H}^{1/2}(\mathbb{R}^d)$ in the following way:

DEFINITION 2.29. Let $\psi \in \dot{H}^{1/2}(\mathbb{R}^d)$. We say that ϕ is a variational solution of (2.12) if there exists $\widetilde{\phi} \in H^1_{0,surf}(\mathcal{S})$ such that $\phi = \widetilde{\phi} + \psi^\dagger$ and

$$\int_{\mathcal{S}} \nabla^{\mu,\gamma}\widetilde{\phi} \cdot P(\Sigma)\nabla^{\mu,\gamma}\varphi = -\int_{\mathcal{S}} \nabla^{\mu,\gamma}\psi^\dagger \cdot P(\Sigma)\nabla^{\mu,\gamma}\varphi$$

for all $\varphi \in H^1_{0,surf}(\mathcal{S})$.

REMARK 2.30. When $\psi \in \dot{H}^1(\mathbb{R}^d)$ then $\psi^\dagger - \psi \in H^1_{0,surf}(\mathcal{S})$, so that Definition 2.29 coincides as expected with Definition 2.22 (see Remark 2.23).

The following proposition ensures the existence and uniqueness of a variational solution for data in $\dot{H}^{1/2}(\mathbb{R}^d)$.

PROPOSITION 2.31. *Let Σ be an admissible diffeomorphism and $\psi \in \dot{H}^{1/2}(\mathbb{R}^d)$. Then there exists a unique variational solution ϕ to (2.12).*

REMARK 2.32. If ψ is changed to $\psi + C$, with $C \in \mathbb{R}$ a constant, then the variational solution ϕ must be changed into $\phi + C$. We could therefore define variational solutions for Dirichlet data in the quotient space, $\psi \in \dot{H}^{1/2}(\mathbb{R})/\mathbb{R}$.

NOTATION 2.33. For all $\psi \in \dot{H}^{1/2}(\mathbb{R}^d)$, we sometimes denote by $\psi^\mathfrak{h} = \phi$ the "harmonic extension" of ψ provided by Proposition 2.31.

PROOF. Since $P(\Sigma)$ is uniformly coercive (see Lemma 2.26), the existence and uniqueness of a solution to the variational equation of Definition 2.29 follows directly from Lax-Milgram's theorem if we can show that $\nabla^{\mu,\gamma}\psi^\dagger \in L^2(\mathcal{S})$. Such a result in proved in the following lemma (we give a more general result for later use). Before stating the lemma, let us introduce the operator $\mathfrak{P}$ defined as

$$\mathfrak{P} = \frac{|D^\gamma|}{(1+\sqrt{\mu}|D^\gamma|)^{1/2}}. \tag{2.14}$$

In Proposition 3.12 we will see that $\mathfrak{P}$ acts on ψ like the square root of the Dirichlet-Neumann operator.

LEMMA 2.34. *Let $s \in \mathbb{R}$, $\psi \in \dot{H}^{s+1/2}(\mathbb{R}^d)$ and let $\psi^\dagger$ be as in Notation 2.28. Then $\nabla^{\mu,\gamma}\psi^\dagger \in H^{s,1}$ and*

$$\|\Lambda^s\nabla^{\mu,\gamma}\psi^\dagger\|_2 \leq C(\chi)\sqrt{\mu}|\mathfrak{P}\psi|_{H^s}, \qquad \|\Lambda^{s-1}\partial_z\nabla^{\mu,\gamma}\psi^\dagger\|_2 \leq C(\chi)\mu|\mathfrak{P}\psi|_{H^s},$$

where $C(\chi)$ is a nonnegative constant depending only on χ.

PROOF OF THE LEMMA. Since $\nabla^{\mu,\gamma} = (\sqrt{\mu}\partial_x, \gamma\sqrt{\mu}\partial_y, \partial_z)^T$ (when $d = 2$, the adaptation to the case $d = 1$ is straightforward), one deduces from the definition of $\psi^\dagger$ that

$$\begin{aligned}\|\Lambda^s\nabla^{\mu,\gamma}\psi^\dagger\|_2 \quad &\leq \quad \sqrt{\mu}\|\Lambda^s\chi(\sqrt{\mu}z|D^\gamma|)\partial_x\psi\|_2 + \sqrt{\mu}\|\Lambda^s\chi(\sqrt{\mu}z|D^\gamma|)\gamma\partial_y\psi\|_2 \\ &+ \quad \sqrt{\mu}\|\Lambda^s\chi'(\sqrt{\mu}z|D^\gamma|)|D^\gamma|\psi\|_2,\end{aligned}$$

and the result easily follows from Lemma 2.20 (with $\lambda_1=\mu$ and $\lambda_2=\gamma$).
The estimate on $\Lambda^{s-1}\partial_z\nabla^{\mu,\gamma}\psi^\dagger$ is obtained in exactly the same way. □

Taking $s=0$ in the lemma gives us the required information to conclude. □

2.3. Regularity estimates

We give here some estimates on the variational solution ϕ to (2.12) provided by Proposition 2.31. For this purpose, we use the functional spaces $L^\infty H^s$ and $H^{s,k}$ introduced in Definition 2.11. Let us also introduce some notations:

NOTATION 2.35. We write, as in Example 2.21,

$$Q(\Sigma)=P(\Sigma)-I.$$

We also use the notation

$$A+\langle B\rangle_{s>t_0+1}=A \quad \text{if} \quad s\leq t_0+1, \quad \text{and} \quad A+\langle B\rangle_{s>t_0+1}=A+B \quad \text{otherwise.}$$

The following proposition[2] gives some information on the horizontal regularity of the solution to (2.12). We recall that $a\vee b=\max\{a,b\}$.

PROPOSITION 2.36. *Let $t_0>d/2$, $s\geq 0$ and Σ be an admissible diffeomorphism satisfying*

$$\frac{1}{k(\Sigma)}+\|Q(\Sigma)\|_{L^\infty H^{t_0}}\leq M_0$$

and also denote by $\mathfrak{C}_s$ any constant of the form (with $Q=Q(\Sigma)$)

$$\mathfrak{C}_s=C(M_0,\|Q\|_{H^{s\vee(t_0+1/2),2}}).$$

Let $\psi\in\dot{H}^{s+1/2}(\mathbb{R}^d)$ and ϕ be the unique variational solution to (2.12) provided by Proposition 2.31. Then,

$$\|\Lambda^s\nabla^{\mu,\gamma}\phi\|_2\leq\sqrt{\mu}\mathfrak{C}_{\min\{s,t_0+3/2\}}\big(|\mathfrak{P}\psi|_{H^s}+\langle\|Q\|_{H^{s,1}}|\mathfrak{P}\psi|_{H^{t_0+3/2}}\rangle_{s>t_0+3/2}\big),$$

where $\mathfrak{P}$ is as in (2.14).
Moreover, if $s\geq -t_0+1$ (which is always the case if $d=2$), the same estimate holds on $\|\nabla^{\mu,\gamma}\phi\|_{H^{s,1}}$.

REMARK 2.37. Since the variational solution can be defined with data in the quotient space, $\psi\in\dot{H}^{s+1/2}(\mathbb{R})/\mathbb{R}$ (see Remark 2.32), one can replace the assumption $\psi\in\dot{H}^{s+1/2}(\mathbb{R})$ by $\psi\in\dot{H}^{s+1/2}(\mathbb{R})/\mathbb{R}$ in the statement of the proposition.

PROOF. The main step in the proof of the proposition is the following lemma. To simplify the text, we do not state the precise assumptions on $\mathbf{g}$. (The trace $\mathbf{g}_{|_{z=-1}}$ must make sense, as well as the quantities involved in the estimate.)

LEMMA 2.38. *Let $t_0>d/2$, $s\geq 0$, and Σ be an admissible diffeomorphism. There exists a unique variational solution $u\in H^1_{0,surf}(\mathcal{S})$ to the boundary value problem*

$$\begin{cases}\nabla^{\mu,\gamma}\cdot P(\Sigma)\nabla^{\mu,\gamma}u=-\nabla^{\mu,\gamma}\cdot\mathbf{g} & \text{in} \quad \mathcal{S},\\ u_{|_{z=0}}=0, \qquad \partial_{\mathbf{n}}u_{|_{z=-1}}=-\mathbf{e_z}\cdot\mathbf{g}_{|_{z=-1}}.\end{cases}\tag{2.15}$$

[2] The proof gives a slightly more precise expression for the quantity $\mathfrak{C}_s$ that appears in the statement, namely,

$$\mathfrak{C}_s=C(M_0,\|Q\|_{H^{s\vee(t_0+1/2),1}},\|Q\|_{L^\infty H^{s\vee(t_0+1/2)-1/2}},\|\partial_z Q\|_{L^\infty H^{s\vee(t_0+1/2)-3/2}}),$$

which is controlled by the expression given in the proposition by the Sobolev embedding of Proposition 2.12.

Moreover, one has

$$\|\nabla^{\mu,\gamma}\Lambda^s u\|_2 \leq \mathfrak{C}_{\min\{s,t_0+3/2\}}\big(\|\mathbf{g}\|_{H^{s,j}} + \big\langle\|\Lambda^s Q\|_2\|\mathbf{g}\|_{H^{t_0+1/2,1}}\big\rangle_{s>t_0+3/2}\big),$$

with $j=0$ *if* $s\leq t_0$ *and* $j=1$ *otherwise.*
Moreover, if $s\geq -t_0+1$*, then the same estimates hold on* $\|\nabla^{\mu,\gamma}u\|_{H^{s,1}}$ *provided that one takes* $j=1$ *for* $0\leq s\leq t_0$ *also.*

PROOF OF LEMMA 2.38. For the sake of clarity, we write P, Q and k instead of $P(\Sigma)$, $Q(\Sigma)$ and $k(\Sigma)$ throughout this proof.
By definition, $u\in H^1_{0,surf}(\mathcal{S})$ is a variational solution to (2.15) if for all $\varphi\in H^1_{0,surf}(\mathcal{S})$, one has (recall that $\partial_{\mathbf{n}}$ is always taken upwards)

$$\begin{aligned}\int_{\mathcal{S}}\nabla^{\mu,\gamma}u\cdot P\nabla^{\mu,\gamma}\varphi &= \int_{\mathcal{S}}(\nabla^{\mu,\gamma}\cdot\mathbf{g})\varphi - \int_{z=-1}\partial_{\mathbf{n}}u\varphi\\ &= -\int_{\mathcal{S}}\mathbf{g}\cdot\nabla^{\mu,\gamma}\varphi\end{aligned}\tag{2.16}$$

and the existence and uniqueness of a variational solution follows from the coercivity of P (see Lemma 2.34) and Lax-Milgram's theorem.
In order to prove the estimates stated in the lemma, let us introduce the smoothed fractional derivative operator Λ^s_δ as follows: for any $\delta>0$, let

$$\Lambda^s_\delta = \Lambda^s\,\chi(\delta\Lambda),$$

where χ is a smooth, even, real valued and compactly supported function equal to 1 in a neighborhood of the origin; Λ^s_δ is thus a Fourier multiplier of symbol $\langle\xi\rangle^s\chi(\delta\langle\xi\rangle)$. Since this symbol is smooth and compactly supported, Λ^s_δ is a smoothing operator; in particular, if $u\in H^1_{0,surf}(\mathcal{S})$ is the variational solution of (2.15), then $(\Lambda^s_\delta)^2u$ also belongs to $H^1_{0,surf}(\mathcal{S})$, so that one can choose $\varphi=(\Lambda^s_\delta)^2u$ as a test function in (2.16). This yields

$$\int_{\mathcal{S}}\nabla^{\mu,\gamma}u\cdot P\nabla^{\mu,\gamma}(\Lambda^s_\delta)^2u = -\int_{\mathcal{S}}\mathbf{g}\cdot\nabla^{\mu,\gamma}(\Lambda^s_\delta)^2u.$$

Since P is symmetric and Λ^s_δ commutes with $\nabla^{\mu,\gamma}$ and is symmetric for the scalar product of $L^2(\mathbb{R}^d)$, we deduce that

$$\int_{\mathcal{S}}\Lambda^s_\delta(P\nabla^{\mu,\gamma}u)\cdot\nabla^{\mu,\gamma}\Lambda^s_\delta u = -\int_{\mathcal{S}}\Lambda^s_\delta\mathbf{g}\cdot\nabla^{\mu,\gamma}\Lambda^s_\delta u$$

and thus that (just note that $[\Lambda^s_\delta,P]=[\Lambda^s_\delta,Q]$),

$$\int_{\mathcal{S}}P\nabla^{\mu,\gamma}\Lambda^s_\delta u\cdot\Lambda^s_\delta\nabla^{\mu,\gamma}u = -\int_{\mathcal{S}}[\Lambda^s_\delta,Q]\nabla^{\mu,\gamma}u\cdot\nabla^{\mu,\gamma}\Lambda^s_\delta u - \int_{\mathcal{S}}\Lambda^s_\delta\mathbf{g}\cdot\nabla^{\mu,\gamma}\Lambda^s_\delta u.$$

It follows from the coercivity of P (see Lemma 2.26) that

$$k\|\nabla^{\mu,\gamma}\Lambda^s_\delta u\|_2 \leq \|[\Lambda^s_\delta,Q]\nabla^{\mu,\gamma}u\|_2 + \|\Lambda^s_\delta\mathbf{g}\|_2.$$

We first distinguish two cases for small values of s:
- *Case 1:* $0\leq s\leq t_0$. Using the commutator estimate provided by Corollary B.17(1), we get

$$k\|\nabla^{\mu,\gamma}\Lambda^s_\delta u\|_2 \leq \|\Lambda^s_\delta\mathbf{g}\|_2 + \|Q\|_{L^\infty H^{t_0}}\|\Lambda^{s-\underline{\delta}}\nabla^{\mu,\gamma}u\|_2,$$

for some $\underline{\delta}>0$, and therefore, with $\epsilon=\underline{\delta}$,

$$\|\nabla^{\mu,\gamma}\Lambda^s_\delta u\|_2 \leq M_0\big(\|\Lambda^s\mathbf{g}\|_2 + \|\Lambda^{s-\epsilon}\nabla^{\mu,\gamma}u\|_2\big).$$

- *Case 2:* $t_0 \le s \le t_0 + 3/2$. The commutator estimate provided by Corollary B.17(2) gives, if we apply it with $T_0 = t_0 + 1/2$ rather than t_0,

$$k\|\nabla^{\mu,\gamma}\Lambda^s_\delta u\|_2 \le \|\Lambda^s_\delta \mathbf{g}\|_2 + \|\Lambda^{s\vee T_0} Q\|_2 \|\nabla^{\mu,\gamma} u\|_{L^\infty H^{\min\{s-\underline{\delta},T_0\}}}.$$

Moreover, we can take any $\underline{\delta} \in (0,1)$, such that $\underline{\delta} < T_0 - d/2 = t_0 - d/2 + 1/2$, so that we can choose $\underline{\delta} > 1/2$. We now need the estimate

$$\|\nabla^{\mu,\gamma} u\|_{L^\infty H^{\min\{s-\underline{\delta},t_0+1/2\}}} \le \mathfrak{C}_s\big(\|\mathbf{g}\|_{H^{s,1}} + \|\Lambda^{s-\epsilon}\nabla^{\mu,\gamma} u\|_2\big),$$

where $\epsilon = \underline{\delta} - 1/2 > 0$. For the sake of clarity, this estimate is proved in Lemma 2.39 below (take $r = \min\{s - \underline{\delta}, t_0 + 1/2\}$ in this lemma). This yields

$$\|\nabla^{\mu,\gamma}\Lambda^s_\delta u\|_2 \le \mathfrak{C}_s\big(\|\mathbf{g}\|_{H^{s,1}} + \|\Lambda^{s-\epsilon}\nabla^{\mu,\gamma} u\|_2\big).$$

It follows from these two cases, and after letting $\delta \to 0$ and using a finite induction on s, that

$$\forall 0 \le s \le t_0 + 3/2, \qquad \|\nabla^{\mu,\gamma}\Lambda^s u\|_2 \le \mathfrak{C}_s \|\mathbf{g}\|_{H^{s,j}}, \tag{2.17}$$

with $j = 0$ if $0 \le s \le t_0$ and $j = 1$ otherwise.
For higher values of s, let us consider one more case:
- *Case 3:* $t_0 + 3/2 \le s$. This time the commutator estimate of Corollary B.16(2) gives

$$k\|\nabla^{\mu,\gamma}\Lambda^s_\delta u\|_2 \lesssim \|\mathbf{g}\|_{H^{s,0}} + \|Q\|_{L^\infty H^{t_0+1}}\|\Lambda^{s-1}\nabla^{\mu,\gamma} u\|_2 + \|\Lambda^s Q\|_2 \|\nabla^{\mu,\gamma} u\|_{L^\infty H^{t_0}}.$$

Letting $\delta \to 0$ and using a finite induction on s therefore yields, with the help of (2.17) and Lemma 2.39 below, that

$$\|\nabla^{\mu,\gamma}\Lambda^s u\|_2 \le \mathfrak{C}_{t_0+3/2}\big(\|\mathbf{g}\|_{H^{s,1}} + \|\Lambda^s Q\|_2 \|\mathbf{g}\|_{H^{t_0+1/2,1}}\big). \tag{2.18}$$

Before we go further, let us prove the following lemma used in the proof above.

LEMMA 2.39. *Under the same assumptions as in Lemma* 2.38, *one has, for all* $1 - t_0 \le r \le t_0 + 1/2$ *:*

$$\|\nabla^{\mu,\gamma} u\|_{L^\infty H^r} \le \mathfrak{C}_{r+1/2}\big(\|\mathbf{g}\|_{H^{r+1/2,1}} + \|\Lambda^{r+1/2}\nabla^{\mu,\gamma} u\|_2\big).$$

PROOF OF LEMMA 2.39. We use Proposition 2.12 to get

$$\|\nabla^{\mu,\gamma} u\|_{L^\infty H^r} \lesssim \|\nabla^{\mu,\gamma} u\|_{H^{r+1/2,1}};$$

since $\gamma \le 1$, we get

$$\|\Lambda^{r-1/2}\partial_z \nabla^{\mu,\gamma} u\|_2 \le \sqrt{\mu}\|\Lambda^{r+1/2}\partial_z u\|_2 + \|\Lambda^{r-1/2}\partial_z^2 u\|_2,$$

so that

$$\|\nabla^{\mu,\gamma} u\|_{L^\infty H^r} \lesssim \|\Lambda^{r+1/2}\nabla^{\mu,\gamma} u\|_2 + \|\Lambda^{r-1/2}\partial_z^2 u\|_2. \tag{2.19}$$

In order to get an upper bound on $\|\Lambda^{r-1/2}\partial_z^2 u\|_2$, we use the equation (2.15) satisfied by u. If we decompose the $(d+1)\times(d+1)$ matrix Q into

$$Q = \begin{pmatrix} Q_d & \mathbf{q} \\ \mathbf{q}^T & q_{d+1} \end{pmatrix},$$

this gives

$$\begin{aligned} (1 + q_{d+1})\partial_z^2 u &= -\nabla^{\mu,\gamma}\cdot \mathbf{g} - \sqrt{\mu}\nabla^\gamma \cdot \big((I_d + Q_d)\sqrt{\mu}\nabla^\gamma u + \mathbf{q}\partial_z u\big) \\ &\quad -(\partial_z \mathbf{q})\cdot \sqrt{\mu}\nabla^\gamma u - \mathbf{q}\cdot\sqrt{\mu}\nabla^\gamma \partial_z u + \partial_z q_{d+1}\partial_z u. \end{aligned} \tag{2.20}$$

Since $1 + q_{d+1} \geq k$ (from the coercivity of P), Corollary B.6 allows us to bound $\|\Lambda^{r-1/2}\partial_z^2 u\|_2$ from above (up to a multiplicative constant M_0) by the $H^{r-1/2,0}$-norm of the r.h.s of (2.20).
There are two kinds of terms that depend on u in the r.h.s. of (2.20): those (e.g., $\mathbf{q}\cdot\sqrt{\mu}\nabla^\gamma\partial_z u$) that do not imply derivatives of the coefficients of Q, and those (e.g., $\partial_z\mathbf{q}\cdot\sqrt{\mu}\nabla^\gamma u$) that involve such derivatives. The former are controlled through Corollary B.5(1) by $\|Q\|_{L^\infty H^{t_0}}\|\Lambda^{r+1/2}\nabla^{\mu,\gamma}u\|_2$ while Corollary B.5(3) (with $s = r-1/2$, $s_1 = r-\epsilon$, $s_2 = r\vee t_0 - 1/2$, with $\epsilon > 0$ and $t_0 - \epsilon - d/2 > 0$) allows us to control the latter by $\|Q\|_{H^{r\vee t_0+1/2,1}}\|\nabla^{\mu,\gamma}u\|_{L^\infty H^{r-\epsilon}}$. Therefore, we get that $\|\Lambda^{r-1/2}\partial_z^2 u\|_2$ is bounded from above by

$$M_0\big(\|\mathbf{g}\|_{H^{r+1/2,1}} + \|\Lambda^{r+1/2}\nabla^{\mu,\gamma}u\|_2 + \|Q\|_{H^{r\vee t_0+1/2,1}}\|\nabla^{\mu,\gamma}u\|_{L^\infty H^{r-\varepsilon}}\big).$$

Together with (2.19), this yields

$$\|\nabla^{\mu,\gamma}u\|_{L^\infty H^r} \leq \mathfrak{C}_{r+1/2}\big(\|\mathbf{g}\|_{H^{r+1/2,1}} + \|\Lambda^{r+1/2}\nabla^{\mu,\gamma}u\|_2 + \|\nabla^{\mu,\gamma}u\|_{L^\infty H^{r-\varepsilon}}\big);$$

by induction (after remarking that $\|\nabla^{\mu,\gamma}u\|_{L^\infty H^{t_0}} < \infty$ — see also (2.21) below), this shows that $\|\nabla^{\mu,\gamma}u\|_{L^\infty H^r} < \infty$. Now, noting that for any $\delta > 0$ there exists $C_\delta > 0$ such that.

$$\|\nabla^{\mu,\gamma}u\|_{L^\infty H^{r-\varepsilon}} \leq C_\delta\|\nabla^{\mu,\gamma}u\|_{L^\infty H^{r-1/2}} + \delta\|\nabla^{\mu,\gamma}u\|_{L^\infty H^r},$$

we get

$$\|\nabla^{\mu,\gamma}u\|_{L^\infty H^r} \leq \mathfrak{C}_{r+1/2}\big(\|\mathbf{g}\|_{H^{r+1/2,1}} + \|\Lambda^{r+1/2}\nabla^{\mu,\gamma}u\|_2 + \|\nabla^{\mu,\gamma}u\|_{L^\infty H^{r-1/2}}\big).$$

Therefore, the result follows if we are able to prove that

$$\|\nabla^{\mu,\gamma}u\|_{L^\infty H^{r-1/2}} \leq \mathfrak{C}_{r+1/2}\big(\|\mathbf{g}\|_{H^{r,1}} + \|\Lambda^r\nabla^{\mu,\gamma}u\|_2\big), \tag{2.21}$$

which we will now do. Proceeding as above (with $r-1$ instead of r), it is enough to prove such a control for $\|\Lambda^{r-1}\partial_z^2 u\|_2$ for which we use again (2.20), Corollary B.6 and the product estimates of Corollary B.5 (this time with $s = r-1$, $s_1 = r\vee t_0 - 1$, $s_2 = r$) to get

$$\begin{aligned}\|\Lambda^{r-1}\partial_z^2 u\|_2 &\leq M_0\big(\|\mathbf{g}\|_{H^{r,1}} + (\|Q\|_{L^\infty H^{r\vee t_0}} + \|\partial_z Q\|_{L^\infty H^{r\vee t_0-1}})\|\Lambda^r\nabla^{\mu,\gamma}u\|_2\big)\\ &\leq \mathfrak{C}_{r+1/2}\big(\|\mathbf{g}\|_{H^{r,1}} + \|\Lambda^r\nabla^{\mu,\gamma}u\|_2\big),\end{aligned} \tag{2.22}$$

which ends the proof. □

The main estimate of Lemma 2.38 corresponds exactly to (2.17) and (2.18). In order to prove that the same estimate also holds on $\|\nabla^{\mu,\gamma}u\|_{H^{s,1}}$ when $s-1 \geq -t_0$, we need an estimate of $\|\Lambda^{s-1}\partial_z\nabla^{\mu,\gamma}u\|_2$ and more specifically of $\|\Lambda^{s-1}\partial_z^2 u\|_2$. For $s \leq t_0$, this is provided by (2.22). For $t_0 < s \leq t_0 + 3/2$, we proceed as for (2.22) with $s_1 = t_0 \vee (s-1)$ and $s_2 = (t_0+1/2)\vee s - 1$ in Corollary B.5(3) to get

$$\begin{aligned}\|\Lambda^{s-1}\partial_z^2 u\|_2 &\leq M_0\big(\|\mathbf{g}\|_{H^{s,1}} + \|Q\|_{L^\infty H^{t_0\vee(s-1)}}\|\Lambda^s\nabla^{\mu,\gamma}u\|_2\\ &\quad + \|Q\|_{H^{(t_0+1/2)\vee s,1}}\|\nabla^{\mu,\gamma}u\|_{L^\infty H^{t_0\vee(s-1)}}\big)\\ &\leq \mathfrak{C}_s\|\mathbf{g}\|_{H^{s,1}},\end{aligned}$$

where we used Lemma 2.39 and (2.17) to derive the second inequality. Finally, for $s > t_0 + 3/2$, we estimate the components of (2.20) with the first two product estimates of Corollary B.5. For instance, for the component $\partial_z\mathbf{q}\cdot\sqrt{\mu}\nabla^\gamma u$, we get

$$\begin{aligned}\|\Lambda^{s-1}\partial_z\mathbf{q}\cdot\sqrt{\mu}\nabla^\gamma u\|_2 &\lesssim \|\partial_z\mathbf{q}\|_{L^\infty H^{t_0}}\|\Lambda^{s-1}\nabla^{\mu,\gamma}u\|_2 + \|\Lambda^{s-1}\partial_z\mathbf{q}\|_2\|\nabla^{\mu,\gamma}u\|_{L^\infty H^{t_0}}\\ &\leq \mathfrak{C}_{t_0+3/2}\big(\|\mathbf{g}\|_{H^{s,1}} + \|Q\|_{H^{s,1}}\|\mathbf{g}\|_{H^{t_0+1/2,1}}\big),\end{aligned}$$

where we used the estimate on $\|\Lambda^{s-1}\nabla^{\mu,\gamma}u\|_2$ already proved and Lemma 2.39 to obtain the second inequality. □

In order to prove Proposition 2.36, let $u = \phi - \psi^\dagger$ and $\mathbf{g} = P\nabla^{\mu,\gamma}\psi^\dagger$. Before applying Lemma 2.38, let us prove that for all $s \geq 0$,

$$\|\Lambda^s\mathbf{g}\|_2 \leq \sqrt{\mu}\mathfrak{C}_{t_0+1/2}|\mathfrak{P}\psi|_{H^s} + \big\langle\sqrt{\mu}\|\Lambda^s Q\|_2|\mathfrak{P}\psi|_{H^{t_0+1/2}}\big\rangle_{s>t_0+1/2}. \tag{2.23}$$

Using the product estimate of Corollary B.5(2), we get that for all $s \geq 0$,

$$\|\Lambda^s\mathbf{g}\|_2 \lesssim \big(1 + \|Q\|_{L^\infty H^{t_0}}\big)\|\nabla^{\mu,\gamma}\psi^\dagger\|_{H^{s,0}} + \big\langle\|\Lambda^s Q\|_2\|\nabla^{\mu,\gamma}\psi^\dagger\|_{L^\infty H^{t_0}}\big\rangle_{s>t_0}.$$

Since $\|\nabla^{\mu,\gamma}\psi^\dagger\|_{L^\infty H^{t_0}} \lesssim \|\nabla^{\mu,\gamma}\psi^\dagger\|_{H^{t_0+1/2,1}}$, we get from Lemma 2.34 that (2.23) is satisfied for $0 \leq s \leq t_0$ and $s \geq t_0 + 1/2$. For $t_0 \leq s \leq t_0 + 1/2$, the result follows by interpolation.
Applying Lemma 2.38 also requires the proof that for all $s \geq t_0$,

$$\|\Lambda^{s-1}\partial_z\mathbf{g}\|_2 \leq \sqrt{\mu}\mathfrak{C}_{\min\{s,t_0+3/2\}}\big(|\mathfrak{P}\psi|_{H^s} + \big\langle\|Q\|_{H^{s,1}}|\mathfrak{P}\psi|_{H^{t_0+3/2}}\big\rangle_{s>t_0+3/2}\big), \tag{2.24}$$

which is done below. Since we have

$$\partial_z\mathbf{g} = (\partial_z Q)\nabla^{\mu,\gamma}\psi^\dagger + P\partial_z\nabla^{\mu,\gamma}\psi^\dagger, \tag{2.25}$$

this is done exactly as for the control of $\|\Lambda^{s-1}\partial_z^2 u\|_2$ at the end of the proof of Lemma 2.39.

In order to prove the last point of the proposition, we still need an estimate $\|\Lambda^{s-1}\partial_z\nabla^{\mu,\gamma}\phi\|_2$. From Lemma 2.38, the only thing left to prove is an upper bound on $\|\Lambda^{s-1}\partial_z\mathbf{g}\|_2$ for $0 \leq s \leq t_0 + 1/2$ and $s \geq -t_0 + 1$. We get from (2.25) and the product estimates of Corollary B.5(3) in Appendix B (with $s_1 = s - 1/2$ and $s_2 = t_0 - 1/2$) for the first component of $\partial_z\mathbf{g}$ and Corollary B.5(2) in Appendix B for the second one, that

$$\begin{aligned}\|\Lambda^{s-1}\partial_z\mathbf{g}\|_2 &\leq \|\Lambda^{t_0-1/2}\partial_z Q\|_2\|\nabla^{\mu,\gamma}\psi^\dagger\|_{L^\infty H^{s-1/2}} \\ &\quad +(1 + \|Q\|_{L^\infty H^{t_0}})\|\Lambda^{s-1}\partial_z\nabla^{\mu,\gamma}\psi^\dagger\|_2 \\ &\leq \sqrt{\mu}\mathfrak{C}_{t_0+1/2}|\mathfrak{P}\psi|_{H^s},\end{aligned}$$

where Proposition 2.12 and Lemma 2.34 have been used to derive the second inequality. The end of the proof is then straightforward. □

In the following corollary, we give a regularity estimate on ϕ in the particular case when Σ is the regularizing diffeomorphism given by Example 2.15.

COROLLARY 2.40. *Let $t_0 > d/2$ and $s \geq 0$. Also let $\zeta, b \in H^{s+1/2} \cap H^{t_0+1}(\mathbb{R}^d)$ be such that (2.6) is satisfied and let Σ be as in Example 2.15.*
Then, for all $\psi \in \dot{H}^{s+1/2}(\mathbb{R}^d)$, the unique variational solution ϕ to (2.12) satisfies

$$\begin{aligned}&\forall 0 \leq s \leq t_0 + 3/2, \quad \|\Lambda^s\nabla^{\mu,\gamma}\phi\|_2 \leq \sqrt{\mu}M(s + 1/2)|\mathfrak{P}\psi|_{H^s}, \\ &\forall t_0 + 3/2 \leq s, \quad \|\Lambda^s\nabla^{\mu,\gamma}\phi\|_2 \leq \sqrt{\mu}M\big(|\mathfrak{P}\psi|_{H^s} + |(\varepsilon\zeta, \beta b)|_{H^{s+1/2}}|\mathfrak{P}\psi|_{H^{t_0+3/2}}\big),\end{aligned}$$

where $M(s)$ and M are as in (2.10) and $\mathfrak{P}$ as in (2.14).
Moreover, if $s \geq -t_0 + 1$, then the same estimates hold on $\|\nabla^{\mu,\gamma}\phi\|_{H^{s,1}}$.

REMARK 2.41. If one chooses the nonregularizing diffeomorphism of Example 2.14, small adaptations of the proof show that if $\zeta, b \in H^{s+1} \cap H^{t_0+2}(\mathbb{R}^d)$, the result still holds provided that the estimate is replaced by

$$\|\Lambda^s\nabla^{\mu,\gamma}\phi\|_2 \leq \sqrt{\mu}M\big(|\mathfrak{P}\psi|_{H^s} + \langle|(\varepsilon\zeta, \beta b)|_{H^{s+1}}|\mathfrak{P}\psi|_{H^{t_0+1}}\rangle_{s>t_0+1}\big)$$

(and a similar bound on $\|\nabla^{\mu,\gamma}\phi\|_{H^{s,1}}$ if $s \geq -t_0+1$), which costs half a derivative more for ζ.

REMARK 2.42. The corollary shows that for all $0 \leq s \leq t_0 + 1/2$, one has $\|\Lambda^s \nabla^{\mu,\gamma}\psi^{\mathfrak{h}}\|_2 \leq \sqrt{\mu}M_0|\mathfrak{P}\psi|_{H^s}$ (recall that the "harmonic extension" $\psi^{\mathfrak{h}}$ of ψ is defined in Notation 2.33). In the next chapter (Proposition 3.19), we show that the reverse inequality also holds.

PROOF. One can write $\Sigma(X,z) = (X, z+\sigma(X,z))$ with

$$\sigma(X,z) = (\varepsilon\zeta^{(\delta)}(X,z) - \beta b_{(\delta)}(X))z + \varepsilon\zeta^{(\delta)}(X,z).$$

We then use the formulas for $Q(\Sigma)$ given in Example 2.21 and the bound on $1/k(\Sigma)$ provided by Remark 2.27, respectively. We also know from Proposition 2.16 that Σ is admissible and that $1 + \partial_z\sigma \geq c_0$, with $c_0 = \frac{h_{min}}{2}$, if (2.6) is satisfied and $\delta > 0$ is chosen small enough.
A direct use of Corollary B.6 (Appendix B) thus gives

$$\|Q(\Sigma)\|_{H^{s,1}} \leq M_0 \|\nabla^{\mu,\gamma}\sigma\|_{H^{s,1}}, \tag{2.26}$$

and since Σ is a regularizing diffeomorphism (the second point of Definition 2.17), we also have

$$\|\nabla^{\mu,\gamma}\sigma\|_{H^{s,1}} \leq M_0(|\varepsilon\zeta|_{H^{s+1/2}} + |\beta b|_{H^{s+1/2}}). \tag{2.27}$$

We can therefore deduce that $\|Q(\Sigma)\|_{H^{s,1}} \leq M_0|(\varepsilon\zeta, \beta b)|_{H^{s+1/2}}$ and $\mathfrak{C}_s \leq M(s + 1/2)$, so that the result follows from Proposition 2.36. □

2.4. Strong solutions to the Laplace equation

The variational solution considered in Corollary 2.40 is in fact a strong solution in the sense that the Neumann condition in (2.12) is not only satisfied in a variational sense, but also in $H^{s-1/2}(\mathbb{R}^d)$, provided that $s \geq -t_0 + 1$. Indeed, under this assumption, one has $\nabla^{\mu,\gamma}\phi \in H^{s,1}$ and therefore (see Proposition 2.12), the trace $\nabla^{\mu,\gamma}\phi_{|_{z=0}}$ is well defined in $H^{s-1/2}(\mathbb{R}^d)$. Moreover, since one has $P(\Sigma)_{|_{z=0}} = I + Q(\Sigma)_{|_{z=0}}$ with $Q(\Sigma)_{|_{z=0}} \in H^{s-1/2} \cap H^{t_0}(\mathbb{R}^d)$, we get by a standard product estimate that

$$\partial_{\mathbf{n}}\phi_{|_{z=0}} := \mathbf{e}_z \cdot P(\Sigma)_{|_{z=0}} \nabla^{\mu,\gamma}\phi_{|_{z=0}}$$

belongs to $H^{s-1/2}(\mathbb{R}^d)$.

REMARK 2.43. The assumptions used in the above analysis are rather weak and do not imply in particular that $\phi \in \dot{H}^2(\mathcal{S})$, which is the classical framework that defines strong solutions. In order to obtain such a regularity on ϕ, Corollary 2.40 shows that it is necessary to assume that $\zeta, b \in H^{t_0+1}(\mathbb{R}^d)$ and $\psi \in \dot{H}^{3/2}(\mathbb{R}^d)$.

Similarly, if we choose to work with the exact Laplace equation (2.7) in the fluid domain Ω rather than the transformed Laplace equation (2.12) in the strip $\mathcal{S}$, the variational solution provided by Proposition 2.9 is a strong solution under the same regularity assumptions as above (i.e., $\zeta, b \in H^{s+1/2} \cap H^{t_0+1}(\mathbb{R}^d)$, $\psi \in \dot{H}^{s+1/2}(\mathbb{R}^d)$ and $s \geq -t_0 + 1$). Indeed, we have

$$\nabla^{\mu,\gamma}\Phi_{|_{z=\zeta}} = I_{\mu,\gamma}(J_\Sigma^{-1})^T I_{\mu,\gamma}^{-1} \nabla^{\mu,\gamma}\phi_{|_{z=0}},$$

which is well defined in $H^{s-1/2}(\mathbb{R}^d)$ since $\nabla^{\mu,\gamma}\phi_{|_{z=0}} \in H^{s-1/2}(\mathbb{R}^d)$ (as seen above) and the coefficients of $(J_\Sigma^{-1})^T_{|_{z=0}}$ all belong to $H^{s-1/2} \cap H^{t_0}(\mathbb{R}^d)$. It follows from a product estimate and the regularity on $\mathbf{n}$ inferred from the regularity of ζ that

$$\partial_{\mathbf{n}}\Phi_{|_{z=\zeta}} := I_{\mu,\gamma}\mathbf{n}\cdot\nabla^{\mu,\gamma}\Phi_{|_{z=\zeta}}$$

is also well defined in $H^{s-1/2}(\mathbb{R}^d)$.

To end this section, let us state the following corollary that extends the comments made in Remark 2.43 to the Laplace equation in the fluid domain.

COROLLARY 2.44. *Let $t_0 > d/2$, $\zeta, b \in H^{t_0+1}(\mathbb{R}^d)$ and $\psi \in \dot{H}^{3/2}(\mathbb{R}^d)$. Then there exists a unique solution $\Phi \in H^2(\Omega)$ to the boundary value problem (2.7).*

2.5. Supplementary remarks

2.5.1. Choice of the diffeomorphism. The regularizing diffeomorphisms used here show that it is important to straighten the fluid domain in a careful way. While we focused here on regularity properties, it is also possible to work with diffeomorphisms that grant the transformed Laplace equation (2.12) some special structure. Such a trick has for instance been used in [**108**]. We also refer to [**181, 277**] for other kinds of regularizing diffeomorphisms.

2.5.2. Nonasymptotically flat bottom and surface parametrizations. Throughout this chapter, we assumed that the bottom and surface parametrizations have Sobolev regularity. We did this for the sake of simplicity, but it is not difficult to handle the case of bottoms that do not vanish at infinity (such as sinusoidal bottoms) and surfaces that are not asymptotical to the axis $z = 0$ (to handle the propagation of bores, for instance). In [**7**], the authors develop a full local well-posedness theory for the water waves equations in uniformly local Sobolev spaces (see §B.4 in Appendix B for the definition and basic properties of these spaces). Let us consider here a more simple situation by assuming that the fluid domain Ω is given by

$$(2.28)\qquad \Omega = \{(X,z) \in \mathbb{R}^d \times \mathbb{R}, -1 + \beta\big(\underline{b}(X) + b(X)\big) < z < \varepsilon(\underline{\zeta}(X) + \zeta(X))\},$$

where $\underline{b}$ and $\underline{\zeta}$ belong to $C_b^\infty(\mathbb{R}^d)$, the space of infinitely smooth functions that are bounded together with all their derivatives. The nonvanishing depth condition is consequently replaced by

$$(2.29)\quad \exists h_{min} > 0, \quad \forall X \in \mathbb{R}^d, \quad 1 + \varepsilon(\zeta(X) + \underline{\zeta}(X)) - \beta(b(X) + \underline{b}(X)) \geq h_{min}.$$

The existence of a variational solution to (2.7) is obtained exactly as in Proposition 2.9; the definition of admissible diffeomorphisms and the derivation of the transformed problem (2.12) are also unaffected provided that one replaces M_0 by $\underline{M}_0$ defined as

$$\underline{M}_0 = C(\frac{1}{h_{min}}, \mu_{max}, |\zeta|_{H^{t_0+1}}, |b|_{H^{t_0+1}}, |\underline{\zeta}|_{W^{1,\infty}}, |\underline{b}|_{W^{1,\infty}}).$$

Definition 2.17 (regularizing diffeomorphisms) can also be easily extended by requiring that the diffeomorphism Σ be of the form

$$\Sigma(X,z) = (X, z + \sigma(X,z) + \underline{\sigma}(X,z))$$

with σ and $\underline{\sigma}$ satisfying

$$\|\nabla^{\mu,\gamma}\sigma\|_{H^{s,0}} \leq \underline{M}_0(|\varepsilon\zeta|_{H^{s+1/2}} + |\beta b|_{H^{s+1/2}}),$$
$$\|\nabla^{\mu,\gamma}\underline{\sigma}\|_{L^\infty_z W^{k,\infty}_X} \leq \underline{M}_0(|\varepsilon\underline{\zeta}|_{W^{k+1,\infty}} + |\beta\underline{b}|_{W^{k+1,\infty}}),$$

where k is the smallest integer larger or equal to s.

EXAMPLE 2.45. Let $\underline{\sigma}(X,z) = \big(\varepsilon\underline{\zeta}(X) - \beta\underline{b}(X)\big)z + \varepsilon\underline{\zeta}(X)$; choosing σ as in Example 2.14 (resp. 2.15), we obtain an admissible (resp. regularizing) diffeomorphism associated to the fluid domain (2.28).

Using the commutator estimates of §B.3 in Appendix B, one can then extend all the results of this chapter to the present situation. For instance, Corollary 2.40 can be adapted as follows, where we use the notations[3]

$$\underline{M} = C(\frac{1}{h_{min}}, \mu_{max}, |\zeta|_{H^{t_0+2}}, |b|_{H^{t_0+2}}, \underline{\zeta}, \underline{b}),$$
$$\underline{M}(s) = C(\frac{1}{h_{min}}, \mu_{max}, |\zeta|_{H^{s\vee(t_0+1)}}, |b|_{H^{s\vee(t_0+1)}}, \underline{\zeta}, \underline{b}).$$

COROLLARY 2.46. *Let $t_0 > d/2$ and $s \geq 0$. Let $\zeta, b \in H^{s+1/2} \cap H^{t_0+1}(\mathbb{R}^d)$, $\underline{\zeta}, \underline{b} \in C^\infty_b(\mathbb{R}^d)$ be such that (2.29) is satisfied. Let Σ be the regularizing diffeomorphism of Example 2.45.*
Then, for all $\psi \in \dot{H}^{s+1/2}(\mathbb{R}^d)$, the unique variational solution ϕ to (2.12) satisfies

$$\forall 0 \leq s \leq t_0 + 3/2, \quad \|\Lambda^s\nabla^{\mu,\gamma}\phi\|_2 \leq \sqrt{\mu}\underline{M}(s+1/2)|\mathfrak{P}\psi|_{H^s},$$
$$\forall t_0 + 3/2 \leq s, \quad \|\Lambda^s\nabla^{\mu,\gamma}\phi\|_2 \leq \sqrt{\mu}\underline{M}\big(|\mathfrak{P}\psi|_{H^s} + |(\varepsilon\zeta,\beta b)|_{H^{s+1/2}}|\mathfrak{P}\psi|_{H^{t_0+3/2}}\big).$$

Moreover, if $s \geq -t_0 + 1$, then the same estimates hold on $\|\nabla^{\mu,\gamma}\phi\|_{H^{s,1}}$.

2.5.3. Rough bottoms. In [**4**] very wild bottoms are considered (the only assumption is that the total water depth exceeds some nonnegative constant). The reason why wild bottoms are not an obstruction to local well-posedness is that their contributions to the surface evolution are infinitely smoothing (from the ellipticity of the potential equation). We refer to Section A.4 in Appendix A for more comments on this point.
However, note that the elliptic estimates derived in the framework of [**4**] cannot be made uniform with respect to μ as done in this chapter. In order to understand the difficulties encountered if one wants to handle simultaneously the shallow water limit and rough bottom, let us consider the two situations described in §1.7:

(1) Nonsmooth topographies. Variational solutions to the potential equation (2.7) still exist with nonsmooth topographies but working with the transformed equation (2.7) is more delicate; in particular, the diffeomorphism used to straighten the fluid domain must necessarily have some smoothing properties in order for the matrix $P(\Sigma)$ of (2.13) to be well defined. The estimates of Proposition 2.36 are not appropriate and should therefore have been improved to allow nonsmooth topography in a shallow water context.

[3]The dependence on $\underline{\zeta}$ and $\underline{b}$ is not made explicit for the sake of clarity. One can easily check that it is enough to assume in Corollary 2.46 that $\underline{\zeta}, \underline{b} \in C^{k+1}_b(\mathbb{R}^d)$, with k the smallest integer strictly larger than s.

(2) Rapidly varying topographies. Let us consider here a bottom parametrization of the form $-1+\beta b(X/\alpha)$, with $\alpha \ll 1$ (in dimensionless variables). When considering such bottoms, the estimates of Proposition 2.36 are uniform with respect to $\mu \in (0,\mu_{max})$ but not with respect to α when $\alpha \to 0$. Consequently, these estimates are useless for rough bottoms. In [**107**], uniform estimates with respect to both α and μ are given in the particular case $\beta=\alpha=\sqrt{\mu}$ (and for periodic bottoms), but with rather low regularity. Dealing with rough bottoms ($\alpha \ll 1$) in shallow water ($\mu \ll 1$) therefore remains an open problem in many aspects, even at the basic level of the potential equation (2.7) (or its transformed version (2.12))

2.5.4. Infinite depth. *For the sake of clarity, we take $\gamma = 1$ here; adaptations to the general case are straightforward.* As explained in Remark 1.13, the equations are nondimensionalized differently in infinite depth and, instead of (2.7), the relevant potential equation becomes

$$\begin{cases} \Delta_{X,z}\Phi = 0 & \text{in } \Omega, \\ \Phi_{\vert z=\epsilon\zeta} = \psi, \end{cases} \tag{2.30}$$

where $0<\epsilon<\epsilon_{max}$ is the *steepness* of the wave, and the fluid domain is now given by

$$\Omega = \{(X,z)\in\mathbb{R}^d\times\mathbb{R}, \quad z<\epsilon\zeta(X)\}.$$

We adapt the Definition 2.7 of variational solutions to the case of infinite depth as follows. In the statement below, $\psi^{ext}\in H^1(\Omega)$ denotes any $H^1(\Omega)$ extension of ψ (choosing a different extension has no incidence).

DEFINITION 2.47. Let $\psi\in H^1(\mathbb{R}^d)$. We say that Φ is a variational solution to (2.7) if there exists $\widetilde{\Phi}\in H^1_{0,surf}(\Omega)$ such that $\Phi=\psi^{ext}+\widetilde{\Phi}$ and

$$\int_\Omega \nabla_{X,z}\widetilde{\Phi}\cdot\nabla_{X,z}\varphi = -\int_\Omega \nabla_{X,z}\psi^{ext}\cdot\nabla_{X,z}\varphi$$

for all $\varphi\in H^1_{0,surf}(\Omega)$.

As for the case of finite depth, on can straighten the fluid domain using admissible (and possibly regularizing) diffeomorphisms. The adaptations of Definitions 2.13 and 2.17 to the case of infinite depth are straightforward. We mainly consider diffeomorphisms of the form $\Sigma(X,z)=(X,z+\sigma(X,z))$.

EXAMPLE 2.48. With the notations of Examples 2.14 and 2.15, we obtain an admissible diffeomorphism by choosing $\sigma(X,z)=\epsilon\zeta(X)$. Choosing $\sigma(X,z)=\epsilon\sigma^{(\delta)}\zeta=\epsilon\chi(\delta z|D|)\zeta$, with $\delta>0$ small enough, one gets a regularizing diffeomorphism.

Using any admissible diffeomorphism Σ, the boundary value problem (2.30) is transformed into

$$\begin{cases} \nabla_{X,z}\cdot P(\Sigma)\nabla_{X,z}\phi = 0 & \text{in} \quad \mathbb{R}^{d+1}_-, \\ \phi_{\vert z=0} = \psi, \end{cases} \tag{2.31}$$

where $\phi=\Phi\circ\Sigma$ and

$$P(\Sigma) = |\det J_\Sigma| J_\Sigma^{-1}(J_\Sigma^{-1})^T. \tag{2.32}$$

The definition of variational solutions to (2.31) can be adapted from Definition 2.22 as it was for Definition 2.47.

In the finite depth situation, the existence of a variational solution for the potential equation (or its transformed version) was a direct consequence of the Lax-Milgram theorem, after noting that the $H^1(\Omega)$ and $\dot{H}^1(\Omega)$ norms define the same topology on $H^1_{0,surf}(\Omega)$. This observation was a consequence of the Poincaré inequality (2.8), which is no longer true in infinite depth. One can, however, note that the following weighted Poincaré inequality holds [4] for $\alpha > 1$,

$$\forall u \in \mathcal{D}(\mathbb{R}^{d+1}_-), \qquad \|u\|_{L^2_\alpha(\mathbb{R}^{d+1}_-)} \leq \mathrm{Cst}\,\|\partial_z u\|_2, \tag{2.33}$$

where $\mathbb{R}^{d+1}_- = \{(X,z) \in \mathbb{R}^{d+1}, z < 0\}$, $L^2_\alpha(\mathbb{R}^{d+1}_-) = L^2(\mathbb{R}^{d+1}_-, \langle z\rangle^{-2\alpha} dX dz)$, and $\langle z\rangle = (1+z^2)^{1/2}$.

In infinite depth, we therefore take a slightly different definition H^1_{surf}. More precisely, we define $H^1_{0,surf}(\mathbb{R}^{d+1}_-)$ as the completion of $\mathcal{D}(\mathbb{R}^{d+1}_-)$ for the norm $\|\nabla_{X,z}u\|_2 + \|u\|_{L^2_\alpha}$. The weighted Poincaré inequality (2.33) then allows us to establish the existence of variational solutions to (2.31) exactly as in the finite depth case (this method can of course be adapted to construct variational solutions to the original equation (2.30)).
As in §2.2.4, we can extend the notion of variational solutions to all $\psi \in H^{1/2}$. All the results in this section hold by replacing μ by 1 and $\mathfrak{P}$ by $|D|^{1/2}$. In particular, the estimates of Lemma 2.34 become $\|\nabla_{X,z}\psi^\dagger\|_{H^{s,1}} \leq \mathrm{Cst}\,\big||D|^{1/2}\psi\big|_{H^s}$. The regularity estimates of §2.3 can then be extended to the case of infinite depth. For instance, Corollary 2.40 can be adapted as follows.

COROLLARY 2.49. *Let $t_0 > d/2$ and $s \geq 0$. Let $\zeta \in H^{s+1/2} \cap H^{t_0+1}(\mathbb{R}^d)$ and $\Sigma = (X, z + \sigma(X,z))$ be the regularizing diffeomorphism of Example 2.48. Then, for all $\psi \in H^{s+1/2}(\mathbb{R}^d)$, the unique variational solution ϕ to (2.31) satisfies*

$$\begin{array}{ll} \forall 0 \leq s \leq t_0 + 3/2, & \|\Lambda^s \nabla_{X,z}\phi\|_2 \leq \tilde{M}(s+1/2)\big||D|^{1/2}\psi\big|_{H^s}, \\ \forall t_0 + 3/2 \leq s, & \|\Lambda^s \nabla_{X,z}\phi\|_2 \leq \tilde{M}\big(\big||D|^{1/2}\psi\big|_{H^s} + |\epsilon\zeta|_{H^{s+1/2}}\big||D|^{1/2}\psi\big|_{H^{t_0+3/2}}\big), \end{array}$$

where $\tilde{M}(s)$ and $\tilde{M}$ are given by

$$\tilde{M} = C(|\zeta|_{H^{t_0+2}}), \qquad M(s) = C(|\zeta|_{H^{s\vee(t_0+1)}}).$$

Moreover, if $s \geq -t_0 + 1$, then the same estimates hold on $\|\nabla_{X,z}\phi\|_{H^{s,1}}$.

REMARK 2.50. In the finite depth case, we assumed that ψ was in the Beppo-Levi space $\dot{H}^{s+1/2}(\mathbb{R}^d)$. The assumption made here (namely, $\psi \in H^{s+1/2}(\mathbb{R}^d)$) is stronger, and requires in particular that ψ be in $L^2(\mathbb{R}^d)$. This assumption can be weakened by assuming that ψ belongs only to the space $\mathring{H}^{s+1/2}(\mathbb{R}^d)$ defined as

$$\mathring{H}^{s+1/2}(\mathbb{R}^d) = \{f \in \mathcal{S}'(\mathbb{R}^d), \quad |D|^{1/2}\psi \in H^s(\mathbb{R}^d)\}. \tag{2.34}$$

[4] Indeed one has

$$\begin{array}{rcl} (1+z^2)^{-\alpha}|u(X,z)|^2 & = & (1+z^2)^{-\alpha}\Big|\displaystyle\int_z^0 \partial_z u(X,z')dz'\Big|^2 \\ & \leq & \dfrac{|z|}{(1+z^2)^\alpha}\displaystyle\int_{-\infty}^0 |\partial_z u|^2(X,z')dz', \end{array}$$

and the result follows upon integrating both sides in X and z.

Since this space is not easy to handle, for simplicity we choose to state the infinite depth extensions of the results proved in finite depth in the simple functional framework $\psi \in H^{s+1/2}(\mathbb{R}^d)$.

REMARK 2.51. As shown in [**4**], it is possible to deal with situations where the depth is finite in some regions (and possibly not a graph) and infinite in others.

2.5.5. Nonhomogeneous Neumann conditions at the bottom. As seen in §1.6 of the introduction, in the case of moving bottoms, the dimensionless equation for the velocity potential becomes

$$\begin{cases} \Delta^{\mu,\gamma}\Phi = 0, & -1+\beta b \le z \le \varepsilon\zeta, \\ \Phi_{|_{z=\varepsilon\zeta}} = \psi, & \sqrt{1+\beta^2|\nabla b|^2}\partial_{\mathbf{n}}\Phi_{|_{z=-1+\beta b}} = \mu\nu\dfrac{\beta}{\varepsilon\delta}\partial_\tau b \end{cases}$$

instead of (2.7). It is therefore necessary to deal with the case of nonhomogeneous Neumann conditions at the bottom. This is done in §A.5.1 (Appendix A).

2.5.6. Analyticity. We prove in Appendix A (Proposition A.5) that the dependence on the surface parametrization of the solution to the (transformed) Laplace equation is analytic.

CHAPTER 3

The Dirichlet-Neumann Operator

This chapter is devoted to the study of the Dirichlet-Neumann operator, which plays an important role in the water waves equations (1.28).
In §3.1, we give a precise definition of this operator and describe some of its basic properties. We then give in §3.2 estimates of its operator norm, which are sharp with respect to the regularity of the surface elevation and with respect to the various dimensionless parameters involved. We show in particular that in the shallow water limit ($\mu \ll 1$), the Dirichlet-Neumann must be seen as an operator of order two (rather than one) if a sharp dependence on μ is required.
The next step, addressed in §3.3, consists of studying the shape derivative of the Dirichlet-Neumann operator (that is, its derivative with respect to the surface and bottom parametrizations ζ and b). We give estimates on the shape derivatives of any order and a useful explicit expression for the first-order derivative of the Dirichlet-Neumann operator with respect to ζ.
In §3.4, we then give some technical results (commutator estimates) that will be needed for the resolution of the water waves equation.

We then derive some properties of the Dirichlet-Neumann operator that are required for the derivation of asymptotic models to the water waves equations. We first give in §3.5 an important qualitative result that makes the link between the Dirichlet-Neumann operator and the vertically averaged velocity. Next, in §3.6 we study the asymptotic properties of the Dirichlet-Neumann operator in two different regimes (shallow water and small amplitude) that cover all the regimes considered in these notes.

In the last section of this chapter (§3.7), we provide some extensions of these results: extensions to nonasymptotically flat bottom and surface parametrizations, infinite depth and rough bottoms. Additional properties of the Dirichlet-Neumann operator (shape analyticity, self-adjointness, invertibility, symbolic analysis) are addressed later in Appendix A, where related operators (the Neumann-Neumann operator, for instance) are also investigated.

Throughout this chapter, we always assume that the dimensionless parameters ε, μ, β, γ satisfy

$$0 \leq \varepsilon, \beta, \gamma \leq 1, \quad 0 \leq \mu \leq \mu_{max}, \tag{3.1}$$

for some $\mu_{max} > 0$ (not necessarily small) and that $\zeta, b \in H^{t_0+1}(\mathbb{R}^d)$ ($t_0 > d/2$) are such that

$$\exists h_{min} > 0, \quad \forall X \in \mathbb{R}^d, \quad 1 + \varepsilon\zeta(X) - \beta b(X) \geq h_{min}. \tag{3.2}$$

We also denote by M_0 and M any constants of the form

$$M_0 = C\big(\frac{1}{h_{min}}, \mu_{max}, |\zeta|_{H^{t_0+1}}, |b|_{H^{t_0+1}}\big), \quad M = C(M_0, |\zeta|_{H^{t_0+2}}, |b|_{H^{t_0+2}}\big), \tag{3.3}$$

and, for all $s \geq 0$, we also define $M(s)$ as

$$M(s) = C(M_0, |\zeta|_{H^s}, |b|_{H^s}). \tag{3.4}$$

3.1. Definition and basic properties

3.1.1. Definition. In Chapter 2 we saw that for all $\zeta, b \in H^{t_0+1}(\mathbb{R}^d)$ satisfying (3.2), it is possible to construct an admissible diffeomorphism in the sense of Definition 2.13 (given by Example 2.14 or 2.15, for instance), mapping the fluid domain Ω onto the flat strip $\mathcal{S} = \mathbb{R}^d \times (-1, 0)$. Using this diffeomorphism, the Laplace equation (2.7) in Ω transforms into a variable coefficients elliptic equation in $\mathcal{S}$, namely,

$$\begin{cases} \nabla^{\mu,\gamma} \cdot P(\Sigma)\nabla^{\mu,\gamma}\phi = 0 & \text{in} \quad \mathcal{S}, \\ \phi_{|z=0} = \psi, & \partial_{\mathbf{n}}\phi_{|z=-1} = 0, \end{cases} \tag{3.5}$$

where

$$P(\Sigma) = |\det J_\Sigma| I^{-1}_{\mu,\gamma} J^{-1}_\Sigma I^2_{\mu,\gamma} (J^{-1}_\Sigma)^T I^{-1}_{\mu,\gamma}, \tag{3.6}$$

and where $I_{\mu,\gamma}$ is defined in (2.1) and J_Σ is the Jacobian matrix of Σ, $J_\Sigma = d_{X,z}\Sigma$; finally, $\partial_{\mathbf{n}}$ stands as usual for the upward conormal derivative

$$\partial_{\mathbf{n}}\phi_{|z=-1} = \mathbf{e}_z \cdot P(\Sigma)\nabla^{\mu,\gamma}\phi_{|z=-1}.$$

We have also seen that for all $\psi \in \dot{H}^{1/2}(\mathbb{R}^d)$, there exists a unique variational solution $\phi \in \dot{H}^1(\mathcal{S})$ to (3.5). As in Notation 2.33, we write $\psi^\mathfrak{h} = \phi$ for this solution. It follows that the mapping

$$G[\Sigma]\psi : \begin{array}{lcl} \dot{H}^{1/2}(\mathbb{R}^d)/\mathbb{R} & \to & \mathbb{R} \\ \varphi & \mapsto & \int_\mathcal{S} \nabla^{\mu,\gamma}\psi^\mathfrak{h} \cdot P(\Sigma)\nabla^{\mu,\gamma}\varphi^\mathfrak{h} \end{array} \tag{3.7}$$

is continuous and belongs therefore to the topological dual $(\dot{H}^{1/2}/\mathbb{R})'$ of $\dot{H}^{1/2}/\mathbb{R}$.

An important observation is that $G[\Sigma]\psi$ does not depend on the admissible diffeomorphismmapping the fluid domain Ω onto the flat strip $\mathcal{S}$.

LEMMA 3.1. *Let $t_0 > d/2$ and $\zeta, b \in H^{t_0+1}(\mathbb{R}^d)$ be such that (3.2) is satisfied. Also let Σ_1 and Σ_2 be two admissible diffeomorphisms, and $\psi \in \dot{H}^{1/2}(\mathbb{R}^d)$. Then one has $G[\Sigma_1]\psi = G[\Sigma_2]\psi$.*

PROOF. Let us denote by $\psi^{\mathfrak{h}_1}$ and $\psi^{\mathfrak{h}_2}$ the unique variational solutions to (3.5) corresponding to $\Sigma = \Sigma_1$ and $\Sigma = \Sigma_2$, respectively. We know that

$$\psi^{\mathfrak{h}_1} \circ \Sigma_1^{-1} = \psi^{\mathfrak{h}_2} \circ \Sigma_2^{-1} = \Phi,$$

where Φ is the unique variational solution to the original Laplace equation (2.7). Consequently, we have $\psi^{\mathfrak{h}_2} = \psi^{\mathfrak{h}_1} \circ (\Sigma_1^{-1} \circ \Sigma_2)$. More generally, for all $\varphi \in \dot{H}^{1/2}(\mathbb{R}^d)$, one has $\varphi^{\mathfrak{h}_2} = \varphi^{\mathfrak{h}_1} \circ (\Sigma_1^{-1} \circ \Sigma_2)$. We deduce that

$$\forall \varphi \in \dot{H}^{1/2}(\mathbb{R}^d)/\mathbb{R}, \qquad \nabla^{\mu,\gamma}\varphi^{\mathfrak{h}_2} = I_{\mu,\gamma} J^T_{\Sigma_1^{-1}\circ\Sigma_2} I^{-1}_{\mu,\gamma} \nabla^{\mu,\gamma}\varphi^{\mathfrak{h}_1} \circ (\Sigma_1^{-1} \circ \Sigma_2),$$

where $I_{\mu,\gamma}$ is as in (2.1), while $J_{\Sigma_1^{-1}\circ\Sigma_2}$ is the Jacobian matrix of $\Sigma_1^{-1} \circ \Sigma_2$. Since $\Sigma_1^{-1} \circ \Sigma_2$ is a diffeomorphism mapping $\mathcal{S}$ onto itself, we therefore deduce from a simple change of variable that

$$\int_\mathcal{S} \nabla^{\mu,\gamma}\psi^{\mathfrak{h}_1} \cdot P(\Sigma_1)\nabla^{\mu,\gamma}\varphi^{\mathfrak{h}_1} = \int_\mathcal{S} \nabla^{\mu,\gamma}\psi^{\mathfrak{h}_2} \cdot \widetilde{P}\nabla^{\mu,\gamma}\varphi^{\mathfrak{h}_2}$$

with

$$\widetilde{P} = I_{\mu,\gamma}^{-1} J_{\Sigma_1^{-1}\circ\Sigma_2}^{-1} I_{\mu,\gamma} P(\Sigma_1)\circ(\Sigma_1^{-1}\circ\Sigma_2) I_{\mu,\gamma} (J_{\Sigma_1^{-1}\circ\Sigma_2}^{-1})^T I_{\mu,\gamma}^{-1} |\det J_{\Sigma_1^{-1}\circ\Sigma_2}|.$$

Since $\Sigma_1 = \Sigma_2 \circ (\Sigma_2^{-1}\circ\Sigma_1)$, we have $J_{\Sigma_1} = J_{\Sigma_2}\circ(\Sigma_2^{-1}\circ\Sigma_1) J_{\Sigma_2^{-1}\circ\Sigma_1}$; replacing J_{Σ_1} by this expression in the definition (2.13) of $P(\Sigma_1)$, we get that $\widetilde{P} = P(\Sigma_2)$, and the proof is complete. □

Using Lemma 3.1, we can give the following definition of the Dirichlet-Neumann operator.

DEFINITION 3.2 (Dirichlet-Neumann operator). Let $t_0 > d/2$. Let $\zeta, b \in H^{t_0+1}(\mathbb{R}^d)$ be such that (3.2) is satisfied. Let also Σ be any admissible diffeomorphism in the sense of Definition 2.13. We define the *Dirichlet-Neumann operator* $\mathcal{G}_{\mu,\gamma}[\varepsilon\zeta, \beta b]$ as

$$\mathcal{G}_{\mu,\gamma}[\varepsilon\zeta,\beta b] : \begin{array}{ccc} \dot{H}^{1/2}(\mathbb{R}^d) & \to & (\dot{H}^{1/2}(\mathbb{R}^d)/\mathbb{R})' \\ \psi & \mapsto & G[\Sigma]\psi, \end{array}$$

where $G[\Sigma]\psi$ is as in (3.7).

Since $H^{1/2}(\mathbb{R}^d)$ is continuously embedded and dense in $\dot{H}^{1/2}(\mathbb{R}^d)$ (see Proposition 2.3(5)), it is possible to identify $\mathcal{G}_{\mu,\gamma}[\varepsilon\zeta,\beta b]\psi$ with an element of $H^{-1/2}(\mathbb{R}^d)$. We will see later in Theorem 3.15 that $\mathcal{G}_{\mu,\gamma}[\varepsilon\zeta,\beta b]$ maps $\dot{H}^{s+1/2}(\mathbb{R}^d)$ into $H^{s-1/2}(\mathbb{R}^d)$ (for $s \geq 0$ — by duality, the Dirichlet-Neumann operator can also be defined on Sobolev spaces of negative index, as explained in Remark 3.17 below).

PROPOSITION 3.3. *Let $t_0 > d/2$ and $\zeta, b \in H^{t_0+1}(\mathbb{R}^d)$ be such that (3.2) is satisfied. Then, for all $\psi \in \dot{H}^{1/2}(\mathbb{R}^d)$, one has $\mathcal{G}_{\mu,\gamma}[\varepsilon\zeta,\beta b]\psi \in H^{-1/2}(\mathbb{R}^d)$.*

REMARK 3.4. It is possible to identify more precisely the topological dual $(\dot{H}^{1/2}(\mathbb{R}^d)/\mathbb{R})'$ as a subspace of $H^{-1/2}(\mathbb{R}^d)$. In [**59**], it is proved that $(\dot{H}^{1/2}(\mathbb{R}^d)/\mathbb{R})'$ can be identified with the subspace $H_0^{-1/2}(\mathbb{R}^d)$ of $H^{1/2}(\mathbb{R}^d)$ defined as

$$H_0^{-1/2}(\mathbb{R}^d) = \{u \in H^{-1/2}(\mathbb{R}^d), \quad \exists \underline{u} \in H^{1/2}(\mathbb{R}^d), \quad u = |D|\underline{u}\}.$$

More precisely, if $l \in (\dot{H}^{1/2}(\mathbb{R}^d)/\mathbb{R})'$, then there is a unique $u_l \in H_0^{-1/2}(\mathbb{R}^d)$, with

$$\forall v \in \dot{H}^{1/2}(\mathbb{R}^d)/\mathbb{R}, \qquad l(v) = \int_{\mathbb{R}^d} \widehat{|D|v}\,\overline{\widehat{\underline{u}_l}}.$$

REMARK 3.5. We explain in §3.7.1 how to construct the Dirichlet-Neumann operator if $\zeta, b \in C_b^1(\mathbb{R}^d)$, and in §3.7.3 in the case of infinite depth. It is also shown in [**7**] that for all $\zeta, b \in H_{ul}^{t_0+1}(\mathbb{R}^d)$ (we recall that H_{ul}^s stands for the uniformly local version of $H^s(\mathbb{R}^d)$; see §B.4 in Appendix B), one can define the Dirichlet-Neumann operator as a linear operator on $H_{ul}^{1/2}(\mathbb{R}^d)$ with values in $H_{ul}^{-1/2}(\mathbb{R}^d)$.

PROOF. Since $H^{1/2}(\mathbb{R}^d)$ is continuously embedded in $\dot{H}^{1/2}(\mathbb{R}^d)/\mathbb{R}$, the fact that $\mathcal{G}\psi$ belongs to $(\dot{H}^{1/2}(\mathbb{R}^d)/\mathbb{R})'$ implies that the restriction $\mathcal{G}\psi_{|_{H^{1/2}}}$ belongs to $H^{1/2}(\mathbb{R}^d)'$. In order to prove that this restriction fully determines $\mathcal{G}\psi$, we have to prove that if $\langle \mathcal{G}\psi, \varphi\rangle = 0$ for all $\varphi \in H^{1/2}(\mathbb{R}^d)$, then $\mathcal{G}\psi = 0$. This is granted by the fact that $H^{1/2}(\mathbb{R}^d)$ is dense in $\dot{H}^{1/2}(\mathbb{R}^d)$, as shown in Proposition 2.3(5). □

Definition 3.2 differs from the definition given in (1.29) of Chapter 1. It is therefore important to show that they coincide (for ζ, b and ψ smooth enough to use Green's formula). In order to prove that this is the case, let us recall that Corollary 2.44 asserts that for all $\zeta, b \in H^{t_0+1}(\mathbb{R}^d)$ and $\psi \in \dot{H}^{3/2}(\mathbb{R}^d)$, there exists a unique strong solution $\Phi \in \dot{H}^2(\mathcal{S})$ (see Definition 2.2 for a definition of this space) to the Laplace equation in the fluid domain, namely,

$$\begin{cases} \Delta^{\mu,\gamma}\Phi = 0 & \text{in } \Omega, \\ \Phi_{|_{z=\varepsilon\zeta}} = \psi, \quad \partial_{\mathbf{n}}\Phi_{|_{z=-1+\beta b}} = 0. \end{cases} \tag{3.8}$$

(Recall that, as always in these notes, $\partial_{\mathbf{n}}$ refers to the upward conormal derivative, so that $\partial_{\mathbf{n}}\Phi_{|_{z=-1+\beta b}} = I_{\mu,\gamma}\mathbf{n} \cdot \nabla^{\mu,\gamma}\Phi_{|_{z=-1+\beta b}}$.)
We can now state the following proposition.

PROPOSITION 3.6. *Let $t_0 > d/2$ and $\zeta, b \in H^{t_0+1}(\mathbb{R}^d)$ be such that (3.2) is satisfied. For all $\psi \in \dot{H}^{3/2}(\mathbb{R}^d)$, one has*

$$\mathcal{G}_{\mu,\gamma}[\varepsilon\zeta, \beta b]\psi = \sqrt{1+\varepsilon^2|\nabla\zeta|^2}\partial_{\mathbf{n}}\Phi_{|_{z=\varepsilon\zeta}}.$$

PROOF. Denoting $\phi = \Phi \circ \Sigma$, where Σ is a regularizing diffeomorphism, we obtain as in §2.4 that $\sqrt{1+\varepsilon^2|\nabla\zeta|^2}\partial_{\mathbf{n}}\Phi_{|_{z=\varepsilon\zeta}} = \partial_{\mathbf{n}}\phi_{|_{z=0}}$ in $H^{1/2}(\mathbb{R}^d)$. (Recall that $\partial_{\mathbf{n}}$ always denotes the conormal derivative, so that it does not have the same expression in both sides of this equality.) Therefore, for all $\varphi \in \mathcal{D}(\mathbb{R}^d)$,

$$(\sqrt{1+\varepsilon^2|\nabla\zeta|^2}\partial_{\mathbf{n}}\Phi_{|_{z=\varepsilon\zeta}}, \varphi) = (\partial_{\mathbf{n}}\phi, \varphi),$$

where on the left-hand-side, functions on $\Gamma^+ = \{(X,z), z = \varepsilon\zeta(X)\}$ are identified with functions on $\mathbb{R}^d$. Under the assumptions made in the statement of the proposition, we have $P(\Sigma)\nabla^{\mu,\gamma}\psi^\mathfrak{h} \in H^1(\mathcal{S})$ and we can therefore apply the divergence theorem to get Green's formula,

$$(\sqrt{1+\varepsilon^2|\nabla\zeta|^2}\partial_{\mathbf{n}}\Phi_{|_{z=\varepsilon\zeta}}, \varphi) = \int_{\mathcal{S}} P(\Sigma)\nabla^{\mu,\gamma}\psi^\mathfrak{h} \cdot \nabla^{\mu,\gamma}\varphi^\mathfrak{h}.$$

The above identity shows that $\mathcal{G}_{\mu,\gamma}[\varepsilon\zeta,\beta b]\psi$ and $\varphi \mapsto (\sqrt{1+\varepsilon^2|\nabla\zeta|^2}\partial_{\mathbf{n}}\Phi_{|_{z=\varepsilon\zeta}}, \varphi)$ coincide as linear forms on $H^{1/2}(\mathbb{R}^d)$. The result follows therefore from Proposition 3.3. □

REMARK 3.7. As shown in the proof, a similar definition can be made in terms of the solution ϕ to (3.5),

$$\mathcal{G}_{\mu,\gamma}[\varepsilon\zeta, \beta b]\psi = \mathbf{e}_z \cdot P(\Sigma)\nabla^{\mu,\gamma}\phi_{|_{z=0}}.$$

Note that we could have defined the Dirichlet-Neumann operator through this formula, which makes sense in $H^{s-1/2}(\mathbb{R}^d)$ as soon as $s \geq 0$ and $s \geq -t_0 + 1$ (see §2.4), instead of using Definition 3.2. Actually, the restriction $s \geq -t_0 + 1$ (which is only a restriction in dimension $d = 1$) can be removed, as proved in [**6**].

NOTATION 3.8. For the sake of clarity, we simply write $\mathcal{G}$ instead of $\mathcal{G}_{\mu,\gamma}[\varepsilon\zeta, \beta b]$ when no confusion is possible.

3.1.2. Basic properties. We begin this section with some basic symmetry properties of the Dirichlet-Neumann operator.

PROPOSITION 3.9. *Let $t_0 > d/2$ and $\zeta, b \in H^{t_0+1}(\mathbb{R}^d)$ be such that (3.2) is satisfied. Then*

(1) *The Dirichlet-Neumann operator is symmetric for the* $\dot{H}^{1/2}(\mathbb{R}^d)' - \dot{H}^{1/2}(\mathbb{R}^d)$*-duality product,*

$$\forall \psi_1, \psi_2 \in \dot{H}^{1/2}(\mathbb{R}^d), \qquad \langle \mathcal{G}\psi_1, \psi_2 \rangle_{(\dot{H}^{1/2})' - \dot{H}^{1/2}} = \langle \mathcal{G}\psi_2, \psi_1 \rangle_{(\dot{H}^{1/2})' - \dot{H}^{1/2}}.$$

(2) *The Dirichlet-Neumann operator is positive on* $\dot{H}^{1/2}(\mathbb{R}^d)$*,*

$$\forall \psi \in \dot{H}^{1/2}(\mathbb{R}^d), \qquad \langle \mathcal{G}\psi, \psi \rangle_{(\dot{H}^{1/2})' - \dot{H}^{1/2}} \geq 0.$$

(3) *One has, for all* $\psi_1, \psi_2 \in \dot{H}^{1/2}(\mathbb{R}^2)$*,*

$$|\langle \mathcal{G}\psi_1, \psi_2 \rangle_{(\dot{H}^{1/2})' - \dot{H}^{1/2}}| \leq \langle \mathcal{G}\psi_1, \psi_1 \rangle_{(\dot{H}^{1/2})' - \dot{H}^{1/2}} \langle \mathcal{G}\psi_2, \psi_2 \rangle_{(\dot{H}^{1/2})' - \dot{H}^{1/2}}.$$

(4) *If* $\psi_1, \psi_2 \in H^{1/2}(\mathbb{R}^d)$*, the* $(\dot{H}^{1/2})' - \dot{H}^{1/2}$*-duality product can be replaced by the* $H^{-1/2} - H^{1/2}$ *duality product in the identities above.*

(5) *Moreover, if* $\mathcal{G}\psi_1, \mathcal{G}\psi_2 \in L^2(\mathbb{R}^d)$*, then the* $H^{-1/2} - H^{1/2}$*-duality product can be replaced by the* L^2*-scalar product.*

REMARK 3.10. We can deduce from Proposition 3.6 that $\mathcal{G}\psi_1 \in H^{1/2}(\mathbb{R}^d)$ for all $\psi_1 \in H^{3/2}(\mathbb{R}^d)$. A consequence of the above proposition is therefore that $\mathcal{G}$ is symmetric on $H^{3/2}(\mathbb{R}^d)$ for the L^2-scalar product. A stronger statement is that it admits a self-adjoint realization on $L^2(\mathbb{R}^d)$ with domain $H^1(\mathbb{R}^d)$, as shown in Proposition A.14 in Appendix A.

NOTATION 3.11. To alleviate the notations we always write $(\mathcal{G}\psi_1, \psi_2)$ instead of $\langle \mathcal{G}\psi_1, \psi_2 \rangle$, even though this notation is abusive outside the framework of the fourth point of Proposition 3.9.

PROOF. By definition of $\mathcal{G}$, one has, for any admissible diffeomorphism Σ,

$$\langle \mathcal{G}\psi_1, \psi_2 \rangle = \int_{\mathcal{S}} P(\Sigma)\nabla^{\mu,\gamma}\psi_1^{\mathfrak{h}} \cdot \nabla^{\mu,\gamma}\psi_2^{\mathfrak{h}} \tag{3.9}$$

$$= \int_{\mathcal{S}} P(\Sigma)^{1/2}\nabla^{\mu,\gamma}\psi_1^{\mathfrak{h}} \cdot P(\Sigma)^{1/2}\nabla^{\mu,\gamma}\psi_2^{\mathfrak{h}}, \tag{3.10}$$

where $P(\Sigma)^{1/2}$ stands for the square root of the positive definite matrix $P(\Sigma)$. (Note that the symmetry in ψ_1 and ψ_2 of the above expression proves the first point of the proposition.) Taking $\psi = \psi_1 = \psi_2$ in (3.10) also shows that $\mathcal{G}$ is positive. Moreover, it follows from the Cauchy-Schwarz inequality that

$$|\langle \mathcal{G}\psi_1, \psi_2 \rangle| \leq \|P(\Sigma)^{1/2}\nabla^{\mu,\gamma}\psi_1^{\mathfrak{h}}\|_2 \, \|P(\Sigma)^{1/2}\nabla^{\mu,\gamma}\psi_2^{\mathfrak{h}}\|_2,$$

which yields the third point of the proposition, since one has

$$\langle \mathcal{G}\psi_1, \psi_1 \rangle = \|P(\Sigma)^{1/2}\nabla^{\mu,\gamma}\psi_1^{\mathfrak{h}}\|_2^2 \tag{3.11}$$

(just take $\psi_1 = \psi_2$ in (3.10)).
Using Proposition 3.3, the last two points do not raise any particular difficulty. □

Together with Proposition 3.9, the following proposition shows that the quantity $\frac{1}{\mu}(\mathcal{G}\psi, \psi)^{1/2}$ defines a (semi)-norm equivalent to $|\mathfrak{P}\psi|_2$, where $\mathfrak{P}$ is the Fourier multiplier defined in (2.14), that is, $\mathfrak{P} = \frac{|D^\gamma|}{(1 + \sqrt{\mu}|D^\gamma|)^{1/2}}$.

PROPOSITION 3.12. *Let $t_0 > d/2$ and $\zeta, b \in H^{t_0+1}(\mathbb{R}^d)$ be such that (3.2) is satisfied. For all $\psi \in \dot{H}^{1/2}(\mathbb{R}^2)$, one has, using Notation 3.11,*

$$(\psi, \frac{1}{\mu}\mathcal{G}\psi) \leq M_0|\mathfrak{P}\psi|_2^2 \quad \text{and} \quad |\mathfrak{P}\psi|_2^2 \leq M_0(\psi, \frac{1}{\mu}\mathcal{G}\psi),$$

where M_0 is given by (3.3).

REMARK 3.13. We show in Proposition 3.18 below that the first estimate of the proposition can be generalized into

$$\forall 0 \leq s \leq t_0 + 1/2, \qquad (\Lambda^s\psi_1, \frac{1}{\mu}\Lambda^s\mathcal{G}\psi_2) \leq M_0|\mathfrak{P}\psi_1|_{H^s}|\mathfrak{P}\psi_2|_{H^s}.$$

REMARK 3.14. Since $\|\nabla^{\mu,\gamma}\psi^{\mathfrak{h}}\|_2 \leq M_0(\mathcal{G}\psi, \psi)$ and $(\mathcal{G}\psi, \psi) \leq M_0\|\nabla^{\mu,\gamma}\psi^{\mathfrak{h}}\|_2$ (this is a consequence of (3.11)), we deduce from the proposition that

$$\|\nabla^{\mu,\gamma}\psi^{\mathfrak{h}}\|_2 \leq \sqrt{\mu}M_0|\mathfrak{P}\psi|_2 \quad \text{and} \quad |\mathfrak{P}\psi|_2 \leq \sqrt{\mu}M_0\|\nabla^{\mu,\gamma}\psi^{\mathfrak{h}}\|_2.$$

Generalizations providing upper and lower bounds for $\|\Lambda^s\nabla^{\mu,\gamma}\psi^{\mathfrak{h}}\|_2$ are provided respectively by Corollary 2.40 and Proposition 3.19 below.

PROOF. Let Σ be an admissible diffeomorphism. The first estimate of the proposition follows directly from (3.11) and Corollary 2.40 (with $s = 0$).

The second estimate is more delicate. Let φ be a smooth function, with compact support in $(-1, 0]$ and such that $\varphi(0) = 1$; also define $v(X, z) = \varphi(z)\mathfrak{P}\psi^{\mathfrak{h}}$ (with $\psi^{\mathfrak{h}}$ defined as in Notation 2.33). Since $v_{|_{z=-1}} = 0$, one can get, after taking the Fourier transform with respect to the horizontal variables,

$$\frac{|\xi^\gamma|^2}{1+\sqrt{\mu}|\xi^\gamma|}|\widehat{\psi}(\xi)|^2 \leq 2\int_{-1}^0 |\partial_z\widehat{v}(\xi, z)||\widehat{v}(\xi, z)|dz.$$

Noting that

$$|\widehat{v}| \leq |\varphi|_\infty|\widehat{\mathfrak{P}\psi^{\mathfrak{h}}}| \quad \text{and} \quad |\partial_z\widehat{v}| \leq |\partial_z\varphi|_\infty|\widehat{\mathfrak{P}\psi^{\mathfrak{h}}}| + |\varphi|_\infty|\partial_z\widehat{\mathfrak{P}\psi^{\mathfrak{h}}}|,$$

one gets

$$\begin{aligned}
\frac{|\xi^\gamma|^2}{1+\sqrt{\mu}|\xi^\gamma|}|\widehat{\psi}(\xi)|^2 &\leq 2|\varphi|_\infty|\partial_z\varphi|_\infty\int_{-1}^0 \frac{|\xi^\gamma|^2}{1+\sqrt{\mu}|\xi^\gamma|}|\widehat{\psi^{\mathfrak{h}}}(\xi, z)|^2dz \\
&\quad +2|\varphi|_\infty^2\int_{-1}^0 \frac{|\xi^\gamma|^2}{1+\sqrt{\mu}|\xi^\gamma|}|\widehat{\psi^{\mathfrak{h}}}(\xi, z)|\,|\partial_z\widehat{\psi^{\mathfrak{h}}}(\xi, z)|dz, \\
&\leq 2|\varphi|_\infty|\partial_z\varphi|_\infty\int_{-1}^0 |\xi^\gamma|^2|\widehat{\psi^{\mathfrak{h}}}(\xi, z)|^2dz \\
&\quad +|\varphi|_\infty^2\int_{-1}^0 \frac{\mu|\xi^\gamma|^4}{(1+\sqrt{\mu}|\xi^\gamma|)^2}|\widehat{\psi^{\mathfrak{h}}}(\xi, z)|^2dz + |\varphi|_\infty^2\int_{-1}^0 \frac{1}{\mu}|\partial_z\widehat{\psi^{\mathfrak{h}}}(\xi, z)|^2dz,
\end{aligned}$$

where Young's inequality has been used to obtain the last line. Noting that $\frac{\mu|\xi^\gamma|^4}{(1+\sqrt{\mu}|\xi^\gamma|)^2} \leq |\xi^\gamma|^2$, one has

$$\frac{|\xi^\gamma|^2}{1+\sqrt{\mu}|\xi^\gamma|}|\widehat{\psi}(\xi)|^2 \lesssim \int_{-1}^0 |\xi^\gamma|^2|\widehat{\psi^{\mathfrak{h}}}(\xi, z)|^2dz + \int_{-1}^0 \frac{1}{\mu}|\partial_z\widehat{\psi^{\mathfrak{h}}}(\xi, z)|^2dz,$$

so that, integrating with respect to ξ, one gets

$$|\mathfrak{P}\psi|_2^2 \lesssim \frac{1}{\mu}\|\nabla^{\mu,\gamma}\psi^{\mathfrak{h}}\|_2^2.$$

Owing to Lemma 2.26, Remark 2.27, and (3.9) (with $v=\psi$), one has

$$\|\nabla^{\mu,\gamma}\psi^{\mathfrak{h}}\|_2^2 \le M_0(\psi,\mathcal{G}\psi),$$

and the proposition follows. $\square$

3.2. Higher order estimates

For flat surface and bottom, it is possible to get $\mathcal{G}$ by simple computations similar to those of §1.3.2; one finds that $\mathcal{G}_{\mu,\gamma}[0,0]=\sqrt{\mu}|D^\gamma|\tanh(\sqrt{\mu}|D^\gamma|)$. It follows in particular that there exists a numerical constant C such that, for $s\ge 0$,

$$\frac{1}{C}|\mathfrak{P}^2\psi|_{H^{s-1/2}} \le \frac{1}{\mu}|\mathcal{G}_{\mu,\gamma}[0,0]\psi|_{H^{s-1/2}} \le C|\mathfrak{P}^2\psi|_{H^{s-1/2}}.$$

Noting that $|\mathfrak{P}u|_{H^{s-1/2}}\lesssim \mu^{-1/4}|u|_{H^s}$ and $|\mathfrak{P}u|_{H^{s-1/2}}\le |u|_{H^{s+1/2}}$ for all $u\in H^{s+1/2}(\mathbb{R}^d)$, we deduce that

$$\begin{aligned} |\mathcal{G}_{\mu,\gamma}[0,0]\psi|_{H^{s-1/2}} &\lesssim \mu^{3/4}|\mathfrak{P}\psi|_{H^s}\lesssim\sqrt{\mu}|\nabla\psi|_{H^{s-1/2}},\\ |\mathcal{G}_{\mu,\gamma}[0,0]\psi|_{H^{s-1/2}} &\lesssim \mu|\mathfrak{P}\psi|_{H^{s+1/2}}\lesssim\mu|\nabla\psi|_{H^{s+1/2}}.\end{aligned}$$

This shows that as μ gets very small (shallow water limit), the operator $\frac{1}{\mu}\mathcal{G}_{\mu,\gamma}[0,0]$ becomes unbounded when seen as a first-order operator. When the boundedness of the family of operators $(\frac{1}{\mu}\mathcal{G}_{\mu,\gamma}[0,0])_{0\le\mu\le\mu_{max}}$ is needed, one has to "lose" one derivative and see $\frac{1}{\mu}\mathcal{G}_{\mu,\gamma}[0,0]$ as a second-order operator.

The following theorem shows that a similar behavior is observed in the general case (nonflat bottom and surface).

THEOREM 3.15. *Let $t_0>d/2$ and $s\ge 0$. Let $\zeta,b\in H^{s+1/2}\cap H^{t_0+1}(\mathbb{R}^d)$ be such that (3.2) is satisfied.*

(1) *The operator $\mathcal{G}$ maps continuously $\dot{H}^{s+1/2}(\mathbb{R}^d)$ into $H^{s-1/2}(\mathbb{R}^d)$ and one has*

$$\begin{aligned} &\forall 0\le s\le t_0+3/2, && |\mathcal{G}\psi|_{H^{s-1/2}}\le\mu^{3/4}M(s+1/2)|\mathfrak{P}\psi|_{H^s}\\ &\forall t_0+3/2\le s, && |\mathcal{G}\psi|_{H^{s-1/2}}\le\mu^{3/4}M\big(|\mathfrak{P}\psi|_{H^s}\\ &&&\qquad +|(\varepsilon\zeta,\beta b)|_{H^{s+1/2}}|\mathfrak{P}\psi|_{H^{t_0+3/2}}\big)\end{aligned}$$

with M and $M(s)$ as in (3.3).

(2) *The operator $\mathcal{G}$ maps continuously $\dot{H}^{s+1}(\mathbb{R}^d)$ into $H^{s-1/2}(\mathbb{R}^d)$ and one has*

$$\begin{aligned} &\forall 0\le s\le t_0+1, && |\mathcal{G}\psi|_{H^{s-1/2}}\le\mu M(s+1)|\mathfrak{P}\psi|_{H^{s+1/2}}\\ &\forall t_0+1\le s, && |\mathcal{G}\psi|_{H^{s-1/2}}\le\mu M\big(|\mathfrak{P}\psi|_{H^{s+1/2}}\\ &&&\qquad +|(\varepsilon\zeta,\beta b)|_{H^{s+1}}|\mathfrak{P}\psi|_{H^{t_0+3/2}}\big).\end{aligned}$$

REMARK 3.16. Noting that $|\mathfrak{P}u|_{H^s}\le\max\{1,\mu^{-1/4}\}|\nabla u|_{H^{s-1/2}}$ (recall that $\gamma\le 1$), the first point of the theorem implies that

$$|\mathcal{G}\psi|_{H^{s-1/2}}\le\sqrt{\mu}M\big(|\nabla\psi|_{H^{s-1/2}}+\langle|(\varepsilon\zeta,\beta b)|_{H^{s+1/2}}|\nabla\psi|_{H^{t_0+1}}\rangle_{s>t_0+3/2}\big).$$

A similar variant also holds for the second point of the theorem.

REMARK 3.17. It follows from Theorem 3.15 that $\mathcal{G}$ can also be defined on Sobolev spaces of negative order. More precisely, the symmetry of $\mathcal{G}$ allows us to define $\mathcal{G}\psi\in H^{-s-1/2}(\mathbb{R}^d)$ for all $\psi\in H^{-s+1/2}(\mathbb{R}^d)$ ($s\ge 0$), through the duality formula

$$\forall\varphi\in H^{s+1/2}(\mathbb{R}^d),\qquad \langle\mathcal{G}\psi,\varphi\rangle:=\langle\psi,\mathcal{G}\varphi,\rangle_{H^{-s+1/2}-H^{s-1/2}}.$$

We also get from Remark 3.16 that

$$\forall 0 \leq s \leq t_0 + 3/2, \qquad |\mathcal{G}\psi|_{H^{-s-1/2}} \leq \sqrt{\mu} M |\psi|_{H^{-s+1/2}}.$$

(Note that the above formula furnishes an estimate on $|\mathcal{G}\psi|_{H^{-s-1/2}}$ far all $s \geq 0$ by choosing t_0 large enough, but that tame estimates like those of Theorem 3.15 or Remark 3.16 cannot be directly obtained by duality methods.)

PROOF. *Throughout this proof, Σ is assumed to be the regularizing diffeomorphism of Example* 2.15.

To evaluate $|\mathcal{G}\psi|_{H^{s-1/2}}$, we consider the quantity $\big(\Lambda^{s-1/2}\mathcal{G}\psi, \varphi\big)$ for all $\varphi \in C_0^\infty(\mathbb{R}^d)$. Using the variational definition (3.7) of $\mathcal{G}$, we have

$$\begin{aligned} \big(\Lambda^{s-1/2}\mathcal{G}\psi, \varphi\big) &= \big(\mathcal{G}\psi, \Lambda^{s-1/2}\varphi\big) \\ &= \int_{\mathcal{S}} \nabla^{\mu,\gamma}\psi^{\mathfrak{h}} \cdot P(\Sigma)\nabla^{\mu,\gamma}(\Lambda^{s-1/2}\varphi)^{\mathfrak{h}}. \end{aligned}$$

Noting that Green's identity implies that $\int_{\mathcal{S}} \nabla^{\mu,\gamma}\psi^{\mathfrak{h}} \cdot P(\Sigma)\nabla^{\mu,\gamma} u = 0$ for all $u \in H^1_{0,surf}(\mathcal{S})$, and since $(\Lambda^{s-1/2}\varphi)^{\mathfrak{h}} - (\Lambda^{s-1/2}\varphi)^{\dagger}$ belongs to $H^1_{0,surf}(\mathcal{S})$ — recall that the extension $(\cdot)^{\dagger}$ is defined in Notation 2.28—we deduce that

$$\big(\Lambda^{s-1/2}\mathcal{G}\psi, \varphi\big) = \int_{\mathcal{S}} \nabla^{\mu,\gamma}\psi^{\mathfrak{h}} \cdot P(\Sigma)\nabla^{\mu,\gamma}(\Lambda^{s-1/2}\varphi)^{\dagger}.$$

Using the symmetry of $P(\Sigma)$ and noting that $\nabla^{\mu,\gamma}(\Lambda^{s-1/2}\varphi)^{\dagger} = \Lambda^{s-1/2}\nabla^{\mu,\gamma}\varphi^{\dagger}$, we deduce further that

$$\big(\Lambda^{s-1/2}\mathcal{G}\psi, \varphi\big) = \int_{\mathcal{S}} \Lambda^{s}\big(P(\Sigma)\nabla^{\mu,\gamma}\psi^{\mathfrak{h}}\big) \cdot \Lambda^{-1/2}\nabla^{\mu,\gamma}\varphi^{\dagger}. \tag{3.12}$$

From the Cauchy-Schwarz inequality and Lemma 2.20, we therefore get

$$\begin{aligned} |(\Lambda^{s-1/2}\mathcal{G}\psi, \varphi)| &\lesssim \sqrt{\mu}|\mathfrak{P}\varphi|_{H^{-1/2}}\|\Lambda^{s}(P(\Sigma)\nabla^{\mu,\gamma}\psi^{\mathfrak{h}})\|_2 \\ &\lesssim \mu^{1/4}|\varphi|_2\|\Lambda^{s}(P(\Sigma)\nabla^{\mu,\gamma}\psi^{\mathfrak{h}})\|_2, \end{aligned}$$

where the last line follows easily from the definition of $\mathfrak{P}$. By duality, it follows that

$$|\mathcal{G}\psi|_{H^{s-1/2}} \lesssim \mu^{1/4}\|\Lambda^{s}(P(\Sigma)\nabla^{\mu,\gamma}\psi^{\mathfrak{h}})\|_2. \tag{3.13}$$

The first point of the theorem therefore follows from (3.13) if we can prove the following estimates

$$\begin{aligned} &\forall 0 \leq s \leq t_0 + 3/2, && \|\Lambda^{s}(P(\Sigma)\nabla^{\mu,\gamma}\psi^{\mathfrak{h}})\|_2 \leq \sqrt{\mu}M(s+1/2)|\mathfrak{P}\psi|_{H^s} \\ &\forall t_0 + 3/2 \leq s, && \|\Lambda^{s}(P(\Sigma)\nabla^{\mu,\gamma}\psi^{\mathfrak{h}})\|_2 \leq \sqrt{\mu}M\big(|\mathfrak{P}\psi|_{H^s} \\ & && \qquad +|(\varepsilon\zeta, \beta b)|_{H^{s+1/2}}|\mathfrak{P}\psi|_{H^{t_0+1}}\big). \end{aligned} \tag{3.14}$$

The following lines are devoted to the proof of these controls. The product estimate of Corollary B.5(2) in Appendix B shows that

$$\begin{aligned} &\forall 0 \leq s \leq t_0, && \|\Lambda^{s}(P(\Sigma)\nabla^{\mu,\gamma}\psi^{\mathfrak{h}})\|_2 \lesssim (1 + \|Q\|_{L^\infty H^{t_0}})\|\Lambda^{s}\nabla^{\mu,\gamma}\psi^{\mathfrak{h}}\|_2 \\ &\forall t_0 + 1/2 \leq s, && \|\Lambda^{s}(P(\Sigma)\nabla^{\mu,\gamma}\psi^{\mathfrak{h}})\|_2 \lesssim (1 + \|Q\|_{L^\infty H^{t_0}})\|\Lambda^{s}\nabla^{\mu,\gamma}\psi^{\mathfrak{h}}\|_2 \\ & && \qquad +(1 + \|\Lambda^{s}Q\|_2)\|\nabla^{\mu,\gamma}\psi^{\mathfrak{h}}\|_{L^\infty H^{t_0}}. \end{aligned}$$

Since $\|Q\|_{L^\infty H^{t_0}} \leq M_0$ and $\|\Lambda^s Q\|_2 \leq M_0|(\varepsilon\zeta, \beta b)|_{H^{s+1/2}}$ (see (2.26)–(2.27)), we can use the continuous embedding $H^{t_0+1/2,1} \subset L^\infty H^{t_0}$ (Proposition 2.12) and Corollary 2.40 to get (3.14). (Note that for $t_0 \leq s \leq t_0 + 1/2$, the result is deduced

from the other cases by interpolation.)
The proof of the second point is quite similar. Rewriting (3.12) as

$$(\Lambda^{s-1/2}\mathcal{G}\psi,\varphi)=\int_S \Lambda^{s+1/2}\big(P(\Sigma)\cdot\nabla^{\mu,\gamma}\psi^{\mathfrak{h}}\big)\Lambda^{-1}\nabla^{\mu,\gamma}\varphi^{\dagger},$$

one deduces that

$$\begin{aligned}|(\Lambda^{s-1/2}\mathcal{G}\psi,\varphi)| &\lesssim \sqrt{\mu}|\mathfrak{P}\varphi|_{H^{-1}}\|\Lambda^{s+1/2}(P(\Sigma)\nabla^{\mu,\gamma}\psi^{\mathfrak{h}})\|_2\\ &\lesssim \sqrt{\mu}|\varphi|_2\|\Lambda^{s+1/2}(P(\Sigma)\psi^{\mathfrak{h}})\|_2,\end{aligned}$$

where we used the identity $|\mathfrak{P}\varphi|_{H^{-1}}\leq|\varphi|_2$. The end of the proof follows the same lines as the proof of point (1). □

In the following proposition, we propose a higher order generalization of the first estimate of Proposition 3.12.

PROPOSITION 3.18. *Let $t_0>d/2$ and $0\leq s\leq t_0+1/2$. Let also $\zeta,b\in H^{t_0+1}(\mathbb{R}^d)$ be such that (3.2) is satisfied. Then for all $\psi_1,\psi_2\in\dot{H}^{s+1/2}(\mathbb{R}^d)$, we have*

$$(\Lambda^s\mathcal{G}\psi_1,\Lambda^s\psi_2)\leq\mu M_0|\mathfrak{P}\psi_1|_{H^s}|\mathfrak{P}\psi_2|_{H^s}.$$

PROOF. Integrating by parts in (3.5), one gets

$$(\Lambda^s\mathcal{G}\psi_1,\Lambda^s\psi_2)=\int_S\Lambda^s(P\nabla^{\mu,\gamma}\psi_1^h)\cdot\Lambda^s\nabla^{\mu,\gamma}\psi_2^{\dagger}. \tag{3.15}$$

It follows that

$$\big|(\Lambda^s\mathcal{G}\psi_1,\Lambda^s\psi_2)\big|\leq\|\Lambda^sP\nabla^{\mu,\gamma}\psi_1^h\|_2\|\Lambda^s\nabla^{\mu,\gamma}\psi_2^{\dagger}\|$$

and the result therefore follows from Corollary 2.40 and (3.14). □

We also know from Corollary 2.40 that

$$\forall 0\leq s\leq t_0+1/2,\qquad \|\Lambda^s\nabla^{\mu,\gamma}\psi^{\mathfrak{h}}\|_2\leq M_0|\mathfrak{P}\psi|_{H^s};$$

the following result shows that the reverse inequality holds.

PROPOSITION 3.19. *Let $t_0>d/2$, $0\leq s\leq t_0+1/2$, and Σ be a regularizing diffeomorphism. For all $\psi\in\dot{H}^{s+1/2}(\mathbb{R}^d)$, we denote by $\psi^{\mathfrak{h}}$ the solution to (2.12). Then*

$$\sqrt{\mu}|\mathfrak{P}\psi|_{H^s}\leq M_0\|\Lambda^s\nabla^{\mu,\gamma}\psi^{\mathfrak{h}}\|_2.$$

The case $s=0$ has already been proved in Remark 3.13; for $s>0$, we note that $|\mathfrak{P}\psi|_{H^s}=|\mathfrak{P}(\Lambda^s\psi)|_2$, so we can deduce from the case $s=0$ that

$$\begin{aligned}\sqrt{\mu}|\mathfrak{P}\psi|_{H^s} &\leq M_0\|\nabla^{\mu,\gamma}(\Lambda^s\psi)^{\mathfrak{h}}\|\\ &\leq M_0\big(\|\nabla^{\mu,\gamma}((\Lambda^s\psi)^{\mathfrak{h}}-\Lambda^s\psi^{\mathfrak{h}})\|_2+\|\Lambda^s\nabla^{\mu,\gamma}\psi^{\mathfrak{h}}\|_2\big).\end{aligned}$$

Noting that $u=(\Lambda^s\psi)^{\mathfrak{h}}-\Lambda^s\psi^{\mathfrak{h}}$ is as in Lemma 2.38 with $\mathbf{g}=[\Lambda^s,P]\nabla^{\mu,\gamma}\psi^{\mathfrak{h}}$, we have

$$\begin{aligned}\|\nabla^{\mu,\gamma}((\Lambda^s\psi)^{\mathfrak{h}}-\Lambda^s\phi)\|_2 &\leq \|\mathbf{g}\|_2\\ &\leq M_0\|\Lambda^s\nabla^{\mu,\gamma}\psi^{\mathfrak{h}}\|_2,\end{aligned} \tag{3.16}$$

where we used (3.14) to derive the second inequality.

3.3. Shape derivatives

Let us denote by $\mathbf{\Gamma}$ the set of all $(\zeta, b) \in H^{t_0+1}(\mathbb{R}^d)$ such that (3.2) is satisfied. Let $0 \leq s \leq t_0 + 1/2$ ($t_0 > d/2$). Given $\psi \in \dot{H}^{s+1/2}(\mathbb{R}^d)$, we are interested here in the mapping

$$\mathcal{G}_{\mu,\gamma}[\varepsilon\cdot, \beta\cdot] : \begin{array}{ccc} \mathbf{\Gamma} \subset H^{t_0+1}(\mathbb{R}^d)^2 & \to & H^{s-1/2}(\mathbb{R}^d), \\ \Gamma = (\zeta, b) & \mapsto & \mathcal{G}_{\mu,\gamma}[\varepsilon\zeta, \beta b]\psi. \end{array} \tag{3.17}$$

NOTATION 3.20. Let $j \in \mathbb{N}$ and $\mathbf{h} = (h_1, \dots, h_j), \mathbf{k} = (k_1, \dots, k_j) \in H^{t_0+1}(\mathbb{R}^d)^j$. We denote by $d^j\mathcal{G}_{\mu,\gamma}[\varepsilon\zeta, \beta b](\mathbf{h}, \mathbf{k})\psi$, or simply $d^j\mathcal{G}(\mathbf{h}, \mathbf{k})\psi$ the j-th order derivative of (3.17) at (ζ, b) in the direction $(\mathbf{h}, \mathbf{k})$.
If the bottom is kept fixed, we denote by $d^j\mathcal{G}_{\mu,\gamma}[\varepsilon\zeta, \beta b](\mathbf{h})\psi$, or simply $d^j\mathcal{G}(\mathbf{h})\psi$ the j-th order partial derivative of (3.17) at ζ in the direction $\mathbf{h}$.

The following theorem shows that (3.17) is analytic[1] and gives a very important explicit formula for the first-order partial derivative with respect to ζ.

THEOREM 3.21. *Let $t_0 > d/2$. Then,*
(1) *For all $0 \leq s \leq t_0 + 1/2$ and $\psi \in \dot{H}^{s+1/2}(\mathbb{R}^d)$, the mapping (3.17) is analytic in the sense of Definition A.1 in Appendix A.*
(2) *Let $\Gamma = (\zeta, b) \in \mathbf{\Gamma}$ and $\psi \in \dot{H}^{3/2}(\mathbb{R}^d)$. Then for all $h \in H^{t_0+1}(\mathbb{R}^d)$, one has*

$$d\mathcal{G}(h)\psi = -\varepsilon\mathcal{G}(h\underline{w}) - \varepsilon\mu\nabla^\gamma \cdot (h\underline{V}),$$

with

$$\underline{w} = \frac{\mathcal{G}\psi + \varepsilon\mu\nabla^\gamma\zeta \cdot \nabla^\gamma\psi}{1 + \varepsilon^2\mu|\nabla^\gamma\zeta|^2} \quad \textit{and} \quad \underline{V} = \nabla^\gamma\psi - \varepsilon\underline{w}\nabla^\gamma\zeta.$$

REMARK 3.22.
(1) As suggested by the notations, $\underline{V}$ and $\underline{w}$ coincide with the horizontal and vertical components of the "velocity" $\mathbf{U} = \nabla_{X,z}\Phi$ (with Φ solving (3.8)),

$$\underline{V} = (\nabla^\gamma\Phi)_{|z=\varepsilon\zeta} \quad \text{and} \quad \underline{w} = (\partial_z\Phi)_{|z=\varepsilon\zeta}.$$

(2) From the explicit expression of $\underline{w}$ and Theorem 3.15, one gets easily that for all $s \geq 0$,

$$|\underline{w}|_{H^{s-1/2}} \leq \mu^{3/4}M\big(|\mathfrak{P}\psi|_{H^s} + \langle|(\varepsilon\zeta, \beta b)|_{H^{s+1/2}}|\mathfrak{P}\psi|_{H^{t_0+3/2}}\rangle_{s>t_0+3/2}\big). \tag{3.18}$$

We refer to Proposition 4.4 below for similar results.

REMARK 3.23. In the second point of the theorem, one has $d\mathcal{G}(h)\psi \in H^{1/2}(\mathbb{R}^d)$. This seems to contradict the fact that both terms in the r.h.s. of the formula lie in $H^{-1/2}(\mathbb{R}^d)$. Of course, this can be explained by a cancellation of the most singular components of both terms.

REMARK 3.24.
(1) One can also find an explicit expression for the shape derivative of the Dirichlet-Neumann operator with respect to the bottom parametrization b and to take into account a nonhomogeneous Neumann boundary condition at the bottom (see **[182]** and Appendix A, §A.5).

---

[1]Since this result is not needed for our purposes, the proof is postponed to Appendix A.

(2) The formula of the theorem can be iterated to find explicit expressions for the higher order shape derivatives $d^j\mathcal{G}(\mathbf{h})\psi$, for $j \leq 2$. The formulas are quite complicated but are simplified when $(\zeta, b) = 0$; one finds, for instance, that

$$\begin{aligned} d^2\mathcal{G}_{\mu,\gamma}[0,0](h_1,h_2) &= \varepsilon^2\mathcal{G}_0(h_2\mathcal{G}_0(h_1\mathcal{G}_0\psi)) + \varepsilon^2\mathcal{G}_0(h_1\mathcal{G}_0(h_2\mathcal{G}_0\psi)) \\ &\quad +\varepsilon^2\mu\Delta^\gamma(h_1h_2\mathcal{G}_0\psi) + \varepsilon^2\mu\mathcal{G}_0(h_1h_2\Delta^\gamma\psi), \end{aligned} \tag{3.19}$$

where $\mathcal{G}_0 = \mathcal{G}_{\mu,\gamma}[0,0]$.

PROOF. (1) The proof of this point is postponed to Theorem A.11 in Appendix A.
(2) *For the sake of simplicity, we use here the diffeomorphism* Σ *of Example* 2.14 *but one can easily adapt the proof if another choice is made.*

We denote[2] by $\mathfrak{A}_\psi(\Gamma) = \phi_\Gamma$ the solution to (3.5) for all $\Gamma \in \mathbf{\Gamma}$; we also denote v the derivative of $\mathfrak{A}_\psi$ at (ζ, b) in the direction $(h, 0)$. Differentiating (3.5) with respect to ζ, we get that

$$\left\{ \begin{array}{l} \nabla^{\mu,\gamma} \cdot P(\Sigma)\nabla^{\mu,\gamma}v = -\nabla^{\mu,\gamma} \cdot P'(h)\nabla^{\mu,\gamma}\phi, \\ v_{|_{z=0}} = 0, \qquad \partial_{\mathbf{n}}v_{|_{z=-1}} = -\mathbf{e}_z \cdot P'(h)\nabla^{\mu,\gamma}\phi_{|_{z=-1}}, \end{array} \right. \tag{3.20}$$

where $P'(h)$ stands for the derivative of the mapping $\Gamma \mapsto P(\Sigma_\Gamma)$ at (ζ, b) in the direction $(h, 0)$.
It might not seem obvious to exploit the fact that v — which is the derivative of the solution to (3.5) with respect to ζ in the direction h — solves the boundary value problem (3.20). However, it is typical in free boundary value problems that the good quantity to work with is not v but w defined as

$$w = v - \frac{\partial_z v}{\partial_z \Sigma_3} d\Sigma_3(h), \tag{3.21}$$

where $\Sigma_3(X, z) = z + \sigma(X, z)$ is the third component of the diffeomorphism used to straighten the domain. This quantity w is known as *Alinhac's good unknown*, and its role in the linearization of free boundary problems was understood in [**10**] in the context of rarefaction waves for quasilinear hyperbolic systems. The interpretation of the results presented here in terms of Alinhac's good unknown is due to Alazard and Métivier [**9**] (see also [**252**]). If Σ is, as here, the trivial diffeomorphism of Example 2.14, then w takes the form

$$w = v - \frac{\partial_z v}{1 + \varepsilon\zeta - \beta b}(z + 1)\varepsilon h.$$

The fact that this choice is relevant is a consequence of the following lemma which shows that $v_a = \frac{\partial_z v}{\partial_z \Sigma_3} d\Sigma_3(h)$ solves (3.20) except for the Dirichlet condition at the surface. We will use it to reduce (3.20) to a boundary value problem on w with homogeneous source term and Neumann condition at the bottom.

LEMMA 3.25. *Let* $v_a = \frac{\varepsilon(z+1)}{1+\varepsilon\zeta-\beta b} h\partial_z\phi$. *One has*

$$\left\{ \begin{array}{l} \nabla^{\mu,\gamma} \cdot P(\Sigma)\nabla^{\mu,\gamma}v_a = -\nabla^{\mu,\gamma} \cdot P'(h)\nabla^{\mu,\gamma}\phi, \\ v_{a\,|_{z=0}} = \frac{\varepsilon h}{1+\varepsilon\zeta-\beta b}\partial_z\phi_{|_{z=0}}, \qquad \partial_{\mathbf{n}}v_{a\,|_{z=-1}} = -\mathbf{e}_z \cdot P'(h)\nabla^{\mu,\gamma}\phi_{|_{z=-1}}, \end{array} \right.$$

[2]We refer to §A.1.1 in Appendix A for a study of the mapping $\mathfrak{A}_\psi$ and in particular of its differentiability.

PROOF OF THE LEMMA. Using the chain rule, and writing $\alpha = \frac{\varepsilon(z+1)}{1+\varepsilon\zeta-\beta b}h$, we compute

$$\begin{aligned}\nabla^{\mu,\gamma}\cdot P\nabla^{\mu,\gamma}v_a &= \alpha\nabla^{\mu,\gamma}\cdot P\partial_z\nabla^{\mu,\gamma}\phi + \nabla^{\mu,\gamma}\alpha\cdot P\nabla^{\mu,\gamma}\partial_z\phi + \nabla^{\mu,\gamma}\cdot(\partial_z\phi P\nabla^{\mu,\gamma}\alpha)\\ &= -\alpha\nabla^{\mu,\gamma}\cdot\partial_z P\nabla^{\mu,\gamma}\phi + \nabla^{\mu,\gamma}\alpha\cdot P\nabla^{\mu,\gamma}\partial_z\phi + \nabla^{\mu,\gamma}\cdot(\partial_z\phi P\nabla^{\mu,\gamma}\alpha),\end{aligned}$$

where we used the fact that $\nabla^{\mu,\gamma}\cdot P\nabla^{\mu,\gamma}\phi = 0$ to derive the second identity. In order to put the first two terms of the r.h.s. in divergence form, we remark that

$$\begin{aligned}&-\alpha\nabla^{\mu,\gamma}\cdot\partial_z P\nabla^{\mu,\gamma}\phi + \nabla^{\mu,\gamma}\alpha\cdot P\nabla^{\mu,\gamma}\partial_z\phi\\ &= -\nabla^{\mu,\gamma}\cdot(\alpha\partial_z P\nabla^{\mu,\gamma}\phi) + \nabla^{\mu,\gamma}\alpha\cdot\partial_z(P\nabla^{\mu,\gamma}\phi)\\ &= -\nabla^{\mu,\gamma}\cdot(\alpha\partial_z P\nabla^{\mu,\gamma}\phi) + \partial_z(P\nabla^{\mu,\gamma}\alpha\cdot\nabla^{\mu,\gamma}\phi) - \partial_z\nabla^{\mu,\gamma}\alpha\cdot P\nabla^{\mu,\gamma}\phi.\end{aligned}$$

Once again using the fact that $\nabla^{\mu,\gamma}\cdot P\nabla^{\mu,\gamma}\phi = 0$, we get that $\partial_z\nabla^{\mu,\gamma}\alpha\cdot P\nabla^{\mu,\gamma}\phi = \nabla^{\mu,\gamma}\cdot(\partial_z\alpha P\nabla^{\mu,\gamma}\phi)$, and therefore

$$\begin{aligned}&-\alpha\nabla^{\mu,\gamma}\cdot\partial_z P\nabla^{\mu,\gamma}\phi + \nabla^{\mu,\gamma}\alpha\cdot P\nabla^{\mu,\gamma}\partial_z\phi\\ &= -\nabla^{\mu,\gamma}\cdot(\alpha\partial_z P\nabla^{\mu,\gamma}\phi) + \partial_z(P\nabla^{\mu,\gamma}\alpha\cdot\nabla^{\mu,\gamma}\phi) - \nabla^{\mu,\gamma}\cdot(\partial_z\alpha P\nabla^{\mu,\gamma}\phi)\\ &= -\nabla^{\mu,\gamma}\cdot(\partial_z(\alpha P)\nabla^{\mu,\gamma}\phi) + \nabla^{\mu,\gamma}\cdot(\mathbf{e}_z\otimes(P\nabla^{\mu,\gamma}\alpha)\nabla^{\mu,\gamma}\phi),\end{aligned}$$

where for all pair of vectors $\mathbf{u}$, $\mathbf{v}$, $\mathbf{u}\otimes\mathbf{v}$ stands for the matrix $\mathbf{u}\mathbf{v}^T$. We have therefore proved that

$$\nabla^{\mu,\gamma}\cdot P\nabla^{\mu,\gamma}v_a = \nabla^{\mu,\gamma}\cdot\widetilde{P}\nabla^{\mu,\gamma}\phi,$$

with

$$\widetilde{P} = -\partial_z(\alpha P) + \mathbf{e}_z\otimes(P\nabla^{\mu,\gamma}\alpha) + (P\nabla^{\mu,\gamma}\alpha)\otimes\mathbf{e}_z.$$

We therefore need to prove that $\widetilde{P} = -P'(h)$, which can be done by brute computations from the explicit formula for $P(\Sigma)$ given in Example 2.21.
In order to conclude the proof, we need to check the Neumann condition. One has

$$\begin{aligned}\partial_{\mathbf{n}}v_a{}_{|z=-1} &= \mathbf{e}_z\cdot P\nabla^{\mu,\gamma}(\alpha\partial_z\phi)_{|z=-1}\\ &= \mathbf{e}_z\cdot(P\nabla^{\mu,\gamma}\alpha\otimes\mathbf{e}_z)\nabla^{\mu,\gamma}\phi_{|z=-1},\end{aligned}$$

where we used the fact that $\alpha_{|z=-1} = 0$. The result therefore follows from the observation that $(P\nabla^{\mu,\gamma}\alpha)\otimes\mathbf{e}_z{}_{|z=-1} = \widetilde{P}_{|z=-1}$. ☐

Let $w = v - v_a$. Using the lemma, w solves the boundary value problem

$$\begin{cases}\nabla^{\mu,\gamma}\cdot P(\Sigma)\nabla^{\mu,\gamma}w = 0,\\ w_{|z=0} = -\frac{\varepsilon h}{1+\varepsilon\zeta-\beta b}\partial_z\phi_{|z=0}, \qquad \partial_{\mathbf{n}}w_{|z=-1} = 0.\end{cases}$$

By definition of the Dirichlet-Neumann operator, it follows that

$$\mathbf{e}_z\cdot P(\Sigma)\nabla^{\mu,\gamma}(v-v_a)_{|z=0} = -\mathcal{G}\Big(\frac{\varepsilon h}{1+\varepsilon\zeta-\beta b}\partial_z\phi_{|z=0}\Big),$$

and therefore,

$$\mathbf{e}_z\cdot P(\Sigma)\nabla^{\mu,\gamma}v_{|z=0} = \mathbf{e}_z\cdot P(\Sigma)\nabla^{\mu,\gamma}v_a{}_{|z=0} - \mathcal{G}\Big(\frac{\varepsilon h}{1+\varepsilon\zeta-\beta b}\partial_z\phi_{|z=0}\Big). \tag{3.22}$$

Now, differentiating the identity $\mathcal{G}\psi = \mathbf{e}_z\cdot P(\Sigma)\nabla^{\mu,\gamma}\phi_{|z=0}$ with respect to ζ, we get

$$d\mathcal{G}(h)\psi = \mathbf{e}_z\cdot P'(h)\nabla^{\mu,\gamma}\phi_{|z=0} + \mathbf{e}_z\cdot P(\Sigma)\nabla^{\mu,\gamma}v_{|z=0},$$

which, together with (3.22), yields

$$d\mathcal{G}(h)\psi = \mathbf{e}_z\cdot P'(h)\nabla^{\mu,\gamma}\phi_{|z=0} + \mathbf{e}_z\cdot P(\Sigma)\nabla^{\mu,\gamma}v_a{}_{|z=0} - \mathcal{G}\Big(\frac{\varepsilon h}{1+\varepsilon\zeta-\beta b}\partial_z\phi_{|z=0}\Big).$$

Replacing $\mathcal{G}\psi = \mathbf{e}_z \cdot P(\Sigma)\nabla^{\mu,\gamma}\phi_{|_{z=0}}$ in the definition of $\underline{w}$ given in the statement of the theorem, one easily gets that $\underline{w} = \frac{1}{1+\varepsilon\zeta-\beta b}\partial_z\phi_{|_{z=0}}$, so that the above formula becomes

$$d\mathcal{G}(h)\psi = \mathbf{e}_z \cdot P'(h)\nabla^{\mu,\gamma}\phi_{|_{z=0}} + \mathbf{e}_z \cdot P(\Sigma)\nabla^{\mu,\gamma}v_{a\,|_{z=0}} - \varepsilon\mathcal{G}(h\underline{w}).$$

The theorem therefore follows from the identity

$$-\varepsilon\mu\nabla^\gamma \cdot (h\underline{V}) = \mathbf{e}_z \cdot P'(h)\nabla^{\mu,\gamma}\phi_{|_{z=0}} + \mathbf{e}_z \cdot P(\Sigma)\nabla^{\mu,\gamma}v_{a\,|_{z=0}} \tag{3.23}$$

that we will now prove. With Φ the solution of the boundary value problem (3.8), one can deduce from Remark 3.22 that

$$\nabla^\gamma \cdot (h\underline{V}) = \nabla^\gamma h \cdot \underline{V} + h\big(\Delta^\gamma\Phi + \varepsilon\nabla^\gamma\zeta \cdot \nabla^\gamma\partial_z\Phi\big)_{|_{z=\varepsilon\zeta}}.$$

Moreover, since $\mu\Delta^\gamma\Phi = -\partial_z^2\Phi$ and $\nabla^\gamma\partial_z\Phi_{|_{z=\zeta}} = \nabla^\gamma\underline{w} - \varepsilon\nabla^\gamma\zeta(\partial_z^2\Phi)_{|_{z=\zeta}}$, we deduce that

$$\begin{aligned}
-\varepsilon\mu\nabla^\gamma \cdot (h\underline{V}) &= -\varepsilon\mu\nabla^\gamma h \cdot \underline{V} + \varepsilon h(1+\varepsilon^2\mu|\nabla\zeta|^2)(\partial_z^2\Phi)_{|_{z=\varepsilon\zeta}} - \varepsilon^2\mu h\nabla^\gamma\zeta \cdot \nabla^\gamma\underline{w} \\
&= -\varepsilon\mu\nabla^\gamma h \cdot \underline{V} + \varepsilon h\frac{1+\varepsilon^2\mu|\nabla\zeta|^2}{(1+\varepsilon\zeta-\beta b)^2}(\partial_z^2\phi)_{|_{z=0}} - \varepsilon^2\mu h\nabla^\gamma\zeta \cdot \nabla^\gamma\underline{w}.
\end{aligned}$$

Using the explicit expression of $P(\Sigma)$ given in Example 2.21, one can check that the r.h.s. of this identity matches the r.h.s. of (3.22) and the proof is complete. □

Corollary 3.26 is a direct corollary of Theorem 3.21. (We give a direct proof of this corollary when the bottom is flat.)

COROLLARY 3.26. *Under the same assumptions and notations as in the second point of Theorem* 3.21, *one has, for all* $1 \leq j \leq d$,

$$\partial_j(\mathcal{G}\psi) = \mathcal{G}(\partial_j\psi - \varepsilon\underline{w}\partial_j\zeta) - \varepsilon\mu\nabla^\gamma \cdot (\underline{V}\partial_j\zeta) + d\mathcal{G}(0,\partial_j b)\psi.$$

REMARK 3.27.

(1) For time-dependent functions, a similar linearization formula holds if we differentiate with respect to time.
(2) A generalization of this corollary to derivative of higher order is given in Proposition 4.5.

PROOF. Using Notations 3.8 and 3.20, noting that

$$\partial_j\mathcal{G}\psi = d\mathcal{G}(\partial_j\zeta)\psi + \mathcal{G}\partial_j\psi + d\mathcal{G}(0,\partial_j b)\psi,$$

and replacing the first term of the r.h.s. by the formula given in the second point of Theorem 3.21, gives the result. It is, however, instructive to give a direct proof of the corollary when the bottom is flat.
With Φ the solution of the boundary value problem (3.8), we use Remark 3.22 to get

$$\mathcal{G}\psi = -\varepsilon\mu\underline{V} \cdot \nabla^\gamma\zeta + \underline{w}$$

and consequently

$$\begin{aligned}
\partial(\mathcal{G}\psi) &= -\varepsilon\mu\partial(\underline{V} \cdot \nabla^\gamma\zeta) + \partial\underline{w} \\
&= -\varepsilon\mu\nabla^\gamma \cdot (\partial\zeta\underline{V}) + \varepsilon\mu\partial\zeta\nabla^\gamma \cdot \underline{V} - \varepsilon\mu\partial\underline{V} \cdot \nabla^\gamma\zeta + \partial\underline{w}.
\end{aligned}$$

Noting that

$$\begin{aligned}\nabla^\gamma\cdot\underline{V} &= (\nabla^\gamma\cdot\nabla^\gamma\Phi)_{|_{z=\varepsilon\zeta}}+\varepsilon\nabla^\gamma\zeta\cdot(\partial_z\nabla^\gamma\Phi)_{|_{z=\varepsilon\zeta}},\\ \partial\underline{V} &= (\partial\nabla^\gamma\Phi)_{|_{z=\varepsilon\zeta}}+\varepsilon\partial\zeta(\partial_z\nabla^\gamma\Phi)_{|_{z=\varepsilon\zeta}},\\ \partial\underline{w} &= (\partial\partial_z\Phi)_{|_{z=\varepsilon\zeta}}+\varepsilon\partial\zeta(\partial_z^2\Phi)_{|_{z=\varepsilon\zeta}},\end{aligned}$$

we then get

$$\partial(\mathcal{G}\psi)=-\varepsilon\mu\nabla^\gamma\cdot(\partial\zeta\underline{V})+\varepsilon\partial\zeta(\Delta^{\mu,\gamma}\Phi)_{|_{z=\varepsilon\zeta}}-\varepsilon\mu(\nabla^\gamma\partial\Phi)_{|_{z=\varepsilon\zeta}}\cdot\nabla^\gamma\zeta+(\partial_z\partial\Phi)_{|_{z=\varepsilon\zeta}}.$$

Since $\Delta^{\mu,\gamma}\Phi=0$, we deduce

$$\partial(\mathcal{G}\psi)=-\varepsilon\mu\nabla^\gamma\cdot(\partial\zeta\underline{V})+\sqrt{1+\varepsilon^2|\nabla\zeta|^2}\partial_{\mathbf{n}}(\partial\Phi)_{|_{z=\varepsilon\zeta}},$$

where, as always, $\partial_{\mathbf{n}}$ stands for the conormal derivative. Noting now that $(\partial\Phi)_{|_{z=\varepsilon\zeta}}=\partial\psi-\underline{w}\partial\zeta$ and that $\Delta^{\mu,\gamma}(\partial\Phi)=0$ and $\partial_{\mathbf{n}}(\partial\Phi)_{|_{z=-1}}=0$ (since the bottom is flat), we get

$$\sqrt{1+\varepsilon^2|\nabla\zeta|^2}\partial_{\mathbf{n}}(\partial\Phi)_{|_{z=\varepsilon\zeta}}=\mathcal{G}(\partial\psi-\varepsilon\partial\zeta\underline{w}),$$

and the result follows easily. □

Because of the cancellation property mentioned in Remark 3.23, Theorem 3.21 is not adapted to give estimates on the shape derivatives of the Dirichlet-Neumann operator. Such estimates are given in the proposition below, where we use Notation 2.10 (i.e., $a\vee b=\max\{a,b\}$).

The first two estimates of the proposition are generalizations of the two kinds of estimates of Theorem 3.15 that emphasized the singular behavior of the shallow water limit $\mu\ll 1$. In these estimates, a lot of regularity is required on $(\mathbf{h},\mathbf{k})$ even if $s=0$ (namely, H^{t_0+1} in this case). The last two estimates show that it is possible to choose one coordinate of $\mathbf{h}$ and $\mathbf{k}$ on which only an $H^{s+1/2}$ or H^{s+1}-regularity is required. As a result, there is a stronger requirement on the regularity of ψ.

PROPOSITION 3.28. *Let $t_0>d/2$ and $(\zeta,b)\in H^{t_0+1}$ be such that (3.2) is satisfied. Then*

(1) *For all $0\leq s\leq t_0+1/2$ and $\psi\in\dot{H}^{s+1/2}(\mathbb{R}^d)$,*

$$|d^j\mathcal{G}(\mathbf{h},\mathbf{k})\psi|_{H^{s-1/2}}\leq M_0\mu^{3/4}\prod_{m=1}^{j}|(\varepsilon h_m,\beta k_m)|_{H^{t_0+1}}|\mathfrak{P}\psi|_{H^s}.$$

(2) *For all $0\leq s\leq t_0$ and $\psi\in\dot{H}^{s+1}(\mathbb{R}^d)$,*

$$|d^j\mathcal{G}(\mathbf{h},\mathbf{k})\psi|_{H^{s-1/2}}\leq M_0\mu\prod_{m=1}^{j}|(\varepsilon h_m,\beta k_m)|_{H^{t_0+1}}|\mathfrak{P}\psi|_{H^{s+1/2}}.$$

(3) *For all $0\leq s\leq t_0$ and $\psi\in\dot{H}^{t_0+1}(\mathbb{R}^d)$,*

$$\begin{aligned}|d^j\mathcal{G}(\mathbf{h},\mathbf{k})\psi|_{H^{s-1/2}} &\leq M_0\mu^{3/4}|(\varepsilon h_1,\beta k_1)|_{H^{s+1/2}}\\ &\quad\times\prod_{m>1}^{j}|(\varepsilon h_m,\beta k_m)|_{H^{t_0+1}}|\mathfrak{P}\psi|_{H^{t_0+1/2}}.\end{aligned}$$

(4) *For all* $0 \leq s \leq t_0 - 1/2$ *and* $\psi \in \dot{H}^{t_0+1}(\mathbb{R}^d)$,

$$\begin{aligned} |d^j\mathcal{G}(\mathbf{h},\mathbf{k})\psi|_{H^{s-1/2}} &\leq M_0\mu|(\varepsilon h_1, \beta k_1)|_{H^{s+1}} \\ &\times \prod_{m>1}^{j} |(\varepsilon h_m, \beta k_m)|_{H^{t_0+1}} |\mathfrak{P}\psi|_{H^{t_0+1/2}}. \end{aligned}$$

PROOF. *Throughout this proof,* Σ *is assumed to be the regularizing diffeomorphism of Example* 2.15.
(1) We recall that $\mathfrak{A}_\psi(\Gamma) = \phi_\Gamma$ denotes the solution to (3.5) for all $\Gamma \in \mathbf{\Gamma}$. Differentiating (3.12) with respect to $\Gamma = (\zeta, b)$, we deduce by the same duality argument as in the proof of the first point of Theorem 3.15 that

$$|d^j\mathcal{G}(\mathbf{h},\mathbf{k})\psi|_{H^{s-1/2}} \lesssim \mu^{1/4}\sum \|\Lambda^s\big(d^{j_1}P(\mathbf{h},\mathbf{k})_I\nabla^{\mu,\gamma}d^{j_2}\mathfrak{A}_\psi(\Gamma)(\mathbf{h},\mathbf{k})_{II}\big)\|_2,$$

the summation being taken over all integers j_1 and j_2 such that $j_1 + j_2 = j$, and on all the $2j_1$ and $2j_2$ uplets $(\mathbf{h},\mathbf{k})_I$ and $(\mathbf{h},\mathbf{k})_{II}$ whose coordinates form a permutation of the coordinates of $(\mathbf{h},\mathbf{k})$. The product estimate of Corollary B.5(2) in Appendix B therefore gives that for all $0 \leq s \leq t_0$,

$$\begin{aligned} |d^j\mathcal{G}(\mathbf{h},\mathbf{k})\psi|_{H^{s-1/2}} &\lesssim \mu^{1/4}\sum \|d^{j_1}P(\mathbf{h},\mathbf{k})_I\|_{L^\infty H^{t_0}}\|\Lambda^s\nabla^{\mu,\gamma}d^{j_2}\mathfrak{A}_\psi(\Gamma)(\mathbf{h},\mathbf{k})_{II}\|_2 \\ &\lesssim \mu^{1/4}M_0\prod_{m\in I}|(\varepsilon h_m, \beta k_m)|_{H^{t_0+1}}\|\Lambda^s\nabla^{\mu,\gamma}d^{j_2}\mathfrak{A}_\psi(\Gamma)(\mathbf{h},\mathbf{k})_{II}\|_2, \end{aligned}$$

where we used Corollaries B.5 and B.6 (see Appendix B), and the explicit expression of P given in Example 2.21 to derive the second inequality. We can therefore use Proposition A.7 (postponed to Appendix A for the sake of clarity) to get the result for $0 \leq s \leq t_0$.
For $s = t_0 + 1/2$, Corollary B.5(2) in Appendix B, and the continuous embedding $H^{t_0+1/2,1} \subset L^\infty H^{t_0}$ yield

$$|d^j\mathcal{G}(\mathbf{h},\mathbf{k})\psi|_{H^{t_0}} \lesssim \mu^{1/4}\sum \|d^{j_1}P(\mathbf{h},\mathbf{k})_I\|_{H^{t_0+1/2,1}}\|\nabla^{\mu,\gamma}d^{j_2}\mathfrak{A}_\psi(\Gamma)(\mathbf{h},\mathbf{k})_{II}\|_{H^{t_0+1/2,1}}.$$

Since $\|d^{j_1}P(\mathbf{h},\mathbf{k})_I\|_{H^{t_0+1/2,1}} \lesssim M_0\prod_{m\in I}|(\varepsilon h_m, \beta k_m)|_{H^{t_0+1}}$ (a consequence of Corollaries B.5 and B.6 in Appendix B, and of the regularizing properties of Σ), the result also follows for $s = t_0 + 1/2$ from Proposition A.7 in Appendix A.
The case $t_0 \leq s \leq t_0 + 1/2$ follows as usual by interpolation.
(2) The difference with (2) is exactly the same as the difference between the second point of Theorem 3.15 and the first one.
(3) Let us distinguish two cases:
- *If* (h_1, k_1) *belongs to* $(\mathbf{h},\mathbf{k})_{II}$, then we proceed exactly as above but use Proposition A.8 (Appendix A) instead of Proposition A.7 to control $\|\Lambda^s\nabla^{\mu,\gamma}d^{j_2}\mathfrak{A}_\psi(\Gamma)(\mathbf{h},\mathbf{k})_{II}\|_2$.
- *If* (h_1, k_1) *belongs to* $(\mathbf{h},\mathbf{k})_I$, then we replace the second inequality in the proof of (1) by

$$|d^j\mathcal{G}(\mathbf{h},\mathbf{k})\psi|_{H^{s-1/2}} \leq \mu^{1/4}\sum \|\Lambda^s d^{j_1}P(\mathbf{h},\mathbf{k})_I\|_2\|\nabla^{\mu,\gamma}d^{j_2}\mathfrak{A}_\psi(\Gamma)(\mathbf{h},\mathbf{k})_{II}\|_{L^\infty H^{t_0}}.$$

We use Lemma A.10 in Appendix A to control the first term of the r.h.s. while the second term can be controlled by Proposition A.7 and the continuous embedding $H^{t_0+1/2,1} \subset L^\infty H^{t_0}$.
(4) We adapt the proof of (3) as we adapted the proof of (1) to get (2). □

Let us give a last proposition on the shape derivatives, which states that a property similar to the first estimate of Proposition 3.12 also holds for $d^j\mathcal{G}(\mathbf{h},\mathbf{k})$.

PROPOSITION 3.29. *Let $t_0 > d/2$, $0 \leq s \leq t_0 + 1/2$ and $\Gamma = (\zeta, b) \in \mathbf{\Gamma}$. Then for all $\psi_1, \psi_2 \in \dot{H}^{s+1/2}(\mathbb{R}^d)$, one has*

$$\big|\big(\Lambda^s d^j \mathcal{G}(\mathbf{h},\mathbf{k})\psi_1, \Lambda^s \psi_2\big)\big| \leq \mu M_0 \prod_{m=1}^{j} \big|(\varepsilon h_m, \beta k_m)\big|_{H^{t_0+1}} |\mathfrak{P}\psi_1|_{H^s} |\mathfrak{P}\psi_2|_{H^s}.$$

PROOF. Differentiating (3.15) with respect to ζ, we get

$$\big(\Lambda^s d^j \mathcal{G}(\mathbf{h},\mathbf{k})\psi_1, \Lambda^s \psi_2\big) = \sum \int_{\mathcal{S}} \Lambda^s (d^{j_1} P(\mathbf{h},\mathbf{k})_I \nabla^{\mu,\gamma} d^{j_2} \mathfrak{A}_\psi(\mathbf{h},\mathbf{k})_{II}) \cdot \Lambda^s \nabla^{\mu,\gamma} \psi_2^\dagger,$$

where $\mathfrak{A}_\psi$ is as defined in (A.1) in Appendix A, and where the summation is taken over all integers j_1 and j_2 such that $j_1 + j_2 = j$, and on all the $2j_1$ and $2j_2$ uplets $(\mathbf{h},\mathbf{k})_I$ and $(\mathbf{h},\mathbf{k})_{II}$ whose coordinates form a permutation of the coordinates of $(\mathbf{h},\mathbf{k})$. We then conclude, as in the proof of Proposition 3.18, after using the control on

$$\|\Lambda^s (d^{j_1} P(\mathbf{h},\mathbf{k})_I \nabla^{\mu,\gamma} d^{j_2} \mathfrak{A}_\psi(\mathbf{h},\mathbf{k})_{II})\|_2$$

established in the proof of the first point of Proposition 3.28. □

3.4. Commutator estimates

Propositions 3.9 and 3.12 allow one to control $(u, \mathcal{G}v)$ in general. However, it is sometimes necessary to have more precise estimates, when u and v have some special structure that can be exploited. We have seen that in the case of flat surface and bottom ($\zeta = b = 0$), one has $\mathcal{G}_{\mu,\gamma}[0,0] = \sqrt{\mu}|D^\gamma| \tanh(\sqrt{\mu}|D^\gamma|)$; this operator obviously commutes with ∇^γ, and it is therefore easy to check that

$$\big((\underline{V} \cdot \nabla^\gamma u), \frac{1}{\mu}\mathcal{G}_{\mu,\gamma}[0,0]u\big) \lesssim |\underline{V}|_{W^{1,\infty}} |\mathfrak{P}u|_2^2.$$

The following proposition shows that this identity still holds true for nonflat surface and bottom.

PROPOSITION 3.30. *Let $t_0 > d/2$ and $\zeta, b \in H^{t_0+2}(\mathbb{R}^d)$ be such that (3.2) is satisfied. For all $\underline{V} \in H^{t_0+1}(\mathbb{R}^2)^2$ and $u \in H^{1/2}(\mathbb{R}^2)$, one has*

$$\big((\underline{V} \cdot \nabla^\gamma u), \frac{1}{\mu}\mathcal{G}u\big) \leq M\, |\underline{V}|_{W^{1,\infty}} |\mathfrak{P}u|_2^2.$$

REMARK 3.31. Let χ be a smooth, positive, compactly supported function defined in $\mathbb{R}$ and equal to 1 in a neighborhood of the origin. For all $0 < \iota \leq 1$, we can define a mollifier J^ι by $J^\iota = \chi(\iota|D|)$. A straightforward adaptation of the proof shows that one also has

$$\big(J^\iota(\underline{V} \cdot \nabla^\gamma u), \frac{1}{\mu}\mathcal{G}u\big) \leq M\, |\underline{V}|_{W^{1,\infty}} |\mathfrak{P}u|_2^2.$$

PROOF. First note that, owing to Green's identity,

$$\big((\underline{V} \cdot \nabla^\gamma u), \mathcal{G}u\big) = \int_{\mathcal{S}} P(\Sigma)\nabla^{\mu,\gamma} u^\mathfrak{h} \cdot \nabla^{\mu,\gamma}(\underline{V} \cdot \nabla^\gamma u^\mathfrak{h}), \tag{3.24}$$

so that, with $Q(\Sigma) = P(\Sigma) - I$,

$$\begin{aligned}\big((\underline{V} \cdot \nabla^\gamma u), \mathcal{G}u\big) &= \int_{\mathcal{S}} P(\Sigma)\nabla^{\mu,\gamma} u^\mathfrak{h} \cdot [\nabla^{\mu,\gamma}, \underline{V} \cdot \nabla^\gamma] u^\mathfrak{h} \\ &+ \int_{\mathcal{S}} \nabla^{\mu,\gamma} u^\mathfrak{h} \cdot [Q(\Sigma), (\underline{V} \cdot \nabla^\gamma)]\nabla^{\mu,\gamma} u^\mathfrak{h} + \int_{\mathcal{S}} \nabla^{\mu,\gamma} u^\mathfrak{h} \cdot (\underline{V} \cdot \nabla^\gamma) P(\Sigma)\nabla^{\mu,\gamma} u^\mathfrak{h}.\end{aligned} \tag{3.25}$$

Integrating by parts, one finds

$$\int_{\mathcal{S}} \nabla^{\mu,\gamma} u^{\mathfrak{h}} \cdot (\underline{V} \cdot \nabla^{\gamma}) P(\Sigma) \nabla^{\mu,\gamma} u^{\mathfrak{h}}$$

$$= -\int_{\mathcal{S}} \big((\mathrm{div}_{\gamma} \underline{V}) + \underline{V} \cdot \nabla^{\gamma}\big) \nabla^{\mu,\gamma} u^{\mathfrak{h}} \cdot P(\Sigma) \nabla^{\mu,\gamma} u^{\mathfrak{h}} \tag{3.26}$$

$$= -\int_{\mathcal{S}} (\mathrm{div}_{\gamma} \underline{V}) \nabla^{\mu,\gamma} u^{\mathfrak{h}} \cdot P \nabla^{\mu,\gamma} u^{\mathfrak{h}} - \int_{\mathcal{S}} [\underline{V} \cdot \nabla^{\gamma}, \nabla^{\mu,\gamma}] u^{\mathfrak{h}} \cdot P \nabla^{\mu,\gamma} u^{\mathfrak{h}}$$

$$-\int_{\mathcal{S}} \nabla^{\mu,\gamma} (\underline{V} \cdot \nabla^{\gamma} u^{\mathfrak{h}}) \cdot P \nabla^{\mu,\gamma} u^{\mathfrak{h}}. \tag{3.27}$$

From (3.24), (3.25), (3.26), and (3.27), one therefore gets

$$\big((\underline{V} \cdot \nabla^{\gamma} u), \mathcal{G} u\big) = \int_{\mathcal{S}} P \nabla^{\mu,\gamma} u^{\mathfrak{h}} \cdot [\nabla^{\mu,\gamma}, \underline{V} \cdot \nabla^{\gamma}] u^{\mathfrak{h}}$$

$$+ \frac{1}{2} \int_{\mathcal{S}} \nabla^{\mu,\gamma} u^{\mathfrak{h}} \cdot [Q, (\underline{V} \cdot \nabla^{\gamma})] \nabla^{\mu,\gamma} u^{\mathfrak{h}} - \frac{1}{2} \int_{\mathcal{S}} (\mathrm{div}_{\gamma} \underline{V}) \nabla^{\mu,\gamma} u^{\mathfrak{h}} \cdot P \nabla^{\mu,\gamma} u^{\mathfrak{h}}.$$

Noting that $[\nabla^{\mu,\gamma}, \underline{V} \cdot \nabla^{\gamma}] = \begin{pmatrix} (\nabla^{\gamma} \underline{V}_1) \sqrt{\mu} \partial_x + (\nabla^{\gamma} \underline{V}_2) \gamma \sqrt{\mu} \partial_y \\ 0 \end{pmatrix}$, one deduces easily that

$$\big((\underline{V} \cdot \nabla^{\gamma} u), \mathcal{G} u\big) \lesssim |\underline{V}|_{W^{1,\infty}} (1 + \|Q(\Sigma)\|_{W^{1,\infty}}) \|\nabla^{\mu,\gamma} u^{\mathfrak{h}}\|_2^2,$$

and the results follow from Proposition 2.36 (with $s = 0$). □

We finally give some control on the commutator $[\Lambda^s, \mathcal{G}]$. We treat only the case $0 \leq s \leq t_0$; tame estimates can also be obtained for $s \geq t_0$ **[12]** but are not needed here.

PROPOSITION 3.32. *Let $t_0 > d/2$ and $0 \leq s \leq t_0$. Let $\zeta, b \in H^{t_0+2}(\mathbb{R}^d)$ be such that (3.2) is satisfied. Then, for all $v \in \dot{H}^{s+1/2}(\mathbb{R}^d)$,*

$$\big|[\Lambda^s, \frac{1}{\mu} \mathcal{G}] v\big|_2 \leq M |\mathfrak{P} v|_{H^s}.$$

REMARK 3.33. Since $\mathfrak{P}$ is an operator of order $1/2$, the estimate given in the proposition shows that $[\Lambda^s, \mathcal{G}]$ behaves as an operator of order $s + 1/2$ while one would have expected an operator of order s. The reason for this poor precision is, of course, because we need a good dependence on μ (see the comments at the beginning of §3.2). Replacing (3.29) in the proof below by $\|\Lambda^{-1/2} \nabla^{\mu,\gamma} u^{\dagger}\|_2 \lesssim \mu^{1/4} |u|_2$, one gains half a derivative in v in the estimate of the proposition; the result is a singularity of order $O(\mu^{-1/4})$ as $\mu \to 0$.

PROOF. First note that for all $u \in \mathfrak{S}(\mathbb{R}^2)$,

$$\big(u, [\mathcal{G}, \Lambda^s] v\big) = \big(u, \mathcal{G} \Lambda^s v\big) - \big(\Lambda^s u, \mathcal{G} v\big).$$

Since $\Lambda^s u^{\dagger}_{|_{z=0}} = \Lambda^s u$ (we use here Notation 2.28), it follows from Green's identity that

$$\big(u, [\mathcal{G}, \Lambda^s] v\big) = \int_{\mathcal{S}} \nabla^{\mu,\gamma} u^{\dagger} \cdot P \nabla^{\mu,\gamma} (\Lambda^s v)^{\mathfrak{h}} - \int_{\mathcal{S}} P \nabla^{\mu,\gamma} v^{\mathfrak{h}} \cdot \nabla^{\mu,\gamma} \Lambda^s u^{\dagger}$$

$$= \int_{\mathcal{S}} \Lambda^{-1} \nabla^{\mu,\gamma} u^{\dagger} \cdot \Lambda^1 \big(P \nabla^{\mu,\gamma} \big((\Lambda^s v)^{\mathfrak{h}} - \Lambda^s v^{\mathfrak{h}}\big) - [\Lambda^s, Q] \nabla^{\mu,\gamma} v^{\mathfrak{h}}\big), \tag{3.28}$$

with $P = P(\Sigma) = I + Q$. Since, by Lemma 2.34,

$$\|\Lambda^{-1}\nabla^{\mu,\gamma}u^\dagger\|_2 \lesssim \sqrt{\mu}|\mathfrak{P}u|_{H^{-1}} \lesssim \sqrt{\mu}|u|_2, \tag{3.29}$$

we deduce from the Cauchy-Schwarz inequality that

$$\begin{aligned}(u,[\mathcal{G},\Lambda^s]v) \lesssim \sqrt{\mu}|u|_2 \Big(& \|\Lambda[\Lambda^s,Q]\nabla^{\mu,\gamma}v^{\mathfrak{h}}\|_2 \\ & +M(1+1\vee t_0)\|\Lambda\nabla^{\mu,\gamma}\big((\Lambda^s v)^{\mathfrak{h}} - \Lambda^s v^{\mathfrak{h}}\big)\|_2\Big),\end{aligned} \tag{3.30}$$

which motivates the following lemma:

LEMMA 3.34. *One has*

$$\|\Lambda^1\nabla^{\mu,\gamma}\big((\Lambda^s v)^{\mathfrak{h}} - \Lambda^s v^{\mathfrak{h}}\big)\|_2 \le M_0\|\Lambda^1[\Lambda^s,Q]\nabla^{\mu,\gamma}v^{\mathfrak{h}}\|_2.$$

PROOF. Just note that $w := (\Lambda^s v)^{\mathfrak{h}} - \Lambda^s v^{\mathfrak{h}}$ solves

$$\begin{cases} \nabla^{\mu,\gamma}\cdot P(\Sigma)\nabla^{\mu,\gamma}w = \nabla^{\mu,\gamma}\cdot \mathbf{g}, \\ w_{|z=\varepsilon\zeta} = 0, \qquad \partial_n w_{|z=-1} = -\mathbf{e_z}\cdot\mathbf{g}_{|z=-1}, \end{cases}$$

with $\mathbf{g} = [\Lambda^s, Q]\nabla^{\mu,\gamma}v^{\mathfrak{h}}$, and use Lemma 2.38. □

With the help of the lemma, one deduces from (3.30) that

$$(u,[\mathcal{G},\Lambda^s]v) \le \sqrt{\mu}M(1+1\vee t_0)\big\|\Lambda^1[\Lambda^s,Q]\nabla^{\mu,\gamma}v^{\mathfrak{h}}\big\|_2|u|_2,$$

and thus, owing to Proposition B.8(3) in Appendix B (with $r=1$),

$$(u,\Lambda^1[\mathcal{G},\Lambda^s]v) \le \sqrt{\mu}M|u|_2\|\Lambda^s\nabla^{\mu,\gamma}v^{\mathfrak{h}}\|_2.$$

The result therefore follows from Corollary 2.40, and a duality argument. □

3.5. The Dirichlet-Neumann operator and the vertically averaged velocity

The vertically averaged horizontal component of the velocity $\overline{V}_{\mu,\gamma}[\varepsilon\zeta,\beta b]\psi$ (denoted $\overline{V}$ when no confusion is possible) is an important quantity in shallow water theory. It is naturally defined in terms of the (nondimensionalized) velocity potential Φ solution of (1.30), namely,

$$\begin{cases} \Delta^{\mu,\gamma}\Phi = 0, & -1+\beta b \le z \le \varepsilon\zeta, \\ \Phi_{|z=\varepsilon\zeta} = \psi, & \sqrt{1+\beta^2|\nabla b|^2}\partial_{\mathbf{n}}\Phi_{|z=-1+\beta b} = 0 \end{cases}$$

by the formula

$$\overline{V} = \overline{V}_{\mu,\gamma}[\varepsilon\zeta,\beta b]\psi = \frac{1}{h}\int_{-1+\beta b}^{\varepsilon\zeta}\nabla^\gamma\Phi(\cdot,z)dz \qquad (h = 1+\varepsilon\zeta-\beta b). \tag{3.31}$$

Equivalently, $\overline{V}$ can be defined in terms of the straightened potential $\phi = \Phi\circ\Sigma$; see (3.35) below. The following proposition makes the link between the Dirichlet-Neumann operator $\mathcal{G}\psi$ and this averaged velocity $\overline{V}$.

PROPOSITION 3.35. *Let $t_0 > d/2$ and $\zeta, b \in H^{t_0+1}(\mathbb{R}^d)$ be such that (3.2) is satisfied. Also let $\psi \in \dot{H}^{3/2}(\mathbb{R}^d)$ and $\Phi \in \dot{H}^2(\Omega)$ be the solution to (2.7) provided by Corollary 2.44 and $\overline{V}$ be given by (3.31). The following relation holds*

$$\mathcal{G}\psi = -\mu\nabla^\gamma\cdot(h\overline{V}).$$

REMARK 3.36. This result can be generalized to the case of a nonhomogeneous Neumann boundary condition at the bottom (this is necessary to handle moving bottoms; see §1.6). We still use formula (3.31) to define $\overline{V}$, but with Φ solving the following boundary value problem instead of (1.30),

$$\begin{cases} \Delta^{\mu,\gamma}\Phi = 0, & -1+\beta b \leq z \leq \varepsilon\zeta, \\ \Phi_{|_{z=\varepsilon\zeta}} = \psi, & \sqrt{1+\beta^2|\nabla b|^2}\partial_{\mathbf{n}}\Phi_{|_{z=-1+\beta b}} = B, \end{cases}$$

for some B not necessarily identically equal to zero. (See §A.5.1 in Appendix A for the analysis of this system.) The identity given in the proposition is then generalized into

$$\sqrt{1+\varepsilon^2|\nabla\zeta|^2}\partial_{\mathbf{n}}\Phi_{|_{z=\varepsilon\zeta}} = B - \mu\nabla^\gamma\cdot(h\overline{V}).$$

PROOF. Let us note that, by Green's identity, the following relation holds for all test function $\varphi \in \mathcal{D}(\mathbb{R}^d)$,

$$\int_{\mathbb{R}^d} \mathcal{G}\psi\varphi = \int_\Omega \nabla^{\mu,\gamma}\Phi\cdot\nabla^{\mu,\gamma}\varphi.$$

Since $\partial_z\varphi = 0$, we deduce that

$$\begin{aligned} \int_{\mathbb{R}^d} \mathcal{G}\psi\varphi &= \mu\int_\Omega \nabla^\gamma\Phi\cdot\nabla^\gamma\varphi \\ &= \mu\int_{\mathbb{R}^d} \nabla^\gamma\varphi\cdot\int_{-1+\beta b}^{\varepsilon\zeta} \nabla^\gamma\Phi dz dX \\ &= \mu\int_{\mathbb{R}^d} \nabla^\gamma\varphi\cdot(h\overline{V})dX. \end{aligned}$$

A simple integration by parts then shows that $\int \mathcal{G}\psi\varphi = -\mu\int\nabla^\gamma\cdot(h\overline{V})\varphi$ for all test function φ, so that the result follows. □

3.6. Asymptotic expansions

This subsection is devoted to the asymptotic expansion of the DN operator $\mathcal{G}\psi(=\mathcal{G}_{\mu,\gamma}[\varepsilon\zeta,\beta b]\psi)$ in terms of one or several of the parameters ε, μ, γ and β. We consider two cases that cover all the physical regimes described in the introduction.

3.6.1. Asymptotic expansion in shallow-water ($\mu \ll 1$). In shallow water, that is, when $\mu \ll 1$, the (straightened) Laplace equation (3.5) reduces at first order to the ODE $\partial_z^2\phi = 0$. This fact can be exploited to find an approximate solution ϕ_{app} to (3.5) by a standard BKW expansion. We also recall (see Proposition 3.35) that

$$\mathcal{G}\psi = -\mu\nabla^\gamma\cdot(h\overline{V}),$$

where $\overline{V}$ is the vertically averaged horizontal component of the velocity (see (3.31) and Proposition 3.35). The strategy we adopt here to construct an asymptotic expansion of $\overline{V}$, and thus of $\mathcal{G}\psi$, is to replace $\overline{V}$ by the vertically averaged velocity corresponding to the approximate solution ϕ_{app} of (3.5).

In order to state the main result of this section, we need to define the second-order linear operator $\mathcal{T}[h,b]$ as

(3.32)

$$\mathcal{T}[h,b]V := -\frac{1}{3h}\nabla^\gamma(h^3\nabla^\gamma\cdot V) + \frac{1}{2h}\big[\nabla^\gamma(h^2\nabla^\gamma b\cdot V) - h^2\nabla^\gamma b\nabla^\gamma\cdot V\big] + \nabla^\gamma b\nabla^\gamma b\cdot V,$$

with $h = 1+\varepsilon\zeta-\beta b$.

PROPOSITION 3.37. *Let* $t_0 > d/2$, $s \geq 0$, $n \in \mathbb{N}$, *and* $\zeta, b \in H^{s+2+2n} \cap H^{t_0+2}(\mathbb{R}^d)$ *be such that (3.2) is satisfied. Let* $\psi \in \dot{H}^{s+3+2n}(\mathbb{R}^d)$, *and* $\overline{V}$ *be as in (3.31).*
One can construct a sequence $(\overline{V}_j)_{0\leq j\leq n}$ *such that* $|\overline{V}_j|_{H^s} \leq M(s+2j)|\nabla^\gamma\psi|_{H^{s+2j}}$ $(0 \leq j \leq n)$ *and*

$$|\overline{V} - \sum_{j=0}^{n} \mu^j \overline{V}_j|_{H^s} \leq \mu^{n+1} M(s+2+2n)|\nabla^\gamma\psi|_{H^{s+2+2n}},$$

with $M(s+2+2n)$ *as defined in (3.4). In particular, one can take*

$$\overline{V}_0 = \nabla^\gamma\psi \quad \text{and} \quad \overline{V}_1 = -\mathcal{T}[h, \beta b]\nabla^\gamma\psi.$$

REMARK 3.38. It is worth emphasizing the fact that the constant $M(s+2+2n)$ that appears in the above proposition is uniform with respect to all the parameter ε, μ, β and γ that satisfy (3.1). If further assumptions are made (for instance, $\varepsilon = O(\mu)$), then the asymptotic expansion of $\overline{V}$ can be simplified.

REMARK 3.39. From the relation $\mathcal{G}\psi = -\mu\nabla^\gamma \cdot (h\overline{V})$ and the proposition, it is straightforward to deduce an asymptotic expansion of $\mathcal{G}\psi$,

$$\mathcal{G}\psi = -\mu\nabla^\gamma \cdot (h\nabla^\gamma\psi) + \mu^2\nabla^\gamma \cdot (h\mathcal{T}[h, \beta b]\nabla^\gamma\psi) + O(\mu^3).$$

REMARK 3.40. A direct corollary of the proposition (with $n = 0$) is that the mapping

$$\overline{V}_{\mu,\gamma}[\varepsilon\zeta, \beta b] : \quad \dot{H}^{s+3}(\mathbb{R}^d) \to H^s(\mathbb{R}^d)^d$$

is well defined and that one has

$$|\overline{V}_{\mu,\gamma}[\varepsilon\zeta, \beta b]\psi|_{H^s} \leq M(s+2)|\nabla^\gamma\psi|_{H^{s+2}}.$$

REMARK 3.41. The proof of the proposition (see Lemma 3.42 below) shows that the velocity potential does not depend on z at leading order and that at first order in μ, it depends quadratically on z. A consequence is the fact that at leading order, the horizontal velocity does not depend on z and therefore that $\overline{V}$ coincides with the horizontal velocity evaluated at any level line of the fluid domain. In physical terms, keeping only the leading order terms corresponds to the *hydrostatic approximation.* See Figures 3.1 and 3.2 for a graphical illustration of the asymptotic approximation of the velocity field.

PROOF. We assume throughout this proof that Σ is the trivial diffeomorphism given by Example 2.14, that is, $\Sigma(X, z) = (X, z + \sigma(X, z))$ with

$$\sigma(X, z) = (\varepsilon\zeta - \beta b)z + \varepsilon\zeta.$$

We construct in the following lemma an approximate solution ϕ_{app} to the exact solution ϕ of the potential equation (3.5) under the form

$$\phi_{app}(X, z) = \sum_{j=0}^{n} \mu^j \phi_j(X, z). \tag{3.33}$$

LEMMA 3.42. *Let* $n \in \mathbb{N}$. *There exists a sequence* $(\phi_j)_{0\leq j\leq n}$ *such that for all* $0 \leq j \leq n$, $\phi_j(X, z)$ *is a polynomial of order* $2j$ *in* z *and such that* ϕ_{app} *as given by (3.33) satisfies*

$$\begin{cases} \nabla^{\mu,\gamma} \cdot P(\Sigma)\nabla^{\mu,\gamma}\phi_{app} = \mu^{n+1}R_\mu, \\ \phi_{app\,|z=0} = \psi, \qquad \partial_{\mathbf{n}}\phi_{app\,|z=-1} = \mu^{n+1}r_\mu, \end{cases}$$

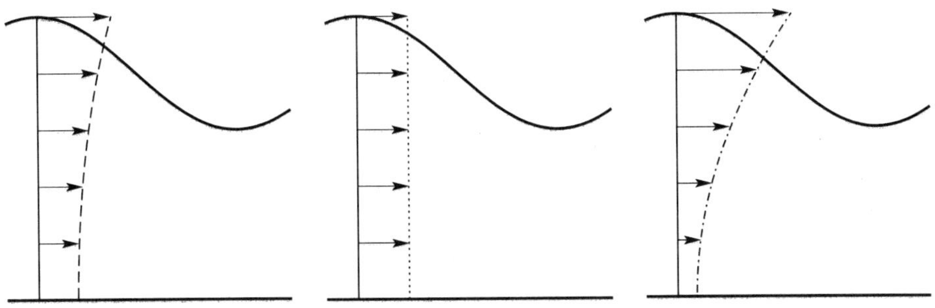

FIGURE 3.1. Shallow water approximation of the horizontal velocity field $V(x_0, \cdot)$ in the fluid domain when $\mu = 1$. Exact velocity field (dash), zero order approximation (dots) and first-order approximation (dash-dots).

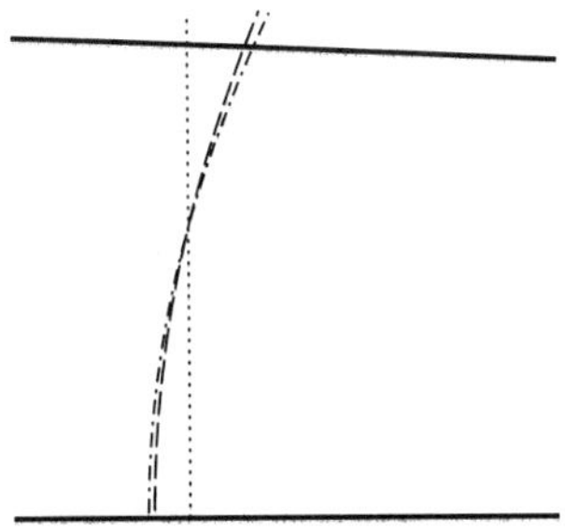

FIGURE 3.2. Shallow water approximation of the horizontal velocity field $V(x_0, \cdot)$ in the fluid domain when $\mu = 0.1$ (zoom). Exact velocity field (dash), zero order approximation (dots) and first-order approximation (dash-dots).

with

$$\|\Lambda^s R_\mu\|_2 + |r_\mu|_{H^s} \leq M(s+2n+1)|\nabla^\gamma \psi|_{H^{s+2n+1}}.$$

Moreover, a possible choice for ϕ_0 *and* ϕ_1 *is given by*

$$\begin{aligned}\phi_0 &= \psi,\\ \phi_1 &= -h^2(\frac{z^2}{2}+z)\Delta^\gamma\psi + zh\beta\nabla^\gamma b\cdot\nabla^\gamma\psi.\end{aligned}$$

PROOF OF LEMMA 3.42. Let us note (see Example 2.21) that the operator $\nabla^{\mu,\gamma} \cdot P(\Sigma)\nabla^{\mu,\gamma}$ can be written under the form

$$\nabla^{\mu,\gamma} \cdot P(\Sigma)\nabla^{\mu,\gamma} = \frac{1}{h}\partial_z^2 + \mu A(\nabla^\gamma, \partial_z),$$

with $h = 1 + \varepsilon\zeta - \beta b$ and

$$A(\nabla^\gamma, \partial_z)\bullet = \nabla^\gamma \cdot (h\nabla^\gamma \bullet) + \partial_z(\frac{|\nabla^\gamma \sigma|^2}{h}\partial_z \bullet) - \nabla^\gamma \cdot (\nabla^\gamma \sigma \partial_z \bullet) - \partial_z(\nabla^\gamma \sigma \cdot \nabla^\gamma \bullet).$$

Similarly, the conormal derivative operator at the bottom can be written as

$$(\partial_{\mathbf{n}} \bullet)_{|_{z=-1}} = \frac{1}{h}(\partial_z \bullet)_{|_{z=-1}} + \mu(\mathbf{b}(\nabla^\gamma, \partial_z)\bullet)_{|_{z=-1}},$$

with

$$\mathbf{b}(\nabla^\gamma, \partial_z) = -\beta\nabla^\gamma b \cdot \nabla^\gamma + \beta^2 \frac{|\nabla^\gamma b|^2}{h}\partial_z.$$

We then replace ϕ by ϕ_{app} in (3.5) and choose the ϕ_j $(0 \leq j \leq n)$ in order to cancel the leading order terms of expansion in μ of the residual. This yields the induction relation

$$\frac{1}{h}\partial_z^2\phi_0 = 0, \qquad \phi_{0\,|_{z=0}} = \psi, \qquad \frac{1}{h}\partial_z\phi_{0\,|_{z=-1}} = 0,$$

and

$$\begin{cases} \frac{1}{h}\partial_z^2\phi_j = -A(\nabla^\gamma, \partial_z)\phi_{j-1}, \\ \phi_{j\,|_{z=0}} = 0, \qquad \frac{1}{h}\partial_z\phi_{j\,|_{z=-1}} = -\mathbf{b}(\nabla^\gamma, \partial_z)\phi_{j-1\,|_{z=-1}}, \end{cases} \qquad (1 \leq j \leq n).$$

Equivalently, one has

$$\phi_0 = \psi \quad \text{and} \quad \phi_j = h\int_z^0\int_{-1}^{z'} A(\nabla^\gamma, \partial_z)\phi_{j-1}dz''dz' - zh(\mathbf{b}(\nabla^\gamma, \partial_z)\phi_{j-1})_{|_{z=-1}},$$

for $1 \leq j \leq n$. The expressions ϕ_0 and ϕ_1 given in the statement of the lemma follow easily from this induction relation.
By construction, ϕ_{app} also satisfies the boundary value problem given in the lemma, with

$$R_\mu = A(\nabla^\gamma, \partial_z)\phi_n \quad \text{and} \quad r_\mu = (\mathbf{b}(\nabla^\gamma, \partial_z)\phi_n)_{|_{z=-1}}. \tag{3.34}$$

The estimates on $\|\Lambda^s R_\mu\|_2$ and $|r_\mu|_{H^s}$ can easily be proved by induction. □

Noting that the definition (3.31) of $\overline{V}$ can be written in terms of ϕ (rather than Φ) using a simple change of variables, we get

$$\overline{V} = \frac{1}{h}\int_{-1}^0 \big[h\nabla^\gamma\phi - (z\nabla^\gamma h + \varepsilon\nabla^\gamma\zeta)\partial_z\phi\big]dz, \tag{3.35}$$

so that if we define

$$\overline{V}_{app} = \sum_{j=0}^n \mu^j \overline{V}_j, \quad \text{with} \quad \overline{V}_j = \int_{-1}^0 \big[\nabla^\gamma\phi_j - \frac{1}{h}(z\nabla^\gamma h + \varepsilon\nabla^\gamma\zeta)\partial_z\phi_j\big]dz, \tag{3.36}$$

we get

$$\overline{V} - \overline{V}_{app} = \int_{-1}^0 \big[\nabla^\gamma u - \frac{1}{h}(z\nabla^\gamma h + \varepsilon\nabla^\gamma\zeta)\partial_z u\big]dz, \tag{3.37}$$

with $u = \phi - \phi_{app}$. We therefore deduce from Proposition B.4 in Appendix B that

$$|\overline{V} - \overline{V}_{app}|_{H^s} \leq M(s+1)(\|\Lambda^s\nabla^\gamma u\|_2 + \|\Lambda^s\partial_z u\|_2),$$

and since $\|\Lambda^s\nabla^\gamma u\|_2 \leq \|\Lambda^{s+1}\partial_z u\|_2$ by the Poincaré inequality (2.8), we get

$$|\overline{V} - \overline{V}_{app}|_{H^s} \leq M(s+1)\|\Lambda^{s+1}\nabla^{\mu,\gamma}u\|_2. \tag{3.38}$$

In order to control the r.h.s. of (3.38), we need the following lemma (recall that the spaces $H^{s,k}$ were introduced in Definition 2.11).

LEMMA 3.43. *Let $h \in H^{s,0}$, $g \in H^s(\mathbb{R}^d)$ and $u \in H^{s+1,1}$ $(s \geq 0)$ solve the boundary value problem*

$$\begin{cases} \nabla^{\mu,\gamma} \cdot P(\Sigma)\nabla^{\mu,\gamma} u = h, \\ u_{|_{z=0}} = 0, \qquad \partial_{\mathbf{n}} u_{|_{z=-1}} = g. \end{cases}$$

Then one has

$$\|\Lambda^s \nabla^{\mu,\gamma} u\|_2 \leq M(s+1)\big(\|\Lambda^s h\|_2 + |g|_{H^s}\big).$$

PROOF OF LEMMA 3.43. Let us first consider the case $s = 0$. Multiplying the equation by u and integrating by parts, one gets easily (using the coercivity of $P(\Sigma)$ provided by Lemma 2.26),

$$\begin{aligned} k(\Sigma)\|\nabla^{\mu,\gamma} u\|_2^2 &\leq \int_{\mathcal{S}} hu + \int_{\mathbb{R}^d} gu_{|_{z=-1}} \\ &\leq \|h\|_2\|u\|_2 + |g|_2 |u_{|_{z=-1}}|_2. \end{aligned}$$

Since we get from the Poincaré inequality (2.8) that

$$\|u\|_2 \leq \|\partial_z u\|_2 \leq \|\nabla^{\mu,\gamma} u\|_2$$

and moreover that

$$|u_{|_{z=-1}}|_2 = \Big|\int_{-1}^{0} \partial_z u\Big|_2 \leq \|\partial_z u\|_2 \leq \|\nabla^{\mu,\gamma} u\|_2,$$

it is easy to deduce (recall that $1/k(\Sigma) \leq M$ by Remark 2.27) that

$$\|\nabla^{\mu,\gamma} u\|_2 \leq M(\|h\|_2 + |g|_2),$$

which proves the lemma in the case $s = 0$.
When $s > 0$, note that $v = \Lambda^s u$ can be decomposed into $v = v_1 + v_2$, where v_1 and v_2 solve

$$\begin{cases} \nabla^{\mu,\gamma} \cdot P(\Sigma)\nabla^{\mu,\gamma} v_1 = \Lambda^s h, \\ v_{1\,|_{z=0}} = 0, \quad \partial_{\mathbf{n}} v_{1\,|_{z=-1}} = \Lambda^s g, \end{cases} \qquad \begin{cases} \nabla^{\mu,\gamma} \cdot P(\Sigma)\nabla^{\mu,\gamma} v_2 = -\nabla^{\mu,\gamma} \cdot \mathbf{g}, \\ v_{2\,|_{z=0}} = 0, \quad \partial_{\mathbf{n}} v_{2\,|_{z=-1}} = -\mathbf{e}_z \cdot \mathbf{g}_{|_{z=-1}}, \end{cases}$$

with $\mathbf{g} = [\Lambda^s, P(\Sigma)]\nabla^{\mu,\gamma} u$. Using the case $s = 0$ to control v_1 and Lemma 2.38 to control v_2, we get

$$\begin{aligned} \|\Lambda^s \nabla^{\mu,\gamma} u\|_2 &\leq \|\nabla^{\mu,\gamma} v_1\|_2 + \|\nabla^{\mu,\gamma} v_2\|_2 \\ &\leq M(\|\Lambda^s h\|_2 + |g|_{H^s} + \|\mathbf{g}\|_2). \end{aligned}$$

Since $\|\mathbf{g}\|_2 \leq M(s+1)\|\Lambda^{s-1}\nabla^{\mu,\gamma} u\|_2$ (see Appendix B, Corollary B.16), we thus get

$$\|\Lambda^s \nabla^{\mu,\gamma} u\|_2 \leq M(s+1)(\|\Lambda^s h\|_2 + |g|_{H^s} + \|\Lambda^{s-1}\nabla^{\mu,\gamma} u\|_2),$$

and a continuous induction (together with the estimate already obtained in the case $s = 0$) yields the result. □

Using Lemma 3.43 with $u = \phi - \phi_{app}$ and $h = -\mu^{n+1} R_\mu$, $g = -\mu^{n+1} r_\mu$, we deduce from (3.38) and Lemma 3.42 that

$$|\overline{V} - \overline{V}_{app}|_{H^s} \leq \mu^{n+1} M(s+2+2n)|\nabla^\gamma \psi|_{H^{s+2+2n}}.$$

The proposition thus follows from the explicit computation of $\overline{V}_0$ and $\overline{V}_1$ using the explicit expressions of ϕ_0 and ϕ_1 given in Lemma 3.42). □

3.6.2. Asymptotic expansion for small amplitude waves ($\varepsilon \ll 1$). The expansions of the Dirichlet-Neumann operator for small amplitude waves have been developed in [**111, 112**]. This method is very efficient to compute the formal expansion in terms of the Dirichlet-Neumann $\mathcal{G}_{\mu,\gamma}[0,0]$ of the undisturbed domain. Instead of adapting it in the present case to give uniform estimates on the truncation error, we propose a very simple method based on Theorem 3.21. We use the notation

$$\mathcal{G}_0 = \mathcal{G}_{\mu,\gamma}[0,0] = \sqrt{\mu}|D^\gamma|\tanh(\sqrt{\mu}|D^\gamma|\cdot) \tag{3.39}$$

and consider the case here of flat bottoms ($b = 0$). As expected, this quantity is the first-order approximation of the Dirichlet-Neumann operator for small amplitude waves. For the next order corrections, we need to define

$$\mathcal{G}_1 = -\mathcal{G}_0(\zeta(\mathcal{G}_0\cdot)) - \mu\nabla^\gamma\cdot(\zeta\nabla^\gamma\cdot) \tag{3.40}$$

$$\mathcal{G}_2 = \mathcal{G}_0(\zeta\mathcal{G}_0(\zeta\mathcal{G}_0\cdot)) + \mu\frac{1}{2}\Delta^\gamma(\zeta^2\mathcal{G}_0\cdot) + \mu\frac{1}{2}\mathcal{G}_0(\zeta^2\Delta^\gamma\cdot). \tag{3.41}$$

(Note that higher order expansion can easily be deduced with the same method.) The dependence on $k = 0,1$ of the estimates presented below should not be surprising: it is another manifestation of the fact that "an extra $\sqrt{\mu}$ costs half a derivative" (see the discussion at the beginning of §3.2).

PROPOSITION 3.44. *Let $t_0 > d/2$, $s \geq 0$ and $k = 0,1$. Let $b = 0$ and $\zeta \in H^{s+(k+1)/2} \cap H^{t_0+2}(\mathbb{R}^d)$ be such that (3.2) is satisfied and $\psi \in \dot{H}^{s+k/2}(\mathbb{R}^d)$. For $n = 0,1,2$, we get*

$$\big|\mathcal{G}\psi - \sum_{j=0}^{n}\varepsilon^j\mathcal{G}_j\psi\big|_{H^{s-1/2}} \leq \varepsilon^{n+1}\mu^{\frac{3+k}{4}}M(s+(k+1)/2)|\mathfrak{P}\psi|_{H^{s+k/2}},$$

where $M(s+(k+1)/2)$ is as defined in (3.4), while $\mathcal{G}_0$, $\mathcal{G}_1$ and $\mathcal{G}_2$ are given in (3.39), (3.40), and (3.41).

REMARK 3.45. We refer to §3.7.4 below for an extension to nonflat bottoms.

REMARK 3.46. We do not consider in these notes the case $\mu \to \infty$ (we recall that we always assume $\mu \leq \mu_{max}$). However, the above expansion remains true as $\mu \to \infty$, provided that ε is sufficiently small. More precisely, it is possible to give a control of the error in terms of the steepness $\epsilon = \varepsilon\sqrt{\mu}$ (see [**12**]). See Proposition 3.52 below for the version of this result in infinite depth.

REMARK 3.47. A look at the proof shows that the dependence on ζ in the right-hand-side is actually a dependence on $\varepsilon\zeta$; that is, one could replace $|\zeta|_{H^s}$ by $|\varepsilon\zeta|_{H^s}$ in the definition (3.4) of $M(s)$ for this proposition.

PROOF. We prove only the case $n = 2$; the modifications to get the cases $n = 0,1$ are straightforward.
A third-order Taylor expansion of $\mathcal{G}\psi$ gives

$$\begin{aligned}\mathcal{G}\psi &= \mathcal{G}_0\psi + d\mathcal{G}_{\mu,\gamma}[0,0](\zeta)\psi + \frac{1}{2}d^2\mathcal{G}_{\mu,\gamma}[0,0](\zeta,\zeta)\psi \\ &\quad + \int_0^1 \frac{1}{2}(1-z)^2 d^3\mathcal{G}_{\mu,\gamma}[\varepsilon z\zeta,0](\zeta,\zeta,\zeta)\psi dz.\end{aligned}$$

Using Theorem 3.21, we compute

$$d\mathcal{G}_{\mu,\gamma}[0,0](\zeta)\psi\cdot\zeta = -\varepsilon\mathcal{G}_0\big(\zeta(\mathcal{G}_0\psi)\big) - \varepsilon\mu\nabla^\gamma\cdot(\zeta\nabla^\gamma\psi),$$

while for the explicit expression of $d^2\mathcal{G}_{\mu,\gamma}[0,0](\zeta,\zeta)\psi$ we use (3.19). Moreover, for all $z\in[-1,0]$, the first and second points of Proposition 3.28 show that

$$|d^3\mathcal{G}_{\mu,\gamma}[\varepsilon z\zeta,0](\zeta,\zeta,\zeta)\psi|_{H^{s-1/2}}\leq \varepsilon^3\mu^{\frac{3+k}{4}}M(s+(k+1)/2)|\mathfrak{P}\psi|_{H^{s+k/2}},$$

with $k=0,1$, and the result follows easily. □

3.7. Supplementary remarks

3.7.1. Nonasymptotically flat bottom and surface parametrizations. The results of this chapter can be extended to the framework of §2.5.2, where the fluid domain is not necessarily asymptotically flat, namely,

$$\Omega=\{(X,z)\in\mathbb{R}^d\times\mathbb{R},-1+\beta\big(\underline{b}(X)+b(X)\big)<z<\varepsilon(\underline{\zeta}(X)+\zeta(X))\},\tag{3.42}$$

where $\underline{b}$ and $\underline{\zeta}$ belong to $C_b^\infty(\mathbb{R}^d)$ and satisfy

$$\exists h_{min}>0,\quad \forall X\in\mathbb{R}^d,\quad 1+\varepsilon(\zeta(X)+\underline{\zeta}(X))-\beta(b(X)+\underline{b}(X))\geq h_{min}.\tag{3.43}$$

Using the results proved in §2.5.2, and in particular Corollary 2.46, we can, for instance, generalize Theorem 3.15 as in the theorem below; we recall the notation

$$\underline{M}=C(\frac{1}{h_{min}},\mu_{max},|\zeta|_{H^{t_0+2}},|b|_{H^{t_0+2}},\underline{\zeta},\underline{b}),$$
$$\underline{M}(s)=C(\frac{1}{h_{min}},\mu_{max},|\zeta|_{H^{s\vee(t_0+1)}},|b|_{H^{s\vee(t_0+1)}},\underline{\zeta},\underline{b}).$$

(As for Corollary 2.46, it is possible to be more precise about the dependence on $\underline{\zeta}$ and $\underline{b}$ of these constants).

THEOREM 3.48. *Let $t_0>d/2$ and $s\geq 0$. Let $\zeta,b\in H^{s+1/2}\cap H^{t_0+1}(\mathbb{R}^d)$ and $\underline{\zeta},\underline{b}\in C_b^\infty(\mathbb{R}^d)$ be such that (3.43) is satisfied.*

(1) *The operator $\mathcal{G}=\mathcal{G}_{\mu,\gamma}[\varepsilon(\zeta+\underline{\zeta}),\beta(b+\underline{b})]$ maps continuously $\dot{H}^{s+1/2}(\mathbb{R}^d)$ into $H^{s-1/2}(\mathbb{R}^d)$ and one has*

$$\begin{aligned}\forall 0\leq s\leq t_0+3/2,\qquad & |\mathcal{G}\psi|_{H^{s-1/2}}\leq\mu^{3/4}\underline{M}(s+1/2)|\mathfrak{P}\psi|_{H^s}\\ \forall t_0+3/2\leq s,\qquad & |\mathcal{G}\psi|_{H^{s-1/2}}\leq\mu^{3/4}\underline{M}\big(|\mathfrak{P}\psi|_{H^s}\\ & \qquad +|(\varepsilon\zeta,\beta b)|_{H^{s+1/2}}|\mathfrak{P}\psi|_{H^{t_0+3/2}}\big).\end{aligned}$$

(2) *The operator $\mathcal{G}=\mathcal{G}_{\mu,\gamma}[\varepsilon(\zeta+\underline{\zeta}),\beta(b+\underline{b})]$ maps continuously $\dot{H}^{s+1}(\mathbb{R}^d)$ into $H^{s-1/2}(\mathbb{R}^d)$ and one has*

$$\begin{aligned}\forall 0\leq s\leq t_0+1,\qquad & |\mathcal{G}\psi|_{H^{s-1/2}}\leq\mu\underline{M}(s+1)|\mathfrak{P}\psi|_{H^{s+1/2}}\\ \forall t_0+1\leq s,\qquad & |\mathcal{G}\psi|_{H^{s-1/2}}\leq\mu\underline{M}\big(|\mathfrak{P}\psi|_{H^{s+1/2}}\\ & \qquad +|(\varepsilon\zeta,\beta b)|_{H^{s+1}}|\mathfrak{P}\psi|_{H^{t_0+3/2}}\big).\end{aligned}$$

The other results in this chapter can be adapted to the present framework in the same way and without any particular difficulty.

3.7.2. Rough bottoms. *For the sake of clarity, we set $\gamma=1$ here.* While the results of **[4]** allow one to define the Dirichlet-Neumann operator even if very wild bottoms are considered (provided that the total water depth exceeds some nonnegative constant), they do not provide uniform estimates in the shallow water regime ($\mu\ll 1$). There are very few rigorous, and only partial, results in the two kinds of "rough" topographies discussed in §1.7 and §2.5.3.

As previously explained, "rough" topographies do not satisfy the standard shallow water assumptions (the typical horizontal scale for the bottom is not large compared to the depth), and the shallow water expansions of §3.6.1 cannot be used. Let us, for instance, consider a bottom Γ^- given by

$$\Gamma_- = \{z = -1 + \beta b(X/\alpha)\}.$$

When $\beta = \sqrt{\mu}$ (medium amplitude topography) and $\varepsilon = 1$ (large amplitude waves), and if $\alpha = 1$, one can apply Proposition 3.37 and Remark 3.39 to get

$$\mathcal{G}_\mu[\zeta, \beta b]\psi \sim -\mu\nabla\cdot\big((1+\zeta-\beta b)\nabla\psi\big) + O(\mu^2), \tag{3.44}$$

where the residual terms involve, among other terms, derivatives of the bottom parametrization b.

If we now consider the "rough" situation corresponding to $\alpha = \sqrt{\mu}$, then each derivative of b in the residual costs $1/\alpha = 1/\sqrt{\mu}$, and there is no reason to believe that the above asymptotic expansion remains valid. It is proved in [**107**] that it must indeed be replaced[3] by

$$\mathcal{G}_\mu[\zeta, \beta b]\psi \sim -\mu\nabla\cdot(h_0\nabla\psi) - \mu\nabla\psi\cdot\nabla_Y \mathrm{sech}(h_0(X)|D_Y|)b_{|_{Y=X/\alpha}} + O(\mu^{1+3/8}), \tag{3.45}$$

where b is considered as a function of the fast variable Y and h_0 is the "homogenized" water depth, $h_0 = 1+\zeta$. Quite obviously, this formula does not correspond to (3.44) with b replaced by $b(\cdot/\alpha)$; the conclusion is that *the shallow water and homogenization limits do not commute.*

The above regime is only a particular example; many other regimes of physical interest remain to be investigated. Moreover, the asymptotic expansion (3.45) holds only in L^2-norm while the well-posedness theory for the water waves equations developed in the next chapter requires much more. It is of course possible to consider homogenization issues for the Dirichlet-Neumann operator independently of the shallow water limit, as in [**79**] for instance.

3.7.3. Infinite depth. *As in* §2.5.4, *we take* $\gamma = 1$ *here for the sake of clarity.* We show here how to adapt the results of this chapter to the infinite depth case for which we recall that the relevant potential equation is

$$\begin{cases} \Delta_{X,z}\Phi = 0 & \text{in } \Omega, \\ \Phi_{|_{z=\epsilon\zeta}} = \psi, \end{cases} \tag{3.46}$$

where the fluid domain is now given by

$$\Omega = \{(X,z) \in \mathbb{R}^d \times \mathbb{R}, \quad z < \epsilon\zeta(X)\},$$

and the *steepness* of the waves ϵ satisfies $0 < \epsilon < \epsilon_{max}$. For regular enough solutions of (3.46), the Dirichlet-Neumann operator in infinite depth is defined as

$$\mathcal{G}[\epsilon\zeta]\psi = \sqrt{1+\epsilon^2|\nabla\zeta|^2}\partial_{\mathbf{n}}\Phi_{|_{z=\epsilon\zeta}}.$$

As shown in this chapter for the case of finite depth, it is convenient to give a variational definition of $\mathcal{G}[\epsilon\zeta]\psi$. Using an admissible diffeomorphism Σ, as in Example 2.48, we instead work with the transformed potential $\phi = \Phi\circ\Sigma$, solving

$$\begin{cases} \nabla_{X,z}\cdot P(\Sigma)\nabla_{X,z}\phi = 0 & \text{in} \quad \mathbb{R}^{d+1}_-, \\ \phi_{|_{z=0}} = \psi, \end{cases} \tag{3.47}$$

where $P(\Sigma)$ is given by (2.32). For all $\psi \in H^{1/2}(\mathbb{R}^d)$, we explained in §2.5.4 how to construct a variational solution $\phi \in \dot{H}^1(\mathbb{R}^{d+1}_-)$ to (3.47). As in the finite depth

---

[3] A more general formula allowing for ζ and ψ to also depend on the fast variable $Y = X/\alpha$ is provided in [**107**].

case, we denote by $\psi^{\mathfrak{h}}$ this solution[4].
As we did in §3.1 for the case of finite depth, we define, for all $\zeta \in H^{t_0+1}(\mathbb{R}^d)$ ($t_0 > d/2$), the Dirichlet-Neumann operator in infinite depth $\mathcal{G}[\epsilon\zeta]$ as the linear mapping

$$\mathcal{G}[\epsilon\zeta] : \begin{array}{ccc} H^{1/2}(\mathbb{R}^d) & \to & H^{-1/2}(\mathbb{R}^d), \\ \psi & \mapsto & \mathcal{G}[\epsilon\zeta]\psi, \end{array} \tag{3.48}$$

with $\mathcal{G}[\epsilon\zeta]\psi$ defined by duality,

$$\forall \varphi \in H^{1/2}(\mathbb{R}^d), \qquad \langle \mathcal{G}[\epsilon\zeta]\psi, \varphi \rangle_{H^{-1/2}\times H^{1/2}} = \int_{\mathbb{R}^{d+1}_-} \nabla_{X,z}\psi^{\mathfrak{h}} \cdot P(\Sigma)\nabla_{X,z}\varphi^{\mathfrak{h}}.$$

The basic properties of Proposition 3.9 can be straightforwardly adapted. For the results of this chapter involving the operator $\mathfrak{P}$ one must carry out the same modifications as in §2.5.4, basically replacing $\mathfrak{P}$ by $|D|^{1/2}$ and setting μ to 1. For instance, the inequalities stated in Proposition 3.12 now become

$$(\psi, \mathcal{G}[\epsilon\zeta]\psi) \le \tilde{M}_0 \big||D|^{1/2}\psi\big|_2^2 \quad \text{and} \quad \big||D|^{1/2}\psi\big|_2^2 \le \tilde{M}_0(\psi, \mathcal{G}[\epsilon\zeta]\psi),$$

where $\tilde{M}_0 = C(\epsilon_{max}, |\zeta|_{H^{t_0+1}})$.
The higher order estimates of §3.2 can be adapted in a similar way. For instance, the infinite depth version of Theorem 3.15 is given below.

THEOREM 3.49. *Let $t_0 > d/2$, $s \ge 0$ and $0 < \epsilon < \epsilon_{max}$. Let also $\zeta \in H^{s+1/2} \cap H^{t_0+1}(\mathbb{R}^d)$.*
The operator $\mathcal{G}[\epsilon\zeta]$ maps continuously $H^{s+1/2}(\mathbb{R}^d)$ into $H^{s-1/2}(\mathbb{R}^d)$ and one has

$$\begin{array}{ll} \forall 0 \le s \le t_0 + 3/2, & |\mathcal{G}[\epsilon\zeta]\psi|_{H^{s-1/2}} \le \tilde{M}(s+1/2)\big||D|^{1/2}\psi\big|_{H^s} \\ \forall t_0 + 3/2 \le s, & |\mathcal{G}[\epsilon\zeta]\psi|_{H^{s-1/2}} \le \tilde{M}\big(\big||D|^{1/2}\psi\big|_{H^s} \\ & \qquad + |\epsilon\zeta|_{H^{s+1/2}}\big||D|^{1/2}\psi\big|_{H^{t_0+3/2}}\big), \end{array}$$

where we recall that $\tilde{M}(s)$ and $\tilde{M}$ are given by

$$\tilde{M} = C(\epsilon_{max}, |\zeta|_{H^{t_0+2}}), \qquad M(s) = C(\epsilon_{max}, |\zeta|_{H^{s\vee(t_0+1)}}).$$

REMARK 3.50.
(1) In finite depth, the Dirichlet-Neumann operator is constructed on the Beppo-Lévi space $\dot{H}^{1/2}(\mathbb{R}^d)/\mathbb{R}$ and not only on $H^{1/2}(\mathbb{R}^d)$. As we discussed in §2.5.4, a natural placeholder for the Beppo-Lévi space $\dot{H}^{1/2}(\mathbb{R}^d)$ in infinite depth is the space $\mathring{H}^{1/2}(\mathbb{R}^d)$ defined in (2.34). However, this is not a very convenient space to work with (especially in dimension $d = 1$) and the construction by duality of the Dirichlet-Neumann operator looks delicate in this space, while it is trivial if we restrict ourselves to $H^{1/2}(\mathbb{R}^d)$. This is why we stick to this framework here.
(2) As noted in Remark 3.7, one can use a direct definition (rather than a definition by duality) of the Dirichlet-Neumann operator, namely,

$$\mathcal{G}[\epsilon\zeta]\psi = \mathbf{e}_z \cdot P(\Sigma)\nabla\phi_{|_{z=0}},$$

which makes sense in $H^{s-1/2}(\mathbb{R}^d)$ as soon as $s \ge 0$ and $s \ge -t_0 + 1$. Using this direct definition, it is possible to define $\mathcal{G}[\epsilon\zeta]$ on $\mathring{H}^{s+1/2}$ for s large enough,

$$\forall s \ge (-t_0+1) \vee 0, \qquad \mathcal{G}[\epsilon\zeta] : \mathring{H}^{s+1/2}(\mathbb{R}^d) \to H^{s-1/2}(\mathbb{R}^d).$$

[4] As shown in §2.5.4, one has $\|\nabla_{X,z}\psi^{\mathfrak{h}}\|_2 \le \text{Cst}\,\big||D|^{1/2}\psi\big|_2$.

The shape derivative formula of Theorem 3.21 can also be adapted[5] to the infinite depth case: for all $\zeta \in H^{t_0+1}(\mathbb{R}^d)$, $\psi \in \mathring{H}^{3/2}(\mathbb{R}^d)$, and all $h \in H^{t_0+1}(\mathbb{R}^d)$, one has

$$d_\zeta \mathcal{G}[\epsilon\cdot](h)\psi = -\epsilon \mathcal{G}[\epsilon\zeta](h\underline{w}) - \epsilon\nabla\cdot(h\underline{V}),$$

with

$$\underline{w} = \frac{\mathcal{G}[\epsilon\zeta]\psi + \epsilon\nabla\zeta\cdot\nabla\psi}{1+\epsilon^2|\nabla\zeta|^2} \quad \text{and} \quad \underline{V} = \nabla\psi - \epsilon\underline{w}\nabla\zeta.$$

The regularity estimates on shape derivatives of Proposition 3.28 can also be adapted in infinite depth. Without any difficulty, one gets the following proposition.

PROPOSITION 3.51. *Let $t_0 > d/2$ and $\zeta \in H^{t_0+1}(\mathbb{R}^{d+1})$. Then, with $\tilde{M}_0 = C(\epsilon_{max}, |\zeta|_{H^{t_0+1}})$,*
(1) *For all $0 \le s \le t_0 + 1/2$ and $\psi \in H^{s+1/2}(\mathbb{R}^d)$,*

$$|d^j\mathcal{G}(\mathbf{h},\mathbf{k})\psi|_{H^{s-1/2}} \le \epsilon^j \tilde{M}_0 \prod_{m=1}^{j} |h_m|_{H^{t_0+1}} \big||D|^{1/2}\psi\big|_{H^s}.$$

(2) *For all $0 \le s \le t_0$ and $\psi \in H^{t_0+1}(\mathbb{R}^d)$,*

$$|d^j\mathcal{G}(\mathbf{h},\mathbf{k})\psi|_{H^{s-1/2}} \le \epsilon^j \tilde{M}_0 |h_1|_{H^{s+1/2}} \prod_{m>1}^{j} |h_m|_{H^{t_0+1}} \big||D|^{1/2}\psi\big|_{H^{t_0+1/2}}.$$

The results of Propositions 3.29, 3.30, and 3.32 can also be established in infinite depth (replacing $\mathfrak{P}$ by $|D|^{1/2}$, μ by 1, and ε by ϵ) in the same way and we therefore do not need to provide more details on this point. Let us conclude this section with the infinite depth version of Proposition 3.44 on small amplitude asymptotic expansions[6] of the Dirichlet-Neumann operator. As in the finite depth case, the proof relies on a simple Taylor expansion together with the estimates on shape derivatives of Proposition 3.51.

PROPOSITION 3.52. *Let $t_0 > d/2$, $s \ge 0$, and $0 < \epsilon < \epsilon_{max}$. Let $\zeta \in H^{s+1/2} \cap H^{t_0+2}(\mathbb{R}^d)$ and $\psi \in H^s(\mathbb{R}^d)$. For $n = 0, 1, 2$, we get*

$$\Big|\mathcal{G}[\epsilon\zeta]\psi - \sum_{j=0}^{n} \epsilon^j \mathcal{G}_j\psi\Big|_{H^{s-1/2}} \le \varepsilon^{n+1}\tilde{M}(s+1/2)\big||D|^{1/2}\psi\big|_{H^{s+k/2}},$$

where $\tilde{M}(s+1/2)$ is as defined in Theorem 3.49 while $\mathcal{G}_0$, $\mathcal{G}_1$ and $\mathcal{G}_2$ are given by

$$\mathcal{G}_0 = |D|, \mathcal{G}_1 = -|D|\big(\zeta(|D|\cdot)\big) - \nabla\cdot(\zeta\nabla\cdot),$$
$$\mathcal{G}_2 = |D|\big(\zeta|D|(\zeta|D|\cdot)\big) + \frac{1}{2}\Delta(\zeta^2|D|\cdot) + \frac{1}{2}|D|(\zeta^2\Delta\cdot).$$

---

[5]The proof is almost the same as the proof for the case of finite depth. One just has to take into account the change in the formula for the diffeomorphism in the definition (3.21) of Alinhac's good unknown. With the most simple choice $\Sigma_3(X,z) = z + \epsilon\zeta$, this yields

$$w = v - \epsilon h\partial_z v.$$

[6]Shallow water expansions are, of course, irrrelevant in infinite depth!

3.7.4. Small amplitude expansions for nonflat bottoms. The analyticity of the Dirichlet-Neumann operator (the first point of Theorem 3.21) is proved in Theorem A.11 in Appendix A.
It is also possible to find an explicit expression for the shape derivative of the Dirichlet-Neumann operator with respect to the bottom parametrization b that complements the shape derivative with respect to ζ given in Theorem 3.21; it is also possible to generalize these formulas to the case of a nonhomogeneous Neumann condition at the bottom; these generalizations are addressed in Appendix A, §A.5.
When $\zeta = b = 0$, the formula for $d\mathcal{G}_{\mu,\gamma}[\varepsilon\zeta, \beta b](0,k)\psi$ simplifies into (see §3.3 for the notations)

$$d\mathcal{G}_{\mu,\gamma}[0,0](0,k)\psi = \mu\beta\mathrm{sech}(\sqrt{\mu}|D^\gamma|)\nabla^\gamma \cdot [k(\mathrm{sech}(\sqrt{\mu}|D^\gamma|)\nabla^\gamma\psi)],$$

which has been used in [**223**], for instance. This formula makes it possible to generalize the expansion formula of Proposition 3.44 to the case of nonflat bottoms,

$$\begin{aligned}\mathcal{G}_{\mu,\gamma}[\varepsilon\zeta,\beta b] \quad &\sim \quad \mathcal{G}_0\psi - \varepsilon\mu\mathcal{G}_1\psi \\ &+\mu\beta\mathrm{sech}(\sqrt{\mu}|D^\gamma|)\nabla^\gamma \cdot [b(\mathrm{sech}(\sqrt{\mu}|D^\gamma|)\nabla^\gamma\psi)].\end{aligned} \tag{3.49}$$

More generally, [**105**] shows how to find explicit expressions for the shape derivative of any order (and with respect to the surface and bottom parametrizations) of $\mathcal{G}$ around $\zeta = 0$ and $b = 0$.

3.7.5. Self-adjointness. Proposition 3.9 shows that $\mathcal{G}$ is symmetric on $H^{3/2}$ for the L^2-scalar product; we show in Proposition A.14 of Appendix A that it admits a self-adjoint realization on $L^2(\mathbb{R}^d)$ with domain $H^1(\mathbb{R}^d)$.

3.7.6. Invertibility. Inverting the Dirichlet-Neumann operator, that is, solving the equation $\mathcal{G}\psi = f$ for ψ, is not always possible. This issue is discussed in Appendix A, §A.3.

3.7.7. Symbolic analysis. We refer to Appendix A, §A.4 for some remarks concerning the symbolic description of the Dirichlet-Neumann operator.

3.7.8. The Neumann-Neumann, Dirichlet-Dirichlet, and Neumann-Dirichlet operators. Several operators related to the Dirichlet-Neumann operator can be defined (we saw in the introduction to §1.6 the relevance of the Neumann-Neumann operator when the bottom is moving). See Appendix A, §A.5 for more details.

CHAPTER 4

Well-posedness of the Water Waves Equations

In this chapter we show the well-posedness of the initial value problem associated to the water waves equations[1]

$$
(4.1)\qquad \begin{cases} \partial_t\zeta - \dfrac{1}{\mu}\mathcal{G}\psi = 0, \\ \partial_t\psi + \zeta + \dfrac{\varepsilon}{2}|\nabla^\gamma\psi|^2 - \dfrac{\varepsilon}{\mu}\dfrac{(\mathcal{G}\psi + \varepsilon\mu\nabla^\gamma\zeta\cdot\nabla^\gamma\psi)^2}{2(1+\varepsilon^2\mu|\nabla^\gamma\zeta|^2)} = 0, \end{cases}
$$

where $\mathcal{G} = \mathcal{G}_{\mu,\gamma}[\varepsilon\zeta, \beta b]$ is as defined in the previous chapter. In view of the subsequent asymptotic analysis (see the following chapters), particular care is paid to get a sharp dependence on the dimensionless parameters ε, μ, β and γ of the existence time and of the size of the solution.

In §4.1, we propose a brief analysis of the linear case that we use to motivate the definition of the energy norm used in the existence proof. By differentiating the equations (4.1), we derive in §4.2 a "quasilinear" system of equations. We then state and prove in §4.3 local existence and stability theorems for the water waves equations (4.1), which give a sharp account of the dependence of the solution on the parameters ε, μ, β and γ. We also relate a technical hyperbolicity condition to the Rayleigh-Taylor stability criterion, stating that the inwards normal derivative of the pressure must be nonnegative at the surface; we then comment on the validity of this condition in the configurations of interest here.

Supplementary results are then provided in §4.4, where we indicate how to handle nonasympotically flat bottoms and surfaces as well as the case of infinite depth. We also comment on other extensions.

N.B. A generalization of the present results in the presence of surface tension is given in Chapter 9.

As in the previous chapters, we always assume that the dimensionless parameters ε, μ, β, γ satisfy

$$
(4.2)\qquad 0 \le \varepsilon, \beta, \gamma \le 1, \quad 0 \le \mu \le \mu_{max},
$$

for some $\mu_{max} > 0$ (not necessarily small) and that $\zeta, b \in H^{t_0+1}(\mathbb{R}^d)$ ($t_0 > d/2$) are such that

$$
(4.3)\qquad \exists h_{min} > 0, \quad \forall X \in \mathbb{R}^d, \quad 1 + \varepsilon\zeta(X) - \beta b(X) \ge h_{min}.
$$

[1] Equation (4.1) uses the "shallow water" nondimensionalization (i.e., we set $\nu = 1$ in (1.28)); we naturally use the "deep water" nondimensionalization when we address the infinite depth case in §4.4.3. See §1.3.3 for comments on the different possible ways to nondimensionalize the equations.

We also denote by M_0 and M any constants of the form

$$M_0 = C\big(\frac{1}{h_{min}}, \mu_{max}, |\zeta|_{H^{t_0+1}}, |b|_{H^{t_0+1}}\big), \qquad (4.4)$$
$$M = C\big(\frac{1}{h_{min}}, \mu_{max}, |\zeta|_{H^{t_0+2}}, |b|_{H^{t_0+2}}\big),$$

and, for all $s \geq 0$, we also define $M(s)$ as

$$M(s) = C\big(M_0, |\zeta|_{H^s}, |b|_{H^s}\big). \qquad (4.5)$$

4.1. Linearization around the rest state and energy norm

In order to motivate the introduction of the energy norm (see (4.8) below), it is instructive to look at the linearization around the rest state $\zeta = 0$, $\psi = 0$. These linearized equations take the form

$$\partial_t U + \mathcal{A}_0 U = 0, \qquad \text{with} \quad \mathcal{A}_0 = \begin{pmatrix} 0 & -\frac{1}{\mu}\mathcal{G}_{\mu,\gamma}[0,\beta b] \\ 1 & 0 \end{pmatrix}, \qquad (4.6)$$

and $U = (\zeta, \psi)^T$.
This system of evolution equations can be made symmetric if we multiply it by the symmetrizer $\mathcal{S}_0$ defined as

$$\mathcal{S}_0 = \begin{pmatrix} 1 & 0 \\ 0 & \frac{1}{\mu}\mathcal{G}_{\mu,\gamma}[0,\beta b] \end{pmatrix}. \qquad (4.7)$$

This suggests that a natural energy for the linearized equations (4.6) is given by

$$(\mathcal{S}_0 U, U) = |\zeta|_2^2 + \big(\frac{1}{\mu}\mathcal{G}_{\mu,\gamma}[0,\beta b]\psi, \psi\big).$$

From Proposition 3.12, we know that the second term in the r.h.s. is uniformly equivalent to $|\mathfrak{P}\psi|_2^2$. It is therefore natural to define the energy $\mathcal{E}^0(U)$ as

$$\mathcal{E}^0(U) = |\zeta|_2^2 + |\mathfrak{P}\psi|_2^2.$$

Higher energy norms for the linearized equations (4.6), are given, for all $N \in \mathbb{N}$, by

$$"\mathcal{E}^N(U)" = \sum_{\alpha\in\mathbb{N}^d, |\alpha|\leq N} |\partial^\alpha \zeta|_2^2 + |\mathfrak{P}\partial^\alpha\psi|_2^2.$$

This quantity, however, is not the best choice to make if one is interested in the study of the full nonlinear equations (4.1). This becomes quite intuitive if we differentiate N times the first equation of (4.1); indeed, Corollary 3.26 shows that for all $\alpha \in \mathbb{N}^d$, $|\alpha| = N$, one has

$$\partial_t \partial^\alpha \zeta + \varepsilon \nabla^\gamma \cdot (\underline{V}\partial^\alpha\zeta) - \frac{1}{\mu}\mathcal{G}(\partial^\alpha\psi - \varepsilon\underline{w}\partial^\alpha\zeta) = \text{lower order terms},$$

with $\underline{w}$ as in the statement of Theorem 3.21 (we refer to Proposition 4.5 below for a more precise statement). This equation has the same structure as the first equation of (4.1), with ζ replaced by $\partial^\alpha\zeta$ and ψ replaced by $\partial^\alpha\psi - \varepsilon\underline{w}\partial^\alpha\zeta$. Therefore, this motivates the following choice for the energy of order $N \geq 1$

$$\mathcal{E}^N(U) = |\mathfrak{P}\psi|_{H^{t_0+3/2}}^2 + \sum_{\alpha\in\mathbb{N}^d, |\alpha|\leq N} |\zeta_{(\alpha)}|_2^2 + |\mathfrak{P}\psi_{(\alpha)}|_2^2, \qquad (4.8)$$

with

$$\zeta_{(\alpha)} = \partial^\alpha\zeta, \qquad \psi_{(\alpha)} = \partial^\alpha\psi - \varepsilon\underline{w}\partial^\alpha\zeta. \qquad (4.9)$$

REMARK 4.1. When $|\alpha|=1$, that is, when $\alpha=\mathbf{e}^j$ $(1\le j\le d)$, then $\psi_{(\mathbf{e}^j)}=\underline{V}_j$, the j-th component of the horizontal velocity evaluated at the interface.

REMARK 4.2. The presence of $|\mathfrak{P}\psi|_{H^{t_0+3/2}}$ in the definition (4.8) is technical. It is necessary to provide a control of ψ and its derivatives in terms of $\zeta_{(\alpha)}$ and $\psi_{(\alpha)}$. See, for instance, the proof of the first point of Lemma 4.6 below.

4.2. Quasilinearization of the water waves equations

4.2.1. Notations and preliminary results. For all $T\ge 0$, we denote by E_T^N the functional space of time-dependent functions associated to the energy (4.8)

$$E_T^N=\{U\in C([0,T];H^{t_0+2}\times \dot{H}^2(\mathbb{R}^d)),\quad \mathcal{E}^N(U(\cdot))\in L^\infty([0,T])\}. \tag{4.10}$$

We also denote by $\mathfrak{m}^N(U)$ any constant of the form

$$\mathfrak{m}^N(U)=C\big(M,|b|_{H^{N+t_0\vee 1+1}},\mathcal{E}^N(U)\big), \tag{4.11}$$

with M as in (4.4).

REMARK 4.3. The dependence on b in $\mathfrak{m}^N(U)$ (and therefore in the estimates derived later in this chapter) is certainly stronger than necessary; since the derivation of asymptotic models (see the following chapters) requires a lot of smoothness on b, and because b is fixed here and does not play any important role in the well-posedness theory of the water waves equations, it is reasonable, however, to work with such a rough precision. See §4.4 for more comments on this point.

We also need to introduce the mappings $\mathcal{w}[\varepsilon\zeta,\beta b]$ and $\mathcal{V}[\varepsilon\zeta,\beta b]$ defined for all $s\ge 0$ and $\zeta\in H^{s+1/2}\cap H^{t_0+2}(\mathbb{R}^d)$ satisfying (4.3) by

$$\mathcal{w}[\varepsilon\zeta,\beta b]:\begin{array}{ccc}\dot{H}^{s+1/2}(\mathbb{R}^d) & \to & H^{s-1/2}(\mathbb{R}^d)\\ \psi & \mapsto & \dfrac{\mathcal{G}\psi+\varepsilon\mu\nabla^\gamma\zeta\cdot\nabla^\gamma\psi}{1+\varepsilon^2\mu|\nabla^\gamma\zeta|^2},\end{array} \tag{4.12}$$

and

$$\mathcal{V}[\varepsilon\zeta,\beta b]:\begin{array}{ccc}\dot{H}^{s+1/2}(\mathbb{R}^d) & \to & H^{s-1/2}(\mathbb{R}^d)^d\\ \psi & \mapsto & \nabla^\gamma\psi-\varepsilon(\mathcal{w}[\varepsilon\zeta]\psi)\nabla^\gamma\zeta,\end{array} \tag{4.13}$$

so that $\underline{w}=\mathcal{w}[\varepsilon\zeta,\beta b]\psi$ and $\underline{V}=\mathcal{V}[\varepsilon\zeta,\beta b]\psi$. It is straightforward, using the product estimate of Proposition B.2 in Appendix B, to get the following proposition.

PROPOSITION 4.4. *It is possible to replace $\mathcal{G}_{\mu,\gamma}[\varepsilon\zeta,\beta b]$ by $\mathcal{w}[\varepsilon\zeta,\beta b]$ in the statements of Theorem* 3.15 *and Proposition* 3.28.

4.2.2. A linearization formula. We can now give an important linearization formula for $\mathcal{G}$. (Note that with $d=1,2$, it is possible to take $N=5$ in the proposition.) We recall that $\varepsilon\vee\beta=\max\{\varepsilon,\beta\}$.

PROPOSITION 4.5. *Let $t_0>d/2$ and $N\in\mathbb{N}$. Moreover, assume that $N\ge t_0+t_0\vee 2+3/2$, $b\in H^{N+1\vee t_0+1}(\mathbb{R}^d)$ and $U=(\zeta,\psi)$ is such that $\mathcal{E}^N(U)<\infty$ and satisfies (4.3).*

(1) *For all $\alpha\in\mathbb{N}^d$ with $1\le|\alpha|\le N$, one has*

$$\frac{1}{\mu}\partial^\alpha(\mathcal{G}\psi)=\frac{1}{\mu}\mathcal{G}\psi_{(\alpha)}+\max\{\varepsilon,\beta\}R_\alpha,\qquad (|\alpha|\le N-1)$$

$$\frac{1}{\mu}\partial^\alpha(\mathcal{G}\psi)=\frac{1}{\mu}\mathcal{G}\psi_{(\alpha)}-\varepsilon\nabla^\gamma\cdot(\zeta_{(\alpha)}\underline{V})+\max\{\varepsilon,\beta\}R_\alpha,\qquad (|\alpha|=N),$$

with $(\zeta_{(\alpha)}, \psi_{(\alpha)})$ as in (4.9) and $\underline{V}$ as in Theorem 3.21, while R_α satisfies

$$|R_\alpha|_2 \leq \mathfrak{m}^N(U).$$

(2) *The dependence of R_α on U is Lipschitz: if R^1_α and R^2_α denote the residuals corresponding to two different U^1 and U^2 satisfying the above assumptions, then*

$$|R^1_\alpha - R^2_\alpha|_2 \leq C\big(\mathfrak{m}^N(U^1), \mathfrak{m}^N(U^2)\big)\, \mathcal{E}^{|\alpha|}(U^1 - U^2).$$

PROOF. We prove the first point of the proposition in the case of a flat bottom ($b = 0$). The general case will be addressed at the end of the proof. We also focus on the most difficult case, namely $|\alpha| = N$. (When $|\alpha| \leq N-1$, it is straightforward that $\varepsilon\nabla^\gamma \cdot (\zeta_{(\alpha)}\underline{V})$ can be put in the residual term.)
Consistent with the notations of §3.3, we denote by $d\mathcal{G}(h)\psi$ the derivative at ζ of the mapping $\zeta' \mapsto \mathcal{G}_{\mu,\gamma}[\varepsilon\zeta', \beta b]\psi$ in the direction h.
The proof requires the following lemma.

LEMMA 4.6. *Let $t_0 > d/2$ and $\zeta \in H^{t_0+2}(\mathbb{R}^d)$ be such that (4.3) is satisfied. Let $r \geq 0$ and $\delta \in \mathbb{N}^d$ be such that $r + |\delta| \leq N - 1$, with N as in the proposition. Then one has*

$$|\mathfrak{P}\psi_{(\delta)}|_{H^1} \leq \mathfrak{m}^N(U) \quad \text{and} \quad |\mathfrak{P}\partial^\delta\psi|_{H^{r+1/2}} \leq \mathfrak{m}^N(U).$$

REMARK 4.7. A straightforward adaptation of the proof of the second estimate shows that the following estimate also holds,

$$|\mathfrak{P}\psi|_{H^N} \leq \mathfrak{m}^N(U)(1 + \varepsilon|\zeta|_{H^{N+1/2}}).$$

PROOF OF THE LEMMA. For the first estimate, we note that

$$\begin{aligned}
|\mathfrak{P}\psi_{(\delta)}|_{H^1} &\leq |\mathfrak{P}\psi_{(\delta)}|_2 + \sum_{\beta\in\mathbb{N}^d, |\beta|=1} |\partial^\beta \mathfrak{P}\psi_{(\delta)}|_2 \\
&\leq |\mathfrak{P}\psi_{(\delta)}|_2 + \sum_{\beta\in\mathbb{N}^d, |\beta|=1} \big(|\mathfrak{P}\psi_{(\delta+\beta)}|_2 + \varepsilon\big|\mathfrak{P}(\partial^\beta \underline{w}\partial^\delta\zeta)\big|_2\big).
\end{aligned}$$

The first component of the r.h.s. is straightforwardly bounded from above by $\mathfrak{m}^N(U)$. For the second component, we note that $|\mathfrak{P}f|_2 \lesssim \mu^{-1/4}|f|_{H^{1/2}}$, so that

$$\big|\mathfrak{P}(\partial^\beta \underline{w}\partial^\delta\zeta)\big|_2 \lesssim \mu^{-1/4}\big|\partial^\beta \underline{w}\partial^\delta\zeta\big|_{H^{1/2}} \lesssim |\mu^{-1/4}\underline{w}|_{H^{t_0+1}}|\zeta|_{H^N},$$

where the second product estimate of Proposition B.2 in Appendix B has been used to derive the last inequality. The result then follows easily from Proposition 4.4.
For the second estimate, we use the assumptions made on r and δ to get

$$|\mathfrak{P}\partial^\delta\psi|_{H^{r+1/2}} \leq \sum_{\beta\in\mathbb{N}^d, |\beta|\leq N-1} |\mathfrak{P}\partial^\beta\psi|_{H^{1/2}}.$$

Moreover, since $\partial^\beta\psi = \psi_{(\beta)} + \varepsilon\underline{w}\partial^\beta\zeta$, we get

$$\begin{aligned}
|\mathfrak{P}\partial^\delta\psi|_{H^{r+1/2}} &\leq \sum_{\beta\in\mathbb{N}^d, |\beta|\leq N-1} \big(|\mathfrak{P}\psi_{(\beta)}|_{H^{1/2}} + \varepsilon|\mathfrak{P}(\underline{w}\partial^\beta\zeta)|_{H^{1/2}}\big) \\
&\leq \sum_{\beta\in\mathbb{N}^d, |\beta|\leq N-1} \big(|\mathfrak{P}\psi_{(\beta)}|_{H^1}\big) + \varepsilon\max\{1, \mu^{-1/4}\}|\underline{w}|_{H^{t_0+1}}|\zeta|_{H^N},
\end{aligned}$$

where we used the identity $|\mathfrak{P}f|_{H^{1/2}} \leq \max\{1, \mu^{-1/4}\}|\nabla f|_2$ to derive the second inequality.

Since $|\underline{w}|_{H^{t_0+1}} \leq \mu^{3/4}M|\mathfrak{P}\psi|_{H^{t_0+3/2}}$ (see (3.18) or Proposition 4.4), the first estimate follows easily from the definition of $\mathfrak{m}^N(U)$ and the first estimate. $\square$

Let us note first that

$$\partial^\alpha \mathcal{G}\psi = \mathcal{G}\partial^\alpha\psi + d\mathcal{G}(\partial^\alpha\zeta)\psi + \varepsilon\mu R_\alpha, \tag{4.14}$$

where $\varepsilon\mu R_\alpha$ is a sum of terms of the form

$$d^j\mathcal{G}(\partial^{\iota^1}\zeta,\dots,\partial^{\iota^j}\zeta)\partial^\delta\psi =: A_{j,\delta,\iota}, \tag{4.15}$$

where $j \in \mathbb{N}$, $\iota = (\iota^1,\dots,\iota^j) \in \mathbb{N}^{dj}$ and $\delta \in \mathbb{N}^d$ satisfy

$$\sum_{i=1}^j |\iota^i| + |\delta| = N, \qquad 0 \le |\delta| \le N-1 \quad \text{and} \quad |\iota^i| < N. \tag{4.16}$$

We therefore turn to control the $A_{j,\delta,\iota}$. Let l be such that $|\iota^l| = \max_{1\le m\le j}|\iota^m|$. Let us distinguish five cases:
(i) The case $|\delta| = j = 1$ and $\iota = \iota^1$ (and thus $|\iota| = N-1$). One then has $A_{j,\delta,\iota} = d\mathcal{G}(\partial^\iota\zeta)\partial^\delta\psi$, and we can therefore use Theorem 3.21 to get

$$\frac{1}{\mu}A_{j,\delta,\iota} = -\varepsilon\frac{1}{\sqrt{\mu}}\mathcal{G}(\partial^\iota\zeta\frac{1}{\sqrt{\mu}}\underline{w}^\delta) - \varepsilon\nabla^\gamma\cdot(\underline{V}^\delta\partial^\iota\zeta),$$

where $\underline{V}^\delta = \boldsymbol{\mathcal{V}}[\varepsilon\zeta]\partial^\delta\psi$ and $\underline{w}^\delta = \boldsymbol{w}[\varepsilon\zeta]\partial^\delta\psi$, the mappings $\boldsymbol{\mathcal{V}}[\varepsilon\zeta]$ and $\boldsymbol{w}[\varepsilon\zeta]$ are as in (4.12)–(4.13). Therefore, we get from Theorem 3.15 and Remark 3.16 that

$$\frac{1}{\mu}|A_{j,\delta,\iota}|_2 \le \varepsilon M|\nabla(\partial^\iota\zeta\frac{1}{\sqrt{\mu}}\underline{w}^\delta)|_2) + \varepsilon|\nabla\cdot(\underline{V}^\delta\partial^\iota\zeta)|_2.$$

We can now deduce with the product estimates of Proposition B.2 (Appendix B) that

$$\frac{1}{\mu}|A_{j,\delta,\iota}|_2 \le \varepsilon M|\zeta|_{H^N}\big(|\frac{1}{\sqrt{\mu}}\underline{w}^\delta|_{H^{t_0+1}} + |\underline{V}^\delta|_{H^{t_0+1}}\big),$$

from which we get, with the help of Proposition 4.4 and the observation that $|\nabla^\gamma\partial^\delta\psi|_{H^{t_0+1}} \le |\mathfrak{P}\partial^\delta\psi|_{H^{t_0+3/2}}$, that

$$\frac{1}{\mu}|A_{j,\delta,\iota}|_2 \le \varepsilon M|\zeta|_{H^N}|\mathfrak{P}\partial^\delta\psi|_{H^{t_0+3/2}}. \tag{4.17}$$

Using Lemma 4.6 (recall that $t_0+3 \le N$) we therefore get

$$|A_{j,\delta,\iota}|_2 \le \varepsilon\mu\mathfrak{m}^N(U). \tag{4.18}$$

(ii) The case $|\delta| \le N-1 \vee t_0 - 3/2$ and $|\iota^l| < N-1$. It follows from the last point of Proposition 3.28 that

$$\frac{1}{\mu}|A_{j,\delta,\iota}|_2 \le \varepsilon M|\partial^{\iota^l}\zeta|_{H^{3/2}}\prod_{m\ne l}|\partial^{\iota^m}\zeta|_{H^{1\vee t_0+3/2}}|\mathfrak{P}\partial^\delta\psi|_{H^{1\vee t_0+1/2}}.$$

From the assumption that $|\iota^l| < N-1$ and the definition of l, we also get that for all $m \ne l$, $|\iota^m| \le [(N-2)/2]$, and thus, from the assumption made on N, we have $|\iota^m| + 1 \vee t_0 + 3/2 \le N$. Using the assumption made here on δ, we deduce from Lemma 4.6 that (4.18) holds.
(iii) The case $N-1 \vee t_0 - 1/2 \le |\delta| \le N-2$. Using the second estimate of Proposition 3.28, we now get

$$|A_{j,\delta,\iota}|_2 \le M\varepsilon^j\mu\prod_{m=1}^j|\partial^{\iota^m}\zeta|_{H^{1\vee t_0+1}}|\mathfrak{P}\partial^\delta\psi|_{H^1}.$$

Since $|\delta|+2\leq N$ and $|\iota^m|+1\vee t_0+1\leq N$, one can conclude, as in the previous case, that (4.18) holds for N, as in the statement of the proposition.
(iv) The case $|\delta|=N-1$. One then has $j=1$ and

$$\begin{aligned} A_{j,\delta,\iota} &= d\mathcal{G}(\partial^\iota\zeta)\partial^\delta\psi \\ &= d\mathcal{G}(\partial^\iota\zeta)\psi_{(\delta)}+\varepsilon d\mathcal{G}(\partial^\iota\zeta)(\underline{w}\partial^\delta\zeta). \end{aligned}$$

From the second point of Proposition 3.28, we deduce that the first component of the r.h.s. satisfies

$$|d\mathcal{G}(\partial^\iota\zeta)\psi_{(\delta)}|_2\leq\varepsilon\mu M|\mathfrak{P}\psi_{(\delta)}|_{H^1}\leq\varepsilon\mu\mathfrak{m}^N(U),$$

where we used Lemma 4.6 to derive the second inequality. For the second component, we use the first point of Proposition 3.28 and Proposition 4.4. We therefore get that (4.18) also holds in this case.
(v) The case $|\delta|=0$ *and* $|\iota^l|=N-1$ *(we take for simplicity* $l=1$*).* Then one has $A_{j,\delta,\iota}=d^2\mathcal{G}(\partial^{\iota^1}\zeta,\partial^{\iota^2}\zeta)\psi$. The third point of Proposition 3.28 provides a good control in terms of regularity but not with respect to the dependence on μ. However, we can remark, after differentiating the formula stated in Theorem 3.21, that

$$\begin{aligned} d^2\mathcal{G}(h_1,h_2)\psi &= -\varepsilon\sqrt{\mu}d\mathcal{G}(h_2)(h_1\frac{1}{\sqrt{\mu}}\underline{w})-\varepsilon\sqrt{\mu}\mathcal{G}(h_1\frac{1}{\sqrt{\mu}}d\underline{w}(h_2)\psi) \\ &- \varepsilon\mu\nabla^\gamma\cdot(h_1 d\underline{V}(h_2)\psi), \end{aligned}$$

where $d\underline{w}(h_2)\psi$ and $d\underline{V}(h_2)\psi$ stand for the derivative at ζ of the mappings $\zeta\mapsto\underline{w}[\varepsilon\zeta]\psi$ and $\zeta\mapsto\underline{V}[\varepsilon\zeta]\psi$. With $h_j=\partial^{\iota^j}\zeta$, one can use the first and third points of Proposition 3.28 to control the first and second term of the r.h.s. respectively; for the third term, we use Proposition 4.4. It follows that (4.18) is satisfied.

We have thus proved that R_α satisfies the desired estimates. The fact that (4.14) coincides with the identity given in the proposition is a consequence of Theorem 3.21.
We now have to show that the first point of the proposition is still valid in the general case of nonflat bottoms. The extra terms to control are of the form[2]

$$d^{j+k}\mathcal{G}(\partial^{\iota^1}\zeta,\dots,\partial^{\iota^j}\zeta;\partial^{\kappa^1}b,\dots,\partial^{\kappa^k}b)\partial^\delta\psi=:B_{j,k,\delta,\iota,\kappa},$$

where $j,k\in\mathbb{N}$, $\iota=(\iota^1,\dots,\iota^j)\in\mathbb{N}^{dj}$, $\kappa=(\kappa^1,\dots,\kappa^k)\in\mathbb{N}^{dk}$ and $\delta\in\mathbb{N}^d$ satisfy

$$\sum_{i=1}^j|\iota^i|+\sum_{i=1}^k|\kappa^i|+|\delta|=N,\qquad\text{and}\qquad\sum_{i=1}^k|\kappa^i|>0.$$

We have to distinguish two cases.
(a) The case $|\iota^1|=N-1$ (and therefore $|\kappa^1|=1$, $\delta=0$). One then has

$$B_{j,k,\delta,\iota,\kappa}=d^2\mathcal{G}(\partial^{\iota^1}\zeta;\partial^{\kappa^1}b)\psi.$$

Proceeding exactly as for case (v) above, we get

$$|B_{j,k,\delta,\iota,\kappa}|_2\leq\max\{\varepsilon,\beta\}\mu\mathfrak{m}^N(U).\tag{4.19}$$

(b) The case $\max_{1\leq i\leq j}|\iota^i|\leq N-2$. It is a direct consequence of Proposition 3.28 that (4.19) also holds in this configuration.

---

[2]Here, $d^{j+k}\mathcal{G}(\partial^{\iota^1}\zeta,\dots,\partial^{\iota^j}\zeta;\partial^{\kappa^1}b,\dots,\partial^{\kappa^k}b)$ stands for $d^{j+k}\mathcal{G}(\mathbf{h},\mathbf{k})$ with $(\mathbf{h},\mathbf{k})_i=(\partial^{\iota^i}\zeta,0)$ if $1\leq i\leq j$ and $(\mathbf{h},\mathbf{k})_i=(0,\partial^{\kappa^i}b)$ if $j+1\leq i\leq j+k$.

We can now address point (2) of the proposition. By definition, $R^1_\alpha - R^2_\alpha$ is a sum of terms of the form $A^1_{j,\delta,\iota} - A^2_{j,\delta,\iota}$, where we used the notation (4.15). The result therefore follows easily from an adaptation of the proof of the first point. □

In the following proposition, we give some control on the $H^{1/2}$-norm of the residual R_α that appears in Proposition 4.5. This result is not needed to derive a priori energy estimates for the water waves equations (4.1) but is necessary to handle the mollification errors that appear in the construction of the solution (see the proof of Theorem 4.16 below).

PROPOSITION 4.8. *Let the assumptions of Proposition* 4.5 *be satisfied and moreover assume that* $\zeta \in L^\infty([0,T]; H^{N+1/2}(\mathbb{R}^d))$. *Then the residuals* R_α *introduced in Proposition* 4.5 *satisfy*

$$\forall 0 \le t \le T, \qquad |R_\alpha(t)|_{H^{1/2}} \le \mathfrak{m}^N(U)\big(\mu^{-1/4} + |\zeta(t)|_{H^{N+1/2}}\big).$$

REMARK 4.9. As noted above, this estimate is not needed to derive energy estimates on the solutions to the exact water waves equations (4.1)—as opposed to the mollified system introduced below in (4.30)—and the singular behavior of this estimate as $\mu \to 0$ is therefore harmless.

PROOF. The proof follows the same lines as the proof of Proposition 4.5. The only difference is that the quantities $A_{j,\delta,\iota}$ and $B_{j,k,\delta,\iota,\kappa}$ must be controlled in $H^{1/2}$ rather than L^2. With exactly the same arguments, for the cases *(i)*, *(ii)*, *(iii)*, and *(v)* we get that

$$\frac{1}{\mu}|A_{j,\delta,\iota}|_{H^{1/2}} \le \varepsilon \mathfrak{m}^N(U)(1 + |\zeta|_{H^{N+1/2}}).$$

In the case *(iv)*, using the same argument as in the proof of Proposition 4.5 would require a control of $|\mathfrak{P}\psi_{(\delta)}|_{H^{3/2}}$, which is not possible since $|\delta| = N-1$ in this case. However, using the first point of Proposition 3.28 rather than the second one to control $d\mathcal{G}(\partial^\iota\zeta)\psi_{(\delta)}$ yields

$$\frac{1}{\mu}|A_{j,\delta,\iota}|_{H^{1/2}} \le \varepsilon\mu^{-1/4}\mathfrak{m}^N(U).$$

We omit the proof of the controls on the $B_{j,k,\delta,\iota,\kappa}$ since they do not cause any particular difficulty. □

4.2.3. The quasilinear system. We show in Proposition 4.10 below that the water waves equations (4.1) can be "quasilinearized". In these quasilinearized equations, a quantity plays a very important role,

$$\underline{\mathfrak{a}} = 1 + \varepsilon\big(\partial_t + \varepsilon\underline{V}\cdot\nabla^\gamma\big)\underline{w}. \tag{4.20}$$

For notational convenience, we also introduce the matricial operators

$$\mathcal{A}[U] = \begin{pmatrix} 0 & -\frac{1}{\mu}\mathcal{G} \\ \underline{\mathfrak{a}} & 0 \end{pmatrix}, \qquad \mathcal{B}[U] = \begin{pmatrix} \varepsilon\nabla^\gamma\cdot(\underline{V}\bullet) & 0 \\ 0 & \varepsilon\underline{V}\cdot\nabla^\gamma \end{pmatrix}. \tag{4.21}$$

We can now state the main result of this section.

PROPOSITION 4.10. *Let* $T > 0$, $t_0 > d/2$ *and* N *be as in Proposition* 4.5. *If* $U = (\zeta,\psi) \in E^N_T$ *and* $b \in H^{N+1\vee t_0+1}(\mathbb{R}^d)$ *satisfy* (4.3) *uniformly on* $[0,T]$ *and solve* (4.1)*, then for all* $\alpha \in \mathbb{N}^d$ *with* $1 \leq |\alpha| \leq N$*, the couple* $U_{(\alpha)} = (\zeta_{(\alpha)}, \psi_{(\alpha)})$ *solves*

$$\partial_t U_{(\alpha)} + \mathcal{A}[U]U_{(\alpha)} = \max\{\varepsilon, \beta\}(R_\alpha, S_\alpha)^T, \qquad (|\alpha| < N),$$
$$\partial_t U_{(\alpha)} + \mathcal{A}[U]U_{(\alpha)} + \mathcal{B}[U]U_{(\alpha)} = \max\{\varepsilon, \beta\}(R_\alpha, S_\alpha)^T, \qquad (|\alpha| = N),$$

where the residuals R_α *and* S_α *satisfy the estimates*

$$|R_\alpha|_2 + |\mathfrak{P} S_\alpha|_2 \lesssim \mathfrak{m}^N(U)$$

uniformly on $[0, T]$.

REMARK 4.11. The "quasilinear" system we refer to is the system of evolution equations on $(U_{(\alpha)})_{0 \leq |\alpha| \leq N}$ (with $U_{(0)} = U$) formed by (4.1) for $\alpha = 0$ and the equations of the proposition for $1 \leq |\alpha| \leq N$.

REMARK 4.12. As for R_α (see the second point of Proposition 4.5), one can check that the dependence of S_α on U is Lipschitz: if S^1_α and S^2_α denote the residuals corresponding to two different U^1 and U^2, then

$$|\mathfrak{P}(S^1_\alpha - S^2_\alpha)|_2 \leq C\big(\mathfrak{m}^N(U^1), \mathfrak{m}^N(U^2)\big)\, \mathcal{E}^{|\alpha|}(U^1 - U^2).$$

PROOF. The first equation is obtained directly by applying ∂^α to the first equation of (4.1) and using Proposition 4.5.

For the second equation, we focus on the most difficult case, namely, $|\alpha| = N$. We first decompose $\alpha = \beta + \kappa \in \mathbb{N}^d$, with $|\kappa| = 1$. Inspired by [**181**], we first apply ∂^κ to the second equation of (4.1) to get the following lemma.

LEMMA 4.13. *The following identity holds*

$$\partial_t \partial^\kappa \psi + \partial^\kappa \zeta + \varepsilon \underline{V} \cdot (\nabla^\gamma \partial^\kappa \psi - \varepsilon \underline{w} \nabla^\gamma \partial^\kappa \zeta) - \frac{\varepsilon}{\mu} \underline{w} \partial^\kappa (\mathcal{G}\psi) = 0, \tag{4.22}$$

where $\underline{V}$ *and* $\underline{w}$ *are as given in Theorem* 3.21.

PROOF. The second equation of (4.1) can be equivalently written under the form

$$\partial_t \psi + \zeta + \frac{\varepsilon}{2}|\nabla^\gamma \psi|^2 - \frac{\varepsilon}{2\mu}(1 + \varepsilon^2 \mu |\nabla^\gamma \zeta|^2)\underline{w}^2 = 0.$$

Applying ∂^κ, we therefore get

$$\partial_t \partial^\kappa \psi + \partial^\kappa \zeta + \varepsilon \nabla^\gamma \psi \cdot \nabla^\gamma \partial^\kappa \psi - \varepsilon^3 \underline{w} \nabla^\gamma \zeta \cdot \underline{w} \nabla^\gamma \partial^\kappa \zeta - \frac{\varepsilon}{\mu}(1 + \varepsilon^2 \mu |\nabla^\gamma \zeta|^2)\underline{w} \partial^\kappa \underline{w} = 0.$$

The result therefore follows easily after noting that

$$\begin{aligned}(1 + \varepsilon^2 \mu |\nabla^\gamma \zeta|^2)\partial^\kappa \underline{w} &= (1 + \varepsilon^2 \mu |\nabla^\gamma \zeta|^2)\partial^\kappa \Big[\frac{\mathcal{G}\psi + \varepsilon \mu \nabla^\gamma \zeta \cdot \nabla^\gamma \psi}{1 + \varepsilon^2 \mu |\nabla^\gamma \zeta|^2}\Big] \\ &= \partial^\kappa (\mathcal{G}\psi) + \varepsilon \mu \nabla^\gamma \zeta \cdot \nabla^\gamma \partial^\kappa \psi + \varepsilon \mu \nabla^\gamma \partial^\kappa \zeta \cdot \nabla^\gamma \psi \\ &\quad - 2\varepsilon^2 \mu \underline{w} \nabla^\gamma \zeta \cdot \nabla^\gamma \partial^\kappa \zeta. \qquad \square\end{aligned}$$

We will now apply ∂^β to (4.22). We will thus find many terms of the form $\partial^\beta(fg)$; in order to check that most of them can be put in the residual term (i.e., that[3] $|\mathfrak{P}\partial^\beta(fg)|_2 \leq \varepsilon \mathfrak{m}^N(U)$), we rely heavily on the product estimate

$$|\mathfrak{P}(fg)|_2 \lesssim |\nabla^\gamma f|_{H^{t_0}} |g|_2 + |f|_\infty |\mathfrak{P} g|_2 \tag{4.23}$$

that stems from the following lemma.

[3] The notation $A \leq \mathfrak{m}^N(U)$ is used as a shortcut for $A(t) \leq \mathfrak{m}^N(U(t))$, for all $0 \leq t \leq T$.

LEMMA 4.14. *Let $f \in H^{t_0+1}(\mathbb{R}^d)$ and $g \in L^2(\mathbb{R}^d)$. Then*

$$\big|[\mathfrak{P}, f]g\big|_2 \lesssim |\nabla^\gamma f|_{H^{t_0}} |g|_2.$$

PROOF OF THE LEMMA. We adapt here the proof of Lemma 5.1 from [**181**]. With $u = [\mathfrak{P}, f]g$, we have

$$\widehat{u}(\xi) = \int_{\mathbb{R}^d} \Big(\frac{|\xi^\gamma|}{(1+\sqrt{\mu}|\xi^\gamma|)^{1/2}} - \frac{|\eta^\gamma|}{(1+\sqrt{\mu}|\eta^\gamma|)^{1/2}} \Big) \widehat{f}(\xi-\eta)\widehat{g}(\eta) d\eta.$$

Noting that there exists a numerical constant c such that $|x/\sqrt{1+\sqrt{\mu}x} - y/\sqrt{1+\sqrt{\mu}y}| \le c|x-y|$ for all $x, y \ge 0$, we deduce that

$$|\widehat{u}(\xi)| \lesssim \int_{\mathbb{R}^d} |\xi^\gamma - \eta^\gamma| |\widehat{f}(\xi-\eta)| |\widehat{g}(\eta)| d\eta.$$

Therefore

$$|u|_2^2 \le |\widehat{\nabla^\gamma f}|_{L^1} |g|_2 \lesssim |\nabla^\gamma f|_{H^{t_0}} |g|_2,$$

and the result is proved. $\square$

For the sake of clarity, we also introduce the notation

$$a \sim b \iff |\mathfrak{P}(a-b)|_2 \le \max\{\varepsilon, \beta\} \mathfrak{m}^N(U). \tag{4.24}$$

We can now state the following lemma.

LEMMA 4.15. *Under the assumptions of the proposition, and with $\alpha = \beta + \kappa$ ($|\kappa| = 1$), the following identities hold,*

$$\begin{aligned} \varepsilon\partial^\beta\big(\underline{V}\cdot(\nabla^\gamma\partial^\kappa\psi - \varepsilon\nabla^\gamma\partial^\kappa\zeta)\big) &\sim \varepsilon\underline{V}\cdot\nabla\psi_{(\alpha)} + \varepsilon^2(\underline{V}\cdot\nabla^\gamma\underline{w})\partial^\alpha\zeta \\ &\quad -\varepsilon^2\underline{V}\cdot\{\partial^\beta, \underline{w}\}\nabla^\gamma\partial^\kappa\zeta, \\ \frac{\varepsilon}{\mu}\partial^\beta\big(\underline{w}\partial^\kappa(\mathcal{G}\psi)\big) &\sim \frac{\varepsilon}{\mu}\underline{w}\partial^\alpha(\mathcal{G}\psi) + \frac{\varepsilon}{\mu}\{\partial^\beta, \underline{w}\}\partial^\kappa(\mathcal{G}\psi), \end{aligned}$$

where $\{\partial^\beta, \underline{w}\} = \sum_{j=1}^d \beta_j \partial_j \underline{w} \partial^{\check{\beta}^j}$, with $\check{\beta}^j = \beta - \mathbf{e}_j$, is the usual Poisson bracket.

PROOF. Let us address the first assertion of the lemma. Since $\varepsilon\partial^\beta\big(\underline{V}\cdot(\nabla^\gamma\partial^\kappa\psi - \varepsilon\underline{w}\nabla^\gamma\partial^\kappa\zeta)\big)$ is a sum of terms of the form $A_{\beta',\beta''} = \varepsilon\partial^{\beta'}\underline{V}\cdot\partial^{\beta''}(\nabla^\gamma\partial^\kappa\psi - \varepsilon\underline{w}\nabla^\gamma\partial^\kappa\zeta)$, with $\beta'+\beta'' = \beta$, and noting that

$$A_{0,\beta} \sim \varepsilon\underline{V}\cdot\nabla^\gamma\psi_{(\alpha)} + \varepsilon^2(\underline{V}\cdot\nabla^\gamma\underline{w})\partial^\alpha\zeta - \varepsilon^2\underline{V}\cdot\{\partial^\beta, \underline{w}\}\nabla^\gamma\partial^\kappa\zeta,$$

we are led to prove that $A_{\beta',\beta''} \sim 0$ if $|\beta''| < N-1$. The most difficult configuration corresponds to $\beta'' = 0$ or $|\beta''| = N-2$, and we thus omit the proof for the intermediate cases. Moreover, since the proof follows the same lines for $\beta'' = 0$ and $|\beta''| = N-2$, we just give it in the latter case. By (4.23), we get

$$|\mathfrak{P}A_{\beta',\beta''}|_2 \le \varepsilon|\partial^{\beta'}\underline{V}|_{H^{t_0+1}}\big(|\partial^{\beta''}g|_2 + |\mathfrak{P}\partial^{\beta''}g|_2\big),$$

with $g = \nabla^\gamma\partial^\kappa\psi - \varepsilon\nabla^\gamma\partial^\kappa\zeta$. Since $|\beta'| = 1$, we deduce easily from the definition of $\underline{V}$, Proposition 4.4, and Lemma 4.6 that $|\partial^{\beta'}\underline{V}|_{H^{t_0+1}} \le \mathfrak{m}^N(U)$; moreover, $\partial^{\beta''}g$ can be put under the form

$$\partial^{\beta''}g = \nabla^\gamma\psi_{(\beta''+\kappa)} - \varepsilon[\partial^{\beta''}, \underline{w}]\nabla^\gamma\partial^\kappa\zeta + \varepsilon(\partial^{\kappa+\beta''}\zeta)\nabla^\gamma\underline{w}.$$

Since the last two components of the r.h.s. are bounded from above in L^2 and $\mathfrak{P}L^2$ norms by $\mathfrak{m}^N(U)$ (this is a consequence of (4.23), Proposition 4.4, and the identity $|\mathfrak{P}f|_2 \leq |\nabla f|_2$), we have obtained that

$$\begin{aligned} |\mathfrak{P}A_{\beta',\beta''}|_2 &\leq \varepsilon\mathfrak{m}^N(U)(1+|\nabla\psi_{(\beta''+\kappa)}|_2+|\mathfrak{P}\nabla\psi_{(\beta''+\kappa)}|_2) \\ &\leq \varepsilon\mathfrak{m}^N(U)(1+|\mathfrak{P}\psi_{(\beta''+\kappa)}|_{H^1}) \\ &\leq \varepsilon\mathfrak{m}^N(U), \end{aligned}$$

where Lemma 4.6 has been used to derive the last inequality.

Let us now prove the second assertion of the lemma. We do the proof only for the case of flat bottoms; the generalization to nonflat topographies is obtained as in the proof of Proposition 4.5 and does not cause any particular difficulty. We proceed as above to check that it is sufficient to prove that $\frac{\varepsilon}{\mu}\partial^{\beta'}\underline{w}\partial^{\beta''+\kappa}(\mathcal{G}\psi) \sim 0$ if $\beta'+\beta''=\beta$ and $|\beta''|<N-2$. We give the proof of the most difficult case, corresponding here to $\beta'=\beta$, $\beta''=0$. Using (4.23) and Theorem 3.15, it is enough to prove that

$$\frac{1}{\sqrt{\mu}}|\mathfrak{P}\partial^\beta\underline{w}|_2 \leq \mathfrak{m}^N(U). \tag{4.25}$$

Recalling that $\underline{w} = w^\pm[\varepsilon\zeta,\beta b]\psi$ (with $b=0$ for flat bottoms) and differentiating this relation with respect to ∂^β yields after multiplication by ε (in order to use the convenient notation $\sim$)

$$\begin{aligned} &(1+\varepsilon^2\mu|\nabla^\gamma\zeta|^2)\frac{\varepsilon}{\sqrt{\mu}}\partial^\beta\underline{w} + 2\varepsilon^2\sqrt{\mu}\nabla^\gamma\partial^\beta\zeta\cdot(\varepsilon\underline{w}\nabla^\gamma\zeta) \\ \sim\ &\frac{\varepsilon}{\sqrt{\mu}}\partial^\beta\mathcal{G}\psi + \varepsilon^2\sqrt{\mu}\nabla^\gamma\zeta\cdot\partial^\beta\nabla^\gamma\psi + \varepsilon^2\sqrt{\mu}\nabla^\gamma\partial^\beta\zeta\cdot\nabla^\gamma\psi. \end{aligned}$$

Moreover, since we can deduce from Theorem 3.21 and Proposition 3.28 that

$$\frac{\varepsilon}{\sqrt{\mu}}\partial^\beta\mathcal{G}\psi \sim \frac{\varepsilon}{\sqrt{\mu}}\mathcal{G}\psi_{(\beta)} - \varepsilon^2\sqrt{\mu}\nabla^\gamma\cdot(\partial^\beta\zeta\underline{V}),$$

we get the formula

$$(1+\varepsilon^2\mu|\nabla^\gamma\zeta|^2)\frac{\varepsilon}{\sqrt{\mu}}\partial^\beta\underline{w} \sim \frac{\varepsilon}{\sqrt{\mu}}\mathcal{G}\psi_{(\beta)} + \varepsilon^2\sqrt{\mu}\nabla^\gamma\zeta\cdot\nabla^\gamma\psi_{(\beta)}.$$

Dividing this formula by ε, it is then easy to deduce from Theorem 3.15 and Lemma 4.6 that (4.25) holds. □

Lemma 4.15 allows one to put most of the terms of the second equation of ∂^β(4.22) in the residual. However, this is not the case for the Poisson brackets that appear in both identities of Lemma 4.15, since they require (when evaluated in $\mathfrak{P}L^2$-norm) a control of $(N+1/2)$ derivatives of ζ. Fortunately, this singularity disappears when one takes the difference of the expressions considered in Lemma 4.15, which is the quantity that appears in ∂^β(4.22). Indeed, one has

$$\begin{aligned} &-\varepsilon^2\underline{V}\cdot\{\partial^\beta,\underline{w}\}\nabla^\gamma\partial^\kappa\zeta - \frac{\varepsilon}{\mu}\{\partial^\beta,\underline{w}\}\partial^\kappa(\mathcal{G}\psi) \\ \sim\ &-\varepsilon\{\partial^\beta,\underline{w}\}\big(\varepsilon\underline{V}\cdot\nabla^\gamma\partial^\kappa\zeta + \frac{1}{\mu}\partial^\kappa(\mathcal{G}\psi)\big) \\ \sim\ &0, \end{aligned}$$

where Theorem 3.21 has been used to derive the second identity.
It follows from this analysis that $\partial^\beta(4.22)$ can be written under the form

$$\partial_t\partial^\alpha\psi + \partial^\alpha\zeta + \varepsilon\underline{V}\cdot(\nabla\psi_{(\alpha)} + \varepsilon\partial^\alpha\zeta\nabla\underline{w}) - \frac{\varepsilon}{\mu}\underline{w}\partial^\alpha(\mathcal{G}\psi) \sim 0.$$

Using the first equation, one can replace $\frac{1}{\mu}\mathcal{G}\psi$ by $\partial_t\zeta$ in this expression,

$$\partial_t\psi_{(\alpha)} + \underline{\mathfrak{a}}\partial^\alpha\zeta + \varepsilon\underline{V}\cdot\nabla^\gamma\psi_{(\alpha)} \sim 0,$$

with $\underline{\mathfrak{a}}$ as in (4.20). This completes the proof of the proposition. □

4.3. Main results

After clarifying our assumptions on the initial condition in §4.3.1, we state our main existence and stability results in §4.3.2 and describe the behavior of the solution in various asymptotic regimes in §4.3.2. The proof of the main theorems occupies §4.3.4, and we end this section with some considerations on the Rayleigh-Taylor stability condition in §4.3.5.

4.3.1. Initial condition. We want to solve (4.1) with initial condition $U^0 = (\zeta^0, \psi^0)$ with U^0 smooth enough. As seen below, the energy estimates used to solve (4.1) require that the quantity $\underline{\mathfrak{a}}$ defined in (4.20) be nonnegative. We therefore have to assume that $\underline{\mathfrak{a}}$ satisfies this assumption at $t = 0$. However, the definition (4.20) of $\underline{\mathfrak{a}}$ involves $\partial_t U$, which is not initially prescribed.
If we write (4.1) under the condensed form

$$\partial_t U + \mathcal{N}(U) = 0,$$

with $\mathcal{N}(U) = (\mathcal{N}_1(U), \mathcal{N}_2(U))^T$ and

$$\mathcal{N}_1(U) = -\frac{1}{\mu}\mathcal{G}\psi, \qquad \mathcal{N}_2(U) = \zeta + \frac{\varepsilon}{2}|\nabla^\gamma\psi|^2 - \frac{\varepsilon}{\mu}\frac{(\mathcal{G}\psi + \varepsilon\mu\nabla^\gamma\zeta\cdot\nabla^\gamma\psi)^2}{2(1+\varepsilon^2\mu|\nabla^\gamma\zeta|^2)}, \tag{4.26}$$

then it is natural to replace $(\partial_t U)_{|t=0}$ by $-\mathcal{N}(U^0)$.
Moreover, since $\partial_t\underline{w} = \boldsymbol{w}[\varepsilon\zeta,\beta b]\partial_t\psi + d\boldsymbol{w}(\partial_t\zeta)\psi$, we consequently replace $(\partial_t\underline{w})_{|t=0}$ by $-\boldsymbol{w}[\varepsilon\zeta^0,\beta b]\mathcal{N}_2(U^0) - d\boldsymbol{w}(\mathcal{N}_1(U^0))\psi^0$. We therefore define $\mathfrak{a}(U)$ as

$$\begin{aligned}\mathfrak{a}(U) \;&=\; 1 + \varepsilon^2\boldsymbol{V}[\varepsilon\zeta,\beta b]\psi\cdot\nabla^\gamma(\boldsymbol{w}[\varepsilon\zeta,\beta b]\psi)\\ &\quad -\varepsilon\boldsymbol{w}[\varepsilon\zeta,\beta b]\mathcal{N}_2(U) - \varepsilon d\boldsymbol{w}(\mathcal{N}_1(U))\psi,\end{aligned} \tag{4.27}$$

so that if U solves (4.1) at time t, then $\mathfrak{a}(U)(t) = \underline{\mathfrak{a}}(t)$. The sign condition $\underline{\mathfrak{a}} > 0$ therefore becomes $\mathfrak{a}(U^0) > 0$ at $t = 0$.

4.3.2. Statement of the theorems. We state here the main result[4] of this chapter, which shows the well-posedness of (4.1) and gives a control of the dependence of the solution with respect to the parameters ε, μ, β and γ satisfying 4.2, namely,

$$0 \le \varepsilon, \beta, \gamma \le 1, \quad 0 \le \mu \le \mu_{max}, \tag{4.28}$$

for some $\mu_{max} > 0$ (not necessarily small). We recall that the functional space E_T^N is defined in (4.10) and that $\varepsilon \vee \beta$ stands for $\max\{\varepsilon, \beta\}$; we also note that for $d = 2$, one can take $t_0 = 3/2$ and $N = 5$ in the statement below.

THEOREM 4.16 (Local existence). *Let $t_0 > d/2$, $N \ge t_0 + t_0 \vee 2 + 3/2$, and $(\varepsilon, \mu, \beta, \gamma)$ satisfy (4.28). Then let $U^0 = (\zeta^0, \psi^0) \in E_0^N$, $b \in H^{N+1\vee t_0+1}(\mathbb{R}^d)$, and moreover assume that*

$$\exists h_{min} > 0, \quad \exists a_0 > 0, \qquad 1 + \varepsilon\zeta^0 - \beta b \ge h_{min} \quad \text{and} \quad \mathfrak{a}(U^0) \ge a_0, \tag{4.29}$$

with $\mathfrak{a}(U^0)$ as defined in (4.27).
Then there exists $T > 0$ and a unique solution $U \in E^N_{T/\varepsilon\vee\beta}$ to (4.1) with initial data U^0. Moreover

$$\frac{1}{T} = c^1_{WW} \quad \text{and} \quad \sup_{t \in [0, \frac{T}{\varepsilon \vee \beta}]} \mathcal{E}^N(U(t)) = c^2_{WW},$$

with $c^j_{WW} = C(\mathcal{E}^N(U^0), \mu_{max}, \frac{1}{h_{min}}, \frac{1}{a_0}, |b|_{H^{N+1\vee t_0+1}})$ *for* $j = 1, 2$.

REMARK 4.17. We refer to §4.3.5 for a physical interpretation of the assumption $\mathfrak{a}(U^0) > 0$ and for considerations regarding its validity. From a mathematical viewpoint, the water waves equations are not strictly hyperbolic (their symbol has a double eigenvalue with a Jordan block), and the condition $\underline{\mathfrak{a}} > 0$ is the Levy-condition on the subprincipal symbol that is required for well-posedness.

Theorem 4.16 is complemented by the following results that show the stability of the solution with respect to perturbations. (The solution U and the time T that appear in the statement below are those furnished by Theorem 4.16.)

THEOREM 4.18 (Stability). *Let the assumptions of Theorem 4.16 be satisfied and moreover assume that there exists $U^{app} = (\zeta^{app}, \psi^{app}) \in E^N_{T/\varepsilon\vee\beta}$ such that*

$$\inf_{0 \le (\varepsilon\vee\beta)t \le T} \inf_{x \in \mathbb{R}^d} 1 + \varepsilon\zeta^{app} - \beta b > 0 \quad \text{and} \quad \partial_t U^{app} + \mathcal{N}(U^{app}) = r,$$

[4]Shinbrot [**300**] and Kano and Nishida [**198**] gave the first well-posedness results for the water waves equations for analytic data. The well-posedness of the water waves equations within a Sobolev class goes back to Nalimov [**265**], Yosihara [**330**], and Craig [**101**] in the one-dimensional case $d = 1$ and with smallness assumptions on the initial data. Beale, Hou, and Lowengrub [**20**] identified the relevance of the Rayleigh-Taylor condition $\mathfrak{a}(U) > 0$ for the well-posedness theory (see §4.3.5 for more comments on this point), and showed that the linearized equations are well posed if this condition holds. The full (nonlinear) well-posedness is due to Wu in infinite depth in dimension $d = 1$ [**326**] and $d = 2$ [**327**]. The result presented here essentially comes from [**218**], where the case of finite depth was treated, and [**12**], where the uniformity of the existence time with respect to the parameters $(\varepsilon, \beta, \gamma, \mu)$ was established; it also includes improvements due to Iguchi [**179**], and from [**221**], which deals with the two fluids generalization of the water waves problem.

with $r = (r^1, r^2)$ *and* $(r^1, \mathfrak{P} r^2) \in L^\infty([0, \frac{T}{\varepsilon \vee \beta}]; H^N(\mathbb{R}^d)^2)$.
We also denote by c_{app} *a constant such that*

$$\sup_{0 \le (\varepsilon \vee \beta) t \le T} \mathcal{E}^N(U^{app}(t)) \le c_{app}.$$

Then, the error $\mathfrak{e} = U - U^{app}$ *satisfies for all* $0 \le (\varepsilon \vee \beta)t \le T$,

$$\mathcal{E}^{N-1}(\mathfrak{e}(t)) \le C(c_{WW}, c_{app})(\mathcal{E}^{N-1}(\mathfrak{e}_{|_{t=0}}) + t|(r_1, \mathfrak{P} r_2)|_{L^\infty([0,t];H^N)}).$$

4.3.3. Asymptotic regimes. Corollary 4.22 below can be used to exhibit a typical scale for the existence time of the solution when stronger conditions than (4.28) are imposed on the parameters ε, β, γ and μ. These stronger conditions correspond to the *asymptotic regimes* already mentioned in §1.5.

DEFINITION 4.19 (Asymptotic regime). We use the term *asymptotic regime* to refer to any subset $\mathcal{A}$ of the admissible set of parameters

$$\mathcal{A} \subset \{(\varepsilon, \beta, \gamma, \mu), \quad 0 \le \varepsilon, \beta, \gamma \le 1, \quad 0 \le \mu \le \mu_{max}\}.$$

NOTATION 4.20. We often write $\mathfrak{p} = (\varepsilon, \beta, \gamma, \mu)$. We also denote by $|\mathcal{A}|$ the diameter of $\mathcal{A}$; quite obviously, $|\mathcal{A}| \le \max\{1, \mu_{max}\}$.

REMARK 4.21. It is of course possible to take $\mathcal{A}$ equal to the whole set of admissible parameters or inversely to one single point of this set.

In the following corollary to Theorem 4.16, we show that to any asymptotic regime $\mathcal{A}$, one can associate a typical time scale over which all the solutions $U_\mathfrak{p}$, with $\mathfrak{p} \in \mathcal{A}$, are known to exist. This will play an important role in the derivation and justification of asymptotic systems (as $\mu \to 0$ for instance) in Chapter 5.

COROLLARY 4.22. *Let* $t_0 > d/2$, $N \ge t_0 + t_0 \vee 2 + 3/2$, *and* $\mathcal{A}$ *be an asymptotic regime. Also let* $(U^0_\mathfrak{p})_{\mathfrak{p} \in \mathcal{A}} \in E^N_0$ *with* $\sup_{\mathfrak{p} \in \mathcal{A}} \mathcal{E}^N(U^0_\mathfrak{p}) < \infty$ *and moreover assume that*

$$\exists h_{min} > 0, \quad \exists a_0 > 0, \quad \forall \mathfrak{p} \in \mathcal{A}, \qquad 1 + \varepsilon \zeta^0_\mathfrak{p} - \beta b \ge h_{min} \quad \text{and} \quad \mathfrak{a}(U^0_\mathfrak{p}) \ge a_0.$$

Then there exists $T > 0$ *such that for all* $\mathfrak{p} \in \mathcal{A}$, *there exists a unique solution* $U_\mathfrak{p} \in E^N_{T/\varepsilon \vee \beta}$ *to* (4.1) *with initial data* U^0. *Moreover,*

$$\frac{1}{T} = c^1_\mathcal{A} \quad \text{and} \quad \sup_{\mathfrak{p} \in \mathcal{A}} \sup_{t \in [0, \frac{T}{\varepsilon \vee \beta}]} \mathcal{E}^N(U_\mathfrak{p}(t)) = c^2_\mathcal{A},$$

with $c^j_\mathcal{A} = C(\sup_{\mathfrak{p} \in \mathcal{A}} \mathcal{E}^N(U^0_\mathfrak{p}), |\mathcal{A}|, \frac{1}{h_{min}}, \frac{1}{a_0}, |b|_{H^{N+1 \vee t_0 + 1}})$ *for* $j = 1, 2$.

PROOF. The result follows directly from Theorem 4.16 since the existence time provided by this theorem depends only on the energy of the initial condition, μ_{max} (straightforwardly replaced here by $|\mathcal{A}|$), and $1/h_{min}$ and $1/a_0$. In particular, there is no singular dependence of this existence time on the parameters $\mathfrak{p} \in \mathcal{A}$. □

We end this section with typical examples of asymptotic regimes that will play an important role in the asymptotic description of the water waves equations (4.1) in Chapter 5.

EXAMPLE 4.23 (Shallow water regime). The shallow water regime corresponds to waves with wavelengths much larger than the depth ($\mu \ll 1$) and with possibly large amplitude and large topography variations (no smallness assumption is made on ε nor β). We consequently define the *shallow water regime* as

$$\mathcal{A}_{SW} = \{(\varepsilon,\beta,\gamma,\mu),\quad 0\le\mu\le\mu_0,\quad 0\le\varepsilon\le 1,\quad 0\le\beta\le 1,\quad 0\le\gamma\le 1\},$$

for some $0<\mu_0\le\mu_{max}$. Note that we do not require that $\mu_0\ll 1$ in the definition of $\mathcal{A}_{SW}$, because the smallness on μ is not necessary for Theorem 4.16 to hold. The smallness of μ is only required to obtain small approximation errors for the various asymptotic models derived in the next chapter (these errors are typically of size $O(\mu)$ or $O(\mu^2)$).

EXAMPLE 4.24 (Long wave regime). The so-called *long wave regime* falls into the range of the shallow water regime of Example 4.23, but it is further assumed that the waves have small amplitude in the sense that $\varepsilon = O(\mu)$. Depending on the assumptions made on the topography parameter β, we distinguish long waves over strong or weak topographies

$$\begin{aligned}
\mathcal{A}_{LW,top+} &= \{(\varepsilon,\beta,\gamma,\mu),\quad 0\le\mu\le\mu_0,\quad 0\le\varepsilon\lesssim\mu,\quad 0\le\beta\le 1,\quad 0\le\gamma\le 1\},\\
\mathcal{A}_{LW,top-} &= \{(\varepsilon,\beta,\gamma,\mu),\quad 0\le\mu\le\mu_0,\quad 0\le\varepsilon\lesssim\mu,\quad 0\le\beta\lesssim\mu,\quad 0\le\gamma\le 1\}.
\end{aligned}$$

EXAMPLE 4.25 (KdV and Kadomtsev-Petviashvili regime). Long waves for which dispersive and nonlinear effects are in perfect balance (that is, $\varepsilon=\mu$) are particularly worthy of interest. In dimension $d=1$ (or $\gamma=0$) and for flat bottoms (or $\beta=0$), we will refer to this regime as the *Korteweg-de Vries regime*,

$$\mathcal{A}_{KdV} = \{(\varepsilon,\beta,\gamma,\mu),\quad 0\le\mu\le\mu_0,\quad \varepsilon=\mu,\quad \beta=0,\quad \gamma=0\}.$$

A two-dimensional generalization is obtained by considering weakly transverse waves ($\gamma=\sqrt{\varepsilon}$); the assumptions on ε, β and μ are otherwise the same as in the KdV regime. This is the *Kadomtsev-Petviashvili regime*

$$\mathcal{A}_{KP} = \{(\varepsilon,\beta,\gamma,\mu),\quad 0\le\mu\le\mu_0,\quad \varepsilon=\mu,\quad \beta=0,\quad \gamma=\sqrt{\mu}\}.$$

Of course, weakly transverse variants of other regimes can be considered (replace $0\le\gamma\le 1$ by $\gamma=\sqrt{\varepsilon}$ in Example 4.23, for instance).

4.3.4. Proof of Theorems 4.16 and 4.18. For the sake of clarity, we write $\epsilon=\max\{\varepsilon,\beta\}$ throughout this proof.

4.3.4.1. *The mollified quasilinear system.* Let $\chi:\mathbb{R}\to\mathbb{R}$ be a smooth, compactly supported function equal to one in a neighborhood of the origin. For all $0<\iota<1$, we denote by J^ι the mollifier $J^\iota=\chi(\iota|D|)$. Also let $0<\nu<1$ and consider the following mollified version of the water waves equations (4.1),

$$(4.30)\qquad \begin{cases} \partial_t\zeta - \frac{1}{\mu}J^\iota\mathcal{G}\psi = 0,\\ \partial_t\psi + J^\iota\zeta + \varepsilon\frac{1}{2}J^\iota\Big(\big|\nabla^\gamma\psi\big|^2 - \frac{\varepsilon}{\mu}\frac{(\mathcal{G}\psi+\varepsilon\mu\nabla^\gamma\zeta\cdot\nabla^\gamma\psi)^2}{2(1+\varepsilon^2\mu|\nabla^\gamma\zeta|^2)^2}\Big) = -\varepsilon\nu J^\iota\Lambda\zeta. \end{cases}$$

From standard results on ODEs, (4.30) has a unique maximal solution $U^{\iota,\nu}=(\zeta^{\iota,\nu},\psi^{\iota,\nu})$ with initial data (ζ^0,ψ^0) on a time interval $[0,T^{\iota,\nu}_{max})$. For all $\alpha\in\mathbb{N}^d$, $1\le|\alpha|\le N$, let $U^{\iota,\nu}_{(\alpha)}=(\zeta^{\iota,\nu}_{(\alpha)},\psi^{\iota,\nu}_{(\alpha)})$, with

$$\zeta^{\iota,\nu}_{(\alpha)}=\partial^\alpha\zeta^{\iota,\nu},\qquad \psi^{\iota,\nu}_{(\alpha)}=\partial^\alpha\psi^{\iota,\nu}-\varepsilon\underline{w}^{\iota,\nu}\partial^\alpha\zeta^{\iota,\nu},$$

and $\underline{w}^{\iota,\nu} = \mathcal{w}[\varepsilon\zeta^{\iota,\nu}, \beta b]\psi^{\iota,\nu}$. Proceeding exactly as we did for the proof of Proposition 4.10, we can check that $U^{\iota,\nu}_{(\alpha)}$ solves[5]

$$\partial_t U_{(\alpha)} + J^\iota\big(\mathcal{A}[U] + \varepsilon\nu\mathcal{C}(D)\big)U_{(\alpha)} + J^\iota\mathcal{B}[U]U_{(\alpha)} = \epsilon\begin{pmatrix} J^\iota R_\alpha \\ J^\iota S_\alpha + S'_\alpha\end{pmatrix} \tag{4.31}$$

with

$$\mathcal{C}(D) = \begin{pmatrix} 0 & 0 \\ \Lambda & 0 \end{pmatrix} \tag{4.32}$$

and where, as in Proposition 4.10, the operator $\mathcal{B}[U]$ must be removed if $|\alpha| < N$. The quantities R_α and S_α satisfy the same estimates as in Proposition 4.10, and S'_α is given by

$$S'_\alpha = -(1-J^\iota)\big[\partial_t \underline{w}\zeta_{(\alpha)}\big] + [\underline{w}, J^\iota]\partial^\alpha\mathcal{G}\psi.$$

Our strategy is to derive energy estimates for the system of evolution equations on $(U^{\iota,\nu}_{(\alpha)})_{0\le|\alpha|\le N}$ formed by (4.30) and (4.31). After showing that these energy estimates do not depend on the mollifying parameter ι, we will construct by completeness a solution $(U^{\nu}_{(\alpha)})_{0\le|\alpha|\le N}$ to the limit system corresponding to $\iota = 0$ (that is, $J^\iota = 1$) on a time interval $[0, \frac{1}{\epsilon}T]$, with T independent of ι and ν. Taking the limit $\nu \to 0$ then provides us with a solution to (4.1). We then prove the uniqueness and stability of this solution.

REMARK 4.26. We introduced the quantity $\nu J^\iota\Lambda\zeta$ in the r.h.s. of (4.30) because it gives an extra control of half a derivative on $\zeta_{(\alpha)}$ that we use to control the mollifying error S'_α. Since $S'_\alpha = 0$ when $\iota = 0$, this additional term is then no longer necessary to get the energy estimates, and it is therefore possible to take the limit $\nu \to 0$.

4.3.4.2. *Symmetrizer and energy.* A symmetrizer for the "quasilinear" system $(4.31)_\alpha$ is given by

$$\begin{aligned} \mathcal{S}^\nu[U] &= \operatorname{diag}(\underline{\mathfrak{a}}, \frac{1}{\mu}\mathcal{G}) + \varepsilon\nu\operatorname{diag}(\Lambda, 0) \\ &=: \mathcal{S}[U] + \varepsilon\nu\operatorname{diag}(\Lambda, 0). \end{aligned}$$

It is therefore natural to introduce the energies $\mathcal{F}^{[l]}(U)$, for all $0 \le l \le N$,

$$\mathcal{F}^{[l]}(U) = \sum_{|\alpha|\le l}\mathcal{F}^\alpha(U), \tag{4.33}$$

with

$$\begin{aligned} (\alpha \neq 0) \qquad \mathcal{F}^\alpha(U) &= \frac{1}{2}\big(\mathcal{S}[U]U_{(\alpha)}, U_{(\alpha)}\big) \\ &= \frac{1}{2}\big(\underline{\mathfrak{a}}\zeta_{(\alpha)}, \zeta_{(\alpha)}\big) + \frac{1}{2}\big(\frac{1}{\mu}\mathcal{G}\psi_{(\alpha)}, \psi_{(\alpha)}\big) \\ (\alpha = 0) \qquad \mathcal{F}^0(U) &= \frac{1}{2}|\zeta|^2_{H^{t_0+3/2}} + \frac{1}{2}\big(\Lambda^{t_0+3/2}\psi, \frac{1}{\mu}\mathcal{G}\Lambda^{t_0+3/2}\psi\big), \end{aligned}$$

and where we recall that $U_{(\alpha)} = (\zeta_{(\alpha)}, \psi_{(\alpha)})^T$.
The following lemma shows that $\mathcal{F}^{[N]}(U)$ is uniformly equivalent to the energy $\mathcal{E}^N(U)$ previously introduced in (4.8).

---

[5]For the sake of clarity, we write $\zeta_{(\alpha)}$ and $\psi_{(\alpha)}$ instead of $\zeta^{\iota,\nu}_{(\alpha)}$ and $\psi^{\iota,\nu}_{(\alpha)}$ when no confusion is possible.

LEMMA 4.27. *Let $T > 0$ and assume that U solves (4.30) on $[0, \frac{1}{\epsilon}T]$ and that*

$$1 + \varepsilon\zeta - \beta b \geq \frac{1}{2}h_{min} \quad \text{and} \quad \underline{\mathfrak{a}} \geq \frac{1}{2}a_0 \quad \text{on} \quad [0, \frac{T}{\epsilon}].$$

Then the following identities hold, for all $0 \leq j \leq N$,

$$\mathcal{E}^j(U) \leq (M + \frac{4}{a_0})\mathcal{F}^{[j]}(U) \quad \text{and} \quad \mathcal{F}^{[j]}(U) \leq \mathfrak{m}^N(U)\mathcal{E}^j(U),$$

where $\mathcal{E}^N(U)$ and $\mathfrak{m}^N(U)$ are defined in (4.8) and (4.11).

PROOF OF THE LEMMA. The first estimate follows from the assumption $\underline{\mathfrak{a}} \geq a_0/2$ and Proposition 3.12. For the second estimate, we again use Proposition 3.12 and the observation that $\underline{\mathfrak{a}} \leq \mathfrak{m}^N(U)$, which follows easily from the fact that $\underline{\mathfrak{a}} = \mathfrak{a}(U)$ with $\mathfrak{a}(U)$ as in (4.20). □

4.3.4.3. *Energy estimates. We assume throughout this section that U solves (4.30) on $[0, \frac{1}{\epsilon}T]$ for some $T > 0$ and that*

$$1 + \varepsilon\zeta - \beta b \geq \frac{1}{2}h_{min} \quad \text{and} \quad \underline{\mathfrak{a}} \geq \frac{1}{2}a_0 \quad \text{on} \quad [0, \frac{T}{\epsilon}].$$

We first consider the case $\alpha = 0$. We take the L^2-scalar product of (4.30) with $(\Lambda^{2(t_0+3/2)}\zeta, \Lambda^{(t_0+3/2)}\frac{1}{\mu}\mathcal{G}\Lambda^{(t_0+3/2)}\psi)^T$; this gives

$$\begin{aligned} \frac{d}{dt}\mathcal{F}^0(U) &= \big([\mathcal{G}, \Lambda^{t_0+3/2}]\psi, \Lambda^{t_0+3/2}\zeta\big) \\ &- \big(\Lambda^{t_0+3/2}(\mathcal{N}_2(U) - \zeta), \frac{1}{\mu}\mathcal{G}\Lambda^{t_0+3/2}\psi\big) - \varepsilon\nu\big(\Lambda^{t_0+5/2}\zeta, \frac{1}{\mu}\mathcal{G}\Lambda^{t_0+3/2}\psi\big) \end{aligned} \tag{4.34}$$

with $\mathcal{N}_2(U)$ as in (4.26). Let us now address the control of the three terms in the r.h.s.

- *Control of* $\big([\mathcal{G}, \Lambda^{t_0+3/2}]\psi, \Lambda^{t_0+3/2}\zeta\big)$. One has

$$\begin{aligned} \big|\big([\mathcal{G}, \Lambda^{t_0+3/2}]\psi, \Lambda^{t_0+3/2}\zeta\big)\big| &\leq \big|[\mathcal{G}, \Lambda^{t_0+3/2}]\psi\big|_2 |\zeta|_{H^{t_0+3/2}} \\ &\leq \epsilon\mu\mathfrak{m}^N(U), \end{aligned} \tag{4.35}$$

where we used Proposition 3.32 to derive the last inequality.

- *Control of* $\big(\Lambda^{t_0+3/2}(\mathcal{N}_2(U) - \zeta), \frac{1}{\mu}\mathcal{G}\Lambda^{t_0+3/2}\psi\big)$. Using Proposition 3.12 and recalling that $|\mathfrak{P}f|_{H^{t_0+3/2}} \leq |f|_{H^{t_0+5/2}}$, one gets

$$\begin{aligned} \big(\Lambda^{t_0+3/2}(\mathcal{N}_2(U) - \zeta), \frac{1}{\mu}\mathcal{G}\Lambda^{t_0+3/2}\psi\big) &\leq \big|\mathcal{N}_2(U) - \zeta\big|_{H^{t_0+5/2}} |\mathfrak{P}\psi|_{H^{t_0+3/2}} \\ &\leq \varepsilon MC(|\mathfrak{P}\psi|^2_{H^{t_0+3}}, |\zeta|^2_{H^{t_0+7/2}}), \end{aligned}$$

where the last inequality stems from the definition (4.26) of $\mathcal{N}_2(U)$ and Theorem 3.15. Using Lemma 4.6 and the assumption made on N, we deduce that

$$\big(\Lambda^{t_0+3/2}(\mathcal{N}_2(U) - \zeta), \frac{1}{\mu}\mathcal{G}\Lambda^{t_0+3/2}\psi\big) \leq \varepsilon\mathfrak{m}^N(U). \tag{4.36}$$

- *Control of* $\varepsilon\nu\big(\Lambda^{t_0+5/2}\zeta, \frac{1}{\mu}\mathcal{G}\Lambda^{t_0+3/2}\psi\big)$. It is a direct consequence of Proposition 3.12 that

$$\varepsilon\nu\big(\Lambda^{t_0+5/2}\zeta, \frac{1}{\mu}\mathcal{G}\Lambda^{t_0+3/2}\psi\big) \leq \varepsilon M|\mathfrak{P}\zeta|_{H^{t_0+5/2}}|\mathfrak{P}\psi|_{H^{t_0+3/2}}.$$

Since $|\mathfrak{P}\zeta|_{H^{t_0+5/2}} \leq |\zeta|_{H^{t_0+7/2}}$, and Lemma 4.6 controls $|\mathfrak{P}\psi|_{H^{t_0+3/2}}$, we therefore get

$$\varepsilon\nu\big(\Lambda^{t_0+5/2}\zeta, \frac{1}{\mu}\mathcal{G}\Lambda^{t_0+3/2}\psi\big) \leq \varepsilon\mathfrak{m}^N(U). \tag{4.37}$$

We can now deduce from (4.34), (4.35), (4.36), and (4.37) that

$$\frac{d}{dt}\mathcal{F}^0(U) \leq \epsilon\mathfrak{m}^N(U). \tag{4.38}$$

We will now study the case $\alpha \neq 0$. Taking the L^2-scalar product of (4.31) with $\mathcal{S}^\nu U_{(\alpha)} = \mathcal{S}^\nu[U]U_{(\alpha)}$ yields

$$\begin{aligned}\big(\mathcal{S}^\nu\partial_t U_{(\alpha)}, U_{(\alpha)}\big) \quad &+ \quad \big(\mathcal{S}^\nu J^\iota\mathcal{B}U_{(\alpha)}, U_{(\alpha)}\big)\\ &= \quad \epsilon\big(J^\iota(R_\alpha, S_\alpha + S'_\alpha)^T, \mathcal{S}^\nu U_{(\alpha)}\big),\end{aligned}$$

where we used the fact that $\big(J^\iota(\mathcal{A}+\varepsilon\nu\mathcal{C})U_{(\alpha)}, \mathcal{S}U_{(\alpha)}\big) = 0$. By definition of $\mathcal{F}^\alpha(U)$ we therefore get

$$\frac{d}{dt}\big(\mathcal{F}^\alpha(U) + \varepsilon\nu|\zeta_{(\alpha)}|^2_{H^{1/2}}\big) = A_1 + A_2 + B_1 + B_2, \tag{4.39}$$

with (recall that $\mathcal{S}^\nu = \mathcal{S} + \varepsilon\nu\mathrm{diag}(|D|, 0)$)

$$A_1 = \frac{1}{2}\big([\partial_t, \mathcal{S}]U_{(\alpha)}, U_{(\alpha)}\big), \qquad A_2 = -\big(\mathcal{S}^\nu J^\iota\mathcal{B}U_{(\alpha)}, U_{(\alpha)}\big),$$

and

$$\begin{aligned}B_1 \quad &= \quad \epsilon\big(J^\iota(R_\alpha, S_\alpha)^T, \mathcal{S}^\nu U_{(\alpha)}\big),\\ B_2 \quad &= \quad \epsilon\big(S'_\alpha, \frac{1}{\mu}\mathcal{G}\psi_{(\alpha)}\big).\end{aligned}$$

We now provide some control on the different components of the r.h.s. of (4.39).

- *Control of* A_1. We can decompose $A_1 = A_{11} + A_{12}$ with

$$A_{11} = \big((\partial_t\underline{\mathfrak{a}})\zeta_{(\alpha)}, \zeta_{(\alpha)}\big), \qquad A_{12} = \big(\frac{1}{\mu}[\partial_t, \mathcal{G}]\psi_{(\alpha)}, \psi_{(\alpha)}\big).$$

Proceeding as in the proof of Lemma 4.27, we get that $|\partial_t\underline{\mathfrak{a}}| \leq \varepsilon\mathfrak{m}^N(U)$ from which we deduce that $|A_{11}| \leq \varepsilon\mathfrak{m}^N(U)$. The same estimate holds for A_{12} as a consequence of Proposition 3.29 and the observation that $[\partial_t, \mathcal{G}]\psi_{(\alpha)} = d\mathcal{G}(\partial_t\zeta)\psi_{(\alpha)}$. We therefore get

$$|A_1|_2 \leq \varepsilon\mathfrak{m}^N(U). \tag{4.40}$$

- *Control of* A_2. We can first note that $A_2 = A_{21} + A_{22} + A_{23}$ with

$$\begin{aligned}A_{21} \quad &= \quad -\varepsilon\big(\underline{\mathfrak{a}}J^\iota\nabla^\gamma\cdot(\zeta_{(\alpha)}\underline{V}), \zeta_{(\alpha)}\big),\\ A_{22} \quad &= \quad -\varepsilon\big(\frac{1}{\mu}\mathcal{G}J^\iota\underline{V}\cdot\nabla^\gamma\psi_{(\alpha)}, \psi_{(\alpha)}\big),\\ A_{23} \quad &= \quad -\varepsilon\nu\big(\Lambda J^\iota\nabla^\gamma\cdot(\zeta_{(\alpha)}\underline{V}), \zeta_{(\alpha)}\big).\end{aligned}$$

Noting that

$$(\underline{\mathfrak{a}}J^\iota\nabla^\gamma\cdot(\underline{V}\bullet)) + (\underline{\mathfrak{a}}J^\iota\nabla^\gamma\cdot(\underline{V}\bullet))^* = [\underline{\mathfrak{a}}, J^\iota]\underline{V}\cdot\nabla^\gamma - [\underline{V}\cdot\nabla^\gamma, J^\iota\underline{\mathfrak{a}}] + \underline{\mathfrak{a}}J^\iota(\nabla^\gamma\underline{V}),$$

and controlling the commutator estimates with J^ι with Corollary B.9 (Appendix B), we get that $A_{21} \leq \varepsilon\mathfrak{m}^N(U)$. The fact that A_{22} satisfies a similar estimate is a direct consequence of Proposition 3.30 and Remark 3.31. For A_{23}, we integrate by parts to get

$$\begin{aligned}|A_{23}| &= \frac{1}{2}\varepsilon\nu\big(\zeta_{(\alpha)}, (\nabla^\gamma\cdot\underline{V})J^\iota\Lambda\zeta_{(\alpha)}\big)\\ &= \frac{1}{2}\varepsilon\nu\big(\Lambda^{1/2}\zeta_{(\alpha)}, \Lambda^{-1/2}(\nabla^\gamma\cdot\underline{V})J^\iota\Lambda\zeta_{(\alpha)}\big).\end{aligned}$$

Using the Cauchy-Schwarz inequality and Proposition B.2 (Appendix B), we deduce that

$$|A_{23}| \leq \varepsilon\mathfrak{m}^N(U) + \varepsilon\nu|\zeta_{(\alpha)}|^2_{H^{1/2}}.$$

We therefore obtain

$$|A_2|_2 \leq \varepsilon\mathfrak{m}^N(U) + \varepsilon\nu|\zeta_{(\alpha)}|^2_{H^{1/2}}. \tag{4.41}$$

- *Control of* B_1. We first decompose $B_1 = B_{11} + B_{12}$ with

$$\begin{aligned}B_{11} &= \epsilon\big(J^\iota(R_\alpha, S_\alpha)^T, \mathcal{S}U_{(\alpha)}\big),\\ B_{12} &= \varepsilon\nu\epsilon\big(J^\iota R_\alpha, \Lambda\zeta_{(\alpha)}\big).\end{aligned}$$

Using the bounds established on R_α and S_α in Proposition 4.10 and in Notation 3.11, there is no problem to control B_{11}, and Proposition 4.8 can be used to control B_{12},

$$|B_{11}| \leq \epsilon\mathfrak{m}^N(U), \qquad |B_{12}| \leq \epsilon\mathfrak{m}^N(U)\big(\varepsilon\nu\mu^{-1/4} + \varepsilon\nu|\zeta|^2_{H^{N+1/2}}\big)$$

and therefore

$$|B_1| \leq \epsilon\mathfrak{m}^N(U)\big(1 + \nu\mu^{-1/4} + \varepsilon\nu|\zeta|^2_{H^{N+1/2}}\big). \tag{4.42}$$

- *Control of* B_2. We can deduce from Proposition 3.12 that

$$|B_2| \leq \epsilon M|\mathfrak{P}S'_\alpha|_2|\mathfrak{P}\psi_{(\alpha)}|_2.$$

Recalling that $S'_\alpha = -(1-J^\iota)\big[\partial_t\underline{w}\zeta_{(\alpha)}\big] + [\underline{w}, J^\iota]\partial^\alpha\mathcal{G}\psi$, and observing that $|\mathfrak{P}S'_\alpha|_2 \leq \mu^{-1/4}|S'_\alpha|_{H^{1/2}}$, it follows that

$$|B_2| \leq \epsilon M\big(|\mu^{-1/4}\partial_t\underline{w}|_{H^{t_0}}|\zeta|_{H^{N+1/2}} + |\mu^{-1/4}\underline{w}|_{H^{t_0+1}}|\mathcal{G}\psi|_{H^{N-1/2}}\big)|\mathfrak{P}\psi_{(\alpha)}|_2,$$

where we used Proposition B.2 to control the first term and Corollary B.9 to control the second one. Proceeding as in §4.3.1 to control $|\partial_t\underline{w}|_{H^{t_0}}$ in terms of $\mathfrak{m}^N(U)$, and using Proposition 4.4 and Theorem 3.15 to control $|\underline{w}|_{H^{t_0}}$ and $|\mathcal{G}\psi|_{H^{N-1/2}}$ respectively, we therefore get

$$|B_2| \leq \epsilon\big(|\zeta|_{H^{N+1/2}} + |\mathfrak{P}\psi|_{H^N}\big)\mathfrak{m}^N(U),$$

From which we can deduce, with the help of Remark 4.7, that

$$|B_2| \leq \epsilon\mathfrak{m}^N(U)\big(1 + |\zeta|_{H^{N+1/2}}\big). \tag{4.43}$$

Gathering all the information from the above estimates, we deduce from (4.39) that for $\alpha \neq 0$

$$\frac{d}{dt}\big(\mathcal{F}^\alpha(U^{\iota,\nu}) + \varepsilon\nu|\zeta_{(\alpha)}|_{H^{1/2}}\big) \leq \epsilon\mathfrak{m}^N(U^{\iota,\nu})(1 + \nu\mu^{-1/4} + |\zeta|_{H^{N+1/2}}). \tag{4.44}$$

Summing (4.44) for all $\alpha \in \mathbb{N}^d$ with $1 \leq |\alpha| \leq N$ and (4.38) for $\alpha = 0$, we deduce that

$$\frac{d}{dt}\mathcal{F}_\nu^{[N]}(U^{\iota,\nu}) \leq \epsilon\mathfrak{m}^N(U^{\iota,\nu})(1 + \nu\mu^{-1/4} + |\zeta|_{H^{N+1/2}}),$$

where we used the notation

$$\mathcal{F}_\nu^{[N]}(U^{\iota,\nu}) = \mathcal{F}^{[N]}(U^{\iota,\nu}) + \varepsilon\nu|\zeta|_{H^{N+1/2}}. \tag{4.45}$$

We can now use Lemma 4.27 to get

$$\frac{d}{dt}\mathcal{F}_\nu^{[N]}(U^{\iota,\nu}) \leq \epsilon C\big(M, \nu\mu^{-1/4}, (\varepsilon\nu)^{-1}, \mathcal{F}_\nu^{[N]}(U^{\iota,\nu}), \frac{1}{a_0}\big). \tag{4.46}$$

4.3.4.4. *Construction of a solution.* We first show that it is possible to find $T^\nu > 0$ *independent of* ι such that $T^\nu/\epsilon < T_{max}^{\iota,\nu}$ for all $0 < \iota, \nu < 1$ and such that $U^{\iota,\nu}$ satisfies the following identities

$$1 + \varepsilon\zeta^\iota - \beta b \geq \frac{1}{2}h_{min} \quad \text{and} \quad \underline{\mathfrak{a}}^\iota \geq \frac{1}{2}a_0 \quad \text{on} \quad [0, \frac{T^\nu}{\epsilon}]. \tag{4.47}$$

We therefore define T^* as

$$T^* = \sup\{t > 0, \quad \frac{t}{\epsilon} \leq T_{\max}^\iota \quad \text{and} \quad (4.47) \quad \text{holds on} \quad [0, \frac{1}{\epsilon}t]\}.$$

(By continuity, the existence of T^* is ensured by the assumptions made in the statement of the proposition.)
From the energy estimate (4.46) we know that there exists $T^\nu > 0$ that depends on $\mathcal{F}_\nu^{[N]}(U^0)$, $\dfrac{1}{h_{min}}$, $\dfrac{1}{a_0}$, $\nu\mu^{-1/4}$ and $(\varepsilon\nu)^{-1}$ but not on ι and such that

$$\mathcal{F}_\nu^{[N]}(U^{\iota,\nu}(t)) \leq C(T^\nu, \mathcal{F}_\nu^{[N]}(U^0)) \quad \text{for all} \quad t \in [0, \frac{1}{\epsilon}\min\{T^*, T^\nu\}]. \tag{4.48}$$

On this time interval, one has

$$\begin{aligned} 1 + \varepsilon\zeta^{\iota,\nu} - \beta b &= (1 + \varepsilon\zeta^0 - \beta b) + \varepsilon\int_0^t \partial_t\zeta^{\iota,\nu} \\ &\geq h_{min} - \min\{T^*, T^\nu\} \sup_{t\in[0,\frac{1}{\epsilon}\min\{T^*,T^\nu\}]} |\partial_t\zeta^{\iota,\nu}|_\infty. \end{aligned}$$

Since $|\partial_t\zeta^{\iota,\nu}|_\infty \lesssim \mathcal{F}_\nu^{[N]}(U^{\iota,\nu}) \leq C(T^\nu, \mathcal{F}_\nu^{[N]}(U^0))$, we deduce (taking a smaller T^ν if necessary) that $1 + \varepsilon\zeta^{\iota,\nu} - \beta b \geq \frac{1}{2}h_{min}$.
With a similar argument, we can show that $\underline{\mathfrak{a}}^\iota \geq \frac{1}{2}a_0$ on $[0, \frac{1}{\epsilon}\min\{T^*, T^\nu\}]$ (for a smaller T^ν if necessary), and we therefore can conclude that $T^* \geq T$. In other words, for all $0 < \iota < 1$ the $U^{\iota,\nu}$ exist (at least) on the same time interval $[0, T^\nu/\epsilon]$ that does not depend on ι. Moreover, it follows from (4.48), the definition (4.45) of $\mathcal{F}_\nu^{[N]}(U)$, and Lemma 4.27 that for all $t \in [0, \frac{1}{\varepsilon\sqrt{\beta}}T^\nu]$,

$$\mathcal{E}^N(U^{\iota,\nu}(t)) + \varepsilon\nu|\zeta^{\iota,\nu}(t)|_{H^{N+1/2}}^2 \leq C\big(T^\nu, \mathcal{E}^N(U^0) + \varepsilon\nu|\zeta^0|_{H^{N+1/2}}^2\big). \tag{4.49}$$

The next step is to prove that for all $0 < \nu < 1$, the sequence $(U^{\iota,\nu})_{0<\iota<1}$ converges to a solution U^ν of the system obtained by setting $\iota = 0$ (or equivalently $J^\iota = 1$ in (4.30)),

$$\begin{cases} \partial_t\zeta - \dfrac{1}{\mu}\mathcal{G}\psi = 0, \\ \partial_t\psi + \zeta + \varepsilon\dfrac{1}{2}\Big(|\nabla^\gamma\psi|^2 - \dfrac{\varepsilon}{\mu}\dfrac{(\mathcal{G}\psi + \varepsilon\mu\nabla^\gamma\zeta\cdot\nabla^\gamma\psi)^2}{2(1+\varepsilon^2\mu|\nabla^\gamma\zeta|^2)^2}\Big) = -\varepsilon\nu\Lambda\zeta. \end{cases} \tag{4.50}$$

A crucial step is the following lemma, which will also prove useful for the proof of the uniqueness of the solutions to (4.1). (Note that, as discussed in [**4**], standard compactness arguments do not work here because the Dirichlet-Neumann operator is nonlocal.)

LEMMA 4.28. *For all* $0<\nu<1$, $(U^{\iota,\nu})_{0<\iota<1}$ *is a Cauchy sequence, as* $\iota\to 0$, *in* $C([0,T^\nu/\epsilon];H^2\times\dot H^2)$.

PROOF. Throughout this proof, we use the notations $\mathcal{G}^j=\mathcal{G}_{\mu,\gamma}[\varepsilon\zeta^j,\beta b]$ ($j=1,2$) and $\underline{U}:=U^2-U^1$. Recalling that by convention $\psi_{(0)}=\psi$ and defining

$$\forall U=(\zeta,\psi),\qquad \underline{\mathcal{E}}^2(U)=\sum_{\alpha\in\mathbb{N}^d,|\alpha|\leq 2}|\partial^\alpha\zeta|_2^2+|\mathfrak{P}\psi_{(\alpha)}|_2^2,$$

(the difference with $\mathcal{E}^2(U)$ as defined in (4.8) is the absence of the term $|\mathfrak{P}\psi|_{H^{t_0+3/2}}$), we can note that owing to (4.49), one has

$$\begin{aligned}|\underline{U}|^2_{H^2\times\dot H^2}&\sim |\underline{\zeta}|^2_{H^2}+|\mathfrak{P}\underline{\psi}|^2_{H^{3/2}}\\ &\leq C\big(T,\mathcal{E}^N(U^0)+\varepsilon\nu|\zeta^0|^2_{H^{N+1/2}}\big)\underline{\mathcal{E}}^2(\underline{U}).\end{aligned}$$

So it is enough to prove that $(U^{\iota,\nu})_{0<\iota<1}$ is a Cauchy sequence for $\underline{\mathcal{E}}^2(\cdot)^{1/2}$. To achieve this, the key point is to show that for all $\alpha\in\mathbb{N}^d$, $1\leq|\alpha|\leq 2$ (the modifications in the case $\alpha=0$ are left to the reader), the quantity $\underline{U}_{(\alpha)}$ solves a system of the form

$$\left\{\begin{array}{l}\partial_t\underline{\zeta}_{(\alpha)}+\varepsilon J^{\iota_2}\nabla^\gamma\cdot(\underline{\zeta}_{(\alpha)}\underline{V}^2)-\frac{1}{\mu}J^{\iota_2}\mathcal{G}^2\underline{\psi}_{(\alpha)}=J^{\iota_2}F_\alpha+\partial^\alpha A_{\iota^1,\iota^2},\\ \partial_t\underline{\psi}_{(\alpha)}+J^{\iota^2}\big((\mathfrak{a}^2+\varepsilon\nu\Lambda)\underline{\zeta}_{(\alpha)}\big)+\varepsilon J^{\iota^2}(\underline{V}^2\cdot\nabla^\gamma\underline{\psi}_{(\alpha)})=J^{\iota_2}G_\alpha+\partial^\alpha B_{\iota^1,\iota^2},\end{array}\right.\tag{4.51}$$

where $\underline{V}^2$ and $\underline{\mathfrak{a}}^2$ are as in Theorem 3.21 and (4.20) respectively, with ζ, ψ replaced by ζ^2, ψ^2, and where

$$|F_\alpha|_2+\sqrt{\varepsilon\nu}|F_\alpha|_{H^{1/2}}+|\mathfrak{P}G_\alpha|_2\leq\widetilde{C}\left(\underline{\mathcal{E}}^2(\underline{U})+\varepsilon\nu|\underline{\zeta}|^2_{H^{5/2}}\right)^{1/2},\tag{4.52}$$

with $\widetilde{C}=C(\frac{1}{\mu},T,\mathcal{E}^N(U^0)+|\zeta^0|^2_{H^{N+1/2}})$, and

$$|\partial^\alpha A_{\iota_1,\iota_2}|_2+\sqrt{\varepsilon\nu}|\partial^\alpha A_{\iota_1,\iota_2}|_{H^{1/2}}+|\mathfrak{P}\partial^\alpha B_{\iota_1,\iota_2}|_2\leq|\iota^2-\iota^1|\widetilde{C}.\tag{4.53}$$

For the sake of simplicity, we give only the proof in the case of flat bottoms. For the general case, the same kind of adaptations as in the proof of Proposition 4.5 must be performed.

Step 1. Proof of the bound on F_α. Proceeding as in the proof of Proposition 4.5, one has $F_\alpha=F_{\alpha,1}+F_{\alpha,2}+F_{\alpha,3}+F_{\alpha,4}$, with

$$\begin{aligned}F_{\alpha,1}&=\frac{1}{\mu}\sum_{\alpha^1+\alpha^2=\alpha,|\alpha^1|=|\alpha^2|=1}*\,d^2\mathcal{G}^2(\partial^{\alpha^1}\underline{\zeta}^2,\partial^{\alpha^2}\zeta^2)\underline{\psi},\\ F_{\alpha,2}&=\frac{1}{\mu}\sum_{\alpha^1+\alpha^2=\alpha,|\alpha^1|=|\alpha^2|=1}*\,d\mathcal{G}^2(\partial^{\alpha^1}\zeta_2)\partial^{\alpha^2}\underline{\psi},\\ F_{\alpha,3}&=-\nabla^\gamma\cdot\big(\zeta^1_{(\alpha)}(\underline{V}^2-\underline{V}^1)\big),\\ F_{\alpha,4}&=-(\mathcal{G}^2-\mathcal{G}^1)\psi^1_{(\alpha)},\end{aligned}$$

where $*$ stands for numerical coefficients of no importance. Using Proposition 3.28 and the bounds on U^1 and U^2 furnished by (4.49), we easily check that $F_{\alpha,1}$ and $F_{\alpha,2}$ satisfy the bound (4.52). The same is also true for $F_{\alpha,3}$ as a consequence of the identity $|\underline{V}^2-\underline{V}^1|_2\leq C(1/\mu,\mathfrak{m}^N(U^1),\mathfrak{m}^N(U^2))|\mathfrak{P}\underline{\psi}|_{H^{3/2}}$ that follows easily from

the definition $\underline{V}$. The result on $F_{\alpha,4}$ stems from Proposition 3.28.
Step 2. Proof of the bound on G_α. An adaptation similar to the one made in the proof of Proposition 4.10 yields the result.
Step 3. Proof of the bound on A_{ι^1,ι^2} and B_{ι^1,ι^2}. One has

$$(A_{\iota^1,\iota^2}, B_{\iota^1,\iota^2})^T = -(J^{\iota^2} - J^{\iota^1})\big(\mathcal{N}(U^1) + \varepsilon\nu\mathcal{C}(D)U^1\big),$$

where $\mathcal{N}(U^1)$ and $\mathcal{C}(D)$ are as in (4.26) and (4.32) respectively. The desired upper bound therefore follows from (4.49) and the observation that $|(J^{\iota^2} - J^{\iota^1})f|_{H^s} \lesssim |\iota^2 - \iota^1||f|_{H^{s+1}}$.
Step 4. Energy estimate. Multiplying (4.51) by $S^\nu\underline{U}_{(\alpha)}$ (for $1 \leq |\alpha| \leq 2$) yields as in §4.3.4.3 above that

$$\begin{aligned}\frac{d}{dt}\big(\mathcal{F}^\alpha(\underline{U}) + \varepsilon\nu|\partial^\alpha\underline{\zeta}|^2_{H^{1/2}}\big) \quad &\leq \quad M\big(|F_\alpha|_{H^{1/2}_{\varepsilon\nu}} + |\mathfrak{P}G_\alpha|_2 + |A_{\iota_1,\iota_2}|_{H^{1/2}_{\varepsilon\nu}} + |\mathfrak{P}B_{\iota_1,\iota_2}|_2\big)\\ &\qquad \times\big(\underline{\mathcal{E}}^2(\underline{U}) + \varepsilon\nu|\zeta|^2_{H^{5/2}}\big)^{1/2},\end{aligned} \tag{4.54}$$

where we used the notation

$$|f|^2_{H^{1/2}_{\varepsilon\nu}} = |f|_2^2 + \varepsilon\nu|f|^2_{H^{1/2}}.$$

When $\alpha = 0$, we write $\underline{\mathcal{F}}^0(U) = |\zeta|^2_{H^{1/2}_{\varepsilon\nu}} + |\mathfrak{P}\psi|_2^2$ and multiply (4.30) by $(\zeta + \varepsilon\nu\Lambda\zeta, \mathfrak{P}^2\psi)^T$ and integrate to obtain without difficulty that

$$\frac{d}{dt}\big(\underline{\mathcal{F}}^0(\underline{U}) + \varepsilon\nu|\underline{\zeta}|^2_{H^{1/2}}\big) \leq \widetilde{C}\big(\underline{\mathcal{E}}^2(\underline{U}) + \varepsilon\nu|\zeta|^2_{H^{5/2}} + |\iota^2 - \iota^1|\big). \tag{4.55}$$

Defining $\underline{\mathcal{F}}^{[2]}(U) = \underline{\mathcal{F}}^0(\underline{U}) + \sum_{1\leq|\alpha|\leq 2}\mathcal{F}^\alpha(\underline{U})$, and using (4.52), we deduce after summing (4.54) (for $1 \leq |\alpha| \leq 2$) and (4.55) that

$$\begin{aligned}\frac{d}{dt}\big(\underline{\mathcal{F}}^{[2]}(\underline{U}) + \varepsilon\nu|\underline{\zeta}|^2_{H^{5/2}}\big) \quad &\leq \quad \widetilde{C}\,\big(\underline{\mathcal{E}}^2(\underline{U}) + \varepsilon\nu|\underline{\zeta}|^2_{H^{5/2}} + |\iota^2 - \iota^1|\big)\\ &\leq \quad \widetilde{C}\,\big(\underline{\mathcal{F}}^{[2]}(\underline{U}) + |\underline{\zeta}|^2_{H^{5/2}} + |\iota^2 - \iota^1|\big),\end{aligned}$$

where we used Lemma 4.27 to derive the second inequality. The result then follows from Gronwall's inequality and Lemma 4.27. □

Using the lemma and Proposition 2.3, we know that there exists $(\zeta^\nu, \psi^\nu) \in C([0, T^\nu/\epsilon]; H^2 \times \dot{H}^2)$ such that

$$\lim_{\iota\to 0}|\zeta^{\iota,\nu} - \zeta^\nu|_{H^2} = 0 \qquad \text{and} \qquad \lim_{\iota\to 0}|\psi^{\iota,\nu} - \psi^\nu|_{\dot{H}^2} = 0.$$

In order to prove that $U^\nu = (\zeta^\nu, \psi^\nu)$ solves (4.50) (adding to ψ^ν a time-dependent function if necessary), the only nontrivial task is to check that $\mathcal{G}[\varepsilon\zeta^{\iota,\nu}, \beta b]\psi^\nu$ converges to $\mathcal{G}[\varepsilon\zeta^\nu, \beta b]\psi^\nu$ in $C([0, T^\nu/\epsilon]; L^2)$ (for instance). We thus write

$$\begin{aligned}\mathcal{G}[\varepsilon\zeta^{\iota,\nu}, \beta b]\psi^\nu - \mathcal{G}[\varepsilon\zeta^\nu, \beta b]\psi^\nu \quad &= \quad \mathcal{G}[\varepsilon\zeta^{\iota,\nu}, \beta b](\psi^{\iota,\nu} - \psi^\nu)\\ &\qquad + \big(\mathcal{G}[\varepsilon\zeta^\nu, \beta b] - \mathcal{G}[\varepsilon\zeta^{\iota,\nu}, \beta b]\big)\psi^\nu.\end{aligned}$$

Since the sequence $(U^{\iota,\nu})$ is uniformly bounded for the energy $\mathcal{E}^N$ through (4.49), the first term of the r.h.s. converges to 0 in L^2 owing to Theorem 3.15, and the second owing to the third point of Proposition 3.28. It is then easy to check that U^ν solves (4.50), and moreover, one has

$$\mathcal{E}^N(U^\nu(t)) + \varepsilon\nu|\zeta^\nu(t)|^2_{H^{N+1/2}} \leq C\big(T^\nu, \mathcal{E}^N(U^0) + \varepsilon\nu|\zeta^0|^2_{H^{N+1/2}}\big).$$

At this point of the analysis, T^ν may depend on $(\varepsilon\nu)^{-1}$, which prevents us from taking the limit $\nu \to 0$ in the above inequality. Looking carefully at the energy estimates of §4.3.4.3, we can note that this singular dependence comes from the component B_2 alone. Since $B_2 = 0$ when $\iota = 0$, we can remove the dependence of T^ν on $(\varepsilon\nu)^{-1}$. We can use this fact to prove that there exists $\widetilde{T}$ that depends on $\mathcal{F}_\nu^{[N]}(U^0)$, $\dfrac{1}{h_{min}}$, $\dfrac{1}{a_0}$, $\nu\mu^{-1/4}$ *but not on* $(\varepsilon\nu)^{-1}$ such that (4.47) holds on $[0, \widetilde{T}/\epsilon]$. Moreover, (4.49) can be made more precise,

$$\mathcal{E}^N(U^\nu(t)) + \varepsilon\nu|\zeta^\nu(t)|^2_{H^{N+1/2}} \le C\big(\widetilde{T}, \mathcal{E}^N(U^0) + \varepsilon\nu|\zeta^0|^2_{H^{N+1/2}}\big).$$

Since the dependence on $(\varepsilon\nu)^{-1}$ is removed from this energy estimate, we can then proceed as for Lemma 4.28 to prove that $(U^\nu)_{0<\nu<1}$ is a Cauchy sequence in $C([0,\widetilde{T}/\epsilon]; H^2 \times \dot{H}^2)$ and therefore converges (adding a time-dependent function to the velocity potential if necessary) to a solution $U \in C([0,\widetilde{T}/\epsilon]; H^2 \times \dot{H}^2)$ to the water waves equations (4.1). The existence time $\widetilde{T}$ is not the same as the existence time T claimed in the statement of the theorem, since it depends on some extra terms (such as $\nu\mu^{-1/4}$, which is singular as $\mu \to 0$). However, in the limit $\nu \to 0$, these terms disappear and one gets the desired result.

4.3.4.5. *Uniqueness and stability.* Quite obviously, uniqueness follows from the stability property stated in Theorem 4.18 that we will now prove. We recall (see (4.31) with $\iota = \nu = 0$) that for all $\alpha \in \mathbb{N}^d$, $|\alpha| \le N-1$, one has

$$\partial_t U_{(\alpha)} + \mathcal{A}[U]U_{(\alpha)} = \epsilon(R_\alpha, S_\alpha)^T,$$

with $\mathcal{E}^0\big((R_\alpha, S_\alpha)\big) \le c_{WW}$ (we recall that $\mathcal{E}^0\big((\zeta,\psi)\big) = |\zeta^2| + |\mathfrak{P}\psi|_2^2)$. We get similarly that

$$\partial_t U^{app}_{(\alpha)} + \mathcal{A}[U^{app}]U^{app}_{(\alpha)} = \epsilon(R^{app}_\alpha, S^{app}_\alpha)^T + r_{(\alpha)},$$

with $r_{(\alpha)} = (\partial^\alpha r^1, \partial^\alpha r^2 - \varepsilon \underline{w}^{app}\partial^\alpha r^1)^T$ and $\mathcal{E}^0\big((R^{app}_\alpha, S^{app}_\alpha)\big) \le c_{app}$. Subtracting these quantities gives the following equations on $\mathfrak{e}_{(\alpha)} = U_{(\alpha)} - U^{app}_{(\alpha)}$

$$\partial_t \mathfrak{e}_{(\alpha)} + \mathcal{A}[U]\mathfrak{e}_{(\alpha)} = -(\mathcal{A}[U] - \mathcal{A}[U^{app}])U^{app}_{(\alpha)} + \epsilon(R_\alpha - R^{app}_\alpha, S_\alpha - S^{app}_\alpha)^T - r_{(\alpha)}.$$

Using the fact that $|\alpha| \le N-1$, and also using Proposition 3.28 and Lemma 4.6, we can note that

$$\mathcal{E}^0\big(\mathcal{A}[U] - \mathcal{A}[U^{app}])U^{app}_{(\alpha)}\big) \le \epsilon C(c_{WW}, c_{app})\mathcal{E}^{N-1}(\mathfrak{e}),$$

while we deduce from the Lipschitz dependence of (R_α, S_α) on U (the second point of Proposition 4.5 and Remark 4.12) that

$$\mathcal{E}^0\big((R_\alpha - R^{app}_\alpha, S_\alpha - S^{app}_\alpha)\big) \le C(c_{WW}, c_{app})\mathcal{E}^{N-1}(\mathfrak{e}).$$

The result then follows easily with the same energy estimates as in §4.3.4.3.

4.3.5. The Rayleigh-Taylor criterion. In §4.3.5.1 we rewrite the water waves equations (4.1) as a set of evolution equations on ζ and $\underline{V}$, where we recall that $\underline{V}$ and $\underline{w}$ are respectively the horizontal and vertical components of the velocity evaluated at the surface (see Theorem 3.21 and Remark 3.22). From this expression, we deduce the relation between the condition $\mathfrak{a} > 0$ made in the statement of Theorem 4.16 and the classical Rayleigh-Taylor criterion [**308**],

$$\inf_{\mathbb{R}^d}\big(-\partial_z P_{|z=\varepsilon\zeta}\big) > 0, \tag{4.56}$$

where P is the dimensionless pressure (see Remark 1.14). We then make some comments on the validity of (4.56).

4.3.5.1. *Reformulation of the equations.* Let us denote by $\underline{\mathbf{U}} = \mathbf{U}_{|_{z=\varepsilon\zeta}}$ the trace of the velocity field $\mathbf{U}$ at the surface, and $\underline{V}$ and $\underline{w}$ its horizontal and vertical components respectively. As seen in Remark 1.14, the first equation of (4.1) can then be restated as

$$\partial_t\zeta + \varepsilon\underline{V}\cdot\nabla\zeta - \frac{1}{\mu}\underline{w} = 0. \tag{4.57}$$

Let us note that

$$(\partial\mathbf{U})_{|_{z=\zeta}} = \partial\underline{\mathbf{U}} - \varepsilon\partial\zeta(\partial_z\mathbf{U})_{|_{z=\varepsilon\zeta}},$$

with ∂ standing for ∂_j $(1 \leq j \leq d)$ or ∂_t, and that

$$(\nabla P)_{|_{z=\varepsilon\zeta}} = -\varepsilon\nabla\zeta(\partial_z P)_{|_{z=\varepsilon\zeta}}$$

(where we used the fact that the pressure is constant at the surface).
Taking the trace at the surface of Equation (H1)′ in §1.1.2 (or rather, of its dimensionless version given in Remark 1.14) we get

$$\begin{aligned}
\partial_t\underline{V} + \varepsilon(\underline{V}\cdot\nabla)\underline{V} - \varepsilon\big(\partial_t\zeta + \varepsilon\underline{V}\cdot\nabla\zeta - \frac{1}{\mu}\underline{w}\big)(\partial_z\underline{V})_{|_{z=\varepsilon\zeta}} &= \varepsilon\frac{P_0}{\rho a g}(\partial_z P)_{|_{z=\varepsilon\zeta}}\nabla\zeta \\
\partial_t\underline{w} + \varepsilon(\underline{V}\cdot\nabla)\underline{w} - \varepsilon\big(\partial_t\zeta + \varepsilon\underline{V}\cdot\nabla\zeta - \frac{1}{\mu}\underline{w}\big)(\partial_z\underline{w})_{|_{z=\zeta}} &= -\frac{P_0}{\rho a g}(\partial_z P)_{|_{z=\varepsilon\zeta}} - \frac{1}{\varepsilon},
\end{aligned}$$

where $P_0 = \rho g H_0$. Using (4.57), we therefore get that $(\zeta, \underline{V})$ solves

$$\begin{cases} \partial_t\zeta + \varepsilon\underline{V}\cdot\nabla\zeta - \frac{1}{\mu}\underline{w} = 0, \\ \partial_t\underline{V} + \underline{\mathfrak{a}}\nabla\zeta + \varepsilon(\underline{V}\cdot\nabla)\underline{V} = 0, \end{cases} \tag{4.58}$$

with

$$\underline{\mathfrak{a}} = -\varepsilon\frac{P_0}{\rho a g}(\partial_z P)_{|_{z=\varepsilon\zeta}} = 1 + \varepsilon(\partial_t + \underline{V}\cdot\nabla)\underline{w}. \tag{4.59}$$

This new formulation (4.58)–(4.59) of the water-waves equations (4.1) allows us to make the link between (4.56) and the (strict)-hyperbolicity condition $\underline{\mathfrak{a}} > 0$.

PROPOSITION 4.29. *The Rayleigh-Taylor condition* (4.56) *is equivalent to the condition* $\underline{\mathfrak{a}} > 0$ *required in the statement of Theorem* 4.16.

REMARK 4.30. It is worth pointing out that the derivation of (4.58) from the free surface Euler equations does not require that the flow be irrotational.

4.3.5.2. *Comments on the Rayleigh-Taylor criterion* (4.56). We end this section with some comments on the assumption that $\inf_{\mathbb{R}^d}\mathfrak{a}(U^0) > 0$. This assumption was made in the statement of Theorem 4.16 and which, by Proposition 4.29, is equivalent to the assumption that the Rayleigh-Taylor condition (4.56) is initially satisfied.
Ebin [**141**] showed that the water waves equations are ill posed when this condition is not satisfied; the following proposition shows that it is always (uniformly) satisfied in asymptotic regimes for which $\varepsilon\mu$ is small enough.

PROPOSITION 4.31. *Let* $t_0 > d/2$ *and* $\mathcal{A}$ *be an asymptotic regime. Also let* $(U^0_\mathfrak{p})_{\mathfrak{p}\in\mathcal{A}} \in E^N_0$ *with* $\sup_{\mathfrak{p}\in\mathcal{A}}\mathcal{E}^N(U^0_\mathfrak{p}) < \infty$ *and moreover assume that*

$$\exists h_{min} > 0, \qquad \forall \mathfrak{p}\in\mathcal{A}, \qquad 1 + \varepsilon\zeta^0_\mathfrak{p} - \beta b \geq h_{min}.$$

There exists $m_0 = m_0(\sup_{\mathfrak{p}\in\mathcal{A}} \mathcal{E}^N(U^0_{\mathfrak{p}}), \frac{1}{h_{min}})$ *and* $a_0 > 0$ *such that for all* $\mathfrak{p} \in \mathcal{A}$ *such that* $\varepsilon\mu \leq m_0$, *one has* $\mathfrak{a}(U^0_{\mathfrak{p}}) \geq a_0$.

PROOF. Recall first that $\mathfrak{a}(U^0)$ is defined in (4.27) and can be written $\mathfrak{a}(U^0_{\mathfrak{p}}) = 1 + \varepsilon\mathfrak{b}(U^0_{\mathfrak{p}})$, with

$$\mathfrak{b}(U) = \varepsilon\mathcal{V}[\varepsilon\zeta,\beta b]\psi \cdot \nabla^\gamma(\mathcal{w}[\varepsilon\zeta,\beta b]\psi) - \mathcal{w}[\varepsilon\zeta,\beta b]\mathcal{N}_2(U) - d\mathcal{w}(\mathcal{N}_1(U))\psi. \tag{4.60}$$

By Theorem 3.15 and Proposition 4.4, we easily get

$$|\mathfrak{b}|_\infty \leq \mu\mathfrak{m}^N(U^0_{\mathfrak{p}}),$$

and therefore $|\mathfrak{a}(U^0_{\mathfrak{p}})|_\infty \geq 1 - \varepsilon\mu\mathfrak{m}^N(U^0_{\mathfrak{p}})$, from which the result follows directly. □

It turns out that the assumption $\inf_{\mathbb{R}^d} \mathfrak{a}(U) > 0$, or equivalently the Rayleigh-Taylor condition (4.56), is satisfied in much more general configurations. Wu proved in [**326, 327**] that this condition is satisfied by any solution of the water waves equations in infinite depth. Using a maximum principle as in [**327**], we show in the proposition below that this is also true in finite depth, for the case of flat bottoms. For nonflat bottoms, a geometrical condition on the bottom parametrization b is needed, however. This condition requires the introduction of the "anisotropic Hessian" $\mathcal{H}^\gamma_b$ associated to the bottom parametrization b,

$$\mathcal{H}^\gamma_b := \begin{pmatrix} \partial_x^2 b & \gamma\partial_{xy}^2 b \\ \gamma\partial_{xy}^2 b & \gamma^2\partial_y^2 b \end{pmatrix}.$$

We also recall that the velocity potential Φ^0 associated to the initial condition $U^0 = (\zeta^0, \psi^0)$ is found by solving the boundary value problem

$$\begin{cases} \Delta^{\mu,\gamma}\Phi^0 = 0, & -1 + \beta b \leq z \leq \varepsilon\zeta^0, \\ \Phi^0\,_{|_{z=\varepsilon\zeta^0}} = \psi^0, & \partial_{\mathbf{n}}\Phi^0\,_{|_{z=-1+\beta b}} = 0. \end{cases} \tag{4.61}$$

PROPOSITION 4.32. *Let* $t_0 > d/2$ *and* $U^0 \in E^N_0$ *and moreover assume that*

$$\exists h_{min} > 0, \qquad 1 + \varepsilon\zeta^0 - \beta b \geq h_{min}.$$

Then the Rayleigh-Taylor condition (4.56) *is satisfied provided that*

$$-\varepsilon^2\beta\mu\mathcal{H}^\gamma_b(\nabla^\gamma\Phi^0\,_{|_{z=-1+\beta b}}) \leq 1,$$

which is in particular always the case for flat bottoms since $\mathcal{H}^\gamma_b = 0$ *if* $b = 0$.

PROOF. Taking the divergence of the nondimensionalized Euler equations discussed in Remark 1.14, and recalling that $\mathbf{U} = \nabla^{\mu,\gamma}\Phi$, with Φ solving (1.30), we get

$$\begin{aligned} -\frac{P_0}{\rho a g}\Delta^{\mu,\gamma}P &= \frac{\varepsilon}{\mu}\Delta^{\mu,\gamma}|\nabla^{\mu,\gamma}\Phi|^2 \\ &= 2\frac{\varepsilon}{\mu}\Big(\mu(\partial_x U_1)^2 + \gamma^2\mu(\partial_y U_2)^2 + (\partial_z U_3)^2\Big), \end{aligned}$$

where we denoted $\mathbf{U} = (U_1, U_2, U_3)$. (The adaptation to the case $d = 1$ is straightforward.) It follows that P is subharmonic in the fluid domain and therefore attains its minimum on the surface or on the bottom. Let us prove now that under the assumption made in the statement of the proposition, the minimum cannot be attained at the bottom.

By Hopf's maximum principle, we know that the outward conormal derivative must be nonpositive at every point where the minimum is reached. It is therefore enough to prove that $\partial_{\mathbf{n}} P_{|_{z=-1+\beta b}} \leq 0$ (recall that the notation $\partial_{\mathbf{n}}$ stands for the *upward* conormal derivative). We get from Remark 1.14 that

$$-\frac{P_0}{\rho a g}\partial_{\mathbf{n}} P_{|_{z=-1+\beta b}} = \frac{1}{2}\partial_{\mathbf{n}}\big(\varepsilon|\nabla^\gamma\Phi|^2 + \frac{\varepsilon}{\mu}(\partial_z\Phi)^2\big)_{|_{z=-1+\beta b}} + \frac{1}{\varepsilon}(1+\beta^2|\nabla b|^2)^{-1/2}.$$

Differentiating the relation

$$(1+\beta^2|\nabla b|^2)^{1/2}\partial_{\mathbf{n}}\Phi_{|_{z=-1+\beta b}}\big(= -\beta\mu\nabla^\gamma b\cdot\nabla^\gamma\Phi_{|_{z=-1+\beta b}} + \partial_z\Phi_{|_{z=-1+\beta b}}\big) = 0$$

with respect to j ($j = x, y$), one also gets

$$\begin{aligned}(1+\beta^2|\nabla b|^2)^{1/2}\partial_{\mathbf{n}}(\partial_j\Phi)_{|_{z=-1+\beta b}} &= \beta\mu\nabla^\gamma\partial_j b\cdot\nabla^\gamma\Phi_{|_{z=-1+\beta b}}\\ &-\beta\partial_j b(1+\beta^2|\nabla b|^2)^{1/2}\partial_{\mathbf{n}}(\partial_z\Phi)_{|_{z=-1+\beta b}},\end{aligned}$$

so that finally

$$-\frac{P_0}{\rho a g}\partial_{\mathbf{n}} P_{|_{z=-1+\beta b}} = \frac{1}{\varepsilon}(1+\beta^2|\nabla b|^2)^{-1/2}\big(\varepsilon^2\beta\mu\mathcal{H}_b^\gamma(\nabla^\gamma\Phi_{|_{z=-1+\beta b}}) + 1\big).$$

The assumption made in the statement of the proposition directly implies that $\partial_{\mathbf{n}} P_{|_{z=-1+\beta b}} \leq 0$, so that, as previously stated, the minimum of P must be reached at the surface.
At every point where this minimum is achieved, one must have, owing to Hopf's maximum principle, $-\partial_z P > 0$. Since by assumption P is constant at the surface, this means that $-\partial_z P > 0$ *everywhere* on $\mathbb{R}^d$. It is therefore bounded from below by a nonnegative constant on any compact set of $\mathbb{R}^d$. In order to check that this is also the case at infinity, we can equivalently prove, owing to Proposition 4.31, that $\mathfrak{a}(U^0)$ satisfies such a condition. Recalling that $\mathfrak{a}(U^0) = 1 + \varepsilon\mathfrak{b}(U^0)$ with $\mathfrak{b}(U)$, as defined in (4.60), this is a direct consequence of the decay at infinity of $\mathfrak{b}$. □

4.4. Supplementary remarks

4.4.1. Nonasymptotically flat bottom and surface parametrizations. We show here how to extend Theorem 4.16 to the case where the bottom parametrization is $b = \underline{b}$ with $\underline{b}$ belonging to $C_b^\infty(\mathbb{R}^d)$ instead of $H^{N+1\vee t_0+1}(\mathbb{R}^d)$, and the initial condition is of the form $\zeta^0 + \underline{\zeta}^0$ with $\underline{\zeta}^0 \in \dot{H}^\infty(\mathbb{R}^d) = \cap_s \dot{H}^s(\mathbb{R}^d)$. A finite regularity would be enough for $\underline{b}$ and $\underline{\zeta}^0$, but we do not focus on this point for the sake of clarity. In order to state the theorem, let us define $(\underline{\zeta}, \underline{\psi})$ as

$$\underline{\zeta}(t,\cdot) = \cos(t\omega(D^\gamma))\underline{\zeta}^0, \qquad \underline{\psi}(t,\cdot) = \frac{\sin(t\omega(D^\gamma))}{\omega(D^\gamma)}\underline{\zeta}^0, \tag{4.62}$$

with

$$\omega(\xi)^2 = |\xi|\frac{\tanh(\sqrt{\mu}|\xi|)}{\sqrt{\mu}}.$$

In the statement below[6], N is as in Theorem 4.16, and the parameters $(\varepsilon, \mu, \beta, \gamma)$ are assumed to satisfy (4.28).

[6]The constants $\underline{c}_{WW}^j$ depend on $\underline{\zeta}$ and $\underline{b}$ through their $W^{k,\infty}(\mathbb{R}^d)$ norms for k large enough, but, for the sake of clarity, we do not make this dependence explicit.

THEOREM 4.33. *Let* $U^0 = (\zeta^0, \psi^0) \in E_0^N$, *and* $\underline{U}^0 = (\underline{\zeta}^0, 0)$ *with* $\underline{\zeta}^0 \in L^\infty(\mathbb{R}^d) \cap \dot{H}^\infty(\mathbb{R}^d)$. *Also assume that the bottom parametrization is given by* $b = \underline{b} \in C_b^\infty(\mathbb{R}^d)$ *and that the following conditions hold,*

$$(4.63) \qquad \begin{array}{ll} \exists h_{min} > 0, & 1 + \varepsilon(\zeta^0 + \underline{\zeta}^0) - \beta \underline{b} \geq h_{min}, \\ \exists a_0 > 0, & \mathfrak{a}(U^0 + \underline{U}^0) \geq a_0, \end{array}$$

with $\mathfrak{a}(U^0 + \underline{U}^0)$ *as defined in (4.27).*
Then, with $\underline{U} = (\underline{\zeta}, \underline{\psi})$ *given by* (4.62), *there exists* $T > 0$ *and a unique* $U \in E^N_{T/\varepsilon\vee\beta}$ *such that* $U + \underline{U}$ *solves (4.1) with initial data* $U^0 + \underline{U}^0$. *Moreover,*

$$\frac{1}{T} = \underline{c}^1_{WW} \quad \textit{and} \quad \sup_{t\in[0,\frac{T}{\varepsilon\vee\beta}]} \mathcal{E}^N(U(t)) = \underline{c}^2_{WW},$$

with $\underline{c}^j_{WW} = C(\mathcal{E}^N(U^0), \mu_{max}, \frac{1}{h_{min}}, \frac{1}{a_0}, \underline{\zeta}, \underline{\psi}, \underline{b})$ *for* $j = 1, 2$.

REMARK 4.34. With Theorem 4.33, the bottom can be a smooth step function or it can be oscillating. The surface parametrization can be more general than in Theorem 4.16 since it can be a smooth step function, but oscillating functions that do not decay to zero do not belong to $H^N + L^\infty \cap \dot{H}^\infty$ and therefore are not covered by this theorem. Working in the uniformly local Sobolev spaces described in §B.4 (Appendix B) one can handle such situations, as shown in [**7**]. Note, however, that since the Sobolev and uniformly Sobolev norms $|\cdot|_{H^s}$ and $|\cdot|_{H^s_{ul}}$ are not equivalent on $H^s(\mathbb{R}^d)$, the well-posedness result of [**7**] is a different result than, and not an extension of, a local well-posedness result in Sobolev spaces as was shown in Theorem 4.16.

REMARK 4.35. In the statement of the theorem, we did not make explicit the dependence of the constants $\underline{c}^j_{WW}$ ($j = 1, 2$) on $(\underline{\zeta}, \underline{\psi})$. Simple estimates show that this dependence can be controlled by the $L^\infty([0, \frac{T}{\varepsilon\vee\beta}]; L^\infty \cap \dot{H}^k \times \dot{H}^k)$ norm of $(\underline{\zeta}, \underline{\psi})$ (for some $k \in \mathbb{N}$). Since the $L^\infty \cap \dot{H}^k \times \dot{H}^k$- norm of $(\underline{\zeta}(t), \underline{\psi}(t))$ may in general grow linearly with t, the constants $\underline{c}^j_{WW}$ might not be uniformly bounded as $\varepsilon \vee \beta \to 0$, in which case the theorem does not ensure that T is uniformly bounded from below as $\varepsilon \vee \beta \to 0$. A straightforward way to avoid such a scenario is to assume that $\underline{\zeta}^0$ is of size $O(\varepsilon \vee \beta)$. This smallness assumption is not always necessary though to obtain a uniform lower bound for T, and a refined analysis of the dependence of $\underline{c}^j_{WW}$ on $(\underline{\zeta}, \underline{\psi})$ could allow one to remove it in some cases.

PROOF. We indicate here how to adapt the energy estimates of §4.3.4.3 to the present situation. Let us first remark that for all $s \geq 0$ and $U = (\zeta, \psi) \in H^{s+1/2} \times \dot{H}^{s+1/2}(\mathbb{R}^d)$, the quantities

$$\mathcal{G}[\varepsilon(\zeta + \underline{\zeta}), \beta\underline{b}](\psi + \underline{\psi}), \quad \nabla^\gamma(\zeta + \underline{\zeta}), \quad \text{and} \quad \nabla^\gamma(\psi + \underline{\psi})$$

that are involved if one looks for a solution $U + \underline{U}$ of (4.1) are all well defined and belong to $H^{s-1/2}(\mathbb{R}^d)$. This is indeed a direct consequence of the fact[7] that $\underline{U} \in C(\mathbb{R}; C_b(\mathbb{R}^d) \cap \dot{H}^\infty(\mathbb{R}^d))^2$ and of the results on the Dirichlet-Neumann operator on nonasymptotically flat domains provided in §3.7.1.
With the notations introduced in (4.33), we can also use the results of §3.7.1 to get

[7]This stems from (4.62) and the assumption that $\underline{\zeta}^0 \in L^\infty(\mathbb{R}^d) \cap \dot{H}^\infty(\mathbb{R}^d)$.

energy estimates on $\mathcal{F}^\alpha(U+\underline{U})$ when $\alpha \neq 0$, as in §4.3.4.3.
When $\alpha = 0$, however, $\zeta + \underline{\zeta} \notin L^2(\mathbb{R}^d)$ (while we had $\partial^\alpha(\zeta+\underline{\zeta}) \in L^2(\mathbb{R}^d)$ for $\alpha \neq 0$), and the estimate of §4.3.4.3 must be adapted. We first write (4.1) under the form

$$\left\{\begin{array}{l} \partial_t \zeta - \frac{1}{\mu}\underline{\mathcal{G}}\psi = -(\partial_t \underline{\zeta} - \frac{1}{\mu}\underline{\mathcal{G}}\underline{\psi}) \\ \partial_t \psi + \zeta + \frac{\varepsilon}{2}|\nabla^\gamma(\psi+\underline{\psi})|^2 - \frac{\varepsilon}{\mu}\frac{(\underline{\mathcal{G}}(\psi+\underline{\psi}) + \varepsilon\mu\nabla^\gamma(\zeta+\underline{\zeta})\cdot\nabla^\gamma(\psi+\underline{\psi}))^2}{2(1+\varepsilon^2\mu|\nabla^\gamma(\zeta+\underline{\zeta})|^2)} \\ \qquad = -(\partial_t\underline{\psi} + \underline{\zeta}), \end{array}\right. \tag{4.64}$$

where $\underline{\mathcal{G}} = \mathcal{G}_{\mu,\gamma}[\varepsilon(\zeta+\underline{\zeta}), \beta\underline{b}]$, and with the initial condition $U^0 = (\zeta^0, \psi^0)$. Since by definition $\underline{U} = (\underline{\zeta}, \underline{\psi})$ solves

$$\left\{\begin{array}{l} \partial_t \underline{\zeta} - \frac{1}{\mu}\mathcal{G}_0\underline{\psi} = 0 \\ \partial_t\underline{\psi} + \underline{\zeta} = 0, \end{array}\right.$$

with $\mathcal{G}_0 = \mathcal{G}_{\mu,\gamma}[0,0]$, (4.64) can be put under the form

$$\left\{\begin{array}{l} \partial_t \zeta - \frac{1}{\mu}\underline{\mathcal{G}}\psi = \frac{1}{\mu}(\mathcal{G}_{\mu,\gamma}[\varepsilon(\zeta+\underline{\zeta}),\beta\underline{b}] - \mathcal{G}_{\mu,\gamma}[0,0])\underline{\psi} \\ \partial_t\psi + \zeta + \frac{\varepsilon}{2}|\nabla^\gamma(\psi+\underline{\psi})|^2 - \frac{\varepsilon}{\mu}\frac{(\underline{\mathcal{G}}(\psi+\underline{\psi}) + \varepsilon\mu\nabla^\gamma(\zeta+\underline{\zeta})\cdot\nabla^\gamma(\psi+\underline{\psi}))^2}{2(1+\varepsilon^2\mu|\nabla^\gamma(\zeta+\underline{\zeta})|^2)} = 0. \end{array}\right.$$

Now, we can use the second point of Proposition 3.28 (with the adaptations to nonasymptotically flat bottoms and surface parametrizations presented in §3.7.1) to get that

$$\left|\frac{1}{\mu}(\mathcal{G}_{\mu,\gamma}[\varepsilon(\zeta+\underline{\zeta}),\beta\underline{b}] - \mathcal{G}_{\mu,\gamma}[0,0])\underline{\psi}\right|_2 \leq (\varepsilon\vee\beta)\underline{M}|\mathfrak{P}\underline{\psi}|_{H^1},$$

with

$$\underline{M} = C(\frac{1}{h_{min}}, \mu_{max}, |\zeta|_{H^{t_0+2}}, \underline{\zeta}, \underline{b}).$$

The presence of a nonzero right-hand-side in the equation on ζ can therefore be handled easily to get an estimate on $|\zeta|_2$, as in §4.3.4.3. An estimate on $|\mathfrak{P}\psi|_2$ is obtained in a similar way. □

4.4.2. Rough bottoms. As shown in [**4**] no smoothness is required on the bottom parametrization to get local well-posedness of the water waves equations, provided that the water depth is always bounded from below by a positive constant. (By standard ellipticity results, the contribution of the bottom to the Dirichlet-Neumann operator is infinitely smooth.) As explained in §3.7.2, this existence result is not compatible with the shallow water limit. See also the comments in §A.1 (Appendix A) on the symbolic analysis of the Dirichlet-Neumann operator. Even at the formal level, some regularity on b is also needed to write the asymptotic systems (see chapter 5).

Theorem 4.33 can be used to provide a local in time solution of the water waves equations for a highly oscillating bottom (the second kind of "rough bottoms" mentioned in §1.7). However, the existence time provided by Theorem 4.33 is not uniform with respect to the scale of the bottom variations; in particular, to our knowledge, there is not any rigorous homogenization result on the full water waves equations[8].

4.4.3. Very deep water ($\mu \gg 1$) and infinite depth. Let us also stress that, even though no smallness assumption on μ is made in this chapter, we do not allow this parameter to be very large (very deep water). Indeed, the existence time in Theorem 4.16 is of the form

$$\frac{1}{T} = C\big(\mathcal{E}^N(U^0), \mu_{max}, \frac{1}{h_{min}}, \frac{1}{a_0}, |b|_{H^{N+1\vee t_0+1}}\big),$$

and therefore can shrink to zero when μ_{max} (the upper bound for μ) is large. In [**12**], it is shown that the existence time depends on μ through $\epsilon = \varepsilon\sqrt{\mu}$ only, which is the *steepness* of the wave (see §1.5). It is therefore allowable to have large μ provided that ε is small enough to keep this steepness bounded. The analysis is made much simpler, though, if we assume that μ remains bounded, and since this is not a severe limitation to applications in coastal oceanography, we decided to make such an assumption. An important point made by Iguchi in [**181**] is that, in this case, the present approach to the water waves equations does not require the use of a Nash-Moser theorem to construct a solution, as in [**218, 12**].

As frequently demonstrated in the literature, the case of very deep depth is replaced by *infinite* depth. We show here how to adapt Theorem 4.16 to this situation. *As in* §2.5.4 *and* 3.7.3, *we take* $\gamma = 1$ *for the sake of clarity.*
The first step is to write the water waves equations in infinite depth. One cannot directly take $\mu = \infty$ in (1.28); however, if we replace $\mathcal{G}_{\mu,\gamma}$ by $\tilde{\mathcal{G}}_{\mu,\gamma}$ as defined in Remark 1.13, the water waves equations in infinite depth are obtained[9] by taking formally $\mu = \infty$ on the resulting formulation. This yields

$$\left\{\begin{array}{l} \partial_t\zeta - \mathcal{G}[\epsilon\zeta]\psi = 0, \\ \partial_t\psi + \zeta + \frac{\epsilon}{2}|\nabla\psi|^2 - \epsilon\frac{(\mathcal{G}[\epsilon\zeta]\psi + \epsilon\nabla\zeta\cdot\nabla\psi)^2}{2(1+\epsilon^2|\nabla\zeta|^2)} = 0, \end{array}\right. \tag{4.65}$$

where we recall that ϵ is the steepness of the wave (see §1.3.1), while $\mathcal{G}[\epsilon\zeta]$ is defined in §3.7.3.
From the analysis of the Dirichlet-Neumann operator in infinite depth performed in §3.7.3, one is led to replace the energy $\mathcal{E}^N(U)$ defined in (4.8) by the following "infinite depth" version

$$\tilde{\mathcal{E}}^N(U) = \big||D|^{1/2}\psi\big|^2_{H^{t_0+3/2}} + \sum_{\alpha\in\mathbb{N}^d, |\alpha|\leq N} |\zeta_{(\alpha)}|_2^2 + \big||D|^{1/2}\psi_{(\alpha)}\big|_2^2, \tag{4.66}$$

and consequently, the functional space E_T^N must be replaced by

$$\tilde{E}_T^N = \{U \in C([0,T]; H^{t_0+2}(\mathbb{R}^d) \times \mathring{H}^2(\mathbb{R}^d)), \quad \tilde{\mathcal{E}}^N(U(\cdot)) \in L^\infty([0,T])\}, \tag{4.67}$$

---

[8] Due to low regularity, the homogenization result on the Dirichlet-Neumann operator presented in (3.45) of §3.7.2 cannot be used in the well-posedness theory developed in this chapter.

[9] The equations obtained are actually

$$\left\{\begin{array}{l} \partial_t\zeta - 2\pi\mathcal{G}[\epsilon\zeta]\psi = 0, \\ \partial_t\psi + \zeta + 2\pi\frac{\epsilon}{2}|\nabla\psi|^2 - 2\pi\epsilon\frac{(\mathcal{G}[\epsilon\zeta]\psi + \epsilon\nabla\zeta\cdot\nabla\psi)^2}{2(1+\epsilon^2|\nabla\zeta|^2)} = 0, \end{array}\right.$$

Rescaling ζ and t by a factor of $\sqrt{2\pi}$ gives (4.65). The best way to obtain directly (4.65) is to start from the infinite depth version of (1.3) and to nondimensionalize t by $t_0 = \sqrt{L/g}$ and ψ by $\Phi_0 = a_{surf}\sqrt{gL_x}$ (the other quantities being nondimensionalized as in §1.3.3).

where we recall that

$$\mathring{H}^{s+1/2}(\mathbb{R}^d) = \{f \in \mathcal{S}'(\mathbb{R}^d), \quad |D|^{1/2}\psi \in H^s(\mathbb{R}^d)\}.$$

We also define the infinite depth version of the mapping $\mathfrak{a}(\cdot)$ introduced in (4.27)

$$\tilde{\mathfrak{a}}(U) = 1 + \epsilon^2 \mathcal{V}[\epsilon\zeta]\psi \cdot \nabla(\underline{w}[\epsilon\zeta]\psi) - \epsilon \underline{w}[\epsilon\zeta]\tilde{\mathcal{N}}_2(U) - \epsilon d\underline{w}(\tilde{\mathcal{N}}_1(U))\psi, \tag{4.68}$$

where $\mathcal{V}[\epsilon\zeta]$, $\underline{w}[\epsilon\zeta]$ and $\tilde{\mathcal{N}}^j(U)$ are easily adapted to the case of infinite depth from the definition of the corresponding quantities in finite depth.
We can now state the following local well-posedness theorem of the water waves equations in infinite depth (4.65).

THEOREM 4.36. *Let $t_0 > d/2$, $N \geq t_0 + t_0 \vee 2 + 3/2$, and $0 < \epsilon < \epsilon_{max}$. Let $U^0 = (\zeta^0, \psi^0) \in \tilde{E}_0^N$, and moreover assume that*

$$\exists a_0 > 0, \qquad \tilde{\mathfrak{a}}(U^0) \geq a_0, \tag{4.69}$$

with $\tilde{\mathfrak{a}}(U^0)$ as defined in (4.68).
Then there exists $T > 0$ and a unique solution $U \in \tilde{E}^N_{T/\epsilon}$ to (4.65) with initial data U^0. Moreover

$$\frac{1}{T} = \tilde{c}^1_{WW} \quad \text{and} \quad \sup_{t\in[0,\frac{T}{\epsilon}]} \tilde{\mathcal{E}}^N(U(t)) = \tilde{c}^2_{WW},$$

with $\tilde{c}^j_{WW} = C(\tilde{\mathcal{E}}^N(U^0), \epsilon_{max}, \frac{1}{a_0})$ for $j = 1, 2$.

PROOF. The proof is a straightforward adaptation of the proof performed in the finite depth case, using the results of §3.7.3. Note that we assumed that $\psi \in \mathring{H}^{N+1/2}$ and not in $H^{N+1/2}$; however, for $0 < |\alpha| \leq N$, one has $\psi_{(\alpha)} \in H^{1/2}$, and $\mathcal{G}[\epsilon\zeta]\psi_{(\alpha)}$ therefore is well defined owing to Theorem 3.49. When $\alpha = 0$, this theorem cannot be used to define $\mathcal{G}[\epsilon\zeta]\psi$ because ψ is not necessarily in $H^{1/2}(\mathbb{R}^d)$; in this case, we therefore use the second point of Remark 3.50. □

4.4.4. Global well-posedness. For the water waves problem (1.28), the eigenvalues of the linear part of the equations are (with $\gamma = 1$)

$$\omega_\pm(\xi) = \pm\Big(\frac{|\xi|\tanh(\sqrt{\mu}|\xi|)}{\sqrt{\mu}\nu}\Big)^{1/2}.$$

The Hessian of these functions has maximal rank d, and one expects from the stationary phase theorem a time decay in L^∞ norm of order $O(t^{-d/2})$. Since nonlinearities in (1.28) are quadratic, this decay is far from enough to conclude to global existence when the horizontal dimension is $d = 1$. If quadratic nonlinearities could be removed (by a normal form) or if an extra $O(t^{-1/2})$ time decay could be gained (by a null form condition), an existence time of size $O(e^{ct/\varepsilon^2})$ for initial data of size $O(\varepsilon)$ could be expected. Such a result was proved by Wu [**328**] in the case of infinite depth, with methods combining Klainerman's vector fields and a clever change of variables. Note that for technical reasons, the result of [**328**] gives an existence time $O(e^{ct/\varepsilon})$ instead of $O(e^{ct/\varepsilon^2})$.
In horizontal dimension $d = 2$, the L^∞-decay is $O(t^{-1})$, and the situation should be the same as for the quadratic wave equation in dimension 3: a generic existence time of order $O(e^{ct/\varepsilon})$, and global existence if we are able to implement a normal form transform or use a null form condition. In the case of infinite depth, such a result has been proved independently (and under different assumptions and

conclusions) by Wu [**329**], who extended the tools developed in [**328**] to the case $d = 2$, and by Germain, Masmoudi, and Shatah [**156**], who obtained it using their space-time resonance methodology (see [**222**] for a review).

4.4.5. Low regularity. In [**6**], Alazard, Burq, and Zuily addressed the issue, lowering the regularity required on the initial data to ensure local well-posedness. Following their previous approach developed in [**9, 4**], after suitable para-linearizations, the authors arrange the equations into an explicit symmetric system of quasilinear waves equation type; this allows them to solve the Cauchy problem with initial surface elevation $\zeta^0 \in H^{s+1/2}(\mathbb{R}^d)$ and trace of the velocity at the surface $\underline{\mathbf{U}}^0 \in H^s(\mathbb{R}^d)^{d+1}$ with s the classical "quasilinear regularity" $s > 1 + d/2$. In particular, the system can be solved for initial surfaces having unbounded curvature.

Taking their analysis further and following classical works for quasi-linear wave equations, they use Strichartz estimates (see [**5, 77**] for such estimates in the context of the water waves equations) to lower the regularity threshold, allowing, for instance, non-Lipschitz initial velocity fields. An application of this low regularity well-posedness allows them to handle the water waves equations on a canal, using Boussinesq's extension method.

CHAPTER 5

Shallow Water Asymptotics: Systems. Part 1: Derivation

This chapter is devoted to the derivation and rigorous justification of asymptotic models that describe, in various shallow water regimes, the solutions of the water waves equations[1]

$$\left\{\begin{array}{l} \partial_t \zeta - \frac{1}{\mu}\mathcal{G}\psi = 0, \\ \partial_t \psi + \zeta + \frac{\varepsilon}{2}|\nabla^\gamma \psi|^2 - \frac{\varepsilon}{\mu}\frac{(\mathcal{G}\psi + \varepsilon\mu\nabla^\gamma\zeta\cdot\nabla^\gamma\psi)^2}{2(1+\varepsilon^2\mu|\nabla^\gamma\zeta|^2)} = 0. \end{array}\right. \tag{5.1}$$

We recall that there are two main categories of asymptotic regimes:[2] shallow water ($\mu \ll 1$) and deep water ($\mu \not\ll 1$). Within each of these categories, various subregimes can be identified, depending on the assumptions made on the nonlinearity parameter ε, the topography parameter β, and the anisotropy parameter γ.

- The shallow water regime ($\mu \ll 1$). Depending on the size of ε, one can identify the following important subregimes[3] for which a rich variety of asymptotic models can be derived:
  - Large amplitude models. If no assumption is made on the nonlinearity parameter (i.e., $0 \le \varepsilon \le 1$), one obtains at first order (with respect to μ) the Nonlinear Shallow Water (or Saint-Venant) equations, and at second order the Green-Naghdi (or Serre, or fully nonlinear Boussinesq) equations.
  - Long wave models. If in addition it is assumed that $\varepsilon \le \mu$, then the Green-Naghdi equations can be simplified into the Boussinesq equations.
- The deep water regime ($\mu \not\ll 1$). When μ is not small, it is still possible to derive asymptotic analysis if the *steepness* $\epsilon = \varepsilon\sqrt{\mu} = a_{surf}/L_x$ is small.

We consider here shallow water regimes only and postpone the study of the deep water regime to Chapter 8. Together with the derivation of asymptotic models, we prove consistency results which roughly state that the solutions to the water waves equations "almost" solve the asymptotic systems.

After deriving in §5.1 a first series of asymptotic systems (all written in terms of the surface elevation ζ and the vertically averaged horizontal velocity $\overline{V}$), we

[1] Since only shallow water regimes are considered here, we set the parameter ν to 1 in the water waves equations (1.28).

[2] We speak here about direct approximations of the solutions of the water waves equations. The *modulation approximations* described in Chapter 8 can be considered as a third category of asymptotic models.

[3] We do not mention in this introduction the dependence on β and γ, but these refinements are also treated in this chapter.

show in §5.2 that it is possible to derive various classes of asymptotically equivalent models (that couple the surface elevation ζ to a velocity that is not necessarily $\overline{V}$). These models can have much better dispersive properties, and thus a wider range of application in oceanography. We also derive in §5.3 asymptotically equivalent models with better (for mathematical analysis, for instance) nonlinear properties.

It is possible to handle the case where the bottom is not fixed but provided by a (given) time-dependent function; we show in §5.4 how to derive shallow water asymptotic models in this situation.

All the asymptotic models derived in this chapter depend on an analysis of the structure of the pressure and velocity fields in the fluid domain. We present in §5.5 another application of these results by showing how to reconstruct the surface elevation from pressure measurements at the bottom.

Finally, we give some supplementary remarks in §5.6. After giving technical results in §5.6.1, we provide in §5.6.2 some considerations that explain the different possible choices of the velocity unknown to the horizontal velocity evaluated at some level line of the fluid domain. We also give in §5.6.3 a formulation of various shallow water models in $(h, h\overline{V})$ rather than $\zeta, \overline{V}$ variables, and we give in §5.6.4 the version with dimensions of these models. In §5.6.5, we interpret the so-called lake and great lake equations as a rigid lid limit of the nonlinear shallow water and Green-Naghdi equations. Finally, we briefly discuss in §5.6.6 the inclusion of friction effects at the bottom.

N.B. This chapter is devoted to the *derivation* of asymptotic systems for the water waves equations. The *justification* of these models will be performed in Chapter 6, where a brief mathematical description of the equations is also provided.

Throughout this chapter, we assume, as always, that the dimensionless parameters ε, μ, β, and γ satisfy

$$0 \leq \varepsilon, \beta, \gamma \leq 1, \quad 0 \leq \mu < \mu_{max}, \tag{5.2}$$

and that $\zeta, b \in H^{t_0+1}(\mathbb{R}^d)$ $(t_0 > d/2)$ are such that

$$\exists h_{min} > 0, \quad \forall X \in \mathbb{R}^d, \quad 1 + \varepsilon\zeta(X) - \beta b(X) \geq h_{min}. \tag{5.3}$$

We also denote by M_0 and M any constants of the form

$$\begin{aligned} M_0 &= C\big(\frac{1}{h_{min}}, \mu_{max}, |\zeta|_{H^{t_0+1}}, |b|_{H^{t_0+1}}\big), \\ M &= C\big(\frac{1}{h_{min}}, \mu_{max}, |\zeta|_{H^{t_0+2}}, |b|_{H^{t_0+2}}\big), \end{aligned} \tag{5.4}$$

and, for all $s \geq 0$, we also define $M(s)$ and $N(s)$ as

$$M(s) = C(M_0, |\zeta|_{H^s}, |b|_{H^s}) \tag{5.5}$$

$$N(s) = C(M(s), |\nabla^\gamma \psi|_{H^s}) \tag{5.6}$$

5.1. Derivation of shallow water models ($\mu \ll 1$)

This section is devoted to the study of water waves when the shallowness parameter $\mu = H_0^2/L_x^2$ is small (recall that H_0 is the typical depth and L_x the typical wavelength).

The most general situation (when no assumption is made on ε, β and γ) is addressed in §5.1.1, where the the Green-Naghdi equations (also called Serre [**278**] or *fully* nonlinear Boussinesq equations [**321**]) and the Nonlinear Shallow Water equations are derived. The key point is the link between $\mathcal{G}\psi$ and the averaged velocity $\overline{V}$ provided by Proposition 3.35 and an asymptotic expansion of $\nabla^\gamma\psi$ in terms of $\overline{V}$ that we prove here.

We then derive simpler models under additional assumptions on ε. More precisely, we consider in §5.1.2 *medium amplitude* models ($\varepsilon = O(\sqrt{\mu})$) and in §5.1.3 we investigate *small amplitude* models ($\varepsilon = O(\mu)$). For all these regimes, we also take into consideration the size of the bottom variations.

All the models derived in this section can be cast under the form

$$(A) \qquad \begin{cases} \partial_t \zeta + \nabla^\gamma \cdot (h\overline{V}) = 0, \\ \partial_t \overline{V} + \mathcal{N}_{(A)}(\zeta, \overline{V}) = 0, \end{cases}$$

where $\mathcal{N}_{(A)}(\zeta, \overline{V})$ is a nonlinear (pseudo)-differential operator that depends on the asymptotic model (A), and where we recall that the averaged velocity $\overline{V}$ can be built from the solution (ζ, ψ) to the water waves equations (5.1) through (3.31) (see also Remark 3.40). It is convenient to define here the notion of *consistency*.

DEFINITION 5.1. Let $\mathcal{A}$ be an asymptotic regime in the sense of Definition 4.19. We note that the water waves equations are *consistent at order $O(\mu^k)$ with (A) in the regime $\mathcal{A}$* if there exists $n \in \mathbb{N}$ and $T > 0$ such that for all $s \geq 0$ and $\mathfrak{p} = (\varepsilon, \beta, \gamma, \mu) \in \mathcal{A}$,

- There exists a solution $(\zeta, \psi) \in C([0, \frac{T}{\varepsilon\vee\beta}]; H^{s+n} \times \dot{H}^{s+n+1})$ to the water waves equations (5.1).
- With $\overline{V} = \overline{V}_{\mu,\gamma}[\varepsilon\zeta, \beta b]\psi$ as in (3.31), one has

$$\begin{cases} \partial_t \zeta + \nabla^\gamma \cdot (h\overline{V}) = 0, \\ \partial_t \overline{V} + \mathcal{N}_{(A)}(\zeta, \overline{V}) = \mu^k R, \end{cases}$$

with $|R|_{H^s} \leq N(s+n)$ on $[0, \frac{T}{\varepsilon\vee\beta}]$, and $N(s+n)$ as in (5.6).

5.1.1. Large amplitude models ($\mu \ll 1$ and $\varepsilon = O(1)$, $\beta = O(1)$). We consider here two shallow water/large amplitude models that allow large bottom variations (see §1.5). These two models are the Green-Naghdi and the Nonlinear Shallow Water equations, and they differ only by their precision. The corresponding *asymptotic regime* in the sense of Definition 4.19 is

$$\mathcal{A}_{SW} = \{(\varepsilon, \beta, \gamma, \mu), \quad 0 \leq \mu \leq \mu_0, \quad 0 \leq \varepsilon \leq 1, \quad 0 \leq \beta \leq 1, \quad 0 \leq \gamma \leq 1\},$$

for some $0 < \mu_0 \leq \mu_{max}$.

5.1.1.1. *The Nonlinear Shallow Water (NSW) equations.* We show here that the solutions to the water waves equations (5.1) are *consistent*[4] *at order $O(\mu)$* with the NSW (or Saint-Venant[5]) equations (that are usually written in the particular

[4]According to Definition 5.1, this is a convenient way to state the result of Proposition 5.2: the exact solution of the water waves equations solves the NSW equations up to terms of order $O(\mu)$.

[5]After Adhémar Jean Claude Barré de Saint-Venant, who derived them in 1871 [**281, 282**] in the one-dimensional case. He used simple energetic considerations, assuming that the pressure is hydrostatic (i.e. $P(X,z) = P_{atm} - \rho g(z-\zeta)$ in dimensional form); as shown in §5.5 below, this is always true at leading order in the shallow water regime $\mu \ll 1$. Note also that Lagrange had previously derived in 1781 the linear version of these equations [**212**].

case $\varepsilon = \beta = \gamma = 1$), which couple the evolution of the free surface elevation ζ to the evolution of the vertically averaged mean velocity[6] $\overline{V}$:

$$\left\{\begin{array}{l}\partial_t\zeta + \nabla^\gamma\cdot(h\overline{V}) = 0,\\ \partial_t\overline{V} + \nabla^\gamma\zeta + \varepsilon(\overline{V}\cdot\nabla^\gamma)\overline{V} = 0,\end{array}\right. \tag{5.7}$$

where we recall that $h = 1 + \varepsilon\zeta - \beta b$.
The consistency result mentioned above is the following.

PROPOSITION 5.2. *In the shallow water regime $\mathcal{A}_{SW}$, the water waves equations* (5.1) *are consistent at order $O(\mu)$ with the Nonlinear Shallow Water equations* (5.7) *in the sense of Definition* 5.1 *(with $n = 4$).*

REMARK 5.3. Let us insist on the fact that no smallness assumption is required on Proposition 5.2 to hold. The smallness of μ is needed only for the residual to be small.

PROOF. The fact that $(\zeta, \overline{V})$ solves exactly the first equation of the NSW equations is a direct consequence of Proposition 3.35.
In order to check that the second equation is satisfied up to μR with R as in the statement of the proposition, we first need to get an asymptotic expansion of $\nabla^\gamma\psi$ in terms of $\overline{V}$. In order to do so, recall that, owing to Proposition 3.37, one has

$$\overline{V} = \nabla^\gamma\psi + \mu R_1$$

with $|R_1|_{H^s} \leq M(s+2)|\nabla^\gamma\psi|_{H^{s+2}}$. The following lemma gives some control on $\partial_t R_1$.

LEMMA 5.4. *For all $s \geq 0$, one has*

$$|R_1|_{H^s} \leq M(s+2)|\nabla^\gamma\psi|_{H^{s+2}} \quad \text{and} \quad |\partial_t R_1|_{H^s} \leq N(s+4),$$

with $N(s+4)$ as defined in (5.6).

PROOF OF THE LEMMA. The first estimate has already been proved. For the second one, let us set $n = 0$ in Proposition 3.37, so that $\mu^{n+1}R_1 = \overline{V} - \overline{V}_{app}$ (recall that $\overline{V}_{app}$ is given by (3.36)). After differentiating (3.37) with respect to t we thus get that

$$\begin{aligned}\mu^{n+1}\partial_t R_1 &= \int_{-1}^0 (\nabla^\gamma\partial_t u - \frac{1}{h}(z\nabla^\gamma h + \varepsilon\nabla^\gamma\zeta)\partial_z\partial_t u)dz\\ &\quad - \int_{-1}^0 \partial_t(\frac{z\nabla^\gamma h + \varepsilon\nabla^\gamma\zeta}{h})\partial_z u dz.\end{aligned} \tag{5.8}$$

Note now that $v = \partial_t u$ can be decomposed into $v = v_1 + v_2$ with v_1 and v_2 solving

$$\left\{\begin{array}{l}\nabla^{\mu,\gamma}\cdot P(\Sigma)\nabla^{\mu,\gamma}v_1 = -\mu^{n+1}\partial_t R_\mu,\\ v_1\,_{|_{z=0}} = 0, \qquad \partial_{\mathbf{n}} v_1\,_{|_{z=-1}} = -\mu^{n+1}\partial_t r_\mu\end{array}\right.$$

and

$$\left\{\begin{array}{l}\nabla^{\mu,\gamma}\cdot P(\Sigma)\nabla^{\mu,\gamma}v_2 = -\nabla^{\mu,\gamma}\cdot\mathbf{g}\\ v_2\,_{|_{z=0}} = 0, \qquad \partial_{\mathbf{n}} v_2\,_{|_{z=-1}} = -\mathbf{e}_z\cdot\mathbf{g},\end{array}\right.$$

[6] A formulation of the shallow water equations in $(h, h\overline{V})$ variables is given later in §5.6.3.1.

with R_μ and r_μ as in (3.34), and $\mathbf{g} = \partial_t P(\Sigma)\nabla^{\mu,\gamma} u$.
Using Lemma 3.43 to control v_1, we get

$$\begin{aligned}\|\Lambda^s \nabla^{\mu,\gamma} v_1\|_2 &\leq \mu^{n+1}(\|\Lambda^s \partial_t R_\mu\|_2 + |\partial_t r_\mu|_{H^s}) \\ &\leq \mu^{n+1} C(M(s+2n+1), |\partial_t \zeta|_{H^{s+2n+1}\cap H^{t_0+1}})|(\nabla^\gamma \psi, \partial_t \nabla^\gamma \psi)|_{H^{s+2n+1}}.\end{aligned}$$

Using (5.1) to control the time derivatives of ζ and ψ in terms of spacial derivatives, we get easily

$$\|\Lambda^s \nabla^{\mu,\gamma} v_1\|_2 \leq \mu^{n+1} N(s+2n+3). \tag{5.9}$$

In order to control v_2, we use Lemma 2.38 to get

$$\begin{aligned}\|\Lambda^s \nabla^{\mu,\gamma} v_2\|_2 &\leq M(s+1)\|\Lambda^s \mathbf{g}\|_2 \\ &\leq C(M(s+1), |\partial_t \zeta|_{H^{s+1}}, |\partial_t \zeta|_{H^{t_0+1}})\|\Lambda^s \nabla^{\mu,\gamma} u\|_2.\end{aligned}$$

Using Lemma 3.43 to get some control on u and (5.1) to control the time derivatives in terms of spatial ones, we deduce

$$\|\Lambda^s \nabla^{\mu,\gamma} v_2\|_2 \leq \mu^{n+1} N(s+2n+3). \tag{5.10}$$

From (5.9) and (5.10) we deduce the following control on $\partial_t u = v_1 + v_2$,

$$\|\Lambda^s \nabla^{\mu,\gamma} \partial_t u\|_2 \leq \mu^{n+1} N(s+2n+3),$$

so that the result follows from (5.8) and the Poincaré inequality. □

It is now very easy to show that the second of the NSW equations is satisfied up to a term μR with R as in the statement of the proposition. Indeed, one just has to apply ∇^γ to the second equation of (5.1) and to replace $\mathcal{G}\psi$ by $-\mu\nabla^\gamma(h\overline{V})$ and $\nabla^\gamma\psi$ by $\overline{V} - \mu R_1$ in all their occurrences. Denoting by μR all the terms of order $O(\mu)$, the estimates of Lemma 5.4 provide the desired control on R. □

5.1.1.2. *The Green-Naghdi (GN) equations.* We show here that the solutions to the water waves equations (5.1) are consistent at order $O(\mu^2)$ (vs. $O(\mu)$ for the NSW equations (5.7)) with the Green-Naghdi equations[7], which couple the evolution of the free surface elevation ζ to the evolution of the vertically averaged mean velocity[8] $\overline{V}$:

$$\begin{cases} \partial_t \zeta + \nabla^\gamma \cdot (h\overline{V}) = 0, \\ (1+\mu\mathcal{T}[h,\beta b])\partial_t \overline{V} + \nabla^\gamma \zeta + \varepsilon(\overline{V}\cdot\nabla^\gamma)\overline{V} + \mu\varepsilon\big(\mathcal{Q}[h](\overline{V}) + \mathcal{Q}_b[h](\overline{V})\big) = 0, \end{cases} \tag{5.11}$$

where we recall that $h = 1 + \varepsilon\zeta - \beta b$ and where $\mathcal{T} = \mathcal{T}[h,\beta b]$ is defined as

$$\begin{aligned}\mathcal{T}V &= -\frac{1}{3h}\nabla^\gamma(h^3 \nabla^\gamma \cdot V) \\ &\quad + \beta\frac{1}{2h}\big[\nabla^\gamma(h^2 \nabla^\gamma b \cdot V) - h^2 \nabla^\gamma b \nabla^\gamma \cdot V\big] + \beta^2 \nabla^\gamma b \nabla^\gamma b \cdot V,\end{aligned} \tag{5.12}$$

while the quadratic forms $\mathcal{Q}[h]$ and $\mathcal{Q}_b[h]$ are defined as

$$\mathcal{Q}[h](\overline{V}) = -\frac{1}{3h}\nabla^\gamma\big[h^3\big((\overline{V}\cdot\nabla^\gamma)(\nabla^\gamma\cdot\overline{V}) - (\nabla^\gamma\cdot\overline{V})^2\big)\big] \tag{5.13}$$

[7]These equations were first derived in the case $d = 1$ by Serre [**278**] and then rediscovered by Su and Gardner [**305**] and extended to nonflat bottoms in [**296**]. They can also be found in the literature under the name *fully nonlinear Boussinesq equations* [**321**]; these equations are also the same as those derived in [**261**] through Hamilton's principle.

[8]A formulation of the Green-Naghdi equations in $(h, h\overline{V})$ variables is given later in §5.6.3.2.

and

$$\begin{aligned}\mathcal{Q}_b[h](\overline{V}) &= \frac{\beta}{2h}\big[\nabla^\gamma\big(h^2(\overline{V}\cdot\nabla^\gamma)^2 b\big) - h^2\big((\overline{V}\cdot\nabla^\gamma)(\nabla^\gamma\cdot\overline{V}) - (\nabla^\gamma\cdot\overline{V})^2\big)\nabla^\gamma b\big]\\ &\quad+\beta^2\big((\overline{V}\cdot\nabla^\gamma)^2 b\big)\nabla^\gamma b.\end{aligned}\tag{5.14}$$

REMARK 5.5. It is worth noting that the quadratic form $\mathcal{Q}_b[h,\beta b]$ is purely topographical in the sense that it vanishes when the bottom is flat ($b=0$).

NOTATION 5.6. When no confusion is possible, we simply write $\mathcal{T}$, $\mathcal{Q}$ and $\mathcal{Q}_b$ instead of $\mathcal{T}[h,\beta b]$, $\mathcal{Q}[h]$ and $\mathcal{Q}_b[h]$.

REMARK 5.7. An equivalent and often convenient formulation of the GN equations is the following[9]

$$\begin{cases}\partial_t\zeta + \nabla^\gamma\cdot(h\overline{V}) = 0,\\ [I+\mu\mathcal{T}]\big(\partial_t\overline{V} + \varepsilon(\overline{V}\cdot\nabla^\gamma)\overline{V}\big) + \nabla^\gamma\zeta + \mu\varepsilon\mathcal{Q}_1(\overline{V}) = 0,\end{cases}\tag{5.15}$$

with, using the notation $V^\perp = (-V_2, V_1)^T$,

$$\mathcal{Q}_1(V) = -2\mathcal{R}_1(\gamma\partial_1 V\cdot\partial_2 V^\perp + (\nabla^\gamma\cdot V)^2) + \beta\mathcal{R}_2(V\cdot(V\cdot\nabla^\gamma)\nabla^\gamma b),\tag{5.16}$$

and where the first-order differential operator $\mathcal{R}_1 = \mathcal{R}_1[h,\beta b]$ and $\mathcal{R}_2 = \mathcal{R}_2[h,\beta b]$ are given by

$$\mathcal{R}_1 w = -\frac{1}{3h}\nabla^\gamma(h^3 w) - \beta\frac{h}{2}w\nabla^\gamma b, \qquad \mathcal{R}_2 w = \frac{1}{2h}\nabla^\gamma(h^2 w) + \beta w\nabla^\gamma b.$$

Note that $\mathcal{Q}_1(\overline{V})$ contains only second-order derivatives of $\overline{V}$ while $\mathcal{Q}(\overline{V})$ contains third-order derivatives; this proves convenient for the mathematical analysis of the equations [**232, 13, 184, 185**] or their numerical simulation [**70, 41**].

The consistency result for the Green-Naghdi equations is the following.

PROPOSITION 5.8. *In the shallow water regime $\mathcal{A}_{SW}$, the water waves equations* (5.1) *are consistent at order $O(\mu^2)$ with the Green-Naghdi equations* (5.11) *in the sense of Definition* 5.1 *(with $n=6$).*

REMARK 5.9. As with the NSW equations (5.7), no smallness assumption is required on Proposition 5.8 to hold. The smallness of μ is needed only for the residual to be small.

REMARK 5.10. (1) The asymptotic regime (in the sense of Definition 4.19) is exactly the same for the Green-Naghdi and NSW equations. The only difference is the *precision* of the approximation: we have $O(\mu^2)$ consistency for the GN equations and $O(\mu)$ for the NSW equations. The difference in precision is due to a better description of the velocity field (see Remark 3.41).
(2) The NSW equations can be formally recovered from the GN equations by dropping all the $O(\mu)$ terms in (5.11).

PROOF. The proof follows the same lines as the proof of Proposition 5.2. We just have to check that the second equation is satisfied up to $\mu^2 R$ with R as in the statement of the proposition. We use a more precise asymptotic expansion of $\nabla^\gamma\psi$ in terms of $\overline{V}$ than in the proof of Proposition 5.2, namely,

$$\overline{V} = \nabla^\gamma\psi - \mu\mathcal{T}\nabla^\gamma\psi + \mu^2 R_2,$$

[9] One has to note only that $\mathcal{T}V = \mathcal{R}_1(\nabla^\gamma\cdot V) + \beta\mathcal{R}_2(\nabla^\gamma b\cdot V)$ and that $\mathcal{Q}(\overline{V}) + \mathcal{Q}_b(\overline{V}) = \mathcal{T}((\overline{V}\cdot\nabla^\gamma)\overline{V}) + \mathcal{Q}_1(\overline{V})$.

where we know by Proposition 3.37 that $|R_2|_{H^s} \leq M(s+4)|\nabla^\gamma\psi|_{H^{s+4}}$. We thus deduce that

$$\begin{aligned}\nabla^\gamma\psi &= \overline{V} + \mu\mathcal{T}\nabla^\gamma\psi + \mu^2 R_2 \\ &= \overline{V} + \mu\mathcal{T}\overline{V} + \mu^2\widetilde{R},\end{aligned} \tag{5.17}$$

with $\widetilde{R} = R_2 + \mathcal{T}R_1$. The following lemma gives some control on $\widetilde{R}$.

LEMMA 5.11. *For all $s \geq 0$, one has*

$$|\widetilde{R}|_{H^s} \leq M(s+4)|\nabla^\gamma\psi|_{H^{s+4}} \quad \text{and} \quad |\partial_t\widetilde{R}|_{H^s} \leq N(s+6),$$

with $M(s+4)$ and $N(s+6)$ as defined in (5.5) and (5.6) respectively.

PROOF OF THE LEMMA. The estimate on $|\widetilde{R}|_{H^s}$ is a direct consequence of the upper bound given above on $|R_1|_{H^s}$ and $|R_2|_{H^s}$.
Let us now set $n=1$ so that $\mu^{n+1}R_2 = \overline{V} - \overline{V}_{app}$. Proceeding exactly as in the proof of Lemma 5.4, we obtain

$$|\partial_t R_2|_{H^s} \leq N(s+6),$$

and the result follows easily since $\widetilde{R} = R_2 + \mathcal{T}R_1$. □

The end of the proof is obtained as for Proposition 5.2. □

In the case of small topography variations, ($\beta = O(\mu)$),

$$\mathcal{A}_{SW,top-} = \{(\varepsilon,\beta,\gamma,\mu), \quad 0 \leq \mu \leq \mu_0, \quad 0 \leq \varepsilon \leq 1, \quad 0 \leq \beta \lesssim \mu, \quad 0 \leq \gamma \leq 1\},$$

it is possible to replace the Green-Naghdi equations (5.11) by the following simplification[10]

$$\begin{cases} \partial_t\zeta + \nabla^\gamma\cdot(h\overline{V}) = 0, \\ (1 - \frac{\mu}{3h}\nabla^\gamma(h^3\nabla^{\gamma T}))\partial_t\overline{V} + \nabla^\gamma\zeta + \varepsilon(\overline{V}\cdot\nabla^\gamma)\overline{V} + \mu\varepsilon\mathcal{Q}[h](\overline{V}) = 0. \end{cases} \tag{5.18}$$

More precisely, since the $O(\beta\mu)$ and $O(\beta^2)$ terms of (5.11) are now of order $O(\mu^2)$, we can easily deduce from Proposition 5.8 the following corollary.

COROLLARY 5.12. *In the shallow water regime with small topography variations $\mathcal{A}_{SW,top-}$, the water waves equations (5.1) are consistent at order $O(\mu^2)$ with (5.18) in the sense of Definition 5.1 (with $n=6$).*

REMARK 5.13. Proceeding as for (5.15), one can equivalently write (5.18) under the form

$$\begin{cases} \partial_t\zeta + \nabla^\gamma\cdot(h\overline{V}) = 0, \\ [1 - \frac{\mu}{3h}\nabla^\gamma(h^3\nabla^{\gamma T})](\partial_t\overline{V} + \varepsilon(\overline{V}\cdot\nabla^\gamma)\overline{V}) \\ \qquad + \nabla^\gamma\zeta + \mu\varepsilon\frac{2}{3h}\nabla^\gamma\big[h^3(\gamma\partial_1\overline{V}\cdot\partial_2\overline{V}^\perp + (\nabla^\gamma\cdot\overline{V})^2)\big] = 0. \end{cases} \tag{5.19}$$

[10]It is also possible to replace $h = 1+\varepsilon\zeta - \beta b$ by $\tilde{h} = 1+\varepsilon\zeta$ in the second equation of (5.18),

$$\begin{cases} \partial_t\zeta + \nabla^\gamma\cdot(h\overline{V}) = 0, \\ (1 - \frac{\mu}{3\tilde{h}}\nabla^\gamma(\tilde{h}^3\nabla^{\gamma T}))\partial_t\overline{V} + \nabla^\gamma\zeta + \varepsilon(\overline{V}\cdot\nabla^\gamma)\overline{V} + \mu\varepsilon\mathcal{Q}[\tilde{h}](\overline{V}) = 0, \end{cases}$$

in which case the dependence on the topography b only appears in the first equation. The same simplification can be made on (5.19).

5.1.2. Medium amplitude models ($\mu \ll 1$ and $\varepsilon = O(\sqrt{\mu})$). The simplifications that can be brought to the Green-Naghdi equations (5.11) under the assumption $\varepsilon = O(\sqrt{\mu})$ depend on the topography. As for the surface variations, we distinguish three different regimes: large, medium, and small amplitude topography variations.

5.1.2.1. *Large amplitude topography variations: $\beta = O(1)$.* In this case, no significant simplification can be made, and the full equations must be kept.

5.1.2.2. *Medium amplitude topography variations: $\beta = O(\sqrt{\mu})$.* In this asymptotic regime, namely,

$$\mathcal{A}_{SW,med} = \{(\varepsilon, \beta, \gamma, \mu), \quad 0 \leq \mu \leq \mu_0, \quad 0 \leq \varepsilon \lesssim \sqrt{\mu}, \quad 0 \leq \beta \lesssim \sqrt{\mu}, \quad 0 \leq \gamma \leq 1\},$$

one has, with $\mathcal{Q}$ and $\mathcal{Q}_b$ as defined in (5.13) and (5.14),

$$\varepsilon\mu\big(\mathcal{Q}(\overline{V}) + \mathcal{Q}_b(\overline{V})\big) = -\frac{\mu\varepsilon}{3}\nabla^\gamma\big[\big((\overline{V}\cdot\nabla^\gamma)(\nabla^\gamma\cdot\overline{V}) - (\nabla^\gamma\cdot\overline{V})^2\big)\big] + O(\mu^2).$$

(The meaning of the $O(\mu^2)$ in the r.h.s. becomes clear in the statement of the proposition below.) Introducing the notation

$$\mathcal{Q}^0(\overline{V}) = \mathcal{Q}[1](\overline{V}) = -\frac{1}{3}\nabla^\gamma\big[\big((\overline{V}\cdot\nabla^\gamma)(\nabla^\gamma\cdot\overline{V}) - (\nabla^\gamma\cdot\overline{V})^2\big)\big], \tag{5.20}$$

one can thus replace (5.11) by

$$\begin{cases} \partial_t\zeta + \nabla^\gamma\cdot(h\overline{V}) = 0, \\ (I + \mu\mathcal{T})\partial_t\overline{V} + \nabla^\gamma\zeta + \varepsilon(\overline{V}\cdot\nabla^\gamma)\overline{V} + \mu\varepsilon\mathcal{Q}^0(\overline{V}) = 0. \end{cases} \tag{5.21}$$

More precisely, one deduces the following result from Proposition 5.8.

COROLLARY 5.14. *In the shallow water regime with medium surface and topography variations $\mathcal{A}_{SW,med}$, the water waves equations (5.1) are consistent at order $O(\mu^2)$ with (5.21) in the sense of Definition 5.1 (with $n = 6$).*

REMARK 5.15. Some components in the first term $(I + \mu\mathcal{T})\partial_t\overline{V}$ of the second equation of (5.21) are of size $O(\mu^2)$ but they have been kept because discarding them would have damaged the good properties of the operator $I + \mu\mathcal{T}$ (see §5.6.1.2). A stronger smallness assumption on β allows for a simplification that preserves these properties, as shown in §5.1.2.3 below.

5.1.2.3. *Small amplitude topography variations: $\beta = O(\mu)$.* In this asymptotic regime

$$\mathcal{A}_{SW,med-} = \{(\varepsilon, \beta, \gamma, \mu), \quad 0 \leq \mu \leq \mu_0, \quad 0 \leq \varepsilon \lesssim \sqrt{\mu}, \quad 0 \leq \beta \lesssim \mu, \quad 0 \leq \gamma \leq 1\},$$

a simplification of the term $\mu\mathcal{T}[h, \beta b]\partial_t\overline{V}$ that allows us to simplify (5.21) into

$$\begin{cases} \partial_t\zeta + \nabla^\gamma\cdot(h\overline{V}) = 0, \\ \big(1 - \frac{\mu}{3h}\nabla^\gamma(h^3\nabla^\gamma\cdot)\big)\partial_t\overline{V} + \nabla^\gamma\zeta + \varepsilon(\overline{V}\cdot\nabla^\gamma)\overline{V} + \mu\varepsilon\mathcal{Q}^0(\overline{V}) = 0. \end{cases} \tag{5.22}$$

More precisely, we have the following result.

COROLLARY 5.16. *In the shallow water regime with medium surface and small topography variations $\mathcal{A}_{SW,med-}$, the water waves equations (5.1) are consistent at order $O(\mu^2)$ with (5.22) in the sense of Definition 5.1 (with $n = 6$).*

REMARK 5.17. The main interest of (5.22) is that in the one-dimensional case $d = 1$, it admits unidirectional solutions governed by a scalar equation related to the Camassa-Holm equation, as shown in §7.3.

REMARK 5.18. One can replace h by $\tilde{h} = 1+\varepsilon\zeta$ in the term $\big(1-\frac{\mu}{3h}\nabla^\gamma(h^3\nabla^\gamma\cdot)\big)\partial_t\overline{V}$ in the second equation of (5.22) and keep the same precision of the model.

5.1.3. Small amplitude models ($\mu \ll 1$ and $\varepsilon = O(\mu)$). The asymptotic regime $\mu \ll 1$ and $\varepsilon = O(\mu)$ is usually referred to as the *long wave regime*[11]. It is under this regime that the usual Boussinesq systems[12] can be derived. Here again, a discussion on the amplitude of the bottom variations is needed.

5.1.3.1. *Large amplitude topography variations: $\beta = O(1)$.* In the long wave regime with large topography variations

$$\mathcal{A}_{LW,top+} = \{(\varepsilon, \beta, \gamma, \mu), \quad 0 \leq \mu \leq \mu_0, \quad 0 \leq \varepsilon \lesssim \mu, \quad 0 \leq \beta \leq 1, \quad 0 \leq \gamma \leq 1\},$$

in the Green-Naghdi equations (5.11) we can neglect the terms which are of order $O(\mu^2)$ under the extra assumption $\varepsilon = O(\mu)$, yielding the equations

$$\begin{cases} \partial_t\zeta + \nabla^\gamma \cdot (h\overline{V}) = 0, \\ (I + \mu\mathcal{T}_b)\partial_t\overline{V} + \nabla^\gamma\zeta + \varepsilon(\overline{V}\cdot\nabla^\gamma)\overline{V} = 0, \end{cases} \tag{5.23}$$

where $h_b := 1 - \beta b$ and $\mathcal{T}_b = \mathcal{T}[h_b, \beta b]$, with $\mathcal{T}[\cdot,\cdot]$ as defined in (5.12),

$$\mathcal{T}_b V = -\frac{1}{3h_b}\nabla^\gamma(h_b^3\nabla^\gamma\cdot V) + \frac{\beta}{2h_b}\big[\nabla^\gamma(h_b^2\nabla^\gamma b\cdot V) - h_b^2\nabla^\gamma b\nabla^\gamma\cdot V\big] + \beta^2\nabla^\gamma b\nabla^\gamma b\cdot V. \tag{5.24}$$

This model corresponds to the Boussinesq system for nonflat bottoms derived by Peregrine [**275**] (in the case $\gamma = 1$) and therefore will be referred to as the *Boussinesq-Peregrine model.* Note also that it is necessary to assume the positiveness of h_b in order for $\mathcal{T}_b$ to make sense. With the usual procedure, we deduce the following result from Proposition 5.8.

COROLLARY 5.19. *In the long wave regime with large topography variations $\mathcal{A}_{LW,top+}$, and provided that $h_b = 1 - \beta b \geq h_{min}$, the water waves equations (*5.1*) are consistent at order $O(\mu^2)$ with (*5.23*) in the sense of Definition* 5.1 *(with $n = 6$).*

5.1.3.2. *Small amplitude topography variations: $\beta = O(\mu)$.* In the long wave regime with small topography variations

$$\mathcal{A}_{LW,top-} = \{(\varepsilon, \beta, \gamma, \mu), \quad 0 \leq \mu \leq \mu_0, \quad 0 \leq \varepsilon \lesssim \mu, \quad 0 \leq \beta \lesssim \mu, \quad 0 \leq \gamma \leq 1\},$$

---

[11] This denomination, however, is not consensual; the assumption $\mu \ll 1$ (called here *shallow water* assumption) is sometimes called long-wave assumption.

[12] Also called *weakly nonlinear Boussinesq systems* by opposition to the *fully nonlinear Boussinesq—or Green-Naghdi—systems.*

the equations simplify further into the following *Boussinesq system*[13]

$$
\left\{
\begin{array}{l}
\partial_t\zeta + \nabla^\gamma\cdot(h\overline{V}) = 0,\\
(1-\frac{\mu}{3}\Delta^\gamma)\partial_t\overline{V} + \nabla^\gamma\zeta + \varepsilon(\overline{V}\cdot\nabla^\gamma)\overline{V} = 0,
\end{array}
\right.
\tag{5.25}
$$

for which topographic effects play a role only through the presence of $h = 1+\varepsilon\zeta-\beta b$ in the first equation. More precisely, we have the following result.

COROLLARY 5.20. *In the long wave regime with small topography variations $\mathcal{A}_{LW,top-}$, the water waves equations (*5.1*) are consistent at order $O(\mu^2)$ with (*5.25*) in the sense of Definition* 5.1 *(with $n = 6$).*

PROOF. With the procedure used so far, one does not find directly (5.25) but a similar model where the first term in the second equation is $(1-\frac{\mu}{3}\nabla^\gamma\nabla^{\gamma\,T})\partial_t\overline{V}$. In order to get (5.25), one just has to note that $\nabla^\gamma\nabla^{\gamma\,T}\partial_t\overline{V} = \Delta^\gamma\partial_t\overline{V} + O(\mu)$. □

REMARK 5.21. It is of course possible to derive directly (5.21), (5.22), (5.23), and (5.25) from the water waves equation (5.1). One just has to replace $\overline{V}_0$ and $\overline{V}_1$ in Proposition 3.37 by the simplifications obtained using the additional assumptions on ε and β. For instance, to derive (5.25) from (5.1), the following expansion of $\overline{V}$ is used

$$
\overline{V} = \nabla^\gamma\psi + \frac{\mu}{3}\Delta^\gamma\nabla^\gamma\psi + O(\mu^2,\varepsilon\mu,\beta\mu).
$$

REMARK 5.22. The Nonlinear Shallow Water equations (5.7) can also be simplified if we assume that $\varepsilon = O(\mu)$, $\beta = O(\mu)$. One easily gets that

$$
\left\{
\begin{array}{l}
\partial_t\zeta + \nabla^\gamma\cdot\overline{V} = O(\mu,\varepsilon,\beta)\\
\partial_t\overline{V} + \nabla^\gamma\zeta = O(\mu,\varepsilon,\beta).
\end{array}
\right.
$$

Therefore, up to terms of order $O(\mu,\varepsilon,\beta)$, the surface elevation satisfies the wave equation

$$
\partial_t^2\zeta - \Delta^\gamma\zeta = 0.
$$

We recall that this is a dimensionless equation; in variables with dimension (see §5.6.4 below), this equation becomes

$$
\partial_t^2\zeta - gH_0\Delta\zeta = 0,
$$

which is a wave equation at the *shallow water velocity* $\sqrt{gH_0}$ (also called Lagrange velocity since this equation was first derived by Lagrange in 1781).

---

[13]Named after Joseph Boussinesq, who first added dispersive terms to the nonlinear shallow water (Saint-Venant) equations [**47, 48**] (in dimension $d = 1$ and for a flat bottom). The system (5.25) is not exactly the system derived by Boussinesq, which was (with $d = 1$ and $b = 0$)

$$
\left\{
\begin{array}{l}
\partial_t\zeta + \partial_x(h\overline{V}) = 0,\\
\partial_t\overline{V} + \partial_x\zeta + \varepsilon\overline{V}\partial_x\overline{V} + \frac{\mu}{3}\partial_x\partial_t^2\zeta = 0.
\end{array}
\right.
$$

The system (5.25) seems to be due to Whitham [**324**]; both systems are of course equivalent since

$$
-\mu\partial_x^2\partial_t\overline{V} = \mu\partial_x\partial_t^2\zeta + O(\mu^2).
$$

In order to derive this system, Whitham used a formal expansion of the velocity potential as a polynomial in z; we know by Lemma 3.42 that such an expansion is always correct in the shallow water regime $\mu \ll 1$.

5.2. Improving the frequency dispersion of shallow water models

The (linear) dispersive properties of the water waves equations (5.1) depend on their *dispersion relation*, that is, on the set of all $(\omega, \mathbf{k}) \in \mathbb{R}^{1+d}$, such that there exists a plane wave solution $(\zeta_0, \psi_0)^T e^{i(\mathbf{k}\cdot X - \omega t)}$ to the linearization of (5.1) around the rest state $\zeta = 0$, $\nabla^\gamma \psi = 0$, and for flat bottoms. Since $\mathcal{G}_{\mu,\gamma}[0,0] = \sqrt{\mu}|D^\gamma|\tanh(\sqrt{\mu}|D^\gamma|)$, these linearized equations are given by

$$\begin{cases} \partial_t \zeta - |D^\gamma|\frac{1}{\sqrt{\mu}}\tanh(\sqrt{\mu}|D^\gamma|)\psi = 0, \\ \partial_t \psi + \zeta = 0, \end{cases}$$

and the dispersion relation of the (nondimensionalized[14]) water waves equations is thus given by

$$\omega_{WW}(\mathbf{k})^2 = |\mathbf{k}^\gamma|^2 \frac{\tanh(\sqrt{\mu}|\mathbf{k}^\gamma|)}{\sqrt{\mu}|\mathbf{k}^\gamma|}, \tag{5.26}$$

with $\mathbf{k}^\gamma = (\mathbf{k}_1, \gamma\mathbf{k}_2)^T$.

It is also possible to compute the dispersion relation for the Green-Naghdi equations (5.11) —which is of course the same for (5.21), (5.22), and (5.25), since these models do not differ at the linear level for flat bottoms. One finds

$$\omega_{GN}(\mathbf{k})^2 = \frac{|\mathbf{k}^\gamma|^2}{1 + \frac{\mu}{3}|\mathbf{k}^\gamma|^2}. \tag{5.27}$$

Defining the *phase velocities* (or celerity) for the water waves and Green-Naghdi equations as

$$\begin{aligned} c_{WW}(\sqrt{\mu}|\mathbf{k}^\gamma|) &= \frac{\omega_{WW}(\mathbf{k})}{|\mathbf{k}^\gamma|} = \Big(\frac{\tanh(\sqrt{\mu}|\mathbf{k}^\gamma|)}{\sqrt{\mu}|\mathbf{k}^\gamma|}\Big)^{1/2}, \\ c_{GN}(\sqrt{\mu}|\mathbf{k}^\gamma|) &= \frac{\omega_{GN}(\mathbf{k})}{|\mathbf{k}^\gamma|} = \Big(\frac{1}{1+\frac{\mu}{3}|\mathbf{k}^\gamma|^2}\Big)^{1/2}, \end{aligned}$$

one readily checks that the first-order Taylor expansions at the origin of $c_{WW}(\cdot)$ and $c_{GN}(\cdot)$ coincide. Consequently, the dispersive properties of the Green-Naghdi (or Boussinesq) equation will be close to the dispersive properties of the water waves equations for a large range of wave numbers $\mathbf{k}$ when μ is small. As shown in Figure 5.1, a relative error of, say, 10% is made on the phase velocity when $\sqrt{\mu}|\mathbf{k}^\gamma| \geq 1.8$. Since the value of μ is not arbitrarily small but given by the physical situation under consideration, a great deal of the literature in oceanography (e.g. [**269, 161, 246, 69, 19**]) has been devoted to the derivation of models with *improved frequency dispersion*[15]. Presented here are a few techniques to improve the dispersion relation of the Green-Naghdi and Boussinesq equations, without altering their precision (in the sense of consistency).

We first consider the case of small amplitude waves with large topography (Boussinesq-Peregrine) or small topography (Boussinesq) in §5.2.1 and then turn to the technically more involved case of large amplitude waves (Green-Naghdi) in §5.2.2.

---

[14]See (1.22) and (1.23) for the version with dimensions of the dispersion relation and celerity.

[15]The interest for applications in coastal oceanography is obvious: a model will be valid (at least as far as its linear dispersive properties are concerned) much further offshore if the 10% threshold is obtained for larger values of $\sqrt{\mu}|\mathbf{k}^\gamma|$. The model also gives better results when high frequencies are released, for instance, when a wave propagates above an obstacle as in §5.2.3.

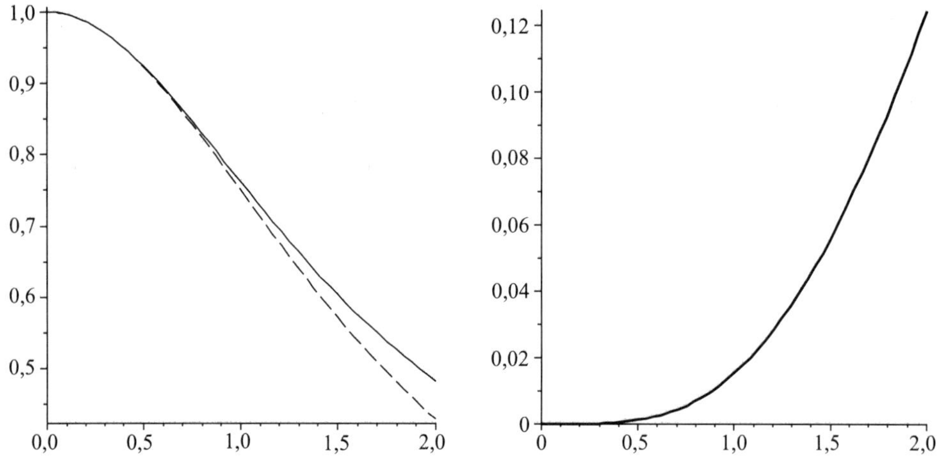

FIGURE 5.1. Left: c_{WW} (solid) and c_{GN} (dashed). Right: relative error $(c_{WW} - c_{GN})/c_{WW}$.

5.2.1. Boussinesq equations with improved frequency dispersion. We show here several techniques to improve the linear frequency dispersion of the Boussinesq-Peregrine system (5.23). In §5.2.1.1, we derive a one-parameter family of Boussinesq systems by using a classical trick, which consists of trading a time derivative of $\overline{V}$ for a space derivative of ζ in the higher order terms; this is a generalization proposed in [**245**] (see also [**19**]) of a technique introduced by Benjamin, Bona, and Mahony [**23**] to derive the BBM equation from the KdV equation (see Remark 7.5).

Another classical technique consists of changing the choice of the velocity unknown; used in a systematic way, as outlined by Bona and Smith [**39**], this allows for the introduction of a second parameter. We propose in §5.2.1.2 a slight adaptation of this technique that allows us to derive a three parameters family of Boussinesq systems.

Finally, we derive in §5.2.1.3 a four parameters family of Boussinesq systems by applying the aforementioned BBM trick on $\partial_t\zeta$. (For the sake of clarity, this last derivation is performed for almost flat bottoms only.)

5.2.1.1. *A first family of Boussinesq-Peregrine systems with improved frequency dispersion.* We prove here that the Boussinesq system with large amplitude topography variations (5.23) is a particular case (corresponding to $\alpha = 1$) of the following one-parameter family of equations

$$(5.28) \qquad \begin{cases} \partial_t\zeta + \nabla^\gamma \cdot (h\overline{V}) = 0, \\ [I + \mu\alpha\mathcal{T}_b]\big(\partial_t\overline{V} + \dfrac{\alpha-1}{\alpha}\nabla^\gamma\zeta\big) + \dfrac{1}{\alpha}\nabla\zeta + \varepsilon(\overline{V}\cdot\nabla^\gamma)\overline{V} = 0, \end{cases}$$

where $h_b := 1 - \beta b$ and $\mathcal{T}_b$ as defined in (5.24), while $\alpha \in \mathbb{R}$ (but only the case $\alpha \geq 1$ yields linearly well-posed models as shown in Remark 5.24). We can note that stronger regularity assumptions[16] are needed on the solution compared to what is required in Corollary 5.19 ($n = 9$ vs. $n = 6$).

[16] These regularity assumptions are certainly not sharp, but since this aspect is not very relevant here, we chose to make the proof as simple as possible.

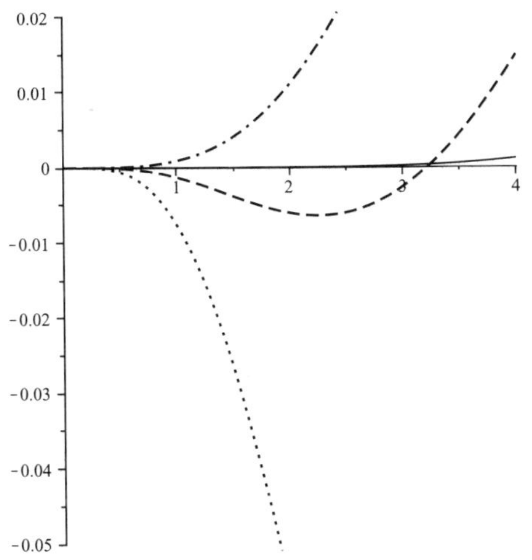

FIGURE 5.2. Relative error $(c_{WW} - c_{XX})/c_{WW}$ with $XX = GN$ (dots), $XX = \alpha$ (dash), $XX = (\alpha, \theta, \delta)$ (dashdot) and $XX = (\alpha, \theta, \delta, \lambda)$ (solid).

COROLLARY 5.23. *In the long wave regime with large topography variations* $\mathcal{A}_{LW,top+}$, *and provided that* $h_b = 1 - \beta b \geq h_{min}$, *the water waves equations* (5.1) *are consistent at order* $O(\mu^2)$ *with* (5.28) *in the sense of Definition* 5.1 *(with* $n = 9$*).*

REMARK 5.24. The dispersion relation and celerity associated to (5.28) are given by

$$\omega_\alpha(\mathbf{k})^2 = |\mathbf{k}^\gamma|^2 \frac{1 + \mu\frac{(\alpha-1)}{3}|\mathbf{k}^\gamma|^2}{1 + \mu\frac{\alpha}{3}|\mathbf{k}^\gamma|^2} \quad \text{and} \quad c_\alpha(\sqrt{\mu}|\mathbf{k}^\gamma|) = \Big(\frac{1 + \mu\frac{(\alpha-1)}{3}|\mathbf{k}^\gamma|^2}{1 + \mu\frac{\alpha}{3}|\mathbf{k}^\gamma|^2}\Big)^{1/2}.$$

The first-order Taylor expansion at the origin of $c_\alpha(\cdot)$ of course matches those of $c_{WW}(\cdot)$ and $c_{GN}(\cdot)$. However, with a good choice of α (oceanographers often take $\alpha = 1.159$, [**19, 80**]), c_α remains close to c_{WW} for larger values of $\sqrt{\mu}\mathbf{k}^\gamma$, as shown in Figure 5.2.

PROOF. We know from Corollary 5.19 that

$$\partial_t \overline{V} = -\nabla^\gamma \zeta + \mu \widetilde{R}, \tag{5.29}$$

with

$$\widetilde{R} = -\frac{\varepsilon}{\mu}(\overline{V} \cdot \nabla^\gamma)\overline{V} - \mu \mathcal{T}_b \partial_t \overline{V} + \mu R$$

(where R is the same as in Corollary 5.19). We therefore can write

$$\partial_t \overline{V} = \alpha \partial_t \overline{V} + (1 - \alpha)\big[- \nabla^\gamma \zeta + \mu \widetilde{R}\big]$$

and therefore

$$(I + \mu \mathcal{T}_b)\partial_t \overline{V} = (I + \mu\alpha \mathcal{T}_b)\partial_t \overline{V} - (1 - \alpha)\mu \mathcal{T}_b \nabla^\gamma \zeta + (1 - \alpha)\mu^2 \mathcal{T}_b \widetilde{R}.$$

Denoting by $\mu^2 R_\alpha$ the residual of the second equation corresponding to the consistency result stated in the corollary, one therefore has $R_\alpha = R - (1 - \alpha)\mathcal{T}_b \widetilde{R}$ and we thus deduce from the definition of $\mathcal{T}_b$ that

$$\begin{aligned} |R_\alpha|_{H^s} &\leq |R|_{H^s} + C(\frac{1}{h_{min}}, |b|_{H^{1+s\vee t_0}})|\widetilde{R}|_{H^{s+2}} \\ &\leq N(s+8) + C(\frac{1}{h_{min}}, |b|_{H^{s+3}})(|\partial_t\overline{V}|_{H^{s+3}} + \sqrt{\mu}|\nabla^\gamma\cdot\partial_t\overline{V}|_{H^{s+3}}), \end{aligned}$$

where we used the bound on R provided by Corollary 5.19. Moreover, since Corollary 5.19 implies that

$$\partial_t\overline{V} = [h_b(I+\mu\mathcal{T}_b)]^{-1}(-h_b\nabla^\gamma\zeta - \varepsilon h_b(\overline{V}\cdot\nabla^\gamma)\overline{V} + \mu h_b R),$$

we can deduce from the properties of $[h_b(I+\mu\mathcal{T}_b)]^{-1}$ stated in Lemma 5.44 (postponed to §5.6.1.1 for the sake of clarity) that

$$|\partial_t\overline{V}|_{H^{s+3}} + \sqrt{\mu}|\nabla^\gamma\cdot\partial_t\overline{V}|_{H^{s+3}} \leq N(s+9),$$

and the result follows. □

5.2.1.2. *A second family of Boussinesq-Peregrine systems with improved frequency dispersion.* The family of equations $(5.28)_\alpha$ furnishes an approximation of the same accuracy (in terms of consistency) as the original Boussinesq equations with large topography variations (5.28).

One more degree of freedom can be achieved by changing the choice of unknown for the velocity. It was noted earlier that it was possible to write Boussinesq systems that differ from those of the family $(5.28)_\alpha$ by replacing the vertically averaged horizontal velocity $\overline{V}$ by another unknown: for instance, Mei and Méhauté [**257**] replaced $\overline{V}$ by the horizontal velocity at the bottom V_{bott}, Peregrine mentioned in [**275**] the possibility of working with the velocity V_0 at $z=0$, and Bona and Smith [**39**] noticed the abundance of possible choices, among which include the velocity evaluated at some level line of the fluid domain. Their work dealt with flat bottoms but has been generalized to nonflat bottoms in [**269**].

In [**269**], Nwogu showed that it was possible to improve the dispersive properties of Boussinesq models by working with the velocity at a certain depth as a dependent variable. This approach was generalized in [**321**] to the fully nonlinear case (Green-Naghdi). When the bottom is not flat, it turns out that in the fully nonlinear case, the standard Green-Naghdi equations written with the average velocity $\overline{V}$ do *not* belong to this new class of fully nonlinear models (see Remark 5.25 below). This is why we use a slightly different approach here, with the introduction of a new dependent variable $V_{\theta,\delta}$ that is not the velocity at a certain depth. The significance is that the computations are somewhat simpler and, more importantly, that the average velocity $\overline{V}$ appears as a particular case ($\theta=\delta=0$). Our choice of dependent variables also differs from the one used in [**227**] and whose purpose is not to improve the dispersive properties but to work with potential variables.
For all $\theta,\delta\geq 0$, we define $V_{\theta,\delta}$ by the relation

$$V_{\theta,\delta} = (I+\mu\theta\mathcal{T}_b)^{-1}(I+\mu\delta\mathcal{T}_b)\overline{V} \tag{5.30}$$

and rewrite the equations with unknown $V_{\theta,\delta}$ rather than $\overline{V}$. The fact that this definition makes sense is shown in Lemma 5.44, which is postponed to §5.6.1.1 for the sake of clarity.

REMARK 5.25. For flat bottoms (or small amplitude topography variations $\beta=O(\mu)$), $V_{\theta,\delta}$ corresponds up to terms of order $O(\mu^2)$ to the horizontal velocity evaluated at the level line

$$\Gamma_{\theta-\delta} = \{h(X)z_{\theta-\delta} + \varepsilon\zeta(X)\},$$

where $z_{\theta-\delta} = -1 + \frac{1}{\sqrt{3}}\sqrt{1+2(\delta-\theta)}$. For large amplitude variations, there is no particular physical meaning to $V_{\theta,\delta}$. We could of course choose to work with $V_{|\Gamma_{\theta-\delta}}$, the horizontal velocity at the level line $z_{\theta-\delta}$ as, for instance, in [**269, 321**]; we decided however to work with $V_{\theta,\delta}$ as defined in (5.30) for two reasons: 1) simplicity 2) the family of models thus obtained contains the equations (5.28) as a particular case (take $\theta = \delta = 0$ in (5.30)), while this is not the case for the other choices of velocity unknown. We refer to §5.6.2.1 for more precision on the relations between $\overline{V}$, $V_{\theta,\delta}$ and $V_{|\theta-\delta}$.

Contrary to the asymptotic models derived so far, the systems presented below couple the surface elevation ζ to the new "velocity" $V_{\theta,\delta}$ (instead of $\overline{V}$). The evolution equation on ζ is therefore no more exact and the asymptotic systems are of the form

$$(\widetilde{A}) \qquad \begin{cases} \partial_t\zeta + \mathcal{N}^1_{(\widetilde{A})}(\zeta, V_{\theta,\delta}) = 0, \\ \partial_t V_{\theta,\delta} + \mathcal{N}^2_{(\widetilde{A})}(\zeta, V_{\theta,\delta}) = 0, \end{cases}$$

where $\mathcal{N}^1_{(\widetilde{A})}(\zeta, V_{\theta,\delta})$ and $\mathcal{N}^2_{(\widetilde{A})}(\zeta, V_{\theta,\delta})$ are nonlinear (pseudo)-differential operators that depend on the asymptotic model (A). Since the first equation is no longer exactly satisfied, we have to generalize the notion of consistency introduced in Definition 5.1.

DEFINITION 5.26. Let $\mathcal{A}$ be an asymptotic regime in the sense of Definition 4.19. The water waves equations are considered *consistent at order* $O(\mu^k)$ *with* $(\widetilde{A})$ *in the regime* $\mathcal{A}$ if there exists $n \in \mathbb{N}$ and $T > 0$ such that for all $s \geq 0$ and $\mathfrak{p} = (\varepsilon, \beta, \gamma, \mu) \in \mathcal{A}$

- There exist a solution $(\zeta, \psi) \in C([0, \frac{T}{\varepsilon\sqrt{\beta}}]; H^{s+n} \times \dot{H}^{s+5})$ to the water waves equations (5.1).
- With $V_{\theta,\delta}$ as in (5.30), one has

$$\begin{cases} \partial_t\zeta + \mathcal{N}^1_{(\widetilde{A})}(\zeta, V_{\theta,\delta}) = \mu^k r, \\ \partial_t\overline{V} + +\mathcal{N}^2_{(\widetilde{A})}(\zeta, V_{\theta,\delta}) = \mu^k R, \end{cases}$$

with $|r|_{H^s} + |R|_{H^s} \leq N(s+n)$ on $[0, \frac{T}{\varepsilon\sqrt{\beta}}]$, and $N(s+n)$ as in (5.6).

The change of unknown (5.30) allows one to generalize the *one-parameter* family of Boussinesq-Peregrine equations $(5.28)_\alpha$ into the following *three-parameter* family of systems depending on $\alpha \in \mathbb{R}$ and $\theta, \delta \geq 0$. (The systems $(5.28)_\alpha$ correspond to the particular case $\theta = \delta = 0$, and (5.23) is recovered by also taking $\alpha = 1$.)

$$(5.31) \quad \begin{cases} [I - \frac{\mu\delta}{3h_b}\nabla^\gamma\cdot(h_b^3\nabla^\gamma\bullet)]\partial_t\zeta + \nabla^\gamma\cdot(h(I+\mu\theta\mathcal{T}_b)V_{\theta,\delta}) + \mu\delta\mathcal{L}_b[h]V_{\theta,\delta} = 0, \\ [I + \mu(\alpha+\theta)\mathcal{T}_b]\big(\partial_t V_{\theta,\delta} + \frac{\alpha+\delta-1}{\alpha+\theta}\nabla^\gamma\zeta\big) \\ \qquad + \frac{1+\theta-\delta}{\alpha+\theta}\nabla^\gamma\zeta + \varepsilon(V_{\theta,\delta}\cdot\nabla^\gamma)V_{\theta,\delta} = 0, \end{cases}$$

where $h_b := 1 - \beta b$ and $\mathcal{T}_b$ is defined in (5.24), while the second-order linear differential operator $\mathcal{L}_b[h]$ is defined as

$$\begin{aligned} \mathcal{L}_b[h]V &= \nabla^\gamma \cdot (hh_b\nabla^\gamma h_b\nabla^\gamma \cdot V) - \frac{1}{3}hh_b\nabla^\gamma h_b\nabla^\gamma\nabla^\gamma \cdot V \\ &\quad -\frac{1}{3h_b}\nabla^\gamma \cdot \left[h_b^3\nabla^\gamma h_b\nabla^\gamma \cdot V + h_b^3\nabla^\gamma(\nabla^\gamma h \cdot V)\right] - \nabla^\gamma \cdot (h\mathcal{T}_b^{II}), \end{aligned} \tag{5.32}$$

where we used the decomposition

$$\mathcal{T}_b = \mathcal{T}_b^I + \mathcal{T}_b^{II} \quad \text{with} \quad \mathcal{T}_b^I = -\frac{1}{3h_b}\nabla^\gamma(h_b^3 \mathrm{div}_\gamma \bullet).$$

(In particular, $\mathcal{T}_b^{II}$ contains at most first-order differential operators.)

COROLLARY 5.27. *In the long wave regime with large topography variations $\mathcal{A}_{LW,top+}$, and provided that $h_b = 1 - \beta b \geq h_{min}$, the water waves equations (5.1) are consistent at order $O(\mu^2)$ with (5.31) in the sense of Definition 5.26 (with $n = 10$).*

REMARK 5.28. The dispersion relation associated to (5.31) is given by

$$\omega_{\alpha,\theta,\delta}(\mathbf{k})^2 = |\mathbf{k}^\gamma|^2 \frac{(1 + \mu\frac{\theta}{3}|\mathbf{k}^\gamma|^2)(1 + \mu\frac{\alpha+\delta-1}{3}|\mathbf{k}^\gamma|^2)}{(1 + \mu\frac{\delta}{3}|\mathbf{k}^\gamma|^2)(1 + \mu\frac{\alpha+\theta}{3}|\mathbf{k}^\gamma|^2)}.$$

The introduction of the two parameters θ and δ allows for more freedom to improve the dispersive properties of the model. For instance, the following values are taken in [**70**],

$$\alpha_{opt} = 1.028, \quad \theta_{opt} = 0.188, \quad \delta_{opt} = 0.112. \tag{5.33}$$

A comparison of the dispersion relation of the corresponding model with the exact (linearized) water waves equations is plotted in Figure 5.2. (The reason it seems worse than c_α on these plots is because these values has been optimized with respect to the group velocity also; see [**70**]).

REMARK 5.29. There is a consequence when one decides to work with such models with improved frequency dispersion. When $\delta \neq 0$, the first equation (conservation of mass) is not exactly satisfied, but only up to $O(\mu^2)$ terms. When $\delta = 0$, this equation is exactly satisfied, but the conservation of energy satisfied by the original model is no longer satisfied if $(\alpha, \theta, \delta) \neq (1, 0, 0)$. It is, however, satisfied up to $O(\mu^2)$ terms, of course (see §6.3.1).

REMARK 5.30. We explain in Remark 5.35 how to derive a *four parameters* generalization of the Boussinesq systems $(5.31)_{\alpha,\theta,\delta}$.

PROOF. After the quite tedious computations, which we omit here, one can note that

$$[I - \mu\delta\frac{1}{3h_b}\nabla^\gamma \cdot (h_b^3\nabla^\gamma \bullet)]\nabla^\gamma \cdot (hV) = \nabla^\gamma \cdot (h(I + \mu\delta\mathcal{T}_b)V) + \mu\delta\mathcal{L}_b[h]V.$$

Using (5.30), we therefore obtain

$$\begin{aligned} [I &- \mu\delta\frac{1}{3h_b}\nabla^\gamma \cdot (h_b^3\nabla^\gamma \bullet)]\nabla^\gamma \cdot (hV) \\ &= \nabla^\gamma \cdot (h(I + \mu\theta\mathcal{T}_b)V_{\theta,\delta}) + \mu\delta\mathcal{L}_b[h]\overline{V} \\ &= \nabla^\gamma \cdot (h(I + \mu\theta\mathcal{T}_b)V_{\theta,\delta}) + \mu\delta\mathcal{L}_b[h]V_{\theta,\delta} - \delta\mu^2 r, \end{aligned}$$

with

$$r = -\frac{1}{\mu}\big(\mathcal{L}_b[h_b]\overline{V} - \mathcal{L}_b[h_b]V_{\theta,\delta}\big).$$

The first equation of (5.31) is therefore obtained after applying $[I - \mu\delta\frac{1}{3h_b}\nabla^\gamma \cdot (h_b^3\nabla^\gamma \bullet)]$ to the first equation of (5.31). For the estimate on $|r|_{H^s}$, we use the definition of $\mathcal{L}_b[h]$ and $V_{\theta,\delta}$ to get

$$\begin{aligned} |r|_{H^s} &\leq M(s+4)|\overline{V}|_{H^{s+4}} \\ &\leq N(s+6), \end{aligned}$$

where we used Proposition 3.37 to derive the last inequality.

For the second equation, we can note that since $\mathcal{T}_b$ does not depend on time, one has

$$(I + \mu\delta\mathcal{T}_b)\partial_t\overline{V} = (I + \mu\theta\mathcal{T}_b)\partial_t V_{\theta,\delta},$$

so that, after being multiplied by $(I + \mu\delta\mathcal{T}_b)$, the second equation of (5.28) can be put under the form

$$(I + \mu(\alpha+\theta)\mathcal{T}_b)\big[\partial_t V_{\theta,\delta} + \frac{\alpha+\delta-1}{\alpha+\theta}\nabla^\gamma\zeta\big] + \frac{1+\theta-\delta}{\alpha+\theta}\nabla^\gamma\zeta + \varepsilon(V_{\theta,\delta}\cdot\nabla^\gamma)V_{\theta,\delta} = \mu^2 R,$$

with

$$\begin{aligned} R &= R_\alpha - \alpha\theta\mathcal{T}_b^2\partial_t V_{\theta,\delta} - (\alpha-1)\delta\mathcal{T}_b^2\nabla^\gamma\zeta \\ &+ \frac{\varepsilon}{\mu}\delta\mathcal{T}_b(\overline{V}\cdot\nabla^\gamma)\overline{V} + \frac{\varepsilon}{\mu^2}\big((V_{\theta,\delta}\cdot\nabla^\gamma)V_{\theta,\delta} - (\overline{V}\cdot\nabla^\gamma)\overline{V}\big), \end{aligned}$$

where $\mu^2 R_\alpha$ is the residual of the velocity equation corresponding to the one parameter improved Boussinesq-Peregrine equations (see Corollary 5.23). Using the definition (5.30) of $V_{\theta,\delta}$, the definition of $\mathcal{T}_b$, and the estimates of Lemma 5.44, we get

$$|R|_{H^s} \leq |R_\alpha|_{H^s} + C(\frac{\varepsilon}{\mu}, M(s+5))\big(1 + |\partial_t\overline{V}|_{H^{s+4}} + |\overline{V}|^2_{H^{s+3}}\big),$$

and the end of the proof follows the same lines as the proof of Corollary 5.23. □

5.2.1.3. *Simplifications for the case of flat or almost flat bottoms.* When the bottom is flat ($\beta = 0$) or its variations are of small amplitude ($\beta = O(\mu)$), then important simplifications can be carried out on (5.31). Simple manipulations then allow us to introduce another parameter λ, thus providing us with a *four parameters* family of Boussinesq systems, namely,

$$(5.34)\qquad \begin{cases} (1-\mu\mathbf{b}\Delta^\gamma)\partial_t\zeta + \nabla^\gamma\cdot(hV_{\theta,\delta}) + \mu\mathbf{a}\Delta^\gamma\nabla^\gamma\cdot V_{\theta,\delta} = 0, \\ (1-\mu\mathbf{d}\Delta^\gamma)\partial_t V_{\theta,\delta} + \nabla^\gamma\zeta + \varepsilon(V_{\theta,\delta}\cdot\nabla^\gamma)V_{\theta,\delta} + \mu\mathbf{c}\Delta^\gamma\nabla^\gamma\zeta = 0, \end{cases}$$

with

$$(5.35)\qquad \mathbf{a} = -\frac{\theta+\lambda}{3}, \quad \mathbf{b} = \frac{\delta+\lambda}{3}, \quad \mathbf{c} = -\frac{\alpha+\delta-1}{3}, \quad \mathbf{d} = \frac{\alpha+\theta}{3}.$$

The corresponding consistency result is given in the following corollary.

COROLLARY 5.31. *In the long wave regime with small topography variations $\mathcal{A}_{LW,top-}$, the water waves equations* (5.1) *are consistent at order $O(\mu^2)$ with* (5.34) *in the sense of Definition* 5.26 *(with $n = 10$).*

PROOF. Since for such regimes one has

$$\mu\mathcal{T}_b = -\frac{\mu}{3}\nabla^\gamma \mathrm{div}_\gamma + O(\mu\beta) \quad \text{and} \quad \mu\frac{1}{3h_b}\nabla^\gamma\cdot(h_b^3\nabla^\gamma \bullet) = \frac{\mu}{3}\Delta^\gamma + O(\mu\beta),$$

it is possible to make such a substitution at every occurrence of $\mu\mathcal{T}_b$ in (5.31), up to a corresponding modification of the residual. Moreover, since $V_{\theta,\delta} = \nabla^\gamma\psi + O(\mu)$, we can write $\mu\nabla^\gamma\mathrm{div}_\gamma V_{\theta,\delta} = \mu\Delta^\gamma V_{\theta,\delta} + O(\mu^2)$. The three parameters family of Boussinesq systems (5.31) therefore can be simplified into

$$\begin{cases}(1-\mu\frac{\delta}{3}\Delta^\gamma)\partial_t\zeta + \nabla^\gamma\cdot(hV_{\theta,\delta}) - \mu\frac{\theta}{3}\Delta^\gamma\nabla^\gamma\cdot V_{\theta,\delta} = O(\mu^2,\varepsilon\mu,\beta\mu),\\ (1-\mu\frac{\alpha+\theta}{3}\Delta^\gamma)\partial_t V_{\theta,\delta} + \nabla^\gamma\zeta - \mu\frac{\alpha+\delta-1}{3}\Delta^\gamma\nabla^\gamma\zeta + \varepsilon(V_{\theta,\delta}\cdot\nabla^\gamma)V_{\theta,\delta} = O(\mu^2,\varepsilon\mu,\beta\mu).\end{cases}$$

Finally, we note that it is possible to introduce a fourth parameter λ by adding $-\mu\frac{\lambda}{3}(\Delta^\gamma\partial_t\zeta - \Delta^\gamma\nabla^\gamma\cdot V_{\theta,\delta})$, which is a $O(\mu^2)$ quantity, to the first equation. □

REMARK 5.32. It is worth noting that in this regime of almost flat bottoms ($\beta = O(\mu)$), the bottom parametrization b appears in the equations (5.34) through $h = 1+\varepsilon\zeta-\beta b$ in the first equation.

REMARK 5.33. The dispersion relation associated to (5.34) is given by

$$\omega_{\alpha,\theta,\delta,\lambda}(\mathbf{k})^2 = |\mathbf{k}^\gamma|^2\frac{(1+\mu\frac{\theta+\lambda}{3}|\mathbf{k}^\gamma|^2)(1+\mu\frac{\alpha+\delta-1}{3}|\mathbf{k}^\gamma|^2)}{(1+\mu\frac{\delta+\lambda}{3}|\mathbf{k}^\gamma|^2)(1+\mu\frac{\alpha+\theta}{3}|\mathbf{k}^\gamma|^2)}.$$

We now have four parameters[17] to improve the dispersive properties of the model. For instance, with

$$\begin{aligned}\alpha &= 1,\\ \theta &= \frac{2}{3}+\frac{\sqrt{133}}{21}-1 \sim 0.216,\\ \delta &= \frac{7}{6}+\frac{\sqrt{805}}{210}-1 \sim 0.302,\\ \lambda &= -\frac{1}{2}-\frac{\sqrt{133}}{21}-\frac{\sqrt{805}}{210}+1 \sim -0.184,\end{aligned}$$

the celerity $c_{\alpha,\theta,\delta,\lambda}(\sqrt{\mu}|\mathbf{k}^\gamma|) = \omega_{\alpha,\theta,\delta,\lambda}(\mathbf{k})/|\mathbf{k}^\gamma|$ is the $(4,4)$ Padé representation of $c_{WW}(\sqrt{\mu}|\mathbf{k}^\gamma|) = (\frac{\tanh(\sqrt{\mu}|\mathbf{k}^\gamma|)}{|\mathbf{k}^\gamma|})^{1/2}$, as in the model derived in [**161**]. As shown in Figure 5.2, the matching with the exact dispersion relation is excellent.

REMARK 5.34. Many of the systems of the family $(5.34)_{\alpha,\theta,\delta,\lambda}$ (with flat bottom) have attracted a lot of interest in the last decades. For example, with $(\mathbf{a},\mathbf{b},\mathbf{c},\mathbf{d}) = (1/3,0,0,0)$ (for instance, take $\alpha=\theta=0$, $\delta=1$ and $\lambda=-1$), one gets the Kaup system [**205**]which is linearly ill posed but integrable. When $\mathbf{b}=\mathbf{d}>0$ and $\mathbf{a}=0$, the corresponding systems are the Bona-Smith models [**39**]. A systematic study of the systems (5.31) was first proposed in [**34, 35**] for $d=1$ and [**36**] for $d=2$ (see also [**73, 68**] for varying bottoms). Note that in [**34, 35, 36**], the coefficients $\mathbf{a}$, $\mathbf{b}$, $\mathbf{c}$ and $\mathbf{d}$ depend on three parameters only, versus four here. The family (5.31) offers therefore some additional possibilities.

[17]Even though we have at our disposal four fitting parameters, the set of coefficients $\mathbf{a}$, $\mathbf{b}$, $\mathbf{c}$ and $\mathbf{d}$ must satisfy the constraint $\mathbf{a}+\mathbf{b}+\mathbf{c}+\mathbf{d}=\frac{1}{3}$ and therefore is a subset of a three-dimensional affine space.

REMARK 5.35. It is of course possible to derive a four parameters family of Boussinesq-Peregrine equations generalizing the three parameters family (5.31) without assuming that the bottom is flat or almost flat. In order to derive (5.34), we added the quantity $\mu\frac{\lambda}{3}\Delta^\gamma\partial_t\zeta + \mu\frac{\lambda}{3}\Delta^\gamma\nabla^\gamma\cdot V_{\theta,\delta} = O(\mu^2)$ to the first equation. If we do not make any smallness assumption on the bottom variations, then one has to add the quantity

$$\mu\frac{\lambda}{3h_b}\nabla^\gamma\cdot(h_b^3\nabla^\gamma\partial_t\zeta) + \mu\frac{\lambda}{3h_b}\nabla^\gamma\cdot(h_b^3\nabla^\gamma\nabla^\gamma\cdot(hV_{\theta,\delta})) = O(\mu^2)$$

to the first equation of (5.31). The complications are only technical.

5.2.2. Green-Naghdi equations with improved frequency dispersion. We show in this section how to improve the dispersive properties of the Green-Naghdi equations (5.11). The techniques are exactly the same as those used above to improve the dispersive properties of the Boussinesq system (5.23), but are technically more involved.

5.2.2.1. *A first family of Green-Naghdi equations with improved frequency dispersion.* We show here that the Green-Naghdi equations (5.11), or its equivalent formulation (5.15), are a particular case (corresponding to $\alpha = 1$) of the following one-parameter family of equations used, for instance, in [**80**]

$$(5.36)\quad \begin{cases} \partial_t\zeta + \nabla^\gamma\cdot(h\overline{V}) = 0, \\ [1+\mu\alpha\mathcal{T}]\big(\partial_t\overline{V} + \varepsilon(\overline{V}\cdot\nabla^\gamma)\overline{V} + \frac{\alpha-1}{\alpha}\nabla^\gamma\zeta\big) + \frac{1}{\alpha}\nabla^\gamma\zeta + \mu\varepsilon\mathcal{Q}_1(\overline{V}) = 0, \end{cases}$$

where $\mathcal{Q}_1(\overline{V})$ is as in (5.16).

COROLLARY 5.36. *In the shallow water regime $\mathcal{A}_{SW}$, the water waves equations (5.1) are consistent at order $O(\mu^2)$ with (5.36) in the sense of Definition 5.1 (with $n = 9$).*

REMARK 5.37. The linear dispersion relation and celerity associated to $(5.36)_\alpha$ are of course the same as those obtained for the Boussinesq systems $(5.28)_\alpha$ (see Remark 5.24), since these two models differ only by nonlinear terms.

PROOF. We know from Proposition 5.8 that

$$\partial_t\overline{V} = -\nabla^\gamma\zeta - \varepsilon(\overline{V}\cdot\nabla^\gamma)\overline{V} + \mu\widetilde{R},$$

with $\widetilde{R} = -\mathcal{T}\big(\partial_t\overline{V} + \varepsilon(\overline{V}\cdot\nabla^\gamma)\overline{V}\big) - \varepsilon\mathcal{Q}_1(\overline{V}) + \mu R$ (where R is the same as in Proposition 5.8). We thus deduce that

$$\partial_t\overline{V} = \alpha\partial_t\overline{V} + (1-\alpha)(-\nabla^\gamma\zeta - \varepsilon(\overline{V}\cdot\nabla^\gamma)\overline{V} + \mu\widetilde{R}),$$

and thus

$$(I+\mu\mathcal{T})\partial_t\overline{V} = (I+\mu\alpha\mathcal{T})\partial_t\overline{V} - \mu(1-\alpha)\mathcal{T}\big(\nabla^\gamma\zeta + \varepsilon(\overline{V}\cdot\nabla^\gamma)\overline{V}\big) + \mu^2(1-\alpha)\mathcal{T}\widetilde{R}.$$

The result therefore follows from Proposition 5.8 with a residual $\mu^2 R_\alpha$ for the velocity equation given by

$$R_\alpha = R - (1-\alpha)\mathcal{T}\widetilde{R}.$$

The estimate on $|R_\alpha|_{H^s}$ is then established as in the proof of Corollary 5.23. □

5.2.2.2. *A second family of Green-Naghdi equations with improved frequency dispersion.* We want to generalize here to the fully nonlinear (Green-Naghdi) case $\varepsilon = O(1)$ the manipulations performed in §5.2.1.2 in the weakly nonlinear (Boussinesq) regime $\varepsilon = O(\mu)$. The natural generalization of the change of unknown (5.30) for the velocity is given, for all $\theta, \delta \geq 0$, by

$$V_{\theta,\delta} = (I + \mu\theta\mathcal{T})^{-1}(I + \mu\delta\mathcal{T})\overline{V}. \tag{5.37}$$

(This is explained in Lemma 5.45, which has been postponed to §5.6.1.2, for the sake of clarity.) The main qualitative difference between (5.37) and (5.30) is that the former is *time dependent* (because $\mathcal{T}$ involves ζ, contrary to $\mathcal{T}_b$).
Another qualitative difference is that in the weakly nonlinear regime $\varepsilon = O(\mu)$, one has

$$\varepsilon(\overline{V} \cdot \nabla^\gamma)\overline{V} = \varepsilon(V_{\theta,\delta} \cdot \nabla^\gamma)V_{\theta,\delta} + O(\mu^2),$$

which is no longer true in the fully nonlinear regime $\varepsilon = O(1)$.
These two qualitative changes induce considerable technical difficulties and we leave to the interested reader the opportunity to adapt the techniques introduced in §5.2.1.2 above (weakly nonlinear case) to the present situation. For relevant references, see [**70, 321**].

5.2.3. The physical relevance of improving the frequency dispersion. Shallow water models can be mathematically justified as $\mu \to 0$. However, from a physical viewpoint, μ has a given value that might not be so small. All the models derived in the previous sections are equivalent as $\mu \to 0$, but behave quite differently when μ is not very small. It is of great physical interest to determine which of these asymptotically equivalent models has the widest range of validity (i.e., still approximates correctly the water waves equations even for not very small values of μ). Indeed, allowing for larger values of μ means that one allows *deeper water* configurations and/or *higher frequency* waves (recall that $\mu = H_0^2/L_x^2$, with H_0 typical depth and L_x typical wavelength).
Looking for models with improved frequency dispersion, as we did in the previous sections, proves to be a useful way to derive such models with an extended range of validity. To illustrate the significance of these manipulations, let us consider a classical benchmark. (See Beji and Battjes [**21**] and Dingemans [**124**]; also see Figure 5.3.) A small-amplitude long wave is generated at the left boundary; when the incident wave encounters the upward part of the bar, it shoals and steepens, which generates higher-harmonics as the nonlinearity increases. These higher-harmonics are then freely released on the downward slope and become deep-water waves behind the bar. One must choose a version of the GN model (with parameters) able to describe correctly these waves; this means that this GN model must have good dispersive properties, even when μ is not very small.

In Figure 5.4, we compare free-surface time-series computed with different GN models and compare them with experimental measurements at several gauges (see Figure 5.3). The interest of the GN with improved frequency dispersion is obvious on these plots. Note that the computations are performed with the numerical code developed in [**70, 41**]. See these references for further information.

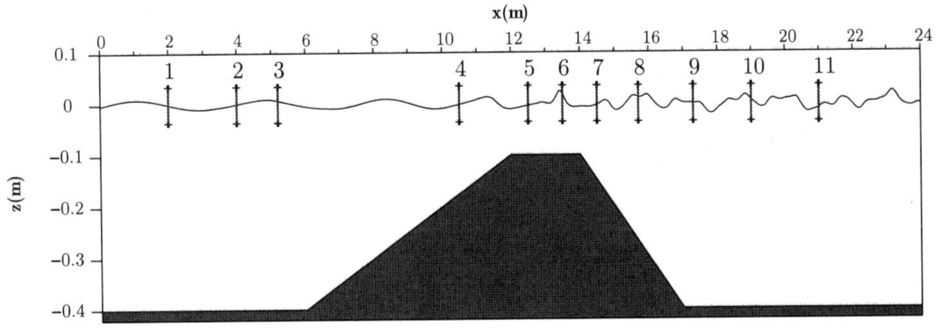

FIGURE 5.3. Experimental set-up and locations of the wave gauges.

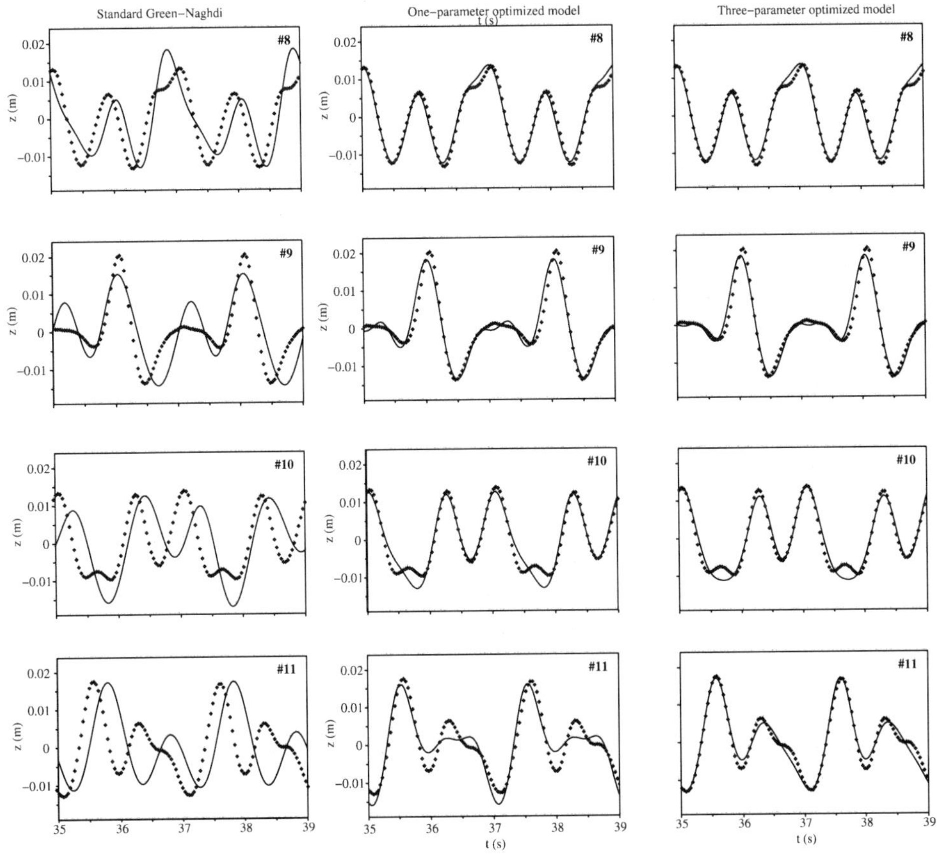

FIGURE 5.4. Comparison of computed free-surface time-series with experimental measurements. From left to right: standard Green-Naghdi equations (5.11), one-parameter Green-Naghdi (5.36), and a three-parameters Green-Naghdi model (see §5.2.2.2).

5.3. Improving the mathematical properties of shallow water models

In the previous section, we showed various techniques to derive asymptotically equivalent models with very good dispersive properties. This was guided by physical motivations (good behavior of the model outside its *a priori* range of validity).

In the same vein, asymptotically equivalent models can be derived for numerical or mathematical purposes. In the latter case, it is of interest to find an asymptotically equivalent model that is trivially well-posed. Though this approach has been used in various regimes (e.g., Boussinesq systems with large topography variations [**68**], $2DH$ Green-Naghdi equations in [**185**]), we discuss here only the case of the Boussinesq systems (5.34) derived in the regime of small amplitude surface and topography variations ($\varepsilon = O(\mu)$, $\beta = O(\mu)$), namely,

$$\left\{ \begin{array}{l} (1 - \mu\mathbf{b}\Delta^\gamma)\partial_t\zeta + \nabla^\gamma \cdot (hV_{\theta,\delta}) + \mu\mathbf{a}\Delta^\gamma\nabla^\gamma \cdot V_{\theta,\delta} = 0, \\ (1 - \mu\mathbf{d}\Delta^\gamma)\partial_t V_{\theta,\delta} + \nabla^\gamma\zeta + \varepsilon(V_{\theta,\delta} \cdot \nabla^\gamma)V_{\theta,\delta} + \mu\mathbf{c}\Delta^\gamma\nabla^\gamma\zeta = 0, \end{array} \right. \tag{5.38}$$

with

$$\mathbf{a} = -\frac{\theta + \lambda}{3}, \quad \mathbf{b} = \frac{\delta + \lambda}{3}, \quad \mathbf{c} = -\frac{\alpha + \delta - 1}{3}, \quad \mathbf{d} = \frac{\alpha + \theta}{3}.$$

It is possible to choose the parameters $\theta \geq 0$, $\delta \geq 0$, $\alpha \in \mathbb{R}$ and $\lambda \in \mathbb{R}$ such that $\mathbf{a} = \mathbf{c}$ and $\mathbf{b} \geq \mathbf{0}$, $\mathbf{d} \geq 0$. For instance,

$$\begin{array}{rcl} (\theta, \delta, \alpha, \lambda) = (2, 2, 2, 1) & \rightsquigarrow & (\mathbf{a}, \mathbf{b}, \mathbf{c}, \mathbf{d}) = (-1, 1, -1, 4/3) \\ (\theta, \delta, \alpha, \lambda) = (0, 0, 1, 0) & \rightsquigarrow & (\mathbf{a}, \mathbf{b}, \mathbf{c}, \mathbf{d}) = (0, 0, 0, 1/3) \\ (\theta, \delta, \alpha, \lambda) = (0, 1/2, 0, -1/2) & \rightsquigarrow & (\mathbf{a}, \mathbf{b}, \mathbf{c}, \mathbf{d}) = (1/6, 0, 1/6, 0). \end{array}$$

(The second example corresponds to (5.25); the third one corresponds to the so-called KdV-KdV system.)
Moreover, since one always has $\mathbf{a} + \mathbf{b} + \mathbf{c} + \mathbf{d} = 1/3$, this motivates the following definition. (Note that the conditions $\mathbf{b} \geq 0$, $\mathbf{d} \geq 0$ ensure the linear well-posedness of the systems.)

DEFINITION 5.38. We call Σ the class of all systems $\Sigma_{a,b,c,d}$ of the form (5.38), and for which $\mathbf{b} \geq 0$, $\mathbf{d} \geq 0$ and $\mathbf{a} = \mathbf{c}$, and

$$\mathbf{a} + \mathbf{b} + \mathbf{c} + \mathbf{d} = 1/3.$$

The systems of the class Σ have a symmetric dispersive part, but their nonlinear part is not symmetric. Finding a symmetrizer that symmetrizes the nonlinear part and also respects the symmetric structure of the dispersive part can prove quite difficult. This is why we instead use the technique from [**36**], which introduces a new velocity unknown $W_{\theta,\delta}$ that symmetrizes the nonlinear terms but leaves the dispersive terms unchanged

$$W_{\theta,\delta} = \sqrt{h}V_{\theta,\delta}. \tag{5.39}$$

This allows us to derive the following *fully symmetric Boussinesq systems* introduced in [**36**]

$$\left\{ \begin{array}{l} (1 - \mu\mathbf{b}\Delta^\gamma)\partial_t\zeta + \nabla^\gamma \cdot (\sqrt{h}W_{\theta,\delta}) + \mu\mathbf{a}\Delta^\gamma\nabla^\gamma \cdot W_{\theta,\delta} = 0, \\ (1 - \mu\mathbf{d}\Delta^\gamma)\partial_t W_{\theta,\delta} + \sqrt{h}\nabla^\gamma\zeta + \mu\mathbf{c}\Delta^\gamma\nabla^\gamma\zeta \\ \qquad + \frac{\varepsilon}{2}\big[(W_{\theta,\delta} \cdot \nabla^\gamma)W_{\theta,\delta} + W_{\theta,\delta}\nabla^\gamma \cdot W_{\theta,\delta} + \frac{1}{2}\nabla^\gamma|W_{\theta,\delta}|^2\big] = 0. \end{array} \right. \tag{5.40}$$

The only difference with respect to (5.38) is the symmetry of the nonlinear terms; in particular, the following definition is not empty.

DEFINITION 5.39. We call Σ' the class of all systems $\Sigma'_{a,b,c,d}$ of the form (5.40), and for which $\mathbf{b} \geq 0$, $\mathbf{d} \geq 0$ and $\mathbf{a} = \mathbf{c}$, and

$$\mathbf{a} + \mathbf{b} + \mathbf{c} + \mathbf{d} = 1/3.$$

The corresponding consistency[18] result is stated in the following corollary.

COROLLARY 5.40. *In the long wave regime with small topography variations* $\mathcal{A}_{LW,top-}$*, the water waves equations* (5.1) *are consistent at order* $O(\mu^2)$ *with* (5.40) *in the sense of Definition* 5.26 *(with* $n = 10$*).*

REMARK 5.41. The systems belonging to the class Σ' introduced in Definition 5.39 are fully symmetric and their mathematical analysis is trivial (see §6.1.3). They will be used in §6.2.3 to justify all the Boussinesq systems for topography variations of small amplitude.
These fully symmetric systems are not exactly the systems derived in [**36**] for which the bottom is assumed to be flat and $\sqrt{h}$ is replaced by $1 + \frac{\varepsilon}{2}\zeta$ (which is true up to $O(\varepsilon^2)$ terms that can be discarded at the precision of the model)

$$(5.41)\quad \begin{cases} (1-\mu\mathbf{b}\Delta^\gamma)\partial_t\zeta + \nabla^\gamma\cdot((1+\frac{\varepsilon}{2}\zeta)W_{\theta,\delta}) + \mu\mathbf{a}\Delta^\gamma\nabla^\gamma\cdot W_{\theta,\delta} = 0,\\ (1-\mu\mathbf{d}\Delta^\gamma)\partial_t W_{\theta,\delta} + (1+\frac{\varepsilon}{2}\zeta)\nabla^\gamma\zeta + \mu\mathbf{c}\Delta^\gamma\nabla^\gamma\zeta \\ \qquad + \frac{\varepsilon}{2}\big[(W_{\theta,\delta}\cdot\nabla^\gamma)W_{\theta,\delta} + W_{\theta,\delta}\nabla^\gamma\cdot W_{\theta,\delta} + \frac{1}{2}\nabla^\gamma|W_{\theta,\delta}|^2\big] = 0. \end{cases}$$

We prefer (5.40) here because it is more elegant in presence of a nonflat topography. (Recall that $h = 1 + \varepsilon\zeta - \beta b$); for the derivation of the KdV approximation in Chapter 7, we will instead work with (5.41).

PROOF. For the first equation, we just have to replace $\sqrt{h}V_{\theta,\delta}$ by $\widetilde{V}_{\theta,\delta}$ in the nonlinear term and to note that since $V_{\theta,\delta} = W_{\theta,\delta} + O(\varepsilon,\beta)$, we can replace $V_{\theta,\delta}$ by $W_{\theta,\delta}$ in the dispersive term; this manipulation changes only the residual terms by a quantity compatible with the estimates on R^1 stated in the corollary.
For the second equation, we can similarly replace $V_{\theta,\delta}$ by $W_{\theta,\delta}$ in the nonlinear and dispersive terms, but not in $\partial_t V_{\theta,\delta}$; for this term, we write

$$\begin{aligned} \partial_t V_{\theta,\delta} + \nabla^\gamma\zeta &= \partial_t\big(\frac{1}{\sqrt{h}}W_{\theta,\delta}\big) + \nabla^\gamma\zeta \\ &= -\frac{\varepsilon\partial_t\zeta}{2h^{3/2}}W_{\theta,\delta} + \frac{1}{\sqrt{h}}\partial_t W_{\theta,\delta} + \nabla^\gamma\zeta \\ &= \frac{1}{\sqrt{h}}\big(\partial_t W_{\theta,\delta} + \frac{\varepsilon}{2}W_{\theta,\delta}\nabla^\gamma\cdot W_{\theta,\delta} + \sqrt{h}\nabla^\gamma\zeta\big) + O(\varepsilon^2,\mu^2,\beta^2), \end{aligned}$$

where we used the first equation to substitute $\partial_t\zeta = -\nabla^\gamma W_{\theta,\delta} + O(\varepsilon,\mu,\beta)$. Multiplying the equation by $\sqrt{h} = 1 + O(\varepsilon,\beta)$, we then get the statement of the theorem after noting that

$$\nabla^\gamma|W_{\theta,\delta}|^2 = 2(\widetilde{V}_{\theta,\delta}\cdot\nabla^\gamma)\widetilde{V}_{\theta,\delta} + O(\mu),$$

which is a consequence of (5.30) and Proposition 3.37. □

5.4. Moving bottoms

We briefly hint here how to derive shallow water asymptotic models from the water waves equations with moving bottoms (1.41) that we can rewrite under the

[18] Definition 5.26 explains the notion of consistency for systems on $(\zeta, V_{\theta,\delta})$, but this notion can be straightforwardly extended to systems on $(\zeta, W_{\theta,\delta})$.

form[19],

$$(5.42)\qquad \begin{cases} \partial_t\zeta - \frac{1}{\mu}\mathcal{G}\psi = \frac{\beta}{\delta\varepsilon}\mathcal{G}^{NN}\partial_\tau b, \\ \partial_t\psi + \zeta + \frac{\varepsilon}{2}|\nabla^\gamma\psi|^2 - \varepsilon\mu\frac{(\frac{1}{\mu}\mathcal{G}\psi + \frac{\beta}{\delta\varepsilon}\mathcal{G}^{NN}\partial_\tau b + \nabla^\gamma(\varepsilon\zeta)\cdot\nabla^\gamma\psi)^2}{2(1+\varepsilon^2\mu|\nabla^\gamma\zeta|^2)} = 0, \end{cases}$$

where $\mathcal{G}$ is the Dirichlet-Neumann operator and $\mathcal{G}^{NN}$ the Neumann-Neumann operator defined in (1.42).
We also recall that $\tau = t/\delta$, where δ defined as

$$\delta = \frac{\text{time scale of the bottom variations}}{\text{typical period of the wave}} = \frac{t_{mb}}{t_0} = \frac{t_{mb}}{L_x/gH_0}.$$

In [**182**], T. Iguchi was interested in modeling the surface perturbation created by a submarine earthquake and therefore investigated the case $\delta \ll 1$ (and $\varepsilon = \beta = 1$).

Let us consider the case here of slower bottom variations corresponding to $\delta = 1$ (in particular, the fast variable $\tau = t/\delta$ coincides with t).
We consider here the same shallow water regime as in §5.1.1

$$\mathcal{A}_{SW} = \{(\varepsilon,\beta,\gamma,\mu),\quad 0\le\mu\le\mu_0,\quad 0\le\varepsilon\le 1,\quad 0\le\beta\le 1,\quad 0\le\gamma\le 1\},$$

for some $0 < \mu_0 \le \mu_{max}$. We also assume that

$$\beta \le \varepsilon$$

so that β/ε can be treated as a $O(1)$ term in the asymptotic expansions. There are, of course, many other regimes worthy of interest and that could be derived with the same strategy.

5.4.1. The Nonlinear Shallow Water equations with moving bottom.

Denoting as usual by $\overline{V}$ the vertically averaged component of the velocity, the first equation of (5.42) can be written in exact form as

$$\partial_t\zeta + \nabla^\gamma\cdot(h\overline{V}) = \frac{\beta}{\varepsilon}\partial_t b,$$

as a consequence of Remark 3.36.
Moreover, since $\mathcal{G}^{NN}\partial_t b = \partial_t b + O(\mu)$ (see §A.5.4), and $\nabla^\gamma\psi = \overline{V} + O(\mu)$ (this follows from (A.27) in Appendix A, since we assumed that $\beta/\varepsilon = O(1)$), we get, after taking the gradient of the second equation in (5.42) and dropping the $O(\mu)$ terms, that the NSW equations (5.7) take the following form when the bottom is moving

$$(5.43)\qquad \begin{cases} \partial_t\zeta + \nabla^\gamma\cdot(h\overline{V}) = \frac{\beta}{\varepsilon}\partial_t b, \\ \partial_t\overline{V} + \nabla^\gamma\zeta + \varepsilon(\overline{V}\cdot\nabla^\gamma)\overline{V} = 0, \end{cases}$$

with $h = 1 + \varepsilon\zeta - \beta b$.

[19]With ν set to 1 because we are interested only in shallow water regimes here.

5.4.2. The Green-Naghdi equations with moving bottom. The Green-Naghdi equations with moving bottom are obtained as the standard Green-Naghdi equations (5.11) with the same adaptation as above due to the moving bottom. The only difference is that the time variations of the bottom now appear in the equation for the velocity; indeed, we know from (A.27) in Appendix A that

$$\overline{V} = \nabla^\gamma \psi + \mu\Big[-\mathcal{T}\nabla^\gamma\psi + \overline{V}_{1,mb} + \varepsilon\nabla^\gamma\zeta\partial_t b\Big] + O(\mu^2),$$

with the "moving bottom" contribution to the $O(\mu)$ component given by

$$\begin{aligned}\overline{V}_{1,mb} &= -\frac{\beta}{\varepsilon}\big(\frac{1}{2}h\nabla^\gamma\partial_t b + \varepsilon\nabla^\gamma\zeta\partial_t b\big)\\ &= -\frac{1}{2}\nabla^\gamma\big(h\frac{\beta}{\varepsilon}\partial_t b\big) - \frac{1}{2}\big(\varepsilon\nabla^\gamma\zeta + \beta\nabla^\gamma b\big)\frac{\beta}{\varepsilon}\partial_t b,\end{aligned}$$

so that (5.17) must be replaced, when the bottom is moving, by

$$\begin{aligned}\nabla^\gamma\psi &= \overline{V} + \mu\Big[\mathcal{T}\nabla^\gamma\psi - \overline{V}_{1,mb}\Big] + O(\mu^2)\\ &= \overline{V} + \mu\Big[\mathcal{T}\overline{V} - \overline{V}_{1,mb}\Big] + O(\mu^2).\end{aligned}$$

This leads to the following generalization of the Green-Naghdi equations (5.11) when the bottom is moving

$$\left\{\begin{array}{l}\partial_t\zeta + \nabla^\gamma\cdot(h\overline{V}) = \frac{\beta}{\varepsilon}\partial_t b,\\ (1+\mu\mathcal{T})\partial_t\overline{V} + \nabla^\gamma\zeta + \varepsilon(\overline{V}\cdot\nabla^\gamma)\overline{V} + \mu\varepsilon\big(\mathcal{Q}(\overline{V}) + \mathcal{Q}_b(\overline{V})\big) = \mu\,\mathbf{mb},\end{array}\right. \tag{5.44}$$

where we recall that $h = 1 + \varepsilon\zeta - \beta b$ and where $\mathcal{T}$, $\mathcal{Q}$ and $\mathcal{Q}_b$ are as in (5.12), (5.13), and (5.14), and where the "moving bottom" term **mb** is defined as

$$\begin{aligned}\mathbf{mb} &= \partial_t\overline{V}_{1,mb} + \varepsilon\nabla^\gamma(\overline{V}\cdot\overline{V}_{1,mb}) + \varepsilon\nabla^\gamma\big((-\nabla^\gamma\cdot(h\overline{V}) + \varepsilon\nabla^\gamma\zeta\cdot\overline{V})\frac{\beta}{\varepsilon}\partial_t b\big)\\ &= \partial_t\overline{V}_{1,mb} - \frac{\varepsilon}{2}\nabla^\gamma\big[\frac{\beta}{\varepsilon}\partial_t b\nabla^\gamma\cdot(h\overline{V}) + \nabla^\gamma\cdot(\frac{\beta}{\varepsilon}\partial_t bh\overline{V})\big].\end{aligned}$$

Using the first equation of (5.44), we deduce that

$$\begin{aligned}\mathbf{mb} = &-\frac{\beta}{2\varepsilon}\big[\nabla^\gamma(h\partial_t^2 b) + (\varepsilon\nabla^\gamma\zeta + \beta\nabla^\gamma b)\partial_t^2 b\big]\\ &-\frac{\beta}{2}\big[\big(\frac{\beta}{\varepsilon}\partial_t b - \nabla^\gamma\cdot(h\overline{V})\big) + \nabla^\gamma\nabla^\gamma\cdot(\partial_t bh\overline{V})\big].\end{aligned} \tag{5.45}$$

5.4.3. A Boussinesq system with moving bottom. In the long wave regime with small topography variation

$$\mathcal{A}_{LW,top-} = \{(\varepsilon,\beta,\gamma,\mu),\quad 0\le\mu\le\mu_0,\quad 0\le\varepsilon\lesssim\mu,\quad 0\le\beta\lesssim\mu,\quad 0\le\gamma\le 1\},$$

and moreover if $\beta\le\varepsilon$, the Green-Naghdi equations with moving bottom (5.44) can be simplified into the following Boussinesq system with moving bottom that generalizes (5.25), and which can be found in [**73**] in its one-dimensional form

$$\left\{\begin{array}{l}\partial_t\zeta + \nabla^\gamma\cdot(h\overline{V}) = \frac{\beta}{\varepsilon}\partial_t b,\\ (1-\frac{\mu}{3}\Delta^\gamma)\partial_t\overline{V} + \nabla^\gamma\zeta + \varepsilon(\overline{V}\cdot\nabla^\gamma)\overline{V} = -\mu\frac{\beta}{2\varepsilon}\nabla^\gamma\partial_t^2 b.\end{array}\right. \tag{5.46}$$

This follows directly after dropping the $O(\mu^2)$ terms in (5.44) and noting that in the asymptotic regime $\mathcal{A}_{LW,top-}$, one has $\mathbf{mb} = -\frac{\beta}{2\varepsilon}\nabla^\gamma\partial_t^2 b + O(\mu)$ (where $\mathbf{mb}$ is defined in (5.45)).

REMARK 5.42. Proceeding as in §5.2, it is possible to derive many Boussinesq systems with moving bottom equivalent to (5.46).

5.5. Reconstruction of the surface elevation from pressure measurements

For experimental measurements of waves, pressure sensors are often used. They measure the pressure[20] at the bottom; the surface elevation is then approximated by a simple hydrostatic reconstruction based on a linear approximation for the pressure. In dimensionless[21] form, this linear approximation is given by

$$-\frac{1}{\varepsilon}\big(P(X,z) - \frac{P_{atm}}{\rho g H_0}\big) = \frac{1}{\varepsilon}(z - \varepsilon\zeta) + \big(1 - \frac{\cosh(\sqrt{\mu}(z+1)|D^\gamma|)}{\cosh(\sqrt{\mu}|D^\gamma|)}\big)\zeta. \tag{5.47}$$

A first-order expansion of this expression with respect to μ shows that the pressure is hydrostatic at leading order. This is classically used to reconstruct the surface elevation from pressure measurements. We show[22] in §5.5.1 that this formula remains valid in a fully nonlinear regime; in §5.5.2, we show that a more precise, nonhydrostatic reconstruction is correct in the (weakly nonlinear) long wave regime.

It is convenient here to work with the nondimensionalized version of the Bernoulli equations (see (H1)$''$ in §1.1.3), namely,

$$\partial_t\Phi + \frac{1}{\varepsilon}z + \frac{\varepsilon}{2}|\nabla^\gamma\Phi|^2 + \frac{\varepsilon}{2\mu}|\partial_z\Phi|^2 = -\frac{1}{\varepsilon}\big(P - \frac{P_{atm}}{\rho g H_0}\big), \tag{5.48}$$

which one can easily deduce from the dimensionless Euler equations provided in Remark 1.14. We also denote by P_{bott} the (dimensionless) pressure measured at the bottom

$$P_{bott} = P_{|_{z=-1+\beta b}}.$$

REMARK 5.43. We deal here with the reconstruction of the surface elevation for time-dependent waves. When dealing with progressive waves, a more precise reconstruction can be performed and an exact formula for the surface elevation can be derived in some cases [**144, 92, 81, 93**].

---

[20] There are also more elaborate sensors called "PUV" sensors that are able to measure the pressure P and the horizontal components (u, v) of the velocity at the bottom. This information on the bottom velocity can be valuable in deriving a reconstruction formula of order $O(\mu^2)$ in the strongly nonlinear regime with strong topography variations ($\varepsilon = O(1)$, $\beta = O(1)$) not investigated here.

[21] The dimensional form of this approximation is

$$-\frac{1}{\rho g}\big(P(X,z) - P_{atm}\big) = z - \zeta + \big(1 - \frac{\cosh((z+H_0)|D|)}{\cosh(H_0|D|)}\big)\zeta.$$

It can easily be derived from the linear analysis in §1.3.2.

[22] We give just a formal proof here; making it rigorous does not cause any particular difficulty.

5.5.1. Hydrostatic reconstruction. Let us consider here the most general shallow water regime (large amplitude and bottom variations)

$$\mathcal{A}_{SW} = \{(\varepsilon, \beta, \gamma, \mu), \quad 0 \le \mu \le \mu_0, \quad 0 \le \varepsilon \le 1, \quad 0 \le \beta \le 1, \quad 0 \le \gamma \le 1\},$$

for some $0 < \mu_0 \le \mu_{max}$. We show here that an order $O(\mu)$ approximation of the surface elevation is provided by

$$\zeta = \frac{1}{\varepsilon}\Big[P_{bott} - \frac{P_{atm}}{\rho g H_0} - 1 + \beta b\Big] + O(\mu), \tag{5.49}$$

which is the standard hydrostatic reconstruction.

Evaluating (5.48) at the bottom, one gets

$$\partial_t \Phi_b + \frac{1}{\varepsilon}(-1 + \beta b) + \frac{\varepsilon}{2}|(\nabla^\gamma \Phi)_b|^2 + \frac{\varepsilon}{2\mu}|(\partial_z \Phi)_b|^2 = -\frac{1}{\varepsilon}\big(P_{bott} - \frac{P_{atm}}{\rho g H_0}\big), \tag{5.50}$$

with the notations

$$\Phi_b = \Phi_{|_{z=-1+\beta b}}, \qquad (\nabla^\gamma \Phi)_b = (\nabla^\gamma \Phi)_{|_{z=-1+\beta b}}, \qquad (\partial_z \Phi)_b = (\partial_z \Phi)_{|_{z=-1+\beta b}}.$$

Using (5.57), these quantities can be expressed in terms of the transformed potential $\phi = \Phi \circ \Sigma$ where, as in §5.6.2.1, Σ is the trivial diffeomorphism; using the notation $\phi_b = \phi_{|_{z=-1}}$, we get

$$\Phi_b = \phi_b, \qquad (\nabla^\gamma \Phi)_b = \nabla \phi_b - \frac{1}{h}(-\nabla^\gamma h + \varepsilon \nabla^\gamma \zeta)\partial_z \phi_{|_{z=-1}}, \qquad (\partial_z \Phi)_b = \frac{1}{h}\partial_z \phi_{|_{z=-1}},$$

and as usual, $h = 1 + \varepsilon\zeta - \beta b$. By Proposition 3.37 and Lemma 3.42, we also know that

$$\phi_b = \psi + O(\mu), \qquad \overline{V} = \nabla^\gamma \psi + O(\mu), \qquad \text{and} \qquad \partial_z \phi_{|_{z=-1}} = O(\mu).$$

Taking the horizontal gradient of (5.50), we therefore get

$$\partial_t \overline{V} + \frac{\beta}{\epsilon}\nabla^\gamma b + \varepsilon(\overline{V} \cdot \nabla^\gamma)\overline{V} = -\frac{1}{\varepsilon}\nabla^\gamma P_{bott} + O(\mu).$$

But we also know from Proposition 5.2 that

$$\partial_t \overline{V} + \varepsilon(\overline{V} \cdot \nabla^\gamma)\overline{V} = -\nabla^\gamma \zeta + O(\mu),$$

so that

$$-\frac{1}{\epsilon}\nabla^\gamma(1 + \varepsilon\zeta - \beta b) = -\frac{1}{\varepsilon}\nabla^\gamma P_{bott} + O(\mu).$$

Integrating this formula in space after noting that (5.50) implies that

$$\lim_{|X| \to \infty} P_{bott}(t, X) - \frac{P_{atm}}{\rho g H_0} = 1$$

(this is a consequence of the assumptions that the fluid is at rest and b vanishes at infinity), we get

$$P_{bott} - \frac{P_{atm}}{\rho g H_0} = 1 + \varepsilon\zeta - \beta b + O(\mu\varepsilon),$$

which coincides, when $b = 0$, with the first-order expansion with respect to μ of the linear formula (5.47); the formula (5.49) follows.

5.5.2. Nonhydrostatic, weakly nonlinear reconstruction. We show here that the standard hydrostatic reconstruction (5.49) of the surface elevation can be made more precise by including the next order terms. For the sake of simplicity, here we investigate only the case of the long-wave regime with small topography variations

$$\mathcal{A}_{LW,top-} = \{(\varepsilon,\beta,\gamma,\mu), \quad 0\leq\mu\leq\mu_0, \quad 0\leq\varepsilon\lesssim\mu, \quad 0\leq\beta\lesssim\mu, \quad 0\leq\gamma\leq 1\}.$$

We show here that in this regime, an order $O(\mu^2)$ approximation of the surface elevation is provided by

$$\zeta = \frac{1}{\varepsilon}(1-\mu\frac{1}{2}\Delta^\gamma)\Big[-1+\beta b+P_{bott}-\frac{P_{atm}}{\rho g H_0}\Big]+O(\mu^2), \tag{5.51}$$

which is a higher order approximation than (5.49).
We proceed as in §5.5.1 but use a higher order expansion of $\nabla^\gamma\Phi_b$ in terms of $\overline{V}$, namely (see (5.59) below),

$$\nabla^\gamma\Phi_b = \overline{V}+\mu\frac{1}{6}\Delta^\gamma\overline{V}+O(\mu^2),$$

where we used the fact that $\varepsilon = O(\mu)$, $\beta = O(\mu)$. Taking the horizontal gradient of (5.50), we therefore get

$$(1+\mu\frac{1}{6}\Delta^\gamma)\partial_t\overline{V}+\frac{\beta}{\epsilon}\nabla^\gamma b+\varepsilon(\overline{V}\cdot\nabla^\gamma)\overline{V} = -\frac{1}{\varepsilon}\nabla^\gamma P_{bott}+O(\mu). \tag{5.52}$$

But we also know from Corollary 5.20 that

$$(1-\frac{\mu}{3}\Delta^\gamma)\partial_t\overline{V}+\nabla^\gamma\zeta+\varepsilon(\overline{V}\cdot\nabla^\gamma)\overline{V} = O(\mu^2).$$

Applying $(1+\mu\frac{1}{2}\Delta^\gamma)$ to this identity yields

$$(1+\mu\frac{1}{6}\Delta^\gamma)\partial_t\overline{V}+\varepsilon(\overline{V}\cdot\nabla^\gamma)\overline{V} = -(1+\mu\frac{1}{2}\Delta^\gamma)\nabla^\gamma\zeta+O(\mu^2),$$

which, together with (5.52), yields

$$-\frac{1}{\epsilon}\nabla^\gamma(1+\varepsilon\zeta-\beta b)-\mu\frac{1}{2}\Delta^\gamma\nabla^\gamma\zeta = -\frac{1}{\varepsilon}\nabla^\gamma P_{bott}+O(\mu^2).$$

Integrating this formula in space and taking into account the boundary conditions at infinity then gives

$$P_{bott}-\frac{P_{atm}}{\rho g H_0} = 1+\varepsilon\zeta-\beta b+\mu\varepsilon\frac{1}{2}\Delta^\gamma\zeta+O(\mu^2\varepsilon),$$

which coincides, when $b=0$, with the second-order expansion with respect to μ of the linear formula (5.47). Rewriting it under the form

$$(1+\mu\frac{1}{2}\Delta^\gamma)\zeta = \frac{1}{\varepsilon}\Big[-1+\beta b+P_{bott}-\frac{P_{atm}}{\rho g H_0}\Big]+O(\mu^2)$$

and applying $(1-\mu\frac{1}{2}\Delta^\gamma)$ to this identity gives (5.51).

5.6. Supplementary remarks

5.6.1. Technical results.

5.6.1.1. *Invertibility properties of $h_b(I+\mu\mathcal{T}_b)$.* The invertibility of $h_b(I+\mu\mathcal{T}_b)$ has been used in the proof Corollary 5.23 and for the definition (5.30) of V_θ. In order to establish this invertibiilty result and to give some precise estimates on the inverse operator, we need the space X^s defined as

$$X^s = \{V \in H^s(\mathbb{R}^d)^d, |V|_{X^s} < \infty\} \quad \text{with} \quad |V|_{X^s}^2 = |V|_{H^s}^2 + \mu|\nabla^\gamma \cdot V|_{H^s}^2,$$

as well as $\nabla^\gamma L^2(\mathbb{R}^d) = \{V = \nabla^\gamma f, \quad f \in L^2(\mathbb{R}^d)\}$. This is done in the following lemma.

LEMMA 5.44. *Let $t_0 > d/2$ and $b \in H^{t_0+1}(\mathbb{R}^d)$ be such that $h_b = 1-\beta b \geq h_{min}$. Then the mapping*

$$h_b(I+\mu\mathcal{T}_b): \qquad X^0 \to L^2(\mathbb{R}^d)^d + \nabla^\gamma L^2(\mathbb{R}^d)$$

is well defined, one-to-one and onto. Moreover, for all $s \geq 0$, if $b \in H^{1+s\vee t_0}(\mathbb{R}^d)$, then

$$\forall W \in H^s(\mathbb{R}^d)^d, \qquad |[h_b(I+\mu\mathcal{T}_b)]^{-1}W|_{X^s} \leq C(\frac{1}{h_{min}}, |b|_{H^{1+s\vee t_0}})|W|_{H^s}.$$

PROOF. The fact that $h_b(I+\mu\mathcal{T}_b)$ maps $X^0(\mathbb{R}^d)^d$ into $L^2(\mathbb{R}^d)^d + \nabla^\gamma L^2$ is a direct consequence of the definition of $\mathcal{T}_b$ and of X^0, and classical product estimates. Note now that one can write

$$h_b\mathcal{T}_b = (S_b^1)^* h_b S_b^1 + (S_b^2)^* h_b S_b^2,$$

with $h_b = 1-\beta b$ and where the first and zero order differential operators S_b^1 and S_b^2 are given by

$$S_b^1 = \frac{h_b}{\sqrt{3}}\mathrm{div}_\gamma - \frac{\sqrt{3}}{2}\beta(\nabla^\gamma b)^T, \qquad S_b^2 = \frac{1}{2}\beta(\nabla^\gamma b)^T.$$

It follows that

$$\begin{aligned}
(5.53)\quad \big(h_b(I+\mu\mathcal{T}_b)V, V\big) &= \big(h_bV, V\big) + \mu\big(h_bS_b^1V, S_b^1V\big) + \mu\big(h_bS_b^2V, S_b^2V\big) \\
(5.54)\quad &\geq h_{min}\big(|V|_2^2 + \mu|S_b^1V|_2^2 + \mu|S_b^2V|_2^2\big),
\end{aligned}$$

from which one easily gets that $(I+\mu\mathcal{T}_b)$ is one-to-one.
Let us now prove that $h_b(I+\mu\mathcal{T}_b)$ is onto, that is, we want to prove that for all $W \in L^2(\mathbb{R}^d)^d$, and all $f \in L^2(\mathbb{R}^d)$, there exist $V \in X^0$ such that $h_b(I+\mu\mathcal{T}_b)V = W + \sqrt{\mu}\nabla^\gamma f$. This is a direct consequence of Lax-Milgram's theorem after the following points have been checked:

(i) The bilinear form $(V^1, V^2) \in X^0 \times X^0 \mapsto (h_b(I+\mu\mathcal{T}_b)V^1, V^2)$ is continuous. Indeed, we easily get from the above that

$$\begin{aligned}
(h_b(I+\mu\mathcal{T}_b)V^1, V^2) &= \big(h_bV^1, V^2\big) + \mu\big(h_bS_b^1V^1, S_b^1V^2\big) + \mu\big(h_bS_b^2V^1, S_b^2V^2\big) \\
&\leq C(|b|_{W^{1,\infty}})|V^1|_{X^0}|V^2|_{X^0}.
\end{aligned}$$

(ii) This bilinear form is also coercive on X^0. We can indeed note that

$$\begin{aligned}|S_b^1 V|^2+|S_b^2 V|^2 &= \big(\frac{h_b^2}{3}(\nabla^\gamma\cdot V)^2-h_b\beta(\nabla^\gamma\cdot V)\nabla^\gamma b\cdot V+\frac{3}{4}\beta^2|\nabla^\gamma b\cdot V|^2\big)\\ &\quad+\frac{1}{4}\beta^2|\nabla^\gamma b\cdot V|^2\\ &= \frac{1}{12}h_b^2(\nabla^\gamma\cdot V)^2+|\frac{h_b}{2}\nabla^\gamma\cdot V-\beta\nabla^\gamma b\cdot V|^2,\end{aligned}$$

so that (5.54) allows us to conclude that

$$|V|_{X^0}^2\leq C(\frac{1}{h_{min}})(h_b(I+\mu\mathcal{T}_b)V,V).$$

(iii) The linear form $V\in X^0\mapsto(V,W+\sqrt{\mu}\nabla^\gamma f)$ is continuous, which is obvious since

$$(V,W+\sqrt{\mu}\nabla^\gamma f)=(V,W)-\sqrt{\mu}(f,\nabla^\gamma\cdot V)\leq|V|_{X^0}(|W|_2+|f|_2).$$

We will now prove the estimates stated in the lemma. We need estimates on $|V|_{X^s}$ where $h_b(I+\mu\mathcal{T}_b)V=W$. Let us first note that if $\widetilde{V}$, $\widetilde{W}$ and $\widetilde{f}$ are such that

$$h_b(I+\mu\mathcal{T}_b)\widetilde{V}=\widetilde{W}+\sqrt{\mu}\nabla^\gamma\widetilde{f}, \tag{5.55}$$

then, proceeding as for points (ii) and (iii) above, one gets

$$|\widetilde{V}|_{X^0}\leq C(\frac{1}{h_{min}})\big(|\widetilde{W}|_2+|\widetilde{f}|_2\big). \tag{5.56}$$

Taking $\widetilde{W}=W$ and $f=0$, this is the estimate we need in the case $s=0$. For the general case $s\geq 0$, we first note (recall that $\mathcal{T}_b=\mathcal{T}[h_b,b]$ with $\mathcal{T}$ as in (3.32)) that

$$\begin{aligned}[\Lambda^s,h_b\mathcal{T}_b]V &= -\frac{1}{3}\nabla^\gamma([\Lambda^s,h_b^3]\nabla^\gamma\cdot V)+\frac{1}{2}\nabla^\gamma([\Lambda^s,h_b^2\nabla^\gamma b]\cdot V)\\ &\quad-\frac{1}{2}[\Lambda^s,h_b^2\nabla^\gamma b]\nabla^\gamma\cdot V+[\Lambda^s,\nabla^\gamma b\otimes\nabla^\gamma b]V,\end{aligned}$$

so that (5.55) also holds true with $\widetilde{V}=\Lambda^s V$, and

$$\begin{aligned}\widetilde{W} &= \Lambda^s W-\frac{\mu}{2}[\Lambda^s,h_b^2\nabla^\gamma b]\nabla^\gamma\cdot V+\mu[\Lambda^s,\nabla^\gamma b\otimes\nabla^\gamma b]V,\\ \widetilde{f} &= -\frac{\sqrt{\mu}}{3}[\Lambda^s,h_b^3]\nabla^\gamma\cdot V+\frac{\sqrt{\mu}}{2}[\Lambda^s,h_b^2\nabla^\gamma b]\cdot V.\end{aligned}$$

In particular, we have

$$|\widetilde{W}|_2+|\widetilde{f}|_2\leq|W|_{H^s}+C(|b|_{H^{1+s\vee t_0}})|V|_{X^s},$$

which allows us to deduce from (5.56) that

$$|V|_{X^s}\leq C(\frac{1}{h_{min}})\big(|W|_{H^s}+C(|b|_{H^{1+s\vee t_0}})|V|_{X^s}\big).$$

The result therefore follows from a straightforward continuous induction. □

5.6.1.2. *Invertibility properties of $h(I+\mu\mathcal{T})$.* The invertibility of $h(I+\mu\mathcal{T})$ has been used in the proof Corollary 5.36 and for the definition (5.37) of $V_{\theta,\delta}$. Using the same notations as in §5.6.1.1 above, we gather the main invertibility properties of $h(I+\mu\mathcal{T})$ in the following lemma.

LEMMA 5.45. *Let $t_0 > d/2$ and $\zeta, b \in H^{t_0+1}(\mathbb{R}^d)$ be such that $h = 1+\varepsilon\zeta-\beta b \geq h_{min}$. Then the mapping*

$$h(I+\mu\mathcal{T}) : \qquad X^0 \to L^2(\mathbb{R}^d)^d + \nabla^\gamma L^2(\mathbb{R}^d)$$

is well defined, one-to-one and onto. Moreover, for all $s \geq 0$, if $\zeta, b \in H^{1+s\vee t_0}(\mathbb{R}^d)$, then

$$\forall W \in H^s(\mathbb{R}^d)^d, \qquad |[h(I+\mu\mathcal{T})]^{-1}W|_{X^s} \leq C(\frac{1}{h_{min}}, |b|_{H^{1+s\vee t_0}}, |\zeta|_{H^{1+s\vee t_0}})|W|_{H^s}.$$

PROOF. The proof follows exactly the same lines as the proof of Lemma 5.45, after noting that

$$h\mathcal{T} = (S^1)^* h S^1 + (S^2)^* h S^2,$$

with $h = 1+\varepsilon\zeta-\beta b$ and where the first and zero order differential operators S^1 and S^2 are given by

$$S^1 = \frac{h}{\sqrt{3}}\mathrm{div}_\gamma - \frac{\sqrt{3}}{2}\beta(\nabla^\gamma b)^T, \qquad S^2 = \frac{1}{2}\beta(\nabla^\gamma b)^T.$$

□

5.6.2. Remarks on the "velocity" unknown used in asymptotic models.

5.6.2.1. *Relationship between the averaged velocity $\overline{V}$ and the velocity at an arbitrary elevation.* For the sake of clarity, we take $\gamma = 1$ throughout this section (fully transverse, or isotropic, waves).

As in Chapter 2, we denote by Φ the velocity in the fluid domain

$$\Omega = \{(X,z), -1+\beta b < z < \varepsilon\zeta\},$$

which solves the Laplace equation (2.7). Taking the trivial diffeomorphism of Example 2.14, namely,

$$\Sigma(X,z) = (X, (1+\varepsilon\zeta(X)-\beta b(X))z + \varepsilon\zeta(X)),$$

we denote $\phi = \Phi\circ\Sigma$ the transformed velocity potential that solves (2.12) in the flat strip $\mathcal{S} = \{-1 < z < 0\}$. By definition of the velocity potential, the velocity field $\mathbf{U} = (V, w)$ is related to Φ through the relation $\mathbf{U} = \nabla_{X,z}\Phi$. We can equivalently define $\mathbf{U}$ in terms of the transformed potential ϕ; differentiating the identity $\phi = \Phi\circ\Sigma$, we indeed get

$$\mathbf{U}\circ\Sigma = (J_\Sigma^T)^{-1}\nabla_{X,z}\phi, \tag{5.57}$$

where, as in Chapter 2, J_Σ is the jacobian matrix associated to the diffeomorphism Σ. One easily computes that

$$(J_\Sigma^T)^{-1} = \frac{1}{h}\begin{pmatrix} hI_{d\times d} & -z\nabla h - \varepsilon\nabla\zeta \\ 0 & 1 \end{pmatrix} \qquad (\text{with } h = 1+\varepsilon\zeta-\beta b),$$

so that the horizontal component of the velocity in the fluid domain is given by

$$V\circ\Sigma = \nabla\phi - \frac{1}{h}(z\nabla h + \varepsilon\nabla\zeta)\partial_z\phi.$$

Recalling that an approximation at order $O(\mu^2)$ of ϕ is given by Lemma 3.42,

$$\phi_{app} = \phi_0 + \mu\phi_1 + O(\mu^2),$$

with

$$\phi_0 = \psi, \quad \phi_1 = -h^2(\frac{z^2}{2} + z)\Delta\psi + zh\beta\nabla b\cdot\nabla\psi,$$

we obtain, after some computations, that

$$\begin{aligned} V\circ\Sigma &= (I-\mu\mathcal{T})\nabla\psi \\ &\quad +\mu(\frac{3}{2}z^2+3z+1)\big[-\frac{1}{3h}\nabla(h^3\Delta\psi)+h\nabla h\Delta\psi\big] \\ &\quad +\mu(2z+1)\big[\frac{\beta}{2h}\nabla(h^2\nabla b\cdot\nabla\psi)-\beta\nabla h\nabla b\cdot\nabla\psi+\frac{\beta}{2}h\nabla b\Delta\psi\big]+O(\mu^2). \end{aligned}$$

Moreover, since $\nabla\psi = (I+\mu\mathcal{T})\overline{V}+O(\mu^2)$, we deduce that

$$\begin{aligned} V\circ\Sigma &= \overline{V}+\mu(\frac{3}{2}z^2+3z+1)\big[-\frac{1}{3h}\nabla(h^3\Delta\psi)+h\nabla h\Delta\psi\big] \\ &\quad +\mu(2z+1)\big[\frac{\beta}{2h}\nabla(h^2\nabla b\cdot\nabla\psi)-\nabla h\nabla b\cdot\nabla\psi\big]+O(\mu^2). \end{aligned}$$

Using the fact that $\nabla\psi = \overline{V}+O(\mu)$ in this formula, we obtain

$$\begin{aligned} V\circ\Sigma &= \overline{V}+\mu(\frac{3}{2}z^2+3z+1)\big[-\frac{1}{3h}\nabla(h^3\nabla\cdot\overline{V})+h\nabla h\nabla\cdot\overline{V}\big] \\ &\quad +\mu(2z+1)\big[\frac{\beta}{2h}\nabla(h^2\nabla b\cdot\overline{V})-\beta\nabla h\nabla b\cdot\overline{V}+\frac{\beta}{2}h\nabla b\nabla\cdot\overline{V}\big]+O(\mu^2). \end{aligned} \tag{5.58}$$

Since by definition $\overline{V} = \frac{1}{h}\int_{-1+\beta b}^{\varepsilon\zeta} V = \int_{-1}^{0} V\circ\Sigma$, the above formula is consistent at order $O(\mu^2)$ if the following two conditions hold,

$$\int_{-1}^{0}(\frac{3}{2}z^2+3z+1)dz = 0 \quad \text{and} \quad \int_{-1}^{0}(2z+1)dz = 0,$$

which is obviously the case.
The formula (5.58) can be considerably simplified in the particular case of small amplitude topography variation ($\beta = O(\mu)$), namely,

$$V\circ\Sigma = \overline{V}+\mu(\frac{3}{2}z^2+3z+1)\big[-\frac{1}{3h}\nabla(h^3\nabla\cdot\overline{V})+h\nabla h\nabla\cdot\overline{V}\big]+O(\mu^2). \tag{5.59}$$

Noting that $(\frac{3}{2}z^2+3z+1)$ vanishes for $z_0 = -1+1/\sqrt{3}$, we can conclude that *for small amplitude topography variations, $\beta = O(\mu)$, the averaged velocity $\overline{V}$ coincides*[23] *with the horizontal velocity V evaluated at the level line $\Gamma_{z_0} = \{h(X)z_0 + \varepsilon\zeta(X), X\in\mathbb{R}^d\}$*. For more general topography variations, (5.58) shows that *this is false in general* since $2z_0+1\neq 0$.

[23] Up to $O(\mu^2)$ terms, of course.

5.6.2.2. *Relationship between $V_{\theta,\delta}$ and the velocity at an arbitrary elevation.* We will now show that it is possible to give a physical meaning to $V_{\theta,\delta}$, as defined in (5.30), in the particular case of small amplitude surface and topography variations ($\varepsilon = O(\mu)$, $\beta = O(\mu)$). In this regime, (5.58) can indeed be simplified into

$$V \circ \Sigma = \overline{V} - \mu(\frac{3}{2}z^2 + 3z + 1)\frac{1}{3}\nabla\nabla \cdot \overline{V} + O(\mu^2).$$

Moreover, since a brief look at (5.24) shows that in this regime, one has $\mathcal{T}_b V = \mathcal{T}_0 V + O(\mu)$ with $\mathcal{T}_0 = -\frac{1}{3}\nabla\nabla^T$, we deduce further that

$$V \circ \Sigma = \big[I + \mu(\frac{3}{2}z^2 + 3z + 1)\mathcal{T}_0\big]\overline{V} + O(\mu^2).$$

We also get from (5.30) that

$$V_{\theta,\delta} = \big[I + \mu\theta\mathcal{T}_0\big]^{-1}\big[I + \mu\delta\mathcal{T}_0\big]\overline{V} = \big[I + \mu(\delta - \theta)\mathcal{T}_0\big]\overline{V} + O(\mu^2).$$

By comparing these two expressions, it is straightforward to obtain that *for small amplitude surface and topography variations, $\varepsilon = O(\mu)$ and $\beta = O(\mu)$, the velocity $V_{\theta,\delta}$ coincides with the horizontal velocity evaluated at the level line $\Gamma_{\theta-\delta} = \{h(X)z_{\theta-\delta} + \varepsilon\zeta(X), X \in \mathbb{R}^d\}$*, where $z_{\theta-\delta}$ is the unique root of $\frac{3}{2}z^2 + 3z + 1 = \delta - \theta$ belonging to $[-1, 0]$, $z_{\theta-\delta} = -1 + \frac{1}{\sqrt{3}}\sqrt{1 + 2(\delta - \theta)}$.

REMARK 5.46. In particular, one can note that when $\delta - \theta = 1$, $V_{\theta,\delta} = \nabla\psi + O(\mu^2)$.

5.6.2.3. *Recovery of the vertical velocity from ζ and $\overline{V}$.* We have seen in §5.6.2.1 how to recover the horizontal velocity at an arbitrary elevation in terms of ζ and $\overline{V}$. We show here now to recover the vertical velocity at an arbitrary elevation. Proceeding as in §5.6.2.1, we get

$$w \circ \Sigma = \frac{1}{h}\partial_z\phi,$$

which, together with Lemma 3.42, gives

$$w \circ \Sigma = -\mu h(z + 1)\Delta\psi + \beta\mu\nabla b \cdot \nabla\psi + O(\mu^2).$$

Replacing $\nabla\psi = \overline{V} + O(\mu)$ in this equation, we obtain

$$w \circ \Sigma = -\mu h(z + 1)\nabla \cdot \overline{V} + \beta\mu\nabla b \cdot \overline{V} + O(\mu^2).$$

5.6.3. Formulation in $(h, h\overline{V})$ variables of shallow water models.

5.6.3.1. *The Nonlinear Shallow Water equations.* After simple computations, the NSW equations (5.7) can be written as a system of two conservation laws on $(h, h\overline{V})$

$$\begin{cases} \partial_t h + \varepsilon\nabla \cdot (h\overline{V}) = 0, \\ \partial_t(h\overline{V}) + \varepsilon\nabla \cdot (h\overline{V} \otimes \overline{V}) + \frac{1}{\varepsilon}h\nabla h = -\frac{\beta}{\varepsilon}h\nabla b, \end{cases} \tag{5.60}$$

where $\overline{V} \otimes \overline{V} = \overline{V}\,\overline{V}^T$.

5.6.3.2. *The Green-Naghdi equations.* We give here the $(h, h\overline{V})$ formulation of the one-parameter family of Green-Naghdi equations (5.36) that can be obtained after simple computations

$$\begin{cases} \partial_t h + \varepsilon\nabla\cdot(h\overline{V}) = 0, \\ \partial_t(h\overline{V}) + \varepsilon\nabla\cdot(h\overline{V}\otimes\overline{V}) + \frac{\alpha-1}{\alpha}h\nabla\zeta \\ \qquad + (I+\mu\alpha h\mathcal{T}\frac{1}{h})^{-1}[\frac{1}{\alpha}h\nabla\zeta + \varepsilon\mu h\mathcal{Q}_1(\overline{V})] = 0, \end{cases} \tag{5.61}$$

with $\varepsilon\zeta = h - 1 + \beta b$.

5.6.4. Equations with dimensions. Going back to variables with dimension by inverting the changes of variables and unknowns performed in §1.3.3, the one-parameter family of Green-Naghdi equations (5.36), or rather its equivalent formulation (5.61), shows

$$\begin{cases} \partial_t h + \nabla\cdot(h\overline{V}) = 0, \\ \partial_t(h\overline{V}) + \frac{\alpha-1}{\alpha}gh\nabla\zeta + \nabla\cdot(h\overline{V}\otimes\overline{V}) \\ \qquad + (I+\alpha h\mathcal{T}\frac{1}{h})^{-1}[\frac{1}{\alpha}gh\nabla\zeta + h\mathcal{Q}_1(\overline{V})] = 0, \end{cases}$$

where the dimensionalized version of the operators $\mathcal{T}$ and $\mathcal{Q}_1$ correspond to (5.12) and (5.16), with $\beta = 1$, and where h now stands for the water height with dimensions

$$h = H_0 + \zeta - b.$$

REMARK 5.47. Dimensional forms of all the asymptotic models derived in this chapter can be obtained in the same way.

5.6.5. The lake and great lake equations. The lake and great lake equations can be formally derived from the nonlinear shallow water and Green-Naghdi equations respectively by assuming that the surface deformations are small enough to replace the kinematic free surface condition by a rigid lid assumption. We briefly comment on these equations and their derivation below and refer to the handbook [**54**] for more details and many variants (including viscous and Coriolis terms, for instance).

5.6.5.1. *The lake equations.* Rescaling the time variable by introducing

$$\tau = \varepsilon t,$$

one can rewrite the nonlinear shallow water equations (5.7) under the form

$$\begin{cases} \varepsilon\partial_\tau\zeta + \nabla^\gamma\cdot(h\overline{V}) = 0, \\ \partial_\tau\overline{V} + \overline{V}\cdot\nabla^\gamma\overline{V} + \frac{1}{\varepsilon}\nabla^\gamma\zeta = 0, \end{cases} \tag{5.62}$$

with $h = 1 + \varepsilon\zeta - \beta b$. Letting ε go to zero, one gets formally the *lake equations*

$$\begin{cases} \nabla^\gamma\cdot(h_b\overline{V}) = 0, \\ \partial_\tau\overline{V} + \overline{V}\cdot\nabla^\gamma\overline{V} + \nabla^\gamma\mathbf{p} = 0, \end{cases} \tag{5.63}$$

where $h_b = 1 - \beta b$ and the "pressure" $\mathbf{p}$ is the Lagrange multiplier associated to the constraint $\nabla^\gamma\cdot(h_b\overline{V}) = 0$; formally, $\nabla^\gamma\mathbf{p}$ is the limit as $\varepsilon\to 0$ of the sequence $(1/\varepsilon\nabla^\gamma\zeta^\varepsilon)_\varepsilon$ where $(\zeta^\varepsilon, \overline{V}^\varepsilon)$ solves (5.62). These equations therefore are a generalization of the standard bidimensional incompressible Euler equations (corresponding to the case $b = 0$). In order to justify rigorously the convergence of solutions of

(5.62) to solutions of (5.63) as $\varepsilon \to 0$, the main difficulty is to establish that the existence time for the solutions of (5.62) is uniform with respect to ε; because of the topography term in the first equations, this limit is singular. This problem has been solved in [**56**] as a particular case of a more general phenomenon, namely, the anelastic limit for Euler-type systems.

REMARK 5.48. The lake equations (5.63) are written on the full horizontal space $\mathbb{R}^d$. When written on a domain $\Omega \subset \mathbb{R}^d$, they must be complemented by the boundary condition

$$h_b \overline{V} \cdot \mathbf{n}_{|\partial\omega} = 0. \tag{5.64}$$

The lake equations (5.63)–(5.64) on a bounded domain Ω have been derived in [**163, 229**]. When h_b is bounded from below by a positive constant, global existence and uniqueness of strong solutions is established with Yudovich's methods in [**229**]. constant on Ω. When the water depth h_b vanishes at the boundary, the situation is more delicate and has been solved more recently in [**55**].

5.6.5.2. *The great lake equations.* With the same rescaling in time as in §5.6.5.1, one formally deduces[24] the following set of equations from the Green-Naghdi equations (5.15)

$$\begin{cases} \nabla \cdot (h_b \overline{V}) = 0, \\ (I + \mu \mathcal{T}_b)\partial_\tau \overline{V} + \nabla^\gamma \big[\mathbf{p} - \frac{1}{2}\overline{V} \cdot \overline{V} + \overline{V} \cdot (1 + \mu \mathcal{T}_b)\overline{V}\big] = 0, \end{cases} \tag{5.65}$$

with $h_b = 1 - \beta b$ and $\mathcal{T}_b$ as in (5.24). These equations are related to the *great lake equations* derived in [**63, 64, 230**]

$$\begin{cases} \nabla \cdot (h_b \overline{V}) = 0, \\ \partial_\tau \widetilde{V} + \overline{V}^\perp \nabla^\gamma \wedge \widetilde{V} + \nabla^\gamma \big[\mathbf{p} - \frac{1}{2}\overline{V} \cdot \overline{V} + \overline{V} \cdot \widetilde{V}\big] = 0, \end{cases} \tag{5.66}$$

with

$$\overline{V}^\perp = (-\overline{V}_2, \overline{V}_1)^T, \qquad \widetilde{V} = (I + \mu \mathcal{T}_b)\overline{V}, \qquad \nabla^\gamma \wedge \widetilde{V} = \partial_x \widetilde{V}_2 - \gamma \partial_y \widetilde{V}_1.$$

The only difference between (5.65) and (5.66) is the term $\overline{V}^\perp \nabla^\gamma \wedge \widetilde{V}$ in the second equation of (5.66), which does not appear in (5.65). This apparent discrepancy can easily be explained. Indeed, if $(\zeta^\varepsilon, \psi^\varepsilon)$ is a solution to the water waves equations (5.1), we know from (5.17) that

$$\nabla^\gamma \psi^\varepsilon = (1 + \mu \mathcal{T})\overline{V}^\varepsilon + O(\mu^2).$$

Applying $\nabla\wedge$ to this equation and formally taking the limit $\varepsilon \to 0$ yields

$$\nabla^\gamma \wedge \widetilde{V} = O(\mu^2).$$

[24]The derivation from the Green-Naghdi equations (5.15) is quite technical. It is much easier to derive (5.65) from (5.1). After rescaling time as in §5.6.5.1, and using Proposition 3.35, we can rewrite (5.1) under the form

$$\begin{cases} \varepsilon \partial_\tau \zeta + \nabla^\gamma \cdot (h\overline{V}) = 0, \\ \partial_\tau \psi + \frac{1}{\varepsilon}\zeta + \frac{1}{2}|\nabla^\gamma \psi|^2 - \mu \dfrac{(-\nabla^\gamma \cdot (h\overline{V}) + \varepsilon \nabla^\gamma \zeta \cdot \nabla^\gamma \psi)^2}{2(1 + \varepsilon^2 \mu |\nabla^\gamma \zeta|^2)} = 0. \end{cases}$$

Applying ∇^γ to the second equation and formally letting ε go to zero directly gives (5.65) up to $O(\mu^2)$ terms since by (5.17) we have $\nabla^\gamma \psi = (1 + \mu \mathcal{T})\overline{V} + O(\mu^2)$.

Therefore, the term $\overline{V}^{\perp}\nabla^{\gamma}\wedge\widetilde{V}$ does not appear in (5.65) because it is of size $O(\mu^2)$ and consequently is considered as a residual term[25]. We can therefore conclude that *the great lake equations can be obtained as the rigid lid limit of the Green-Naghdi equations.*

5.6.6. Bottom friction. For some applications, friction effects are important when the water becomes very shallow—for instance, at the run-up and run-down stages. These effects are very complex, depend on the nature of the bottom (rock, sand, mud, etc.), and involve turbulence phenomena. In order to take these effects into account, a classical quadratic friction term is sometimes added to the right-hand-side of the equation on V [**336, 209, 74**]. For the Green-Naghdi equations, this gives

$$(5.67)\qquad \begin{cases} \partial_t\zeta + \nabla^{\gamma}\cdot(h\overline{V}) = 0, \\ [I+\mu\mathcal{T}]\big(\partial_t\overline{V} + \varepsilon(\overline{V}\cdot\nabla^{\gamma})\overline{V}\big) + \nabla^{\gamma}\zeta + \mu\varepsilon\mathcal{Q}_1(\overline{V}) = -f\varepsilon\mu^{-1/2}\frac{1}{h}|\overline{V}|\overline{V}, \end{cases}$$

where f is a nondimensional friction coefficient. As shown in [**187**] a typical value is $f = 0.04$, used in this context in [**41**], for instance. We also refer to [**155, 249**] as well as [**54**] and references therein for a formal direct derivation of shallow water models from the Navier-Stokes equations with viscous and friction terms. See [**139**] for a brief review of the inclusion of dissipative effects in Boussinesq systems.

[25]This is true in the irrotational case considered in these notes. Without the irrotationality assumption (H3) made in §1.1.1, this term is not necessarily of size $O(\mu^2)$; this is why it has been included in [**63, 64, 230**], where no irrotationality assumption is made.

CHAPTER 6

Shallow Water Asymptotics: Systems. Part 2: Justification

This chapter is devoted to the *full* justification of the asymptotic systems derived in Chapter 5. More precisely, we prove that solutions to the water waves equations exist on the relevant time scale associated to the asymptotic regime under consideration, and we show that these solutions remain close to the approximation furnished by the asymptotic models. In other words, we prove that the solution of the asymptotic models strongly converges to the solution of the water waves equations.

As in numerical analysis, such a convergence result stems from a *consistency* and a *stability* result. Since it is obviously equivalent to prove that the solution of the asymptotic model converges towards the solution of the water waves equations, or that the solution of the water waves equations converges to the solution of the asymptotic model, two strategies are possible:

(1) Prove that the solutions of the water waves equations are *consistent* with the asymptotic model and use a *stability* property of the latter. This is the approach used below for shallow water models

(2) Prove that the solutions of the asymptotic model are *consistent* with the water waves equations and use the *stability* of these equations. This is the approach used in Chapter 8 for deep water (or full dispersion) models.

Many shallow water systems have been derived in the previous chapter, and well-posedness results on the relevant time scale are not known for all of them. Rather than proving such results for all these systems, which would be quite tedious, we show that if they are well-posed, then they necessarily provide good approximations to the water waves equations. These models therefore are *almost fully* justified; the last step to obtain *full* well-posedness, namely, a well-posedness result, can be proved independently and therefore is not addressed here.

We also give some comments on the properties of energy conservation of the shallow water models derived in this chapter and on their Hamiltonian structure.

6.1. Mathematical analysis of some shallow water models

6.1.1. The Nonlinear Shallow Water equations. Let us consider here the Nonlinear Shallow Water equations

$$\left\{\begin{array}{l}\partial_t\zeta+\nabla^\gamma\cdot(h\overline{V})=0,\\ \partial_t\overline{V}+\nabla^\gamma\zeta+\varepsilon(\overline{V}\cdot\nabla^\gamma)\overline{V}=0,\end{array}\right. \tag{6.1}$$

where we recall that $h=1+\varepsilon\zeta-\beta b$. Since (6.1) is a symmetrizable hyperbolic system, the two propositions below are very classical; see [**11, 311**].

PROPOSITION 6.1 (Local existence). *Let $t_0 > d/2$, $s \geq t_0 + 1$ and $\zeta^0, b \in H^s(\mathbb{R}^d)$, $V^0 \in H^s(\mathbb{R}^d)^d$ and assume that*

$$\exists h_{min} > 0, \qquad 1 + \varepsilon\zeta^0 - \beta b \geq h_{min} \quad on \quad \mathbb{R}^d.$$

Then there exists $T_{SW} > 0$ such that the NSW equations (6.1) admit a unique solution $(\zeta, \overline{V}) \in C([0, \frac{T_{SW}}{\varepsilon\vee\beta}]; H^s(\mathbb{R}^d)^{1+d})$ with initial condition $(\zeta^0, \overline{V}^0)$.
Moreover

$$\frac{1}{T_{SW}} = c^1_{SW} \quad and \quad \sup_{0\leq(\varepsilon\vee\beta)t\leq T_{SW}} |\zeta(t)|_{H^s} + |\overline{V}(t)|_{H^s} = c^2_{SW},$$

with $c^j_{SW} = C(\frac{1}{h_{min}}, |\zeta^0|_{H^s}, |\overline{V}^0|_{H^s})$ for $j = 1, 2$.

REMARK 6.2. Note in particular that it is possible to find $T > 0$ such that solutions to (6.1) exist uniformly on $[0, \frac{T}{\varepsilon\vee\beta}]$ for all $(\varepsilon, \beta, \gamma, \mu) \in \mathcal{A}_{SW}$, where we recall that $\mathcal{A}_{SW}$ is the shallow water regime

$$\mathcal{A}_{SW} = \{(\varepsilon, \beta, \gamma, \mu), \quad 0 \leq \mu \leq \mu_0, \quad 0 \leq \varepsilon \leq 1, \quad 0 \leq \beta \leq 1, \quad 0 \leq \gamma \leq 1\},$$

for some $0 < \mu_0 \leq \mu_{max}$.

SKETCH OF PROOF. This result is extremely classical but for the sake of completeness, we briefly sketch the proof. We take here $\gamma = 1$ and a flat bottom ($b = 0$); adaptations to the general case are straightforward.
Multiplying the second equation by $h = 1 + \varepsilon\zeta$, (6.1) can be written as

$$S_0(U)\partial_t U + \sum_{j=1}^d A_j(U)\partial_j U = 0,$$

with $U = \begin{pmatrix} \zeta \\ \overline{V} \end{pmatrix}$ and

$$S_0(U) = \begin{pmatrix} 1 & 0 \\ 0 & (1+\varepsilon\zeta)I_{d\times d} \end{pmatrix}, \qquad A_j(U) = \begin{pmatrix} \overline{V}_j & h\mathbf{e}_j^T \\ h\mathbf{e}_j & \overline{V}_j I_{d\times d} \end{pmatrix}.$$

The "symmetrizer" $S_0(U)$ is positive definite as long as $h = 1 + \varepsilon\zeta$ is strictly positive, and the matrices $A_j(U)$ are symmetric. A solution to (6.1) is generally constructed as the limit of the iterative scheme

$$\forall n \in \mathbb{N}, \qquad S_0(U^n)\partial_t U^{n+1} + \sum_{j=1}^d A_j(U^n)\partial_j U^{n+1} = 0.$$

We do not give details on the implementation of this strategy (for details, we refer readers to classical textbooks such as Chapter III.B.1 of [**11**] or Chapter 16 of [**311**]). We do, however, indicate how to prove the key step, which is the following energy estimate[1]: let $F \in L^\infty([0, T/\varepsilon]; H^s(\mathbb{R}^d))$ and $\underline{U} \in L^\infty([0, T/\varepsilon]; H^{s\vee(t_0+1)}(\mathbb{R}^d)) \cap W^{1,\infty}([0, T/\varepsilon]; H^{t_0}(\mathbb{R}^d))$ be such that

$$\exists c_0(\underline{U}) > 0, \quad \forall 0 \leq t \leq T/\varepsilon, \quad \forall W \in \mathbb{R}^{1+d}, \qquad (S_0(\underline{U})W, W) \geq c_0(\underline{U})|W|^2,$$

[1]The main point worthy of interest here is that the peculiar form of this energy estimate ensures uniform estimates on a time scale $O(1/\varepsilon)$ and not only $O(1)$.

(in the particular case considered here, $c_0(\underline{U}) = \inf_{\mathbb{R}^d}(1+\varepsilon\underline{\zeta})$ is the minimum depth), and let U solve

$$(6.2) \qquad S_0(\underline{U})\partial_t U + \sum_{j=1}^d A_j(\underline{U})\partial_j U = \varepsilon F.$$

Then one has, for all $s \geq 0$,

$$(6.3) \qquad \forall 0 \leq t \leq T/\varepsilon, \qquad |U(t)|_{H^s}^2 \leq e^{\varepsilon\lambda_s t}|U^0|_{H^s}^2 + \varepsilon\nu_s \int_0^t |F(t')|_{H^s}^2 dt',$$

with λ_s and ν_s two constants of the form

$$\lambda_s, \nu_s = C\big(T, \frac{1}{c_0}, |\underline{U}|_{L^\infty([0,T/\varepsilon];H^{s\vee(t_0+1)})}, |\underline{U}|_{W^{1,\infty}([0,T/\varepsilon];H^{t_0})}\big).$$

This energy estimate is proved in two steps.
Step 1. L^2 estimate. In order to prove the energy estimate (6.3) in the case $s = 0$, let us compute

$$\frac{1}{2}\frac{d}{dt}\big(S_0(\underline{U})U, U\big) = \frac{1}{2}\big((\partial_t S_0(\underline{U}))U, U\big) + \big(S_0(\underline{U})\partial_t U, U\big),$$

where we used the symmetry of $S_0(\underline{U})$ for the last term. Replacing $S_0(\underline{U})\partial_t U$ by its expression furnished by the equation (6.2) yields

$$\frac{1}{2}\frac{d}{dt}\big(S_0(\underline{U})U, U\big) = \frac{1}{2}\big((\partial_t S_0(\underline{U}))U, U\big) - \sum_{j=1}^d \big(A_j(\underline{U})\partial_j U, U\big) + \varepsilon(F, U).$$

Integrating by parts in the second component of the right-hand-side, one gets

$$\frac{1}{2}\frac{d}{dt}\big(S_0(\underline{U})U, U\big) = \frac{1}{2}\big([\partial_t S_0(\underline{U}) + \sum_{j=1}^d \partial_j A_j(\underline{U})]U, U\big) + \varepsilon(F, U).$$

Denoting by $c_1(\underline{U}) = C(|\underline{U}|_{W^{1,\infty}([0,T/\varepsilon]\times\mathbb{R}^d)})$ a constant such that[2], uniformly on $[0, T/\varepsilon]$,

$$\big\|\partial_t S_0(\underline{U}) + \sum_{j=1}^d \partial_j A_j(\underline{U})\big\|_{L^2\to L^2} \leq \varepsilon c_1(\underline{U}),$$

we then deduce from the Cauchy-Schwarz inequality that

$$\begin{aligned} \frac{d}{dt}\big(S_0(\underline{U})U, U\big) &\leq \varepsilon c_1(\underline{U})|U|_2^2 + 2\varepsilon|F|_2|U|_2 \\ &\leq \varepsilon\frac{1 + c_1(\underline{U})}{c_0(\underline{U})}\big(S_0(\underline{U})U, U\big) + \varepsilon|F|_2^2. \end{aligned}$$

Integrating this differential inequality yields for all $0 \leq t \leq T/\varepsilon$,

$$\big(S_0(\underline{U})U, U\big) \leq \exp(\varepsilon t c_2(\underline{U}))\big(S_0(\underline{U})U, U\big)_{|_{t=0}} + \varepsilon\int_0^t e^{\varepsilon(t-t')c_2(\underline{U})}|F(t')|_2^2 dt',$$

[2]The fact that ε can be factored out in the right-hand-side comes from the peculiar form of $S_0(\underline{U})$ and $A_j(\underline{U})$ that are $O(\varepsilon)$ perturbations of constant terms.

with $c_2(\underline{U}) = \frac{1+c_1(\underline{U})}{c_0(\underline{U})}$. The energy estimate (6.3) follows easily in the case $s = 0$.
Step 2. H^s estimate. Let us first note that $U_{(s)} = \Lambda^s U$ solves

$$S_0(\underline{U})\partial_t U_{(s)} + \sum_{j=1}^d A_j(\underline{U})\partial_j U_{(s)} = \varepsilon F_{(s)},$$

with

$$\begin{aligned} F_{(s)} &= \Lambda^s F - \frac{1}{\varepsilon}[\Lambda^s, S_0(\underline{U})]\partial_t U - \frac{1}{\varepsilon}\sum_{j=1}^d [\Lambda^s, A_j(\underline{U})]\partial_j U \\ &= \Lambda^s F - \frac{1}{\varepsilon}[\Lambda^s, S_0(\underline{U})]S_0(\underline{U})^{-1}F \\ &\quad + \frac{1}{\varepsilon}\sum_{j=1}^d \Big([\Lambda^s, S_0(\underline{U})]S_0(\underline{U})^{-1}A_j(\underline{U}) - [\Lambda^s, A_j(\underline{U})]\Big)\partial_j U, \end{aligned}$$

where we used (6.2) to derive the second identity.
Using the commutator estimates of Proposition B.8 in Appendix B, we get

$$|F_{(s)}|_2 \le c_3(\underline{U})\big(|F|_{H^s} + |U|_{H^s}\big),$$

where $c_3(\underline{U})$ is of the form $c_3(\underline{U}) = C(\frac{1}{c_0(\underline{U})}, |\underline{U}|_{H^{s\vee(t_0+1)}_{T/\varepsilon}})$. Using the L^2 estimate of Step 1 on the system satisfied by $U_{(s)}$ therefore gives (6.3), after applying Gronwall's lemma. □

We now complement Proposition 6.1 with a result showing that the exact solution it furnishes is stable with respect to perturbations.

PROPOSITION 6.3 (Stability). *Let the assumptions of Proposition 6.1 be satisfied and moreover assume that there exists $(\widetilde{\zeta}, \widetilde{V}) \in C([0, \frac{T_{SW}}{\varepsilon\vee\beta}]; H^s(\mathbb{R}^d)^{d+1})$ such that*

$$\begin{cases} \partial_t\widetilde{\zeta} + \nabla^\gamma\cdot(\widetilde{h}\widetilde{V}) = r, \\ \partial_t\widetilde{V} + \nabla^\gamma\widetilde{\zeta} + \varepsilon(\widetilde{V}\cdot\nabla^\gamma)\widetilde{V} = R, \end{cases} \qquad \text{with} \quad \widetilde{h} = 1 + \varepsilon\widetilde{\zeta} - \beta b,$$

with $\widetilde{R} := (r, R) \in L^\infty([0, \frac{T_{SW}}{\varepsilon\vee\beta}]; H^s(\mathbb{R}^d)^{1+d})$. Then, for all $t \in [0, \frac{T_{SW}}{\varepsilon\vee\beta}]$, the error with respect to the solution $(\zeta, \overline{V})$ given by Proposition 6.1 satisfies

$$|\mathfrak{e}|_{L^\infty([0,t];H^{s-1})} \le C(c^1_{SW}, c^2_{SW}, |\widetilde{\zeta}, \widetilde{V}, \widetilde{R}|_{L^\infty([0,t];H^s)})\big(|\mathfrak{e}_{|_{t=0}}|_{H^{s-1}} + t|\widetilde{R}|_{L^\infty([0,t];H^s)}\big),$$

where $\mathfrak{e} = (\zeta, \overline{V}) - (\widetilde{\zeta}, \widetilde{V})$ and c^1_{SW}, c^2_{SW} are as in Proposition 6.1.

PROOF. The result is a direct and classical consequence of the energy estimate (6.3). As for the proof of Proposition 6.1, we consider $\gamma = 1$ and flat bottoms for the sake of clarity. Subtracting the equations satisfied by $U = (\zeta, \overline{V})$ and $\widetilde{U} = (\widetilde{\zeta}, \widetilde{V})$, we obtain

$$S_0(U)\partial_t\mathfrak{e} + \sum_{j=1}^d A_j(U)\partial_j\mathfrak{e} = \varepsilon F, \tag{6.4}$$

with $S_0(\cdot)$ and $A_j(\cdot)$ as in the proof of Proposition 6.1, while F is given by

$$F = -\frac{1}{\varepsilon}\widetilde{R} - \frac{1}{\varepsilon}(S_0(U) - S_0(\widetilde{U}))\partial_t\widetilde{U} - \frac{1}{\varepsilon}\sum_{j=1}^d (A_j(U) - A_j(\widetilde{U}))\partial_j\widetilde{U}.$$

Since $s-1\geq t_0$, we deduce from Proposition B.2 in Appendix B that

$$|F|_{H^{s-1}}\leq\frac{1}{\varepsilon}|\widetilde{R}|_{H^{s-1}}+C\big(|\widetilde{U}|_{H^s},|\partial_t\widetilde{U}|_{H^{s-1}}\big)|\mathfrak{e}|_{H^{s-1}}.$$

We therefore get from the energy estimate (6.3) applied to (6.4) that

$$\forall 0\leq t\leq T_{SW}/\varepsilon,\qquad |\mathfrak{e}(t)|_{H^{s-1}}\leq\varepsilon c_4\int_0^t\big(\frac{1}{\varepsilon}|\widetilde{R}(t')|_{H^{s-1}}+|\mathfrak{e}(t')|_{H^{s-1}}\big)dt',$$

where $c_4=c_4(U,\widetilde{U})$ is of the form

$$c_4=C\big(T,\frac{1}{c_0(U)},|U,\widetilde{U}|_{L^\infty([0,T/\varepsilon];H^s)},|U,\widetilde{U}|_{W^{1,\infty}([0,T/\varepsilon];H^{s-1})}\big).$$

The result is then a direct consequence of Gronwall's lemma. □

6.1.2. The Green-Naghdi equations. Let us now consider the Green-Naghdi equations

$$(6.5)\qquad\begin{cases}\partial_t\zeta+\nabla^\gamma\cdot(h\overline{V})=0,\\ [1+\mu\mathcal{T}]\big(\partial_t\overline{V}+\varepsilon(\overline{V}\cdot\nabla^\gamma)\overline{V}\big)+\nabla^\gamma\zeta+\mu\varepsilon\mathcal{Q}_1(\overline{V})=0,\end{cases}$$

where we recall that $h=1+\varepsilon\zeta-\beta b$ and that $\mathcal{T}$ and $\mathcal{Q}_1$ are given in (5.12) and (5.16). Local existence and stability are addressed in the following two propositions; their statement requires the introduction of the spaces X^s ($s\geq 0$) defined as

$$(6.6)\quad X^s=\{V\in H^s(\mathbb{R}^d)^d,|V|_{X^s}<\infty\}\quad\text{with}\quad |V|_{X^s}^2=|V|_{H^s}^2+\mu|\nabla^\gamma\cdot V|_{H^s}^2.$$

We also need the space Y^s defined as

$$Y^s=H^s(\mathbb{R}^d)\times X^s.$$

PROPOSITION 6.4 (Local existence). *Let $t_0>d/2$, $P\geq 0$, $s\geq t_0+1$ and $\zeta^0,b\in H^{s+P}(\mathbb{R}^d)$, $V^0\in X^{s+P}$ and assume that*

$$\exists h_{min}>0,\qquad 1+\varepsilon\zeta^0-\beta b\geq h_{min}\quad on\quad\mathbb{R}^d.$$

Then, if P is large enough, there exists $T_{GN}>0$ such that the Green-Naghdi equations (6.5) admit a unique solution $(\zeta,\overline{V})\in C([0,\frac{T_{GN}}{\varepsilon\vee\beta}];Y^s)$ with initial condition $(\zeta^0,\overline{V}^0)$. Moreover,

$$T_{GN}=c_{GN}^1\quad and\quad\sup_{0\leq(\varepsilon\vee\beta)t\leq T_{GN}}|(\zeta(t),\overline{V}(t))|_{Y^s}=c_{GN}^2,$$

where $c_{GN}^j=C\big(\frac{1}{h_{min}},|(\zeta^0,V^0)|_{Y^{s+P}}\big)$ for $j=1,2$.

SKETCH OF PROOF. As noted in Remark 6.6 below, the proof is much simpler in dimension $d=1$. We indicate here how the proof works in this simple configuration, dealt with in [**232, 184**], and refer to [**13**] for the two-dimensional case. For the sake of simplicity, we consider a flat bottom here and $\gamma=1$. The Green-Naghdi equations (6.5) can then be written under the form

$$\big[S_0(U)+\mu S_1(U)\big]\partial_tU+\sum_{j=1}^d A_j(U)\partial_jU+\varepsilon B(U)=0,$$

with $S_0(U)$ and $A_j(U)$ as in the proof of Proposition 6.1, while $S_1(U)$ and $B(U)$ are given by

$$S_1(U)=\begin{pmatrix}0\\ -\frac{1}{3}\partial_x(h^3\partial_x\cdot)\end{pmatrix},\qquad B(U)=\begin{pmatrix}0\\ \mu\frac{2}{3}\partial_x(h^3(\partial_x\overline{v})^2)\end{pmatrix},$$

where we recall that $h = 1 + \varepsilon\zeta$ and that we denote $\overline{V} = \overline{v}$ in dimension $d = 1$. Let us note in particular that for all f smooth enough, and with $\mathbf{e}_2 = (0,1)^T$

$$(6.7)\quad (S_1[U]f\mathbf{e}_2, f\mathbf{e}_2) = \frac{1}{3}\big((1+\varepsilon\zeta)^3\partial_x f, \partial_x f\big) \quad \text{and} \quad (S[U]f\mathbf{e}_2, f\mathbf{e}_2) \sim |f|^2_{X^0},$$

where the constants in the equivalence relation depend only on $|\zeta|_\infty$ and $1/c_0(\underline{U})$ (with $c_0(\underline{U}) = \inf_{\mathbb{R}^d}(1+\zeta)$ here).

As in the proof of Proposition 6.1 we do not give all the details of the proof, but indicate how to derive an energy estimate for the linearized system

$$(6.8)\qquad \big[S_0(\underline{U}) + \mu S_1(\underline{U})\big]\partial_t U + \sum_{j=1}^d A_j(\underline{U})\partial_j U = -\varepsilon B(\underline{U}) + \varepsilon F + \varepsilon\sqrt{\mu}\partial_x g\mathbf{e}_2.$$

This energy estimate (which allows the convergence of an iterative scheme to construct a solution to the Green-Naghdi equations) is the following[3],

$$(6.9)\ \forall 0 \le t \le T/\varepsilon, \qquad |U(t)|^2_{Y^s} \le e^{\varepsilon\lambda_s t}|U^0|^2_{Y^s} + \varepsilon\nu_s\int_0^t \big(|F|^2_{H^s} + |g|^2_{H^s} + |\underline{U}|^2_{Y^s}\big),$$

with λ_s and ν_s two constants of the form

$$\lambda_s, \nu_s = C\big(T, \frac{1}{c_0}, |\underline{U}|_{L^\infty([0,T/\varepsilon];H^{s\vee(t_0+1)})}, |\underline{U}|_{W^{1,\infty}([0,T/\varepsilon];H^{t_0})}\big).$$

Step 1. Y^0 estimate. Denoting $S(\underline{U}) = S_0(\underline{U}) + \mu S_1(\underline{U})$, let us compute

$$\frac{1}{2}\frac{d}{dt}\big(S(\underline{U})U, U\big) = \frac{1}{2}\big((\partial_t S_0(\underline{U})U, U) + \frac{1}{2}\mu([\partial_t, S_1(\underline{U})]U, U) + \big(S(\underline{U})\partial_t U, U\big).$$

Proceeding exactly as in the first step of the proof of the energy estimate (6.3) for the shallow water equations, and using the same notations, we get

$$\begin{aligned} \frac{d}{dt}\big(S(\underline{U})U, U\big) \ &\le\ \varepsilon c_2(\underline{U})\big(S_0(\underline{U})U, U\big) + \varepsilon|F|_2^2 \\ &\quad + \varepsilon\sqrt{\mu}(\partial_x g\mathbf{e}_2, U) + \mu\big([\partial_t, S_1(\underline{U})]U, U\big) - 2\varepsilon\big(B(\underline{U}), U\big). \end{aligned}$$

The last three terms of the r.-h.-s. are specific to the Green-Naghdi equations and we now discuss how to control them.

- Control of $\varepsilon\sqrt{\mu}(\partial_x g\mathbf{e}_2, U)$. An integration by parts gives directly

$$\varepsilon\sqrt{\mu}(\partial_x g\mathbf{e}_2, U) \le \varepsilon|g|_2|\sqrt{\mu}\partial_x\overline{v}|_2.$$

- Control of $\mu\big([\partial_t, S_1(\underline{U})]U, U\big)$. From the definition of $S_1(\underline{U})$, one gets after an integration by parts that

$$\begin{aligned} \mu\big([\partial_t, S_1(\underline{U})]U, U\big) \ &=\ \frac{\mu}{3}\big([\partial_t, (1+\varepsilon\underline{\zeta})^3]\partial_x\overline{v}, \partial_x\overline{v}\big) \\ &\le\ \varepsilon c_1(\underline{U})|\sqrt{\mu}\partial_x\overline{v}|_2^2, \end{aligned}$$

where, as in the proof of Proposition 6.1, $c_1(\underline{U}) = C(|\underline{U}|_{W^{1,\infty}([0,T/\varepsilon\times\mathbb{R}^d])})$.

[3]The most important aspect about this energy estimate is that it gives uniform estimates on a time scale $O(1/\varepsilon)$ and *uniformly with respect to μ.*

- Control of $2\varepsilon\big(B(\underline{U}),U\big)$. From the definition of $B(\underline{U})$, one gets after an integration by parts that

$$\begin{aligned}
-2\varepsilon\big(B(\underline{U}),U\big) &= \frac{4}{3}\varepsilon\mu\big((1+\varepsilon\underline{\zeta})^3(\partial_x\underline{v})^2,\partial_x v\big)\\
&\le 4\varepsilon\mu|\partial_x\underline{v}|_\infty(S_1(\underline{U})\underline{U},\underline{U})^{1/2}(S_1(\underline{U})U,U)^{1/2}\\
&\le \varepsilon c_1(\underline{U})\big(|\underline{v}|^2_{X^0}+(S_1(\underline{U})U,U)\big),
\end{aligned}$$

where we used (6.7) to derive the second inequality.

Gathering these estimates, we deduce that

$$\frac{d}{dt}\big(S(\underline{U})U,U\big)\le\varepsilon c_2(\underline{U})\Big(S(\underline{U})U,U\big)+|F|_2^2+|g|_2+|\underline{v}|^2_{X_0}\Big),$$

where we now use the notation $c_2(\underline{U})$ more generally for any constant of the form $c_2(\underline{U})=C(\frac{1}{c_0(\underline{U})},c_1(\underline{U}))$. By Gronwall's lemma and (6.7), we get the energy estimate (6.9) in the case $s=0$.

Step 2. Y^s estimate. Let us first note that $U_{(s)}=\Lambda^s U$ solves

$$[S_0(\underline{U})+\mu S_1(\underline{U})]\partial_t U_{(s)}+\sum_{j=1}^d A_j(\underline{U})\partial_j U_{(s)}=-\Lambda^s B(\underline{U})+\varepsilon F_{(s)}+\varepsilon\sqrt{\mu}\partial_x g_{(s)}\mathbf{e}_2$$

with $F_{(s)}$, as in the second step of the proof of Proposition 6.1, and

$$g_{(s)}=\Lambda^s g+\frac{1}{3\varepsilon}[\Lambda^s,(1+\varepsilon\underline{\zeta})^3]\sqrt{\mu}\partial_x\partial_t\underline{v}.$$

Let us first note that

$$\begin{aligned}
\big|(\Lambda^s B(\underline{U}),U_{(s)})\big| &= \mu\frac{2}{3}\big|(\Lambda^s(h^3(\partial_x\underline{v})^2,\partial_x U_{(s)})\big|\\
&\le c_1(\underline{U})|\underline{U}|_{Y^s}|U|_{Y^s},
\end{aligned}$$

where we used the product estimate of Remark B.3 in Appendix B to derive the second inequality. The fact that $B(\underline{U})$ is replaced by $\Lambda^s B(\underline{U})$ in the equation for $U_{(s)}$ therefore is transparent in the energy estimates.

We also have

$$\begin{aligned}
|g_{(s)}|_2 &\le |g|_{H^s}+c_3(\underline{U})|\partial_t\underline{v}|_{X^{s-1}}\\
&\le |g|_{H^s}+c_3(\underline{U})|\underline{U}|_{Y^s},
\end{aligned}$$

where $c_3(\underline{U})$ is of the form $c_3(\underline{U})=C(\frac{1}{c_0(\underline{U})},|\underline{U}|_{H^{s\vee(t_0+1)}_{T/\varepsilon}})$. We therefore deduce the desired energy estimate, as in the proof of Proposition 6.1, by using the zero-th order estimate of Step 1 on the equation satisfied by $U_{(s)}$. □

In the statement of the stability property below, we use the notations

$$\widetilde{h}=1+\varepsilon\widetilde{\zeta}-\beta b\quad\text{and}\quad\widetilde{\mathcal{T}}=\mathcal{T}[\widetilde{h},\beta b],$$

with $\mathcal{T}[\widetilde{h},\beta b]$ as defined in (5.12).

PROPOSITION 6.5 (Stability). *Let the assumptions of Proposition 6.4 be satisfied and moreover assume that there exists $(\widetilde{\zeta},\widetilde{V})\in C([0,\frac{T_{GN}}{\varepsilon\vee\beta}];Y^{s+P})$ such that*

$$\begin{cases}
\partial_t\widetilde{\zeta}+\nabla^\gamma\cdot(\widetilde{h}\widetilde{V})=r,\\
[I+\mu\widetilde{\mathcal{T}}]\big(\partial_t\widetilde{V}+\varepsilon(\widetilde{V}\cdot\nabla^\gamma)\widetilde{V}\big)+\nabla^\gamma\widetilde{\zeta}+\mu\varepsilon\mathcal{Q}_1(\widetilde{V})=R,
\end{cases}$$

with $\widetilde{R} := (r, R) \in L^\infty([0, \frac{T_{GN}}{\varepsilon\sqrt{\beta}}]; Y^{s+P})$. *Then one has, for all* $t \in [0, \frac{T}{\varepsilon\sqrt{\beta}}]$,

$$|\mathfrak{e}(t)|_{Y^{s-1})} \leq C(c^1_{GN}, c^2_{GN}, |\widetilde{\zeta}, \widetilde{V}, \widetilde{R}|_{L^\infty([0,t];Y^{s+P})})(|\mathfrak{e}_{|_{t=0}}|_{Y^{s+P}} + t|\widetilde{R}|_{L^\infty([0,t];Y^{s+P})}),$$

where $\mathfrak{e} = (\zeta, \overline{V}) - (\widetilde{\zeta}, \widetilde{V})$ *and* c^1_{GN}, c^2_{GN} *are as in Proposition* 6.4.

SKETCH OF PROOF. As for Proposition 6.3, this is a consequence of the energy estimate, now provided by (6.9) in dimension $d = 1$ (see **[13]** for the general case). □

REMARK 6.6. The proof of Propositions 6.4 and 6.5 can be found in **[13]**. The reason why a regularity of order $s + P$ is required (rather than s) on the initial data is that the proof relies on a Nash-Moser theorem, because the energy estimates exhibit derivative losses[4]. As shown above, in the case $d = 1$, these derivative losses disappear and a standard iterative scheme can be used to construct the solution (we only sketched the proof above; for full details, see **[232]** for the case of flat bottoms and **[184]** for the general case); consequently, one can take $P = 0$ in the statements above. When $d = 2$, it is also possible to derive variants of the Green-Naghdi equations (6.5) that do not exhibit derivative losses and for which the above statements hold with $P = 0$ **[185]**.

6.1.3. The Fully Symmetric Boussinesq systems. Let us now investigate the properties of the fully symmetric Boussinesq systems[5] belonging to the class Σ' introduced in Definition 5.39

$$\text{(6.10)} \qquad \begin{cases} (1 - \mu\mathbf{b}\Delta^\gamma)\partial_t\zeta + \nabla^\gamma \cdot (\sqrt{h}W_{\theta,\delta}) + \mu\mathbf{a}\Delta^\gamma\nabla^\gamma \cdot W_{\theta,\delta} = 0, \\ (1 - \mu\mathbf{d}\Delta^\gamma)\partial_t W_{\theta,\delta} + \sqrt{h}\nabla^\gamma\zeta + \mu\mathbf{c}\Delta^\gamma\nabla^\gamma\zeta \\ \qquad + \frac{\varepsilon}{2}\big[(W_{\theta,\delta} \cdot \nabla^\gamma)W_{\theta,\delta} + W_{\theta,\delta}\nabla^\gamma \cdot W_{\theta,\delta} + \frac{1}{2}\nabla^\gamma|W_{\theta,\delta}|^2\big] = 0, \end{cases}$$

where, by definition of Σ', one has $\mathbf{a} = \mathbf{c}$, $\mathbf{d} \geq 0$, $\mathbf{d} \geq 0$ and

$$\mathbf{a} + \mathbf{b} + \mathbf{c} + \mathbf{d} = 1/3.$$

Without the dispersive terms (i.e., if we take $\mathbf{a} = \mathbf{b} = \mathbf{c} = \mathbf{d} = 0$), (6.10) is a hyperbolic system and local existence and stability results are classical, as for the NSW equations in §6.1 above. It is easy to check that the presence of the dispersive terms does not affect the standard proof sketched in the proof of Proposition 6.1 since the terms are symmetric ($\mathbf{a} = \mathbf{c}$). The local existence results are shown below (some of these systems are in fact globally well-posed as shown in **[146]**, but they are not needed for our present purposes).

PROPOSITION 6.7 (Local existence). *Let* $t_0 > d/2$, $s \geq t_0 + 1$ *and* $\zeta^0, b \in H^s(\mathbb{R}^d)$, $W^0 \in H^s(\mathbb{R}^d)^d$ *and assume that*

$$\exists h_{min} > 0, \qquad 1 + \varepsilon\zeta^0 - \beta b \geq h_{min} \quad on \quad \mathbb{R}^d.$$

Then there exists $T_B > 0$ *such that the Boussinesq equations* (6.10) *admit a unique solution* $(\zeta, W_{\theta,\delta}) \in C([0, \frac{T_B}{\varepsilon\sqrt{\beta}}]; H^s(\mathbb{R}^d)^{1+d})$ *with initial condition* (ζ^0, W^0).

[4]Things are different in dimension $d = 2$ because the X^s norm does not control all the derivatives of order $s + 1$; it controls only the H^s norm of the divergence.

[5]It is straightforward to adapt the results of this section to symmetric systems of the form (5.41).

Moreover,

$$\frac{1}{T_B} = c_B^1 \quad \text{and} \quad \sup_{0\leq(\varepsilon\vee\beta)t\leq T_B} |\zeta(t)|_{H^s} + |W_{\theta,\delta}(t)|_{H^s} = c_B^2,$$

with $c_B^j = C(\frac{1}{h_{min}}, |\zeta^0|_{H^s}, |W^0|_{H^s})$ *for* $j = 1, 2$.

The stability of the solutions to (6.10) is then ensured by the proposition below.

PROPOSITION 6.8 (Stability). *Let the assumptions of Proposition* 6.7 *be satisfied and moreover assume that there exists* $(\widetilde{\zeta}, \widetilde{W}) \in C([0, \frac{T_B}{\varepsilon\vee\beta}]; H^s(\mathbb{R}^d)^{d+1})$ *such that*

$$\begin{cases} (1-\mu\mathbf{b}\Delta^\gamma)\partial_t\zeta + \nabla^\gamma\cdot(\sqrt{h}\widetilde{W}) + \mu\mathbf{a}\Delta^\gamma\nabla^\gamma\cdot\widetilde{W} = r, \\ (1-\mu\mathbf{d}\Delta^\gamma)\partial_t\widetilde{W} + \sqrt{h}\nabla^\gamma\zeta + \mu\mathbf{c}\Delta^\gamma\nabla^\gamma\zeta \\ \qquad + \frac{\varepsilon}{2}\big[(\widetilde{W}\cdot\nabla^\gamma)\widetilde{W} + \widetilde{W}\nabla^\gamma\cdot\widetilde{W} + \frac{1}{2}\nabla^\gamma|\widetilde{W}|^2\big] = R, \end{cases}$$

with $\widetilde{R} := (r, R) \in L^\infty([0, \frac{T_B}{\varepsilon\vee\beta}]; H^s(\mathbb{R}^d)^{1+d})$. *Then, for all* $t \in [0, \frac{T_B}{\varepsilon\vee\beta}]$,

$$|\mathfrak{e}|_{L^\infty([0,t];H^{s-1})} \leq C(c_B^1, c_B^2, |\widetilde{R}|_{L^\infty([0,t];H^s)})\big(|\mathfrak{e}_{|_{t=0}}|_{H^s} + t|\widetilde{R}|_{L^\infty([0,t];H^s)}\big),$$

where $\mathfrak{e} = (\zeta, W_{\theta,\delta}) - (\widetilde{\zeta}, \widetilde{W})$ *and* c_B^1, c_B^2 *are as in Proposition* 6.7.

6.2. Full justification (convergence) of shallow water models

We provide here the justification of three important models: the nonlinear shallow water equations (5.7), the Green-Naghdi equations (5.11), and the fully symmetric Boussinesq equations (5.40). For all these models, we use the first strategy mentioned above. The justification of all the other shallow water models[6] derived in §5.1 could be done following the same lines (provided that local existence/stability result has been established for these models[7] over the relevant time scale $t = O(\frac{1}{\varepsilon\vee\beta})$). This is done in §6.2.4, where the general strategy is explained and the example of Boussinesq systems explored in further detail.

REMARK 6.9. For the sake of simplicity, we take $b \in H^\infty(\mathbb{R}^d)$ in the statements of Theorems 6.10, 6.15, and 6.20. It is not difficult to count the finite number of derivatives actually needed for these theorems to hold (for instance, it is enough to take $b \in H^{N+d/2+3/2}(\mathbb{R}^d)$ in Theorem 6.10).

6.2.1. Full justification of the Nonlinear Shallow Water equations.

We provide here a full justification of the Nonlinear Shallow Water (NSW) equations

$$\begin{cases} \partial_t\zeta + \nabla^\gamma\cdot(h\overline{V}) = 0, \\ \partial_t\overline{V} + \nabla^\gamma\zeta + \varepsilon(\overline{V}\cdot\nabla^\gamma)\overline{V} = 0 \end{cases} \tag{6.11}$$

when the parameters ε, β, γ and μ belong to the *shallow water asymptotic regime*

$$\mathcal{A}_{SW} = \{(\varepsilon,\beta,\gamma,\mu), \quad 0\leq\mu\leq\mu_0, \quad 0\leq\varepsilon\leq 1, \quad 0\leq\beta\leq 1, \quad \gamma\leq 1\}.$$

[6] For instance, the GN equations for almost flat bottoms (5.18) or (5.19); the GN equations with medium amplitude surface deformations and medium amplitude (5.21) or small amplitude (5.22) topography variations; the Boussinesq equations with large (5.23) or small (5.25) topography variations; or the Boussinesq equations with improved frequency dispersion (5.28), (5.31), or (5.34), the Green-Naghdi equations with improved dispersion (5.36), etc.

[7] Such results do not exist yet for the equations mentioned in the previous footnote, though for some of them, such as (5.18), this follows from simple adaptations of [**13**], for instance.

The justification of the NSW equations (and extensions such as the so-called Friedrichs expansion) goes back to Ovsjannikov [**270, 271**], Kano and Nishida [**198**], and Kano [**196**] for analytical data. The result below considers Sobolev data; see [**181, 12**]. What is important in this result is that it is uniform with respect to $(\varepsilon,\beta,\gamma,\mu)\in\mathcal{A}_{SW}$. We also recall that the functional space E_T^N that appears in the statement below has been defined in (4.10).

THEOREM 6.10. *Let $N > d/2+6$ and $\mathfrak{p}=(\varepsilon,\beta,\gamma,\mu)\in\mathcal{A}_{SW}$. Also let $U^0=(\zeta^0,\psi^0)\in E_0^N$ and $b\in H^\infty(\mathbb{R}^d)$, and moreover assume that U_0 satisfies the assumptions of Theorem 4.16.*
Then there exists $T=T_{SW}>0$ that does not depend on $\mathfrak{p}\in\mathcal{A}_{SW}$ and such that
(1) *The water waves equations (5.1) admit a unique solution $U=(\zeta,\psi)\in E^N_{\frac{T}{\varepsilon\vee\beta}}$ with initial data U^0, and to which one associates through (3.31) an averaged velocity $\overline{V}\in C([0,\frac{T}{\varepsilon\vee\beta}];H^{N-3}(\mathbb{R}^d)^d)$.*
(2) *There exists a unique solution $(\zeta_{SW},\overline{V}_{SW})\in C([0,\frac{T}{\varepsilon\vee\beta}];H^{N-3}(\mathbb{R}^d)^{1+d})$ to (6.11) with initial condition $(\zeta^0,\overline{V}^0=\overline{V}_{|_{t=0}})$.*
(3) *The following error estimate holds, for all $0\le(\varepsilon\vee\beta)t\le T$,*

$$|\mathfrak{e}_{SW}|_{L^\infty([0,t]\times\mathbb{R}^d)}\le\mu t C\big(\mathcal{E}^N(U^0),\mu_0,\frac{1}{h_{min}},\frac{1}{a_0},b\big),$$

where $\mathfrak{e}_{SW}=(\zeta,\overline{V})-(\zeta_{SW},\overline{V}_{SW})$ and $\mathcal{E}^N(U^0)$ are as in (4.8).

REMARK 6.11. The assumption that $\mathfrak{a}(U^0)\ge a_0>0$ made in the statement of Theorem 4.16 is automatically satisfied when μ_{max} is small enough, since one has $\mathfrak{a}(U^0)=1+O(\mu)$, which can easily be checked from §4.3.1.

REMARK 6.12. Since the statement of the theorem is uniform with respect to $\mathfrak{p}=(\varepsilon,\beta,\gamma,\mu)\in\mathcal{A}_{SW}$, it is possible, in the spirit of Corollary 4.22, to control the error between families of exact solutions to the water waves equations indexed by $\mathfrak{p}\in\mathcal{A}_{SW}$ and their shallow water approximations.

REMARK 6.13. We have assumed in the theorem that $(\zeta_{SW},\overline{V}_{SW})=(\zeta,\overline{V})$ at $t=0$. If we relax this assumption by taking $(\zeta_{SW},\overline{V}_{SW})_{|_{t=0}}=(\zeta,\overline{V})_{|_{t=0}}+O(\mu)$, then it is easy to check that the error of the approximation becomes $O(\mu(1+t))$ rather than $O(\mu t)$.

REMARK 6.14. The existence time considered in the theorem is of order $O(1/\varepsilon\vee\beta)$; since no smallness assumption is made on ε nor β in the shallow water regime $\mathcal{A}_{SW}$, this existence time is uniformly of order $O(1)$ for all the configuration belongings to this regime. On this time scale, the precision of the NSW equations (6.11) is uniformly of size $O(\mu)$.

PROOF. From the assumption made on N, it is possible to find $t_0>d/2$ such that $N\ge t_0+t_0\vee 2+3/2$, and we therefore can use Theorem 4.16 to get the existence and uniqueness of a solution to (5.1). The existence and regularity of $\overline{V}$ comes from Remark 3.40 and Lemma 4.6.
The second point follows directly from Proposition 6.1 (taking a smaller T_{SW} if necessary).
For the third point, we recall that Proposition 5.2 implies that

$$\begin{cases}\partial_t\zeta+\nabla^\gamma\cdot(h\overline{V})=0,\\ \partial_t\overline{V}+\nabla^\gamma\zeta+\varepsilon(\overline{V}\cdot\nabla^\gamma)\overline{V}=\mu R,\end{cases}$$

with $|R|_{H^s} \leq N(s+4)$ (for all $s \geq 0$). Using Proposition 6.3 and the Sobolev embedding $H^{t_0}(\mathbb{R}^d) \subset L^\infty(\mathbb{R}^d)$ ($d = 1, 2$), we therefore get

$$|\zeta - \zeta_{SW}|_{L^\infty([0,t]\times\mathbb{R}^d)} + |\overline{V} - \overline{V}_{SW}|_{L^\infty([0,t]\times\mathbb{R}^d)} \leq \mu t N(t_0 + 5),$$

where we also used the fact that $|\overline{V}^0|_{H^{t_0+1}} \leq M(t_0+3)|\nabla^\gamma \psi^0|_{H^{t_0+3}}$ (see Remark 3.40). We now note that $|\nabla^\gamma \psi|_{H^{t_0+5}} \leq |\mathfrak{P}\psi|_{H^{t_0+11/2}} \leq \mathfrak{m}^N(U)$ (the last inequality stemming from Lemma 4.6 and the fact that $t_0 + 6 \leq N$, which is made possible for a suitable $t_0 > d/2$ by the assumption that $N > d/2 + 6$). □

6.2.2. Full justification of the Green-Naghdi equations. We will now provide a full justification of the Green-Naghdi equations

$$\begin{cases} \partial_t \zeta + \nabla^\gamma \cdot (h\overline{V}) = 0, \\ [1+\mu\mathcal{T}]\big(\partial_t \overline{V} + \varepsilon(\overline{V}\cdot\nabla^\gamma)\overline{V}\big) + \nabla^\gamma\zeta + \mu\varepsilon\mathcal{Q}_1(\overline{V}) = 0, \end{cases} \tag{6.12}$$

where we recall that $h = 1 + \varepsilon\zeta - \beta b$ and that $\mathcal{T}$ and $\mathcal{Q}_1$ are given in (5.12) and (5.16). The parameters ε, β, γ and μ belong to the same *shallow water asymptotic regime* as the NSW equations in §6.2.1 above

$$\mathcal{A}_{SW} = \{(\varepsilon, \beta, \gamma, \mu), \quad 0 \leq \mu \leq \mu_0, \quad 0 \leq \varepsilon \leq 1, \quad 0 \leq \beta \leq 1, \quad \gamma \leq 1\}.$$

In one space dimension, the Green-Naghdi equations had been justified by Makarenko [**248**] for analytic data and by Li [**232**] for Sobolev data. As for Theorem 6.10 above, the result below is uniform with respect to $(\varepsilon, \beta, \gamma, \mu) \in \mathcal{A}_{SW}$. We also recall that the spaces X^s and Y^s have been introduced in (6.6).

THEOREM 6.15. *Let $P \geq 0$, $N > d/2 + 8 + P$ and $\mathfrak{p} = (\varepsilon, \beta, \gamma, \mu) \in \mathcal{A}_{SW}$. Also let $U^0 = (\zeta^0, \psi^0) \in E_0^N$ and $b \in H^\infty(\mathbb{R}^d)$ satisfy (4.29).*
If P is large enough, there exists $T = T_{GN} > 0$ that does not depend on $\mathfrak{p} \in \mathcal{A}_{SW}$ and such that

(1) *There exists a unique solution $U = (\zeta, \psi) \in E^N_{\frac{T}{\varepsilon\vee\beta}}$ to the water waves equations (5.1) with initial data U^0, and to which one associates through (3.31) an averaged velocity $\overline{V} \in C([0, \frac{T}{\varepsilon\vee\beta}]; X^{N-4})$.*
(2) *There exists a unique solution $(\zeta_{GN}, \overline{V}_{GN}) \in C([0, \frac{T}{\varepsilon\vee\beta}]; Y^{N-P-4})$ to (6.12) with initial condition $(\zeta^0, \overline{V}^0 = \overline{V}_{|_{t=0}})$.*
(3) *The following error estimate holds, for all $0 \leq (\varepsilon \vee \beta)t \leq T$,*

$$|\mathfrak{e}_{GN}|_{L^\infty([0,t]\times\mathbb{R}^d)} \leq \mu^2 t C\big(\mathcal{E}^N(U^0), \mu_0, \frac{1}{h_{min}}, \frac{1}{a_0}, b\big),$$

where $\mathfrak{e}_{GN} = (\zeta, \overline{V}) - (\zeta_{GN}, \overline{V}_{GN})$ and $\mathcal{E}^N(U^0)$ are as in (4.8).

REMARK 6.16. The observations made in Remarks 6.11–6.14 still hold in the present case, with minor modifications.

REMARK 6.17. The reasons why we have to take $N \geq d/2 + 8 + P$ for some $P \geq 0$ large enough are the same as in Proposition 6.5 (see Remark 6.6). In particular, one can take $P = 0$ when $d = 1$.

REMARK 6.18. The NSW equation (6.11) and the Green-Naghdi equations (6.12) are valid in the same asymptotic regime $\mathcal{A}_{SW}$; they differ only by their precision. As shown by Theorem 6.15, the error is of size $O(\mu^2)$ for the Green-Naghdi equations versus $O(\mu)$ for the NSW equations.

REMARK 6.19. In dimension $d = 1$ and for flat bottom, the Green-Naghdi equations (6.5) possess solitary waves[8] of the form

$$\zeta(t,x) = \alpha \mathrm{sech}^2[K(x-ct)], \qquad \overline{v}(t,x) = \frac{c}{1+\varepsilon\zeta}\zeta,$$

with

$$c = \sqrt{1+\varepsilon\alpha} \quad \text{and} \quad K = \sqrt{\frac{3}{4}\frac{\varepsilon\alpha}{\mu(1+\varepsilon\alpha)}}.$$

Though no smallness assumption on ε is necessary to justify the Green-Naghdi equations, Theorem 6.15 can be used to conclude that this explicit solitary wave is a good approximation of the full water waves equations only under the extra assumption that $\varepsilon \lesssim \mu$ (long wave regime). Indeed, if this assumption is not made, then one has $K \to \infty$ as $\mu \to 0$, and the error estimate given in the third point of the theorem becomes useless.

PROOF. The proof of the first point is exactly the same as in the proof of Theorem 6.10. The second point is a direct consequence of Proposition 6.4; for the third point, we follow the same lines as for the proof of Theorem 6.10, using Propositions 5.8 and 6.5 rather than Propositions 5.2 and 6.3. □

6.2.3. Full justification of the Fully Symmetric Boussinesq equations. Finally, we provide a full justification of the fully symmetric Boussinesq systems[9]

$$(6.13) \qquad \begin{cases} (1-\mu\mathbf{b}\Delta^\gamma)\partial_t\zeta + \nabla^\gamma\cdot(\sqrt{h}W_{\theta,\delta}) + \mu\mathbf{a}\Delta^\gamma\nabla^\gamma\cdot W_{\theta,\delta} = 0, \\ (1-\mu\mathbf{d}\Delta^\gamma)\partial_t W_{\theta,\delta} + \sqrt{h}\nabla^\gamma\zeta + \mu\mathbf{c}\Delta^\gamma\nabla^\gamma\zeta \\ \qquad + \frac{\varepsilon}{2}\big[(W_{\theta,\delta}\cdot\nabla^\gamma)W_{\theta,\delta} + W_{\theta,\delta}\nabla^\gamma\cdot W_{\theta,\delta} + \frac{1}{2}\nabla^\gamma|W_{\theta,\delta}|^2\big] = 0, \end{cases}$$

where, by definition of Σ', one has $\mathbf{a} = \mathbf{c}$, $\mathbf{b} \geq 0$, $\mathbf{d} \geq 0$, and

$$\mathbf{a}+\mathbf{b}+\mathbf{c}+\mathbf{d} = 1/3,$$

and where $W_{\theta,\delta}$ is as defined in (5.39).
The parameters ε, β, γ and μ belong to the *long wave regime with almost flat bottom*, which is contained in the shallow water regime. The additional assumption is that ε and β are at most of the same order as μ

$$\mathcal{A}_{LW,top-} = \{(\varepsilon,\beta,\gamma,\mu), \quad 0 \leq \mu \leq \mu_{max}, \quad 0 \leq \varepsilon \lesssim \mu, \quad 0 \leq \beta \lesssim \mu, \quad \gamma \leq 1\}.$$

The result below is uniform with respect to $(\varepsilon,\beta,\gamma,\mu) \in \mathcal{A}_{LW}$. It is important to note that an approximation of order $O(\mu^2)$ of the average velocity $\overline{V}$ is found in terms of $W_{\theta,\delta}$ by inverting the changes of unknown[10] $V_{\theta,\delta}$ (5.39) and (5.30)

$$(6.14) \qquad \overline{V}_B = (1-\mu\frac{\delta}{3}\Delta)^{-1}(1-\mu\frac{\theta}{3}\Delta)\Big(\frac{1}{\sqrt{1+\varepsilon\zeta-\beta b}}W_{\theta,\delta}\Big).$$

Conversely, the initial data for (6.13) is given in terms of the initial averaged velocity $\overline{V}^0$ by

$$W^0 = \sqrt{1+\varepsilon\zeta^0-\beta b}(1-\mu\frac{\theta}{3}\Delta)^{-1}(1-\mu\frac{\delta}{3}\Delta)\overline{V}^0.$$

[8]The linear stability of these solitary waves has been proved in [**233, 234**].

[9]The results of this section also hold if we use fully symmetric systems of the form (5.41) rather than (5.40).

[10]We actually replace (5.30) by $V_{\theta,\delta} = (1-\mu\frac{\theta}{3}\Delta)^{-1}(1-\mu\frac{\delta}{3}\Delta)\overline{V}$, which is equivalent up to $O(\mu^2)$ terms in the long wave regime.

THEOREM 6.20. *Let* $N > d/2 + 14$ *and* $\mathfrak{p} = (\varepsilon, \beta, \gamma, \mu) \in \mathcal{A}_{LW,top-}$. *Let* $U^0 = (\zeta^0, \psi^0) \in E_0^N$ *and* $b \in H^\infty(\mathbb{R}^d)$ *satisfy* (4.29). *Also assume that the coefficients* $\mathbf{a}$, $\mathbf{b}$, $\mathbf{c}$ *and* $\mathbf{d}$ *in* (6.13) *satisfy* $\mathbf{a} = \mathbf{c}$ *and* $\mathbf{d} \geq 0$, $\mathbf{d} \geq 0$.
There exists $T = T_B > 0$ *that does not depend on* $\mathfrak{p} \in \mathcal{A}_{LW,top-}$ *and such that*

(1) *There exists a unique solution* $U = (\zeta, \psi) \in E^N_{\frac{T}{\varepsilon \vee \beta}}$ *to the water waves equations* (5.1) *with initial data* U^0, *and to which one associates through* (3.31) *an averaged velocity* $\overline{V} \in C([0, \frac{T}{\varepsilon\vee\beta}]; H^{N-3})$.
(2) *There exists a unique solution* $(\zeta_B, W_{\theta,\delta}) \in C([0, \frac{T}{\varepsilon\vee\beta}]; H^{N-3})$ *to* (6.13) *with initial condition* (ζ^0, W^0) *to which one associates* $\overline{V}_B$ *through* (6.14).
(3) *The following error estimate holds, for all* $0 \leq (\varepsilon \vee \beta)t \leq T$,

$$|\mathfrak{e}_B|_{L^\infty([0,t]\times\mathbb{R}^d)} \leq \mu^2 t C\big(\mathcal{E}^N(U^0), \frac{\varepsilon}{\mu}, \frac{\beta}{\mu}, \mu_0, \frac{1}{h_{min}}, \frac{1}{a_0}, b\big),$$

where $\mathfrak{e}_B = (\zeta, \overline{V}) - (\zeta_B, \overline{V}_B)$ *and* $\mathcal{E}^N(U^0)$ *are as in* (4.8).

REMARK 6.21. The dependence of the estimate on $\frac{\varepsilon}{\mu}$ and $\frac{\beta}{\mu}$ makes it useless in the shallow water regime $\mathcal{A}_{SW}$ (with large surface and bottom variations) since these quantities can grow to infinity; in the long wave regime with almost flat bottoms $\mathcal{A}_{LW,top-}$ both $\frac{\varepsilon}{\mu}$ and $\frac{\beta}{\mu}$ remain uniformly bounded, and the estimate is relevant.

PROOF. The proof of the first point is exactly the same as in the proof of Theorem 6.10. The second point is a direct consequence of Proposition 6.7; for the third point, we follow the same lines as for the proof of Theorem 6.10, using Corollary 5.40 and Proposition 6.8 rather than Propositions 5.2 and 6.3 to get

$$|\widetilde{\mathfrak{e}}_B|_{L^\infty([0,t];H^{t_0+2})} \leq \mu^2 t C\big(\mathcal{E}^N(U^0), \frac{\varepsilon}{\mu}, \frac{\beta}{\mu}, |\mathcal{A}_{SW}|, \frac{1}{h_{min}}, \frac{1}{a_0}, b\big),$$

where $\widetilde{\mathfrak{e}}_B = (\zeta, W) - (\zeta_B, W_{\theta,\delta})$ and $t_0 > d/2$, and

$$W = \sqrt{1 + \varepsilon\zeta - \beta b}(1 - \mu\frac{\theta}{3}\Delta)^{-1}(1 - \mu\frac{\delta}{3}\Delta)\overline{V}.$$

The estimate on $\mathfrak{e}_B$ provided by the theorem therefore follows from the observation that

$$\overline{V} - \overline{V}_B = (1 - \mu\frac{\delta}{3}\Delta)^{-1}(1 - \mu\frac{\theta}{3}\Delta)\big(\frac{1}{\sqrt{1 + \varepsilon\zeta - \beta b}}(W - W_{\theta,\delta})\big),$$

so that

$$|\mathfrak{e}_B|_{L^\infty([0,t];H^{t_0})} \leq C\big(\frac{1}{h_{min}}, |\zeta|_{L^\infty([0,t];H^N)}, |\widetilde{\mathfrak{e}}|_{L^\infty([0,t];H^{t_0+2})}\big).$$ ☐

6.2.4. (Almost) full justification of other shallow water systems. The full justification results given in §§6.2.1, 6.2.2, and 6.2.3 rely on the well-posedness results of §6.1. It would be time-consuming and boring to prove such well-posedness results for all the systems derived in Chapter 5; instead we will establish slightly weaker results for the systems that have not been fully justified in §6.2.1. Since all these systems are consistent (in the relevant asymptotic regimes) at order $O(\mu^2)$ with the Green-Naghdi equations, we can prove that their solutions, if they exist, provide approximations of order $O(\mu^2 t)$ for $t \in [0, \frac{T}{\varepsilon\vee\beta}]$ to the water waves equations. Denoting by (A) any of these asymptotic models, we use the following strategy:

(1) Solutions[11] to (A) are consistent at order $O(\mu^2)$ with the Green-Naghdi equations in a given asymptotic regime $\mathcal{A}$: such results have been proved in Chapter 5.
(2) These solutions are in a $O(\mu^2 t)$ neighborhood of the exact solution to the Green-Naghdi equations with corresponding initial data: This is a consequence of the stability property of the Green-Naghdi equations (see Proposition 6.5).
(3) Since solutions to the Green-Naghdi are $O(\mu^2 t)$ close to the solutions of the water waves equations (see Theorem 6.15), a triangular inequality allows us to conclude that solutions to (A) are also $O(\mu^2 t)$ close to the solutions of the water waves equations.

When an asymptotic model is justified along these lines, we refer to it as *almost fully justified*. Full justification then follows from a well-posedness result (left to the reader!) for this asymptotic system on the relevant time scale $t \in [0, \frac{T}{\varepsilon \vee \beta}]$.

REMARK 6.22. The constraint that well-posedness must hold for times $t \in [0, \frac{T}{\varepsilon\vee\beta}]$ is important. In the long wave regime ($\varepsilon = O(\mu)$, $\beta = O(\mu)$), for instance, a well-posedness theorem for times $t \in [0, T]$ would miss all the relevant effects of these models (namely, dispersion and nonlinearity). Mathematically speaking, it can be more involved to get this correct time scale. For most of the Boussinesq systems, for instance, only a well-posedness result for $t \in [0, T]$ can be found in the literature (e.g., [**129, 206**]).

As an example, we show here how it is possible to *almost fully* justify all the Boussinesq systems derived for small amplitude topography variations. More precisely, we consider the four parameters family derived in §5.2.1.3 and which contains the systems systematically derived in [**34, 36**] in the case of flat bottoms (and for which this notion of almost full justification has been derived in [**36**])

$$(6.15)\qquad \begin{cases} (1-\mu\mathbf{b}\Delta^\gamma)\partial_t\zeta + \nabla^\gamma\cdot(hV_{\theta,\delta}) + \mu\mathbf{a}\Delta^\gamma\nabla^\gamma\cdot V_{\theta,\delta} = 0, \\ (1-\mu\mathbf{d}\Delta^\gamma)\partial_t V_{\theta,\delta} + \nabla^\gamma\zeta + \varepsilon(V_{\theta,\delta}\cdot\nabla^\gamma)V_{\theta,\delta} + \mu\mathbf{c}\Delta^\gamma\nabla^\gamma\zeta = 0, \end{cases}$$

with $\mathbf{a}$, $\mathbf{b}$, $\mathbf{c}$ and $\mathbf{d}$ given by (5.35) and $V_{\theta,\delta}$ by (5.30). In the statement below, the initial data for (6.13) is naturally given in terms of the initial averaged velocity $\overline{V}^0$ by

$$V^0 = (1-\mu\frac{\theta}{3}\Delta)^{-1}(1-\mu\frac{\delta}{3}\Delta)\overline{V}^0.$$

COROLLARY 6.23. *Let $N > d/2+14$ and $\mathfrak{p} = (\varepsilon,\beta,\gamma,\mu) \in \mathcal{A}_{LW,top-}$. Also let $U^0 = (\zeta^0,\psi^0) \in E_0^N$ and $b \in H^\infty(\mathbb{R}^d)$ satisfy (4.29).*
There exists $T = T_B > 0$ that does not depend on $\mathfrak{p} \in \mathcal{A}_{LW,top-}$ and such that
(1) *There exists a unique solution $U = (\zeta,\psi) \in E^N_{\frac{T}{\varepsilon\vee\beta}}$ to the water waves equations (5.1) with initial data U^0, and to which one associates through (3.31) an averaged velocity $\overline{V} \in C([0, \frac{T}{\varepsilon\vee\beta}]; H^{N-3})$.*

[11] These solutions must exist for $t \in [0, \frac{T}{\varepsilon\vee\beta}]$ uniformly with respect to $(\varepsilon,\mu,\beta,\gamma) \in \mathcal{A}$.

(2) *If* $(\zeta_B, V_B) \in C([0, \frac{T}{\varepsilon\vee\beta}]; H^{N-3})$ *solves* (6.15) *with initial condition* (ζ^0, V^0), *we define* $\overline{V}_B$ *by*

$$\overline{V}_B = (1 - \mu\frac{\delta}{3}\Delta)^{-1}(1 - \mu\frac{\theta}{3}\Delta)V_B$$

and the following error estimate holds, for all $0 \leq (\varepsilon \vee \beta)t \leq T$,

$$|\mathfrak{e}_B|_{L^\infty([0,t]\times\mathbb{R}^d)} \leq \mu^2 t C\big(\mathcal{E}^N(U^0), |(\zeta_B, V_B)|_{L^\infty([0,t];H^{N-6})}, \frac{\varepsilon}{\mu}, \frac{\beta}{\mu}, \mu_0, \frac{1}{h_{min}}, \frac{1}{a_0}, b\big),$$

where $\mathfrak{e}_B = (\zeta, \overline{V}) - (\zeta_B, \overline{V}_B)$ *and* $\mathcal{E}^N(U^0)$ *are as in* (4.8).

REMARK 6.24. We emphasize that contrary to Theorem 6.20, the result provided by (6.23) is not a full justification of the Boussinesq system (6.15); it shows only that *if* solutions to this systems satisfying the assumptions of (2) exist, then they furnish a good approximation to the solution of the water waves equations. To complete the full justification, one needs a local existence theorem for (6.15), which remains to be done in most cases (and which is not always true since some of theses systems are ill-posed). The most complete result in this direction is [**287**].

PROOF. Let $\underline{\alpha}$, $\underline{\theta}$, $\underline{\delta}$, $\underline{\lambda}$ be such that $\underline{\mathbf{b}} \geq 0$, $\underline{\mathbf{d}}$, $\underline{\mathbf{a}} = \underline{\mathbf{c}}$ (where the latter parameters are associated to $\underline{\alpha}$, $\underline{\theta}$, $\underline{\delta}$, $\underline{\lambda}$ through (5.35)) and let

$$W = \sqrt{1 + \varepsilon\zeta_B - \beta b}(1 - \mu\frac{\underline{\theta}}{3}\Delta)^{-1}(1 - \mu\frac{\underline{\delta}}{3}\Delta)(1 - \mu\frac{\delta}{3}\Delta)^{-1}(1 - \mu\frac{\theta}{3}\Delta)V_{\theta,\delta}.$$

In addition, we impose that $\underline{\delta} = 0$ (which is possible as seen in §5.3). It is easy to check that (ζ_B, W) solves a system of the form (6.13) with coefficients $\underline{\mathbf{b}} \geq 0$, $\underline{\mathbf{d}} \geq 0$, $\underline{\mathbf{a}} = \underline{\mathbf{c}}$ — and that therefore belongs to the class Σ' of fully symmetric systems — up to a residual $\mu^2(r, R)$ with

$$|r|_{H^{t_0+1}} + |R|_{H^{t_0+1}} \leq C(\frac{\varepsilon}{\mu}, \frac{\beta}{\mu}, |(\zeta_B, V_{\theta,\delta})|_{H^{t_0+6}}) \qquad (t_0 > d/2).$$

We therefore can deduce from (6.8) that there exists a unique solution $(\underline{\zeta}, \underline{W}) \in C([0, \frac{T}{\varepsilon\vee\beta}]; H^{N-3})$ (taking a smaller T if necessary) to this fully symmetric system with initial condition $(\zeta^0, W^0 = W_{|_{t=0}})$ and that

$$|(\zeta_B, W) - (\underline{\zeta}, \underline{W})|_{L^\infty([0,t];H^{t_0})} \leq \mu^2 t C(\frac{1}{h_{min}}, \frac{\varepsilon}{\mu}, \frac{\beta}{\mu}, |(\zeta_B, V_{\theta,\delta})|_{L^\infty([0,t];H^{t_0+6})})$$

and therefore

$$|(\zeta_B, \overline{V}_B) - (\underline{\zeta}, \underline{\overline{V}}_B)|_{L^\infty([0,t]\times\mathbb{R}^d)} \leq \mu^2 t C(\frac{1}{h_{min}}, \frac{\varepsilon}{\mu}, \frac{\beta}{\mu}, |(\zeta_B, V_{\theta,\delta})|_{L^\infty([0,t];H^{t_0+8})}),$$

where $\underline{\overline{V}}_B$ is deduced from $\underline{W}$ through (6.14). Moreover, since Theorem 6.20 provides us with an estimate of the distance between $(\underline{\zeta}, \underline{\overline{V}}_B)$ and $(\zeta, \overline{V})$, we can conclude with a simple triangle inequality. □

REMARK 6.25. The notion of *almost full justification* is quite robust. It can indeed provide a justification to *ill-posed* models. Take, for instance, the well-known Kaup system (which is a particular example, in dimension $d = 1$ and for a flat bottom, of the Boussinesq systems (6.15)),

$$\begin{cases} \partial_t\zeta + \partial_x\big(1 + \varepsilon\zeta)v\big) + \mu\frac{1}{3}\partial_x^3 v = 0, \\ \partial_t v + \partial_x\zeta + \varepsilon v\partial_x v = 0 = 0. \end{cases} \tag{6.16}$$

This system is obviously ill-posed but nonetheless quite interesting since it possesses solutions that can be computed exactly by inverse scattering techniques [**205**]. The above consideration allow us to conclude that these solutions indeed provide a good approximation to the water waves equations.

6.3. Supplementary remarks

6.3.1. Energy conservation. As seen in Chapter 1, the water waves equations (5.1) conserve the total energy (or Hamiltonian[12]) $\mathcal{E}_{WW}$, which is the sum of the kinematic and potential energy

$$\mathcal{E}_{WW} = \frac{1}{2}\int_{\mathbb{R}^d} \psi \frac{1}{\mu}\mathcal{G}\psi + \frac{1}{2}\int_{\mathbb{R}^d} \zeta^2,$$

and therefore it is natural to check whether the various shallow water models justified in this chapter (as well as others derived in Chapter 5) share a similar property.

6.3.1.1. *Nonlinear Shallow Water equations.* For the Nonlinear Shallow Water equations (5.7), the natural energy is found by making an asymptotic expansion of $\mathcal{E}_{WW}$ in terms of μ and keeping only the leading order term. Recalling that $\frac{1}{\mu}\mathcal{G}\psi = -\nabla^\gamma \cdot (hV)$ (see Proposition 3.35) and that $\nabla^\gamma \psi = \overline{V} + O(\mu)$ (see Proposition 3.37), this leads us to define the energy $\mathcal{E}_{NSW}$ as

$$\mathcal{E}_{NSW} = \frac{1}{2}\int_{\mathbb{R}^d} h|\overline{V}|^2 + \frac{1}{2}\int_{\mathbb{R}^d} \zeta^2 \qquad (h = 1 + \varepsilon\zeta - \beta b).$$

Straightforward computations show that *the nonlinear shallow water equations (5.7) conserve this energy.*

6.3.1.2. *Boussinesq systems.* To obtain the energy associated to the Boussinesq models derived in §5.1.3, which are written in terms of the surface elevation ζ and the averaged velocity $\overline{V}$, we proceed as above but keep the $O(\mu)$ terms in the expansion of $\mathcal{E}_{WW}$. As in §5.1.3, the result depends on the amplitude of the topography variations:

- *Large amplitude topography variations ($\beta = O(1)$).* This corresponds to the Boussinesq-Peregrine model (5.23). One then has $\nabla^\gamma \psi = (I + \mu\mathcal{T}_b)\overline{V} + O(\mu^2)$ and therefore

$$\begin{aligned}\mathcal{E}_{B,top+} \quad := \quad & \frac{1}{2}\int_{\mathbb{R}^d} h|\overline{V}|^2 + \frac{1}{2}\int_{\mathbb{R}^d} \zeta^2 \\ & + \mu \int_{\mathbb{R}^d} h_b\Big(\frac{1}{3}|h_b\nabla^\gamma \cdot \overline{V} - \frac{3}{2}\beta\nabla^\gamma b \cdot \overline{V}|^2 + \frac{\beta^2}{4}(\nabla^\gamma b \cdot \overline{V})^2\Big),\end{aligned}$$

  where we recall that $h_b = 1 - \beta b$.
- *Small amplitude topography variations ($\beta = O(\mu)$).* This corresponds to the Boussinesq system (5.25). The energy is found by dropping the $O(\beta\mu)$ terms in $\mathcal{E}_{B,top+}$,

$$\mathcal{E}_{B,top-} := \frac{1}{2}\int_{\mathbb{R}^d} \big(h|\overline{V}|^2 + \mu\frac{1}{3}|\nabla^\gamma \cdot \overline{V}|^2 + \zeta^2\big).$$

[12]The expression given here corresponds to the dimensionless version of the Hamiltonian H given in Chapter 1.

Contrary to what has been observed with the nonlinear shallow water equations, *these energies are not conserved by the Boussinesq equations* (5.23) *for* $\mathcal{E}_{B,top+}$ *or* (5.25) *for* $\mathcal{E}_{B,top-}$. However, these energies are *almost conserved* by the equations in the sense that

$$\frac{d}{dt}\mathcal{E}_{B,top\pm} = O(\mu^2).$$

The discussion above is absolutely identical if we consider Boussinesq systems with improved frequency dispersion as soon as these models are also written in terms of ζ and $\overline{V}$ (such as the one-parameter family of Boussinesq-Peregrine systems (5.23), for instance). For systems involving another velocity unknown, the definition of the energy must be accordingly modified by writing the energy in terms of this new velocity.
For instance, for the families of Boussinesq-Peregrine systems (5.31) and Boussinesq systems (5.34) for large amplitude and small amplitude topography variations respectively, the velocity unknown is $V_{\theta,\delta}$ defined in (5.30)

$$V_{\theta,\delta} = (I + \mu\theta\mathcal{T}_b)^{-1}(I + \mu\delta\mathcal{T}_b)\overline{V}.$$

The energy corresponding to these models therefore is found by replacing $\overline{V}$ by the above formula in $\mathcal{E}_{B,top+}$ and $\mathcal{E}_{B,top-}$. After dropping $O(\mu^2)$ terms, this leads to

$$\begin{aligned}\mathcal{E}^{\delta,\theta}_{B,top+} &:= \frac{1}{2}\int_{\mathbb{R}^d}\zeta^2 + \frac{1}{2}\int_{\mathbb{R}^d} h|V_{\theta,\delta}|^2\\ &+ \mu(\frac{1}{2}+\theta-\delta)\int_{\mathbb{R}^d}\frac{h_b}{3}|h_b\nabla^\gamma\cdot V_{\theta,\delta} - \frac{3}{2}\beta\nabla^\gamma b\cdot V_{\theta,\delta}|^2 + \beta^2\frac{h_b}{4}(\nabla^\gamma b\cdot V_{\theta,\delta})^2\end{aligned}$$

and

$$\mathcal{E}^{\delta,\theta}_{B,top-} := \frac{1}{2}\int_{\mathbb{R}^d}\big(h|V_{\theta,\delta}|^2 + 2\mu(\frac{1}{2}+\theta-\delta)\frac{1}{3}|\nabla^\gamma\cdot V_{\theta,\delta}|^2\big) + \frac{1}{2}\int_{\mathbb{R}^d}\zeta^2.$$

In general, one cannot expect better than almost conservation of these energies, but some particular choices of the parameters can lead to exact conservation, which is another aspect of the usefulness of the Boussinesq systems with improved frequency dispersion.
For instance, take (5.34) with $\mathbf{a} = 1/3$ and $\mathbf{b} = \mathbf{c} = \mathbf{d} = 0$ (in dimension $d = 1$, this corresponds to the Kaup system,). This system conserves the energy $\mathcal{E}^{1,0}_{B,top-}$ in dimension $d = 1$ (and in dimension $d = 1$ if the velocity unknown has zero horizontal vorticity) — but we recall that it is linearly ill-posed!

Another example of Boussinesq systems that exactly conserve the energy is furnished by the fully symmetric systems of the class Σ' introduced in Definition 5.39. It is not difficult to check that the natural energy derived from $\mathcal{E}_{WW}$ and associated to $(6.10)_{abcd}$ is given by

$$\mathcal{E}_{abcd} = \frac{1}{2}\int_{\mathbb{R}^d}\big(|W_{\theta,\delta}|^2 + \mu\mathbf{b}|\nabla^\gamma\cdot W_{\theta,\delta}|^2 + \mu\mathbf{d}|\nabla^\gamma\zeta|^2\big) + \frac{1}{2}\int_{\mathbb{R}^d}\zeta^2,$$

and that it is exactly conserved.

6.3.1.3. *Green-Naghdi equations.* Proceeding as for the Boussinesq systems but keeping the $O(\varepsilon\mu)$ terms, we are led to define the following energy associated to the Green-Naghdi equations (5.11) or more generally (5.36)

$$\mathcal{E}_{GN} := \frac{1}{2}\int_{\mathbb{R}^d}\zeta^2 + \frac{1}{2}\int_{\mathbb{R}^d} h|\overline{V}|^2 + \mu h\Big(\frac{1}{3}|h\nabla^\gamma\cdot\overline{V} - \frac{3}{2}\beta\nabla^\gamma b\cdot\overline{V}|^2 + \frac{\beta^2}{4}|\nabla^\gamma b\cdot\overline{V}|^2\Big).$$

The Green-Naghdi equations (5.11) is the only system among the one-parameter family $(5.36)_\alpha$ (corresponding to $\alpha = 1$) that conserves exactly this energy. When $\alpha \neq 1$, one only has almost conservation of the energy.

6.3.2. Hamiltonian structure. As noted by Zakharov [**333**] (see Chapter 1), the water waves equations (5.1) can be cast as a Hamiltonian evolutionary system with canonical variables (ζ, ψ) and Hamiltonian

$$\mathcal{E}_{WW} = \frac{1}{2}\int_{\mathbb{R}^d} \psi \frac{1}{\mu}\mathcal{G}\psi + \frac{1}{2}\int_{\mathbb{R}^d} \zeta^2.$$

More precisely, the water waves equations can be written as

$$\partial_t \begin{pmatrix} \zeta \\ \psi \end{pmatrix} = \begin{pmatrix} 0 & I \\ -I & 0 \end{pmatrix} \begin{pmatrix} \partial_\zeta \mathcal{E}_{WW} \\ \partial_\psi \mathcal{E}_{WW} \end{pmatrix}. \tag{6.17}$$

In particular, this quantity is preserved during the motion of the wave; in the previous section, we checked whether the different shallow water asymptotic systems derived in Chapter 5 also preserved this quantity (more exactly, its expansion in the relevant asymptotic regime). Another natural question, investigated here, is to determine whether these systems can also be put under a Hamiltonian form.

The most natural way to derive Hamiltonian systems from the water waves equations is to perform an asymptotic expansion of the Hamiltonian $\mathcal{E}_{WW}$ and to use it instead of $\mathcal{E}_{WW}$ in (6.17). Thus the evolution system obtained is Hamiltonian. This approach has been initiated in [**102**] and further developed in [**105, 104**], for instance. Let us briefly review the outcomes of this method in the cases considered in the previous section.

- *Nonlinear Shallow Water equations.* As shown above, the relevant expansion of $\mathcal{E}_{WW}$ is given by

$$\tilde{\mathcal{E}}_{NSW} = \frac{1}{2}\int_{\mathbb{R}^d} h|\nabla^\gamma \psi|^2 + \zeta^2$$

  (the difference with $\mathcal{E}_{NSW}$ is that we wrote this quantity in terms of the canonical variable ψ rather than $\overline{V}$). Since $\nabla^\gamma\psi = \overline{V} + O(\mu)$ (see Proposition 3.37), $\tilde{\mathcal{E}}_{NSW}$ differs from $\mathcal{E}_{NSW}$ by $O(\mu)$ terms only, which is the order of the neglected terms in the expansion of $\mathcal{E}_{WW}$. Replacing $\mathcal{E}_{WW}$ by $\mathcal{E}_{NSW}$ in (6.17), one obtains

$$\begin{cases} \partial_t\zeta + \nabla^\gamma \cdot (h\nabla^\gamma\psi) = 0, \\ \partial_t\psi + \zeta + \frac{1}{2}|\nabla^\gamma\psi|^2 = 0. \end{cases}$$

  Applying ∇^γ to the second equation and setting $V = \nabla^\gamma\psi$, one recovers the Nonlinear Shallow Water (or Saint-Venant) equations (6.1).
- *Boussinesq systems and Green-Naghdi equations.* One can use the same procedure in the asymptotic regimes corresponding to the various Boussinesq systems and to the Green-Naghdi equations. Unfortunately, the systems thus derived, though Hamiltonian, are (linearly) ill-posed.
  Let us, for instance, consider the case of the long wave regime ($\varepsilon = O(\mu)$), for a flat bottom. Since $V_{1,0} = \nabla^\gamma\psi + O(\mu^2)$ (see Remark 5.46), the quantity $\widetilde{\mathcal{E}}^{1,0}_{B,top-}$ defined as

$$\widetilde{\mathcal{E}}^{1,0}_{B,top-} := \frac{1}{2}\int_{\mathbb{R}^d} \big(h|\nabla^\gamma\psi|^2 - \frac{\mu}{3}(\Delta^\gamma\psi)^2\big) + \frac{1}{2}\int_{\mathbb{R}^d} \zeta^2$$

differs from $\mathcal{E}_{1,0}$ by $O(\mu^2)$ terms only. The Hamiltonian system derived from $\widetilde{\mathcal{E}}^{1,0}_{B,top-}$ with the same procedure as above is exactly, in dimension $d=1$, the Kaup system [**205**] (after differentiating the equation in ψ).

- *Higher order Boussinesq systems.* Since the Hamiltonian systems derived from the shallow water expansions of $\mathcal{E}_{WW}$ at order $O(\mu^2)$ are linearly ill-posed, it is natural to proceed to $O(\mu^3)$ expansions and to investigate the associated Hamiltonian systems. This procedure leads to (linearly well-posed) Boussinesq systems of higher order (involving fifth-order derivatives of ζ and $\nabla^\gamma\psi$). See, for instance, [**102**].

The procedure described above allows one to derive Hamiltonian systems canonically associated to the energy. This does not mean that they are the only Hamiltonian systems among the models derived in Chapter 5. For instance, it has been noticed in [**35**] that the $\mathbf{a}, \mathbf{b}, \mathbf{c}, \mathbf{d}$ systems (5.34) are Hamiltonian in dimension $d=1$ and for flat bottoms when $\mathbf{b}=\mathbf{d}$. More precisely, all these systems can always be written as

$$\partial_t \begin{pmatrix} \zeta \\ v \end{pmatrix} = -\begin{pmatrix} 0 & (1-\mu\mathbf{b}\partial_x^2)^{-1}\partial_x \\ (1-\mu\mathbf{d}\partial_x^2)^{-1}\partial_x & 0 \end{pmatrix} \begin{pmatrix} \partial_\zeta \mathcal{H} \\ \partial_v \mathcal{H} \end{pmatrix}$$

with

$$\mathcal{H} = \frac{1}{2}\int (-\mu\mathbf{c}\zeta_x^2 - \mu\mathbf{a}v_x^2 + hv^2 + \zeta^2).$$

When $\mathbf{b}=\mathbf{d}$, the operator

$$J = -\begin{pmatrix} 0 & (1-\mu\mathbf{b}\partial_x^2)^{-1}\partial_x \\ (1-\mu\mathbf{d}\partial_x^2)^{-1}\partial_x & 0 \end{pmatrix}$$

is skew-adjoint and the evolutionary system is therefore Hamiltonian.

REMARK 6.26. A consequence of this Hamiltonian structure is that the Hamiltonian $\mathcal{H}$ is a conserved quantity when $\mathbf{b}=\mathbf{d}$. More generally, it is shown in [**35**] that

$$\frac{d}{dt}\mathcal{H} = -2\mu(\mathbf{b}-\mathbf{d})\int \zeta_t v_{xt}.$$

When $\mathbf{b}=\mathbf{d}$, one also has conservation of the impulse $I(\zeta,v)$

$$I(\zeta,v) = \int \zeta v + \mu\mathbf{b}\zeta_x v_x.$$

Finally, let us mention that the Green-Naghdi equations (5.11) also have a Hamiltonian structure. We refer to [**63**] for the general case ($d=1,2$, nonflat bottom). In the one-dimensional case $d=1$ and for flat bottom, this Hamiltonian formulation is simpler [**234**]. The Hamiltonian coincides with the energy $\mathcal{E}_{GN}$, but the canonical variables are not (ζ,v) but (ζ,m), where

$$m = \big(h - \frac{\mu}{3}\partial_x(h^3\partial_x\cdot)\big)u.$$

The Green-Naghdi equations (5.11) can then be written as

$$\partial_t \begin{pmatrix} \zeta \\ m \end{pmatrix} = -\begin{pmatrix} 0 & \partial_x(h\cdot) \\ h\partial_x & \partial_x(m\cdot)+m\partial_x \end{pmatrix} \begin{pmatrix} \partial_\zeta \mathcal{E}_{GN} \\ \partial_m \mathcal{E}_{GN} \end{pmatrix}.$$

CHAPTER 7

Shallow Water Asymptotics: Scalar Equations

The various models derived in Chapter 5 differ depending on the assumptions made on the parameters ε, μ, β and γ, and on the accuracy chosen for the description of the waves (for instance, this accuracy is $O(\mu)$ for the shallow water equations (5.7) and $O(\mu^2)$ for the Green-Naghdi equations (5.11)). However, they all coincide at the lower order of precision, that is, if we neglect all the terms of size $O(\varepsilon, \beta, \mu)$; indeed, one then obtains

$$\begin{cases} \partial_t \zeta + \nabla^\gamma \cdot \overline{V} = 0, \\ \partial_t \overline{V} + \nabla^\gamma \zeta = 0, \end{cases} \tag{7.1}$$

which implies that ζ solves a wave equation of speed 1 (when $\gamma = 1$),

$$\partial_t^2 \zeta - \Delta^\gamma \zeta = 0.$$

In dimension $d = 1$, any initial perturbation of the free surface will split up into two counter propagating waves travelling at speed ± 1. The aim of this section is to describe how the wave departs from this dynamics when the next order terms are included, as well as possible weakly transverse perturbations. *Throughout this chapter, we consider only flat bottoms* (see §7.4.3 for comments on nonflat bottoms).

In §7.1, we first consider long waves in one dimension and over flat bottoms

$$\varepsilon = \mu, \qquad \gamma = \beta = 0$$

and prove that the two counterpropagating components can be correctly described by a set of two uncoupled Korteweg-de Vries (KdV) equations. We emphasize the particular role of secular growth effects on the quality of the approximation. More precisely, we show that the coupling terms between the two counterpropagating components may accumulate over large time scales and damage the quality of the approximation. We show in §7.1.6 that it is possible to get rid of these effects if the two components are sufficiently decaying in space.

We then investigate in §7.2 the weakly transverse case. More precisely, we do not assume anymore that the waves are one dimensional, but we impose that transverse variations are much slower than longitudinal ones

$$\varepsilon = \mu, \qquad \gamma = \sqrt{\varepsilon}, \quad \beta = 0.$$

In this configuration, one still observes at leading order a splitting of the initial condition into two counterpropagating waves. The difference with the one-dimensional case is that transverse effects appear in the equations describing the evolution of these components over larger scales: the KdV equation must be replaced by the Kadomtsev-Petviashvili (KP) equations.

In §7.3, we focus on the particular case where one of the two counterpropagating waves is zero. In particular, coupling effects between both components disappear, and this makes it possible to study more nonlinear regimes. In particular, for medium amplitude waves

$$\varepsilon = O(\sqrt{\mu}), \qquad \gamma = \beta = 0,$$

we derive a family of equations that generalize the KdV equations to this regime. We show that this family of equations can be related to the Camassa-Holm and Degasperis-Procesi equations that are two important examples of bi-Hamiltonian equations. In the particular case where $\varepsilon = O(\mu)$, we show that these equations degenerate into the KdV and BBM (Benjamin-Bona-Mahoney) equations.

Additional material is given in Section 7.4. We first give in §7.4.1 a brief history of the KdV equation. We then give in §7.4.2 a well-posedness theorem for Camassa-Holm type equations. Finally, we review in §7.4.3 some possible extensions to nonflat bottoms and, in §7.4.4, we comment on the ability of these scalar models to describe wave breaking.

We also give in §7.4.5 *full dispersion* versions of the main scalar models derived in this chapter.

7.1. The splitting into unidirectional waves in one dimension

We are interested here in the one-dimensional case ($d = 1$); the two-dimensional, weakly transverse case will be addressed in §7.2 below. In one dimension, (7.1) implies that both ζ and $\overline{V}$ satisfy a linear wave equation (with $X = x$)

$$(\partial_t^2 - \partial_x^2)\zeta = 0, \qquad (\partial_t^2 - \partial_x^2)\overline{V} = 0,$$

and the solutions to (7.1) have to be sought as the sum of two counterpropagating waves

$$\zeta(t,x) = f_+(x-t) + f_-(x+t), \qquad \overline{V}(t,x) = f_+(x-t) - f_-(x+t). \tag{7.2}$$

The aim of this section is to make the description provided by (7.2) more precise in the so-called KdV regime $\varepsilon = \mu$, and $\mu \ll 1$. More precisely, we want to describe the decoupling of any initial disturbance into two counterpropagating waves at the same $O(\mu^2)$ precision as the Boussinesq systems derived in §5.1.3.

7.1.1. The Korteweg-de Vries equation. As previously noted, we are interested here in the propagation of one-dimensional waves in the long-wave regime and over flat bottoms. The KdV approximation is classically derived under the assumption that the nonlinearity and shallowness parameters ε and μ are of the same order[1], and therefore we take $\varepsilon = \mu$ for the sake of clarity. This leads us to define the KdV asymptotic regime as

$$\mathcal{A}_{KdV} = \{(\varepsilon, \beta, \gamma, \mu), \quad 0 \le \mu \le \mu_0, \quad \varepsilon = \mu, \quad \beta = \gamma = 0\}. \tag{7.3}$$

We know from the analysis in Chapter 5 that in this configuration, the Boussinesq systems (5.25) or asymptotically equivalent versions such as (5.34) or (5.40) provide the same $O(\mu^2)$ accuracy as the Green-Naghdi equations (5.11) and are much simpler. It therefore is natural to investigate the derivation of the KdV approximation from one of these simple models, and the best choice is certainly to take the most

[1] As shown in §7.3.2 below, this assumption can be relaxed into $\varepsilon = O(\mu)$, in which case the existence time in Theorem 7.1 is T/ε instead of T/μ.

convenient for mathematical analysis, namely, one of the fully symmetric Boussinesq systems (5.41). In the one-dimensional case, these systems can be written (with $\varepsilon = \mu$)

$$\begin{cases} (1-\mu\mathbf{b}\partial_x^2)\partial_t\zeta + \partial_x((1+\frac{\mu}{2}\zeta)w) + \mu\mathbf{a}\partial_x^3 w = 0, \\ (1-\mu\mathbf{d}\partial_x^2)\partial_t w + (1+\frac{\mu}{2}\zeta)\partial_x\zeta + \mu\mathbf{c}\partial_x^3\zeta + \mu\frac{3}{2}w\partial_x w = 0, \end{cases}$$

where[2] $\mathbf{a} = \mathbf{c}$, $\mathbf{b} \geq 0$, $\mathbf{d} \geq 0$, and

$$\mathbf{a} + \mathbf{b} + \mathbf{c} + \mathbf{d} = \frac{1}{3},$$

and where $w = W_{\theta,\delta}$ is as defined in (5.39). The analysis below works with any of these systems, but for the sake of clarity, we work with a so-called KdV/KdV type system which makes the computations easier

$$\begin{cases} \partial_t\zeta + \partial_x((1+\frac{\mu}{2}\zeta)w) + \mu\frac{1}{6}\partial_x^3 w = 0, \\ \partial_t w + (1+\frac{\mu}{2}\zeta)\partial_x\zeta + \mu\frac{1}{6}\partial_x^3\zeta + \mu\frac{3}{2}w\partial_x w = 0, \end{cases} \tag{7.4}$$

which corresponds to $\mathbf{a} = \mathbf{c} = 1/6$ and $\mathbf{b} = \mathbf{d} = 0$ (or $\theta = \alpha = 0$, $\delta = -\lambda = 1/2$).

We define the KdV approximation (ζ_{KdV}, w_{KdV}) to (7.4) as the following slow modulation of the first-order solution (7.2)

$$\begin{aligned} \zeta_{KdV}(t,x) &= f_+(x-t,\mu t) + f_-(x+t,\mu t), \\ w_{KdV}(t,x) &= f_+(x-t,\mu t) - f_-(x+t,\mu t), \end{aligned} \tag{7.5}$$

where $f_+ = f(\tilde{x},\tau)$ and $f_- = f(\tilde{x},\tau)$ solve the KdV equations

$$\partial_\tau f_+ + \frac{1}{6}\partial_{\tilde{x}}^3 f_+ + \frac{3}{2}f_+\partial_{\tilde{x}} f_+ = 0, \qquad \partial_\tau f_- - \frac{1}{6}\partial_{\tilde{x}}^3 f_- - \frac{3}{2}f_-\partial_{\tilde{x}} f_- = 0. \tag{7.6}$$

The remainder of this section is devoted to the justification of this approximation. The KdV equation was derived by Boussinesq [**49**] and Korteweg and de Vries [**211**] (see §7.4.1 for further comments on this point), but the justification program was initiated much later by Craig [**101**] and Kano-Nishida [**199**], who studied the propagation of waves in the case where the left-going wave f_- is identically zero and under some more or less restrictive assumptions (in particular, the time scale considered in [**199**] would correspond to $O(1)$ rather than $O(1/\mu)$ times with our notations). Schneider and Wayne [**291**] then handled the general case; the convergence rate of the approximation was then improved in [**36**].

7.1.2. Statement of the main result. We state here the main result that justifies the KdV approximation (7.5)–(7.6) in the following sense (in the statement below, μ_0 is as in (7.3)).

THEOREM 7.1. *Let* $\zeta^0, w^0 \in H^8(\mathbb{R}^d)$. *There exists* $T > 0$ *such that for all* $0 < \mu < \mu_0$,

(1) *There exists a unique solution* $(\zeta_B, w_B) \in C([0, \frac{T}{\mu}]; H^8(\mathbb{R}))$ *to (7.4) with initial condition* (ζ^0, w^0).

[2]We have seen in §5.3 that there exists admissible choices of the parameters α, λ, δ, and θ leading to such values of $\mathbf{a}$, $\mathbf{b}$, $\mathbf{c}$ and $\mathbf{d}$.

(2) *The KdV approximation (7.5)–(7.6) is well defined in* $C([0,T/\mu]\times\mathbb{R})$ *and for all* $0\le t\le \frac{T}{\mu}$, *one has*

$$|(\zeta_B,w_B)-(\zeta_{KdV},w_{KdV})|_{L^\infty([0,t]\times\mathbb{R})}\le \mu\big(\frac{t}{1+\sqrt{t}}+\mu t\big)C(T,\mu_0,|(\zeta^0,w^0)|_{H^8}),$$

where the initial conditions for $f_\pm$ *in (7.6) are given by* $f_\pm^0=\frac{1}{2}(\zeta^0\pm w^0)$.

Since the solutions to the Boussinesq system (7.4) are known to provide good approximations to the water waves equations (see Theorem 6.20), we can deduce from Theorem 7.1 that this is also the case for the KdV approximation. In the statement below $\mathcal{E}^N(U^0)$ (see (4.8) for the definition) is the natural energy of order N for the data of the water waves equations. In particular, the initial condition $\mathcal{E}^N(U^0)<\infty$ can be replaced[3] by the less sharp but more intuitive condition

$$(\zeta^0,\partial_x\psi^0)\in H^{N+1}(\mathbb{R})\times H^N(\mathbb{R}).$$

COROLLARY 7.2. *Let* $N>14$ *and* $\mathfrak{p}=(\varepsilon,\beta,\gamma,\mu)\in\mathcal{A}_{KdV}$. *Also let* $U^0=(\zeta^0,\psi^0)$ *be such that* $\mathcal{E}^N(U^0)<\infty$ *and satisfies (4.29).*
There exists $T=T_{KdV}>0$ *that does not depend on* $\mathfrak{p}\in\mathcal{A}_{KdV}$ *and such that*
(1) *There exists a unique solution* $U=(\zeta,\psi)\in E^N_{T/\mu}$ *to the water waves equations (5.1) with initial data* U^0.
(2) *With* $w^0=\partial_x\psi^0$ *and* $f_\pm$ *solving (7.6) with initial condition* $f_\pm^0=\frac{1}{2}(\zeta^0\pm w^0)$, *the KdV approximation (7.5) satisfies, for all* $0\le t\le\frac{T}{\mu}$

$$|(\zeta,\partial_x\psi)-(\zeta_{KdV},w_{KdV})|_{L^\infty([0,t]\times\mathbb{R})}\le\mu(1+\sqrt{t})C\big(T,\mathcal{E}^N(U^0),\mu_0,\frac{1}{h_{min}},\frac{1}{a_0}\big),$$

where $\mathcal{E}^N(U^0)$ *is as in (4.8).*

REMARK 7.3. The precision of the KdV approximation is $O(\sqrt{\mu})$ for time scales of order $O(1/\mu)$, which is worse than the precision $O(\mu)$ established in Theorem 6.20 for the solution of the Boussinesq system (7.4) on the same time scale. A qualitative explanation for this difference is that the KdV approximation misses the coupling effects between the two counterpropagating components of the wave; technically, this coupling induces *secular growth phenomena* that are discussed below. In some configurations, these coupling effects remain small and the precision of the KdV approximation can be improved into $O(\mu)$. This is, for instance, the case if the two counterpropagating components decay sufficiently in space (see Corollary 7.12 below). Another example occurs when one of these two components vanishes (e.g., if $\zeta^0=w^0$, then the left-going component remains zero). This configuration will be thoroughly investigated in §7.3.

REMARK 7.4. As stated above, the KdV approximation (7.5)–(7.6) neglects the coupling effects between the counterpropagating components of the wave. These coupling effects are relatively small, because the distance between these two components grows larger and larger. This would not be the case if we were in a periodic setting. It therefore is not surprising that the uncoupled KdV approximation (7.5)–(7.6) is *false* in such a configuration (see [**36**]).

---

[3]This is a consequence of (4.8), (4.9), and Proposition 4.4.

REMARK 7.5. We can write $\zeta_{KdV} = \zeta^+_{KdV} + \zeta^-_{KdV}$, with $\zeta^\pm_{KdV} = f_\pm(x \mp t, \mu t)$. In particular, $\zeta^\pm_{KdV}$ solve

$$\partial_t\zeta^+_{KdV} + \partial_x\zeta^+_{KdV} + \mu(\frac{1}{6}\partial_x^3\zeta^+_{KdV} + \frac{3}{2}\zeta^+_{KdV}\partial_x\zeta^+_{KdV}) = 0,$$
$$\partial_t\zeta^-_{KdV} - \partial_x\zeta^-_{KdV} - \mu(\frac{1}{6}\partial_x^3\zeta^-_{KdV} + \frac{3}{2}\zeta^-_{KdV}\partial_x\zeta^+_{KdV}) = 0.$$

We now introduce $\zeta^\pm_{BBM}$ as solutions to the BBM (after Benjamin-Bona-Mahoney [**23**]) equations

$$(1 - \mu\frac{1}{6}\partial_x^2)\partial_t\zeta^+_{BBM} + \partial_x\zeta^+_{BBM} + \mu\frac{3}{2}\zeta^+_{BBM}\partial_x\zeta^+_{BBM} = 0,$$
$$(1 - \mu\frac{1}{6}\partial_x^2)\partial_t\zeta^-_{BBM} - \partial_x\zeta^-_{KdV} - \mu\frac{3}{2}\zeta^-_{BBM}\partial_x\zeta^+_{BBM} = 0.$$

In particular, one has $\partial_t\zeta^\pm_{BBM} = \mp\partial_x\zeta^\pm_{BBM} + O(\mu)$, so that $\zeta^\pm_{BBM}$ solves the same KdV equations as $\zeta^\pm_{KdV}$ up to a $O(\mu^2)$ residual. It follows that one can replace ζ_{KdV} in Theorem 7.1 and Corollary 7.2 by $\zeta_{BBM} = \zeta^+_{BBM} + \zeta^-_{BBM}$ (and similar substitutions for w). The asymptotic equivalence of the KdV and BBM approximations is the analogue for scalar equations of the equivalence of the various Boussinesq systems derived in §5.2.1. More generally, the right-going component (for instance) can be described with the same accuracy by any of the following evolution equations

$$\partial_t\zeta + \partial_x\zeta + \mu(p\partial_x^3\zeta + (p - \frac{1}{6})\partial^3_{xxt}\zeta) + \frac{3}{2}\varepsilon\zeta\partial_x\zeta = 0 \qquad (p \leq \frac{1}{6}). \tag{7.7}$$

See §7.3.2 for further comments on this point.

7.1.3. BKW expansion. We seek an approximate solution (ζ_{app}, w_{app}) to (7.4) under the form

$$\begin{aligned} \zeta_{app}(t,x) &= \zeta_{(0)}(t,x,\mu t) + \mu\zeta_{(1)}(t,x,\mu t), \\ w_{app}(t,x) &= w_{(0)}(t,x,\mu t) + \mu w_{(1)}(t,x,\mu t), \end{aligned} \tag{7.8}$$

with

$$\begin{aligned} \zeta_{(0)}(t,x,\mu t) = f_+(x-t,\mu t) + f_-(x+t,\mu t), \\ w_{(0)}(t,x,\mu t) = f_+(x-t,\mu t) - f_-(x+t,\mu t), \end{aligned} \tag{7.9}$$

so that

$$\zeta_{KdV}(t,x) = \zeta_{(0)}(t,x,\mu t), \qquad w_{KdV}(t,x) = w_{(0)}(t,x,\mu t),$$

and where the corrector term $(\zeta_{(1)}, w_{(1)})$ is determined below.
The first step consists in plugging the ansatz (7.8) into (7.4) and ordering the residual into powers of μ. More precisely, one finds

$$\left\{ \begin{array}{l} \partial_t\zeta_{app} + \partial_x((1 + \frac{\mu}{2}\zeta_{app})w_{app}) + \mu\frac{1}{6}\partial_x^3 w_{app} = \mu R_1^{(1)} + \mu^2 R_1^\mu, \\ \partial_t w_{app} + (1 + \frac{\mu}{2}\zeta_{app})\partial_x\zeta_{app} + \frac{\mu}{6}\partial_x^3\zeta_{app} + \mu\frac{3}{4}\partial_x w_{app}^2 = \mu R_2^{(1)} + \mu^2 R_2^\mu, \end{array}\right. \tag{7.10}$$

(note that $O(1)$ terms vanish because at leading order (7.5) coincides with (7.2)), with $R_j^{(1)} = R_j^{(1)}(t,x,\mu t)$ given by

$$\begin{aligned} R_1^{(1)} &= (\partial_t\zeta_{(1)} + \partial_x w_{(1)}) + \partial_\tau\zeta_{(0)} + \frac{1}{6}\partial_x^3 w_{(0)} + \frac{1}{2}\partial_x(\zeta_{(0)}w_{(0)}) \\ R_2^{(1)} &= (\partial_t w_{(1)} + \partial_x\zeta_{(1)}) + \partial_\tau w_{(0)} + \frac{1}{6}\partial_x^3\zeta_{(0)} + \frac{1}{4}\partial_x(3w_{(0)}^2 + \zeta_{(0)}^2) \end{aligned}$$

and $R_j^\mu = R_j^\mu(t, x, \mu t)$ given by

$$
(7.11)\qquad \begin{aligned}
R_1^\mu &= \partial_\tau \zeta_{(1)} + \frac{1}{6}\partial_x^3 w_{(1)} + \frac{1}{2}\partial_x(\zeta_{(1)}w_{(0)} + \zeta_{(0)}w_{(1)}) + \frac{\mu}{2}\partial_x(w_{(1)}\zeta_{(1)})\\
R_2^\mu &= \partial_\tau w_{(1)} + \frac{1}{6}\partial_x^3 \zeta_{(1)} + \frac{1}{2}\partial_x(3w_{(0)}w_{(1)} + \zeta_{(0)}\zeta_{(1)}) + \frac{\mu}{4}\partial_x(w_{(1)}^2 + \zeta_{(1)}^2).
\end{aligned}
$$

The strategy is to choose the corrector $(\zeta_{(1)}, w_{(1)})$ so that $R_1^{(1)}(t,x,\mu t) = R_2^{(1)}(t,x,\mu t)$ $= 0$ for all $t \in [0, \frac{T}{\mu}]$ (for some $T > 0$) and $x \in \mathbb{R}$. A sufficient condition is to impose that $R_1^{(1)}(t,x,\tau) = R_2^{(1)}(t,x,\tau) = 0$ for all $t \in [0,\frac{T}{\mu}]$, $x \in \mathbb{R}$ and $\tau \in [0,T]$. This leads to the system of equations

$$
(7.12)\qquad \left\{ \begin{aligned}
\partial_t \zeta_{(1)} + \partial_x w_{(1)} &= F_+(x-t,\tau) + F_-(x+t,\tau),\\
\partial_t w_{(1)} + \partial_x \zeta_{(1)} &= G_+(x-t,\tau) + G_-(x+t,\tau) + G_0(x,t,\tau),
\end{aligned}\right.
$$

with

$$
(7.13)\qquad \begin{aligned}
F_+(\tilde{x},\tau) &= \partial_\tau f_+ + \tfrac{1}{6}\partial_{\tilde{x}}^3 f_+ + f_+\partial_{\tilde{x}} f_+,\\
F_-(\tilde{x},\tau) &= \partial_\tau f_- - \tfrac{1}{6}\partial_{\tilde{x}}^3 f_- - f_-\partial_{\tilde{x}} f_-,
\end{aligned}
$$

and

$$
(7.14)\qquad \begin{aligned}
G_+(\tilde{x},\tau) &= \partial_\tau f_+ + \tfrac{1}{6}\partial_{\tilde{x}}^3 f_+ + 2f_+\partial_{\tilde{x}} f_+,\\
G_-(\tilde{x},\tau) &= -\partial_\tau f_- + \tfrac{1}{6}\partial_{\tilde{x}}^3 f_- + 2f_-\partial_{\tilde{x}} f_-,\\
G_0(x,t,\tau) &= -\partial_x(f_+(x-t,\tau)f_-(x+t,\tau)).
\end{aligned}
$$

7.1.4. Consistency of the approximate solution and secular growth. It seems at this point that the problem is underdetermined. Indeed, for any choice of f_+ and f_- (i.e., if we do not impose that they solve (7.6)) it is possible to find $(\zeta_{(1)}, w_{(1)})$ by solving (7.12). With such a choice, the residuals $R_1^{(1)}$ and $R_2^{(1)}$ vanish in (7.10) and the approximate solution (7.8) is consistent at order $O(\mu^2)$ with the Boussinesq system (7.4). An error estimate similar to those performed for the proof of Theorem 6.20 then shows that the approximation (7.8) and hence the KdV approximation (7.5) remain in an $O(\mu)$ neighborhood of the exact solution to (7.4) for times of order $O(1/\mu)$.

The above reasoning is flawed because it does not take into account *secular growth* effects. Indeed, the solution $(\zeta_{(1)}, w_{(1)})$ to (7.12) grows in general linearly in t and therefore becomes of size $O(1/\mu)$ for times of order $O(1/\mu)$. It follows, for instance, that for such times, the quantities $\mu^2 R_j^\mu$ are of size $O(\mu)$ and not $O(\mu^2)$. The remainder of this section is devoted to the proof of the fact that there is only *one* possibility for $(\zeta_{(1)}, w_{(1)})$ to grow sublinearly in t: that f_+ and f_- solve the KdV equations (7.6).

In order to show this, let us first introduce the quantities u and v defined as

$$
(7.15)\qquad u = \zeta_{(1)} + w_{(1)}, \qquad v = \zeta_{(1)} - w_{(1)},
$$

and note that (7.12) is equivalent to

$$
(7.16)\qquad \left\{ \begin{aligned}
(\partial_t + \partial_x)u &= (F_+ + G_+)(x-t,\tau) + (F_- + G_-)(x+t,\tau) + G_0(x,t,\tau),\\
(\partial_t - \partial_x)v &= (F_+ - G_+)(x-t,\tau) + (F_- - G_-)(x+t,\tau) - G_0(x,t,\tau).
\end{aligned}\right.
$$

The following key lemma shows that the only possibility to avoid linear growth of u or v is to impose that $F_+ + G_+ = F_- - G_- = 0$.

LEMMA 7.6. *Let $c_1 \neq c_2$, $t_0 > 1/2$ and $s \in \mathbb{N}$. Let $h_1, h_{3,1}, h_{3,2} \in H^s(\mathbb{R})$, and also let $h_2 = H_2'$ with $H_2 \in H^{s+1}(\mathbb{R})$. Then the unique solution g vanishing at $t = 0$ to*

$$(\partial_t + c_1\partial_x)g = h_1(x - c_1t) + h_2(x - c_2t) + h_{3,1}(x - c_1t)h_{3,2}(x - c_2t)$$

is given by

$$\begin{aligned} g(t,x) &= th_1(x - c_1t) + \frac{1}{c_1 - c_2}\big(H_2(x - c_2t) - H_2(x - c_1t)\big) \\ &\quad + h_{3,1}(x - c_1t)\int_0^t h_{3,2}(x + (c_1 - c_2)s - c_1t)ds. \end{aligned}$$

In particular, $\lim_{t\to\infty} \frac{1}{t}|g(t,\cdot)|_2 = 0$ if and only if $h_1 = 0$, and then one has

$$\begin{aligned} |g(t,\cdot)|_{H^s} &\leq \min\big\{\frac{2}{|c_1 - c_2|}|H_2|_{H^s}, t|h_2|_{H^s}\big\} \\ &\quad + \mathrm{Cst}\min\{\frac{\sqrt{t}}{\sqrt{|c_1 - c_2|}}|h_{3,1}|_{H^s}|h_{3,2}|_{H^s}, t|h_{3,1}|_{L^\infty\cap H^s}|h_{3,2}|_{L^\infty\cap H^s}\}. \end{aligned}$$

REMARK 7.7. The contribution of the nonlinear term $h_{3,1}(x - c_1t)h_{3,2}(x - c_2t)$ is in fact $o(\sqrt{t})$, which is slightly better than the $O(\sqrt{t})$ given by the lemma (see [**26**]).

PROOF. The explicit formula for g can be checked by direct computation. The second part of the lemma follows from a control on the three components of g. The fact that the component corresponding to the contribution of h_2 is bounded from above by $\frac{2}{|c_1 - c_2|}|H_2|_{H^s}$ follows directly from its explicit expression. The fact that it is also bounded by $t|h_2|_{H^s}$ (which is better than the previous estimate for small values of t only) follows from a direct estimate.
For the contribution corresponding to $h_{3,1}(x - c_1t)h_{3,2}(x - c_2t)$, we use Cauchy-Schwarz to get

$$\Big|\int_0^t h_{3,2}(x + (c_1 - c_2)t' - c_1t)dt'\Big|_{L^\infty_x} \leq \frac{\sqrt{t}}{\sqrt{|c_1 - c_2|}}|h_{3,2}|_2$$

(and straightforward adaptations for higher order derivatives); this gives the $O(\sqrt{t})$ upper bound of the lemma. The $O(t)$ bound also follows from a direct estimate. ☐

As mentioned above, the lemma shows that we have to impose $F_+ + G_+ = F_- - G_- = 0$ to avoid linear secular growth in (7.16). From the definition (7.13)–(7.14) of $F_\pm$ and $G_\pm$, these conditions can be written as

$$\begin{aligned} 2\partial_\tau f_+ + \frac{1}{3}\partial_{\tilde{x}}^3 f_+ + 3f_+\partial_{\tilde{x}} f_+ &= 0, \\ 2\partial_\tau f_- - \frac{1}{3}\partial_{\tilde{x}}^3 f_- - 3f_-\partial_{\tilde{x}} f_- &= 0; \end{aligned}$$

these equations are exactly the KdV equations (7.6). Using the KdV equation to express $\partial_\tau f_\pm$ in terms of spacial derivatives of $f_\pm$, we can note that $F_\pm$ and $G_\pm$ can be written under the form $F_\pm = \partial_{\tilde{x}}\widetilde{F}_\pm$ and $G_\pm = \partial_{\tilde{x}}\widetilde{G}_\pm$, with $\widetilde{F}_\pm, \widetilde{G}_\pm \in H^s(\mathbb{R})$. Lemma 7.6 then provides us with estimates on u and v or equivalently, through

(7.15), on $\zeta_{(1)}$ and $w_{(1)}$,

$$\begin{aligned}|\zeta_{(1)}(t,\cdot,\tau)|_{H^s}+|w_{(1)}(t,\cdot,\tau)|_{H^s} &\lesssim \frac{t}{1+t}|\widetilde{F}_\pm(\cdot,\tau)|_{H^{s+1}}+|\widetilde{G}_\pm(\cdot,\tau)|_{H^{s+1}}\\ &+ \frac{t}{1+\sqrt{t}}|f_+(\cdot,\tau)|_{H^{s+1}}|f_-(\cdot,\tau)|_{H^{s+1}},\end{aligned}$$

where we used the definition (7.14) of G_0 to get the last contribution in the above estimate. From the definition of $\widetilde{F}_\pm$ and $\widetilde{G}_\pm$ we then get

$$|\zeta_{(1)}(t,\cdot,\tau)|_{H^s}+|w_{(1)}(t,\cdot,\tau)|_{H^s} \le \frac{t}{1+\sqrt{t}}C\big(|f_\pm(\cdot,\tau)|_{H^{s+3}}\big). \tag{7.17}$$

We now need the following classical result on the KdV equation[4]. Note that the lemma is stated for the right-going KdV equation (i.e., the equation for f_+) but that the same result of course holds for the left-going equation.

LEMMA 7.8 (see, for instance, [**204**]). *Let $s\ge 1$ and $T>0$. Then for all $f^0\in H^s(\mathbb{R})$, there exists a unique solution $f\in C([0,T];H^s(\mathbb{R}))$ to the initial value problem*

$$\partial_\tau f+\frac{1}{6}\partial_{\tilde{x}}^3 f+\frac{3}{2}f\partial_{\tilde{x}}f=0,\qquad f_{|\tau=0}=f^0,$$

and one has $|f|_{L^\infty([0,T];H^s)}\le C(T,|f^0|_{H^s})$.

Recalling that the initial data for f_+ and f_- are

$$f_{+\ |\tau=0}=\frac{\zeta^0+w^0}{2},\qquad f_{-\ |\tau=0}=\frac{\zeta^0-w^0}{2},$$

we get from (7.17) and Lemma 7.8 that

$$|\zeta_{(1)}(t,\cdot,\tau)|_{H^s}+|w_{(1)}(t,\cdot,\tau)|_{H^s}\le\frac{t}{1+\sqrt{t}}C(\tau,|\zeta^0|_{H^{s+3}},|w^0|_{H^{s+3}}). \tag{7.18}$$

We are now ready to prove that the approximate solution (7.8) is consistent with (7.4) at order $O(\mu^2\sqrt{t})$; the secular growth in time therefore is of rate $\sqrt{t}$, which is acceptable since $O(\mu^2\sqrt{t})$ terms remain correctors of the $O(\mu)$ terms over time scales $O(1/\mu)$.

LEMMA 7.9. *Let $T>0$ and $s\ge 0$, and $\zeta^0,w^0\in H^{s+5}(\mathbb{R}^d)$. Then the approximate solution (ζ_{app},w_{app}) provided by (7.8), (7.9), and (7.12) solves (7.10) with residuals satisfying for all $0\le t\le T/\mu$*

$$\begin{aligned}&R_1^{(1)}=R_2^{(1)}=0,\\ &\mu^2|R_1^\mu(t,\cdot,\mu t)|_{H^s}+\mu^2|R_2^\mu(t,\cdot,\mu t)|_{H^s}\le\mu^2\big(\frac{t}{1+\sqrt{t}}+\mu t\big)C(T,|(\zeta^0,w^0)|_{H^{s+6}}).\end{aligned}$$

PROOF. We have already shown that $R_j^{(1)}=0$ $(j=1,2)$ in (7.10). We therefore are left to control R_j^μ. One has, for all $0\le t\le T/\mu$, and with $U_{(j)}=(\zeta_{(j)},w_{(j)})$ $(j=1,2)$,

$$|R_j^\mu(t,\cdot,\mu t)|_{H^s}\le\sup_{0\le t\le T/\mu}\ \sup_{0\le\tau\le T}|R_j^\mu(t,\cdot,\tau)|_{H^s}$$

[4] We need only a local existence result for regular data; such a result, first proved by Kato [**201**], can be established by a straightforward adaptation of the local well-posedness of hyperbolic systems as for Proposition 6.7. Similarly, global well-posedness is known under weaker assumptions ($s>-3/2$ in [**84**]), but with the assumption $s\ge 1$ we do not need to define weak solutions.

and therefore, using (7.11)

$$|R_j^\mu(t,\cdot,\mu t)|_{H^s} \le C\big(|U_{(0)}(t,\cdot)|_{H_T^{s+1}}\big) \times \big(1+|\partial_\tau U_{(1)}(t,\cdot)|_{H_T^{s+1}} + |U_{(1)}(t,\cdot)|_{H_T^{s+3}} + \mu |U_{(1)}(t,\cdot)|^2_{H_T^{s+1}}\big),$$

where $|f(t,\cdot)|_{H_T^s}$ stands for $|f(t,\cdot)|_{H_T^s} = \sup_{0\le\tau\le T}|f(t,\cdot,\tau)|_{H^s}$. To control the terms involving $U_{(1)}$, it is equivalent, thanks to (7.15), to give estimates on u and v solving (7.16). Estimates on $\partial_\tau U_{(1)}$ are obtained in the same way after differentiating (7.16) with respect to τ. All these estimates are a consequence of Lemma 7.6, so that one gets

$$\begin{aligned}|R_j^\mu(t,\cdot,\mu t)|_{H^s} &\le C\big(|U_{(0)}|_{H_T^{s+6}}\big)\big(\frac{t}{1+\sqrt{t}}+\mu t\big),\\ &\le C\big(\sup_{0\le\tau\le T}|f_\pm(\cdot,\tau)|_{H^{s+6}}\big)\big(\frac{t}{1+\sqrt{t}}+\mu t\big),\end{aligned}$$

where we wrote $U_{(0)}$ in terms of f_+ and f_- through (7.9) and used (7.6) to get the last inequality. A control in terms of ζ^0 and w^0 then follows from Lemma 7.8. □

7.1.5. Proof of Theorem 7.1 and Corollary 7.2. From Lemma 7.9 and Proposition 6.8, we deduce directly that for all $t\in[0,\frac{T}{\mu}]$

$$|(\zeta_B,w_B)-(\zeta_{app},w_{app})|_{L^\infty([0,t];H^{s-1})} \le \mu^2 C\big(T,\frac{1}{h_{min}},|(\zeta^0,w^0)|_{H^{s+6}}\big)\big(\mu t^2+\frac{t^2}{1+\sqrt{t}}\big).$$

Moreover, since we get from (7.8) and (7.18) that

$$\begin{aligned}|(\zeta_{app},w_{app})-(\zeta_{KdV},w_{KdV})|_{L^\infty([0,t];H^{s-1})} &\le \mu|(\zeta_{(1)},w_{(1)}))|_{L^\infty([0,t]\times[0,\mu t]_\tau;H^{s-1})}\\ &\le \mu\frac{t}{1+\sqrt{t}}C(\mu t,|\zeta^0|_{H^{s+2}},|w^0|_{H^{s+2}}),\end{aligned}$$

the result stated in Theorem 7.1 follows from a triangular inequality and the Sobolev embedding $H^1(\mathbb{R})\subset L^\infty(\mathbb{R})$.

In order to prove Corollary 7.2, let us first define $\overline{V}^0$ from $U^0=(\zeta^0,\psi^0)$ through (3.31), and define w_B^0 by

$$w_B^0 = \sqrt{1+\mu\zeta^0}(1-\mu\frac{\theta}{3}\partial_x^2)^{-1}(1-\mu\frac{\delta}{3}\partial_x^2)\overline{V}^0,$$

where $(\alpha,\lambda,\theta,\delta)$ is the set of parameters from which **a**, **b**, **c** and **d** are computed (see §5.3). We can then invoke Theorem 6.20 to assert that

(1) There exists $T>0$ and a unique solution $U=(\zeta,\psi)\in E^N_{T/\mu}$ to the water waves equations (5.1) with initial data U^0 to which one can associate an averaged velocity $\overline{V}$ through (3.31).
(2) There exists a unique solution $(\widetilde{\zeta}_B,\widetilde{w}_B)\in C([0,T/\mu];H^{N-3})$ to (7.4) with initial condition (ζ^0,w_B^0)
(3) The following error estimates hold for all $0\le t\le\frac{T}{\mu}$,

$$|(\zeta,\overline{V})-(\widetilde{\zeta}_B,\overline{V}_B)|_{L^\infty([0,t]\times\mathbb{R})} \le \mu^2 t C\big(T,\mathcal{E}^N(U^0),\mu_0,\frac{1}{h_{min}},\frac{1}{a_0}\big), \tag{7.19}$$

where $\overline{V}_B$ is given by

$$\overline{V}_B = (1-\mu\frac{\delta}{3}\partial_x^2)^{-1}(1-\mu\frac{\theta}{3}\partial_x^2)\big(\frac{1}{\sqrt{1+\mu\widetilde{\zeta}_B}}\widetilde{w}_B\big). \tag{7.20}$$

We also know (see Remark 5.21) that

$$|\overline{V}-\partial_x\psi|_{L^\infty([0,t]\times\mathbb{R})}\leq \mu C\big(T,\mathcal{E}^N(U^0),\mu_0,\frac{1}{h_{min}},\frac{1}{a_0}\big) \tag{7.21}$$

and, after observing that the same upper bound holds on $|\overline{V}_B-\widetilde{w}_B|_{L^\infty([0,t]\times\mathbb{R})}$ (this is a direct consequence of (7.20)) and on $|(\widetilde{\zeta}_B,\widetilde{w}_B)-(\zeta_B,w_B)|_{L^\infty([0,t]\times\mathbb{R})}$ (this is a consequence of Proposition 6.8), we also get

$$|(\widetilde{\zeta}_B,\overline{V}_B)-(\zeta_B,w_B)|_{L^\infty([0,t]\times\mathbb{R})}\leq \mu C\big(T,\mathcal{E}^N(U^0),\mu_0,\frac{1}{h_{min}},\frac{1}{a_0}\big). \tag{7.22}$$

The estimate stated in Corollary 7.2 is then a direct consequence of (7.19), (7.21), (7.22), and Theorem 7.1.

7.1.6. An improvement. There exist various techniques to limit the influence of secular growth. See [**216**] for a detailed investigation of this phenomenon. As an example, we show here that under decay assumptions on the initial conditions we can avoid secular growth in the residual terms and therefore replace the $O(\mu^2(1+\sqrt{t}))$ consistency result of Lemma 7.9 into $O(\mu^2)$. The key point is the following lemma, which is a refinement of Lemma 7.6 by Schneider and Wayne [**288, 291**]. (See also Proposition 3.5 of [**216**] for a slight generalization.)

LEMMA 7.10. *Let $c_1\neq c_2$, $t_0>1/2$. Also let $\alpha>1/2$ and $h_{3,1},h_{3,2}\in L^2(\mathbb{R})$ be such that $\underline{h}_{3,1}=(1+x^2)^\alpha h_{3,1}$ and $\underline{h}_{3,2}=(1+x^2)^\alpha h_{3,2}$ belong to $L^2(\mathbb{R})$. Then the unique solution g vanishing at $t=0$ to*

$$(\partial_t+c_1\partial_x)g=h_{3,1}(x-c_1t)h_{3,2}(x-c_2t)$$

satisfies

$$|g(t,\cdot)|_2\lesssim\frac{1}{|c_1-c_2|}|\underline{h}_{3,1}|_2|\underline{h}_{3,2}|_2.$$

PROOF. From the explicit expression for g given in Lemma 7.6, we get

$$g(t,x)=\underline{h}_{3,1}(x-c_1t)\int_0^t k(x,t,s)\underline{h}_{3,2}(x+(c_1-c_2)s-c_1t)ds$$

with $k(x,t,s)=\langle x-c_1t\rangle^{-\alpha}\langle x+(c_1-c_2)s-c_1t\rangle^{-\alpha}$. An elementary analysis of the function k shows that

$$k(x,t,s)\leq\big[\frac{3}{4}(1+(c_1-c_2)^2s^2)\big]^{-\alpha/2},$$

and therefore one can deduce from the Cauchy-Schwarz inequality that

$$\begin{aligned} g(t,x) &\leq \underline{h}_{3,1}(x-c_1t)\Big(\int_0^t\big[\frac{3}{4}(1+(c_1-c_2)^2s^2)\big]^{-\alpha}ds\Big)^{1/2}\frac{1}{\sqrt{|c_1-c_2|}}|\underline{h}_{3,2}|_2\\ &\lesssim \frac{1}{|c_1-c_2|}\underline{h}_{3,1}(x-c_1t)|\underline{h}_{3,2}|_2, \end{aligned}$$

where we used the fact that $\int_0^\infty(1+s^2)^{-\alpha}<\infty$ when $2\alpha>1$. The estimate of the lemma follows directly. □

Thanks to this lemma, it is possible to avoid the secular growth contribution to the estimate of Theorem 7.1 (or Corollary 7.2). The structure of the proof is exactly the same as for the general case; one just has to complement Lemma 7.6 with Lemma 7.10 and to prove that the KdV equations propagates the spatial decay of the initial data. In order to measure this decay, let us introduce the following weighted spaces, where $s \geq 0$, $k \in \mathbb{N}$, and

$$H^s_{(2k)}(\mathbb{R}) = \{f \in H^{s+2k}(\mathbb{R}), \quad \forall j = 0, \dots k, \quad x^j f \in H^{s+2(k-j)}(\mathbb{R})\} \tag{7.23}$$

endowed with its canonical norm $|f|^2_{H^s_{(k)}} = \sum_{j=0}^k |x^j f|^2_{H^{s+2(k-j)}}$. The following lemma is a slight generalization of the decay estimates that can be found in [**291**].

LEMMA 7.11. *Let $s \geq 1$, $k \in \mathbb{N}$ and $T > 0$. Then for all $f^0 \in H^s_{(2k)} \cap H^{s+2k+1}(\mathbb{R})$ there exists a unique solution $f \in C([0,T]; H^s_{(2k)} \cap H^{s+2k+1}(\mathbb{R}))$ to the initial value problem*

$$\partial_\tau f + \frac{1}{6}\partial^3_{\tilde{x}} f + \frac{3}{2} f \partial_{\tilde{x}} f = 0, \qquad f_{|_{\tau=0}} = f^0,$$

and one has $|f|_{L^\infty([0,T]; H^s_{(2k)} \cap H^{s+2k+1}} \leq C(T, |f^0|_{H^s_{(2k)} \cap H^{s+2k+1}})$.

PROOF. We already know from Lemma 7.8 that there is a unique solution $f \in C([0,T]; H^{s+2k+1})$. Therefore, all that is left to prove is the $H^s_{(2k)}$ estimate. By Duhamel's formula, we can write

$$f(\tau, \cdot) = S(\tau) f^0 + \frac{3}{2}\int_0^\tau S(\tau - \tau') f \partial_{\tilde{x}} f(\tau') d\tau',$$

where $S(\tau) = \exp(-\frac{\tau}{6}\partial^3_{\tilde{x}})$. It follows that

$$|f(\tau, \cdot)|_{H^s_{(2k)}} \leq |S(t) f^0|_{H^s_{(2k)}} + \frac{3}{2}\int_0^\tau |S(\tau - \tau') f \partial_{\tilde{x}} f(\tau')|_{H^s_{(2k)}} d\tau' \tag{7.24}$$

and we now show how to control the two components of the right-hand-side.
- *Control of $|S(\tau) f^0|_{H^s_{(2k)}}$.* By Plancherel's formula, one has

$$\begin{aligned} |S(\tau) f^0|^2_{H^s_{(2k)}} &= \sum_{j=0}^k \int \langle \xi \rangle^{2(s+2k-2j)} \big|\partial^j_\xi (e^{i\frac{\tau}{6}\xi^3} \widehat{f^0}(\xi))\big|^2 d\xi \\ &\leq C(\tau) \sum_{j=0}^k \int \langle \xi \rangle^{2(s+2k-2j)} \sum_{0 \leq l \leq j} \langle \xi \rangle^{4(j-l)} \big|\partial^l_\xi \widehat{f^0}(\xi)\big|^2 d\xi \\ &\leq C(\tau) |f^0|_{H^s_{(2k)}}. \end{aligned}$$

- *Control of $\int_0^\tau |S(\tau - \tau') f \partial_{\tilde{x}} f(\tau')|_{H^s_{(2k)}} d\tau'$.* Proceeding as above, one gets

$$\int_0^\tau |S(\tau - \tau') f \partial_{\tilde{x}} f(\tau')|_{H^s_{(2k)}} d\tau' \leq C(\tau) \int_0^\tau |f(\tau', \cdot)|_{H^s_{(2k)}} |f(\tau', \cdot)|_{H^{s+2k+1}} d\tau',$$

and since we know from Lemma 7.8 that $|f(\tau', \cdot)|_{H^{s+2k+1}} \leq C(\tau, |f^0|_{H^{s+2k+1}})$, we get

$$\int_0^\tau |S(\tau - \tau') f \partial_{\tilde{x}} f(\tau')|_{H^s_{(2k)}} d\tau' \leq C(\tau, |f^0|_{H^{s+2k+1}}) \int_0^\tau |f(\tau', \cdot)|_{H^s_{(2k)}} d\tau'.$$

The result therefore follows from (7.24) and a Gronwall argument. □

We can now discuss the following improvement of Theorem 7.1, which states that the error estimate can be improved from $O(\sqrt{\mu})$ to $O(\mu)$ (for time scales $O(1/\mu)$) if the initial conditions have a strong enough spatial decay.

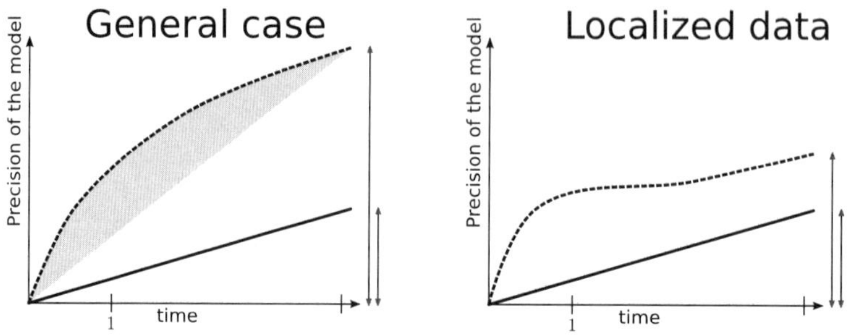

FIGURE 7.1. Precision of the Boussinesq system (7.4) and of the KdV approximation (solid and dashed curve respectively). Left is the general result; right is the improvement of §7.1.6.

COROLLARY 7.12. *Under the assumptions of Theorem 7.1 and under the additional decay assumption that $\zeta^0, w^0 \in H^9_{(2)}(\mathbb{R}^d)$, the error estimate of Theorem 7.1 can be improved into*

$$|(\zeta_B, w_B) - (\zeta_{KdV}, w_{KdV})|_{L^\infty([0,t]\times\mathbb{R})} \leq \mu C(T, \mu_0, |(\zeta^0, w^0)|_{H^9_{(2)}}),$$

for all $t \in [0, T/\mu]$.

REMARK 7.13. It follows of course that the $O(\mu(1+\sqrt{t}))$ error estimate of Corollary 7.2 is improved into a $O(\mu)$ estimate.

REMARK 7.14. A quick examination of the proof shows that one can replace the estimate of Corollary 7.12 by

$$|(\zeta_B, w_B) - (\zeta_{KdV}, w_{KdV})|_{L^\infty([0,t]\times\mathbb{R})} \leq \mu \frac{t}{1+t} C(T, \mu_0, |(\zeta^0, w^0)|_{H^9_{(2)}}),$$

for all $t \in [0, T/\mu]$, which is better for small times.

PROOF. The proof is absolutely similar to the proof of Theorem 7.1, but Lemmas 7.10 and 7.11 allow the growth rate in time of various estimates to improve: terms which were growing as $O(\sqrt{t})$ due to secular growth are now bounded. □

7.2. The splitting into unidirectional waves: The weakly transverse case

The KdV approximation presented in the previous section describes the evolution of one-dimensional surface waves; this means that these waves do not depend on the transverse direction. A natural question therefore is to consider the physically more realistic case of waves that depend on the transverse variable, but with a "weak" dependence, so that transverse variations are less important than longitudinal ones. This corresponds to the Kadomtsev-Petviashvili (KP) regime

$$\mathcal{A}_{KP} = \{(\varepsilon, \beta, \gamma, \mu), \quad 0 \leq \mu \leq \mu_0, \quad \varepsilon = \mu, \quad \gamma = \sqrt{\varepsilon}, \quad \beta = 0\}. \tag{7.25}$$

(Compared to the KdV regime, we assume that $\gamma = \sqrt{\mu}$ rather than $\gamma = 0$.)

As in the one-dimensional case, we can start our analysis from one of the fully symmetric Boussinesq systems (5.41). In the present two-dimensional, weakly transverse case, these systems can be written (with $\varepsilon = \mu$, $\gamma = \sqrt{\varepsilon}$)

$$\begin{cases} (1-\mu\mathbf{b}\Delta^\gamma)\partial_t\zeta + \nabla^\gamma\cdot((1+\frac{\varepsilon}{2}\zeta)W) + \mu\mathbf{a}\Delta^\gamma\nabla^\gamma\cdot W = 0, \\ (1-\mu\mathbf{d}\Delta^\gamma)\partial_t W + (1+\frac{\varepsilon}{2}\zeta)\nabla^\gamma\zeta + \mu\mathbf{c}\Delta^\gamma\nabla^\gamma\zeta \\ \qquad + \frac{\varepsilon}{2}\big[(W\cdot\nabla^\gamma)W + W\nabla^\gamma\cdot W + \frac{1}{2}\nabla^\gamma|W|^2\big] = 0, \end{cases}$$

where $\mathbf{a} = \mathbf{c}$, $\mathbf{b} \geq 0$, $\mathbf{d} \geq 0$ are such that

$$\mathbf{a}+\mathbf{b}+\mathbf{c}+\mathbf{d} = \frac{1}{3},$$

and where $W = (w_1, w_2) = W_{\theta,\delta}$ is as defined in (5.39). As for the one-dimensional case, we choose to work with a KdV/KdV type system that simplifies the computations and corresponds to $\mathbf{a} = \mathbf{c} = 1/6$ and $\mathbf{b} = \mathbf{d} = 0$ in the system above

$$(7.26)\qquad \begin{cases} \partial_t\zeta + \nabla^\gamma\cdot((1+\frac{\varepsilon}{2}\zeta)W) + \mu\frac{1}{6}\Delta^\gamma\nabla^\gamma\cdot W = 0, \\ \partial_t W + (1+\frac{\varepsilon}{2}\zeta)\nabla^\gamma\zeta + \mu\frac{1}{6}\Delta^\gamma\nabla^\gamma\zeta \\ \qquad + \frac{\varepsilon}{2}\big[(W\cdot\nabla^\gamma)W + W\nabla^\gamma\cdot W + \frac{1}{2}\nabla^\gamma|W|^2\big] = 0. \end{cases}$$

Before we define the KP approximation, let us set some notations.

NOTATION 7.15.
(1) For all $s \in \mathbb{R}$, we denote by $\partial_x H^s(\mathbb{R}^d)$ the space of all $f \in H^{s-1}(\mathbb{R}^2)$ such that there exists $\widetilde{f} \in H^s(\mathbb{R}^d)$ satisfying $\partial_x \widetilde{f} = f$. We denote $\partial_x^{-1} f = \widetilde{f}$ and $|f|_{\partial_x H^s} = |\widetilde{f}|_{H^s}$.
(2) We define similarly $\partial_x^2 H^s(\mathbb{R}^2)$.
(3) More generally, ∂_x^{-1} stands for the Fourier multiplier with symbol $-i/\xi_1$, where $\xi = (\xi_1, \xi_2)$ is the dual variable of (x, y).

We define the KP approximation $(\zeta_{KP}, w_{1,KP}, w_{2,KP})$ to (7.26) as the following weakly transverse generalization[5] of the KdV approximation (7.5)

$$(7.27)\qquad \begin{aligned} \zeta_{KP}(t,x,y) &= f_+(x-t,y,\mu t) + f_-(x+t,y,\mu t), \\ w_{1,KP}(t,x,y) &= f_+(x-t,y,\mu t) - f_-(x+t,y,\mu t), \end{aligned}$$

where $f_+ = f(\tilde{x}, y, \tau)$ and $f_- = f(\tilde{x}, y, \tau)$ solve the KP equations[6].

$$(7.28)\qquad \begin{aligned} \partial_\tau f_+ + \frac{1}{2}\partial_{\tilde{x}}^{-1}\partial_y^2 f_+ + \frac{1}{6}\partial_{\tilde{x}}^3 f_+ + \frac{3}{2}f_+\partial_{\tilde{x}} f_+ = 0, \\ \partial_\tau f_- - \frac{1}{2}\partial_{\tilde{x}}^{-1}\partial_y^2 f_- - \frac{1}{6}\partial_{\tilde{x}}^3 f_- - \frac{3}{2}f_-\partial_{\tilde{x}} f_- = 0. \end{aligned}$$

It is worth noting here that there is no ansatz for the transverse component of the velocity w_2 in (7.27), because this component remains small (actually, smaller than the overall precision of the approximation). The remainder of this section is devoted to the justification of the KP approximation. The KP equation was first derived in

[5]One would expect this weakly transverse generalization to depend on y through $\sqrt{\mu}y$ and not y only (see for instance [**225**]). We have a full transverse dependence here because the weak transversality is already taken into account by the nondimensionalization, which is responsible for the presence of the coefficient $\gamma = \sqrt{\mu}$ in (7.26).

[6]In the literature [**208**], these equations are commonly referred to as $KP2$ equations. The $KP1$ equation, with a different sign on the third-order dispersive term, can be derived in the presence of strong surface tension (see §9.3.3).

1970 in [**193**] and in [**2**] in the context of water waves. The first justification of this approximation was done by Kano [**197**] on a too-short time scale ($O(1)$ rather than $O(1/\mu)$). The approximation was then justified for other contexts and toy models [**151, 27, 274**]. Finally, the KP approximation was proved in the water waves case in [**225**], which assumed the existence and control of solutions to the water waves equations. (The fact that this assumption is fulfilled was proved in [**12**] and is a consequence of Chapter 4 in this book.)

7.2.1. Statement of the main result. We state here the main result that justifies the KP approximation (7.27)–(7.28) in the following sense (in the statement below, μ_0 is as in (7.25)).

THEOREM 7.16. *Let $t_0 > 1$, $\zeta^0, w_1^0 \in \partial_x H^{7+t_0}(\mathbb{R}^2)$ and $w_2^0 \in H^{t_0+4}(\mathbb{R}^2)$. Moreover, assume that $\zeta^0, w_1^0 \in \partial_x^2 H^{t_0+3}(\mathbb{R}^2)$. There exists $T > 0$ such that*

(1) *There exists a unique solution $(\zeta_B, w_{1,B}, w_{2,B}) \in C([0, \frac{T}{\mu}]; H^8(\mathbb{R}))$ to (7.26) with initial condition $(\zeta^0, w_1^0, \sqrt{\mu} w_2^0)$.*

(2) *The KP approximation (7.27) is well defined in $C([0, T/\mu] \times \mathbb{R}^2)$ and*

$$|(\zeta_B, w_{1,B}) - (\zeta_{KP}, w_{1,KP})|_{L^\infty([0,\frac{T}{\mu}]\times\mathbb{R}^2)} = o(1) \qquad as \qquad \mu \to 0,$$

where $(\zeta_{KP}, w_{1,KP})$ is given by (7.27) with $f_\pm$ solving (7.28) with initial condition $f_\pm^0 = \frac{1}{2}(\zeta^0 \pm w_1^0)$.

REMARK 7.17. Since the solutions to the Boussinesq system (7.26) are known to provide good approximations to the water waves equations (see Theorem 6.20), we can deduce from Theorem 7.16 that this is also the case for the KP approximation and obtain a corollary to Theorem 7.16 in the spirit of Corollary 7.2.

REMARK 7.18.

(1) The precision of the KP approximation is much worse than in the one-dimensional case (KdV): it is $o(1)$ versus $O(\sqrt{\mu})$ or even $O(\mu)$ (see Corollary 7.12). As shown in [**217**], this poor convergence rate can be improved in some cases (at least at the level of consistency).

(2) Another drawback to the KP approximation is the restrictive zero mean assumptions $\zeta^0, w_1^0 \in \partial_x^2 H^{t_0+3}(\mathbb{R}^2)$. Using the strategy adopted in [**27**] for a toy model, one could weaken this assumption into $\zeta^0, w_1^0 \in \partial_x H^{t_0+3}(\mathbb{R}^2)$, which remains quite restrictive.

(3) The operator $\partial_x^{-1}\partial_y^2$ that appears in the KP equation is typical to the analysis of weakly transverse phenomena. See [**87**] for another occurrence in nonlinear optics.

REMARK 7.19. We can write $\zeta_{KP} = \zeta_{KP}^+ + \zeta_{KP}^-$, with $\zeta_{KP}^\pm = f_\pm(x \mp t, y, \mu t)$. In particular, $\zeta_{KP}^\pm$ solves

$$\partial_t \zeta_{KP}^+ + \partial_x \zeta_{KP}^+ + \mu\big(\frac{1}{2}\partial_x^{-1}\partial_y^2 \zeta_{KP}^+ + \frac{1}{6}\partial_x^3 \zeta_{KP}^+ + \frac{3}{2}\zeta_{KP}^+ \partial_x \zeta_{KP}^+\big) = 0,$$
$$\partial_t \zeta_{KP}^- - \partial_x \zeta_{KP}^- - \mu\big(\frac{1}{2}\partial_x^{-1}\partial_y^2 \zeta_{KP}^- + \frac{1}{6}\partial_x^3 \zeta_{KP}^- + \frac{3}{2}\zeta_{KP}^- \partial_x \zeta_{KP}^+\big) = 0.$$

Proceeding as in the KdV case (see Remark 7.5), one can show that these KP equations can be replaced by any equation of a family of equations, sometimes called KP-BBM (or regularized long wave KP or weakly transverse BBM) equations by analogy with the one-dimensional case, without changing the precision of the approximation. For the right-going component (for instance), this family of equations is given by

$$\partial_t\zeta+\partial_x\zeta+\mu(\frac{1}{2}\partial_x^{-1}\partial_y^2\zeta+p\partial_x^3\zeta+(p-\frac{1}{6})\partial_{xxt}^3\zeta)=0\qquad(p\leq\frac{1}{6}).\tag{7.29}$$

This equation has been studied, for instance, in [**38, 285**] in the case $p=0$.

7.2.2. BKW expansion. We seek an approximate solution (ζ_{app},W_{app}) (with $W_{app}=(w_{1,app},w_{2,app})$) to (7.26) under the form[7]

$$\begin{array}{lcl}\zeta_{app}(t,x)&=&\zeta_{(0)}(t,x,y,\mu t)+\mu\zeta_{(1)}(t,x,y,\mu t),\\ w_{1,app}(t,x)&=&w_{1,(0)}(t,x,y,\mu t)+\mu w_{1,(1)}(t,x,y,\mu t),\\ w_{2,app}(t,x)&=&\mu^{1/2}w_{2,(1/2)}(t,x,y,\mu t),\end{array}\tag{7.30}$$

with

$$\begin{array}{l}\zeta_{(0)}(t,x,y,\mu t)=f_+(x-t,y,\mu t)+f_-(x+t,y,\mu t),\\ w_{1,(0)}(t,x,y,\mu t)=f_+(x-t,y,\mu t)-f_-(x+t,y,\mu t),\end{array}\tag{7.31}$$

so that

$$\zeta_{KP}(t,x)=\zeta_{(0)}(t,x,y,\mu t),\qquad w_{1,KP}(t,x)=w_{1,(0)}(t,x,y,\mu t),$$

and where the corrector terms $w_{2,(1/2)}$ and $(\zeta_{(1)},W_{(1)})$ are determined below. Plugging the ansatz (7.30) into (7.26) and ordering the residual into powers of μ, one finds (recall that $\varepsilon=\mu$ and $\gamma=\sqrt{\mu}$)

$$\begin{cases}\partial_t\zeta\nabla^\gamma\cdot((1+\frac{\varepsilon}{2}\zeta)W)+\mu\frac{1}{6}\Delta^\gamma\nabla^\gamma\cdot W=\mu R_1^{(1)}+\mu^2R_1^\mu,\\ \partial_tW+(1+\frac{\varepsilon}{2}\zeta)\nabla^\gamma\zeta+\mu\frac{1}{6}\Delta^\gamma\nabla^\gamma\zeta\\ \quad+\frac{\varepsilon}{2}[(W\cdot\nabla^\gamma)W+W\nabla^\gamma\cdot W+\frac{1}{2}\nabla^\gamma|W|^2]\\ \quad=\sqrt{\mu}R_2^{(1/2)}+\mu R_2^{(1)}+\mu^{3/2}R_2^{(3/2)}+\mu^2R_2^\mu,\end{cases}\tag{7.32}$$

with $R_2^{(1/2)}=R_2^{(1/2)}(t,x,y,\mu t)$ and $R_j^{(1)}=R_j^{(1)}(t,x,y,\mu t)$ given by

$$\begin{array}{lcl}R_1^{(1)}&=&(\partial_t\zeta_{(1)}+\partial_xw_{1,(1)})+\partial_yw_{2,(1/2)}+\partial_\tau\zeta_{(0)}+\frac{1}{6}\partial_x^3w_{1,(0)}+\frac{1}{2}\partial_x(\zeta_{(0)}w_{1,(0)}),\\ R_2^{(1/2)}&=&\begin{pmatrix}0\\ \partial_tw_{2,(1/2)}+\partial_y\zeta_{(0)}\end{pmatrix},\\ R_2^{(1)}&=&\begin{pmatrix}\partial_tw_{1,(1)}+\partial_x\zeta_{(1)}+\partial_\tau w_{1,(0)}+\frac{1}{6}\partial_x^3\zeta_{(0)}+\frac{1}{4}(\partial_x(3w_{1,(0)}^2+\zeta_{(0)}^2)\\ 0\end{pmatrix}.\end{array}$$

The $O(\mu^{3/2})$ term $R_2^{(3/2)}$ is given by

$$\begin{array}{lcl}R_2^{(3/2)}&=&\begin{pmatrix}0\\ \partial_\tau w_{2,(1/2)}+\frac{1}{6}\partial_x^2\partial_y\zeta_{(0)}\end{pmatrix}\\ &+&\begin{pmatrix}0\\ \frac{1}{2}\zeta_{(0)}\partial_y\zeta_{(0)}+\frac{3}{2}\partial_x(w_{1,(0)}w_{2,(1/2)})+\frac{1}{2}\partial_y(w_{1,(0)})^2\end{pmatrix},\end{array}\tag{7.33}$$

---

[7] We should a priori add $\mu^{1/2}\zeta_{1,(1/2)}(t,x,y,\mu t)$, $\mu^{1/2}w_{1,(1/2)}(t,x,y,\mu t)$ and $\mu w_{2,(1)}$ to the ansatz for ζ_{app} and $w_{1,app}$ respectively, but computations would then show that

$$\begin{cases}\partial_t\zeta_{(1/2)}+\partial_xw_{1,(1/2)}=0,\\ \partial_tw_{1,(1/2)}+\partial_x\zeta_{(1/2)}=0.\end{cases}$$

Therefore $\zeta_{(1/2)}=w_{(1/2)}=0$ if these quantities are initially zero, which is the case by assumption on the initial data. One would further obtain $\partial_tw_{2,(1)}=0$ and therefore $w_{2,(1)}=0$ in accordance with the initial data.

while R_1^μ and R_2^μ are given by

$$\begin{aligned} R_1^\mu &= \partial_\tau \zeta_{(1)} + \frac{1}{6}\Delta^\gamma(\partial_x w_{1,(1)} + \partial_y w_{2,(1/2)}) \\ &\quad + \frac{1}{2}\partial_x\big[\zeta_{(1)}(w_{1,(0)} + \mu w_{1,(1)}) + \zeta_{(0)} w_{1,(1)}\big] + \frac{1}{2}\partial_y\big[(\zeta_{(0)} + \mu\zeta_{(1)})w_{2,(1/2)}\big] \end{aligned} \tag{7.34}$$

and $R_2^\mu = (R_{2,1}^\mu, \sqrt{\mu}R_{2,2}^\mu)^T$ with

$$\begin{aligned} R_{2,1}^\mu &= \partial_\tau w_{1,(1)} + \frac{1}{6}\Delta^\gamma \partial_x \zeta_{(1)} \\ &\quad + \frac{1}{2}\partial_x(3w_{1,(0)}w_{1,(1)} + \zeta_{(0)}\zeta_{1,(1)}) + \frac{\mu}{4}\partial_x(w_{1,(1)}^2 + \zeta_{(1)}^2) \\ &\quad + \frac{1}{2}\partial_y\big(w_{2,(1/2)}\partial_y(w_{1,(0)} + \mu w_{1,(1)})\big) + \frac{1}{2}\partial_x(w_{2,(1/2)})^2 \end{aligned} \tag{7.35}$$

and

$$R_{2,2}^\mu = \sqrt{\mu}\frac{1}{6}\Delta^\gamma\partial_y\zeta_{(1)} + \frac{3}{2}w_{2,(1/2)}\partial_y w_{2,(1/2)} + \frac{1}{2}\partial_y(\zeta_{(0)}\zeta_{(1)} + \mu(\zeta_{(1)})^2). \tag{7.36}$$

We use the same strategy that we used for the KdV equation: we choose the correctors $w_{2,(1/2)}$ and $(\zeta_{(1)}, W_{(1)})$ in such a way that $R_2^{(1/2)}(t,x,\tau) = R_2^{(1)}(t,x,\tau) = 0$ and $R_1^{(1)}(t,x,\tau) = 0$ for all $t \in [0, \frac{T}{\mu}]$, $x \in \mathbb{R}$ and $\tau \in [0,T]$.

- *Cancellation of* $R_2^{(1/2)}(t,x,y,\tau)$. This is equivalent to the equation

$$\partial_t w_{2,(1/2)} + \partial_y \zeta_{(0)} = 0$$

or (using the fact that $w_{2,(1/2)}$ and $\zeta_{(0)}$ vanish at $x = \infty$)

$$\partial_t\partial_x w_{2,(1/2)} + \partial_y\partial_x\zeta_{(0)} = 0.$$

From the definition (7.31) of $\zeta_{(0)}$, we get that $\partial_x\zeta_{(0)} = -\partial_t(f_{+\,|\tilde{x}=x-t}) + \partial_t(f_{-\,|\tilde{x}=x+t})$, so that a time integration yields

$$\begin{aligned} \partial_x w_{2,(1/2)}(t,x,y,\tau) &= (\partial_x w_2^0 - \partial_y f_+^0 + \partial_y f_-^0) \\ &\quad + \partial_y f_+(x-t,y,\tau) - \partial_y f_-(x+t,y,\tau). \end{aligned} \tag{7.37}$$

- *Cancellation of* $R_1^{(1)}(t,x,y,\tau)$ *and* $R_2^{(1)}(t,x,y,\tau)$. From the definition of $R_1^{(1)}$ and $R_2^{(1)}$ this is equivalent to

$$\begin{cases} \partial_t\zeta_{(1)} + \partial_x w_{1,(1)} = F_+(x-t,y,\tau) + F_-(x+t,y,\tau), \\ \partial_t w_{1,(1)} + \partial_x\zeta_{(1)} = G_+(x-t,y,\tau) + G_-(x+t,y,\tau) + G_0(x,t,y,\tau), \end{cases} \tag{7.38}$$

with

$$\begin{aligned} F_+(\tilde{x},y,\tau) &= \partial_\tau f_+ + \partial_x^{-1}\partial_y^2 f_+ + \tfrac{1}{6}\partial_{\tilde{x}}^3 f_+ + f_+\partial_{\tilde{x}} f_+, \\ F_-(\tilde{x},y,\tau) &= \partial_\tau f_- - \partial_x^{-1}\partial_y^2 f_- - \tfrac{1}{6}\partial_{\tilde{x}}^3 f_- - f_-\partial_{\tilde{x}} f_-, \end{aligned} \tag{7.39}$$

where we refer to Notation 7.15 for the meaning of ∂_x^{-1}, and

$$\begin{aligned} G_+(\tilde{x},y,\tau) &= \partial_\tau f_+ + \tfrac{1}{6}\partial_{\tilde{x}}^3 f_+ + 2f_+\partial_{\tilde{x}} f_+, \\ G_-(\tilde{x},y,\tau) &= -\partial_\tau f_- + \tfrac{1}{6}\partial_{\tilde{x}}^3 f_- + 2f_-\partial_{\tilde{x}} f_-, \\ G_0(x,t,y,\tau) &= -\partial_x(f_+(x-t,y,\tau)f_-(x+t,y,\tau)). \end{aligned} \tag{7.40}$$

7.2.3. Consistency of the approximate solution and secular growth. Inspired by the analysis performed in the one-dimensional case (see §7.1.4), we introduce the quantities u and v defined as

$$u = \zeta_{(1)} + w_{1,(1)}, \qquad v = \zeta_{(1)} - w_{1,(1)}, \tag{7.41}$$

and note that (7.38) is equivalent to

$$\begin{cases} (\partial_t+\partial_x)u = (F_+ + G_+)(x-t,y,\tau)+(F_- + G_-)(x+t,y,\tau)+G_0(x,t,y,\tau), \\ (\partial_t-\partial_x)v = (F_+ - G_+)(x-t,y,\tau)+(F_- - G_-)(x+t,y,\tau)-G_0(x,t,y,\tau). \end{cases} \tag{7.42}$$

As for the one-dimensional case, the control of the secular growth is invoked to impose the relations $F_+ + G_+ = F_- - G_- = 0$. For this reason, we need the following two-dimensional generalization of Lemma 7.6 (note that here h_2 contains a component $h_{2,1}$ that is not the derivative of an L^2-function).

LEMMA 7.20. *Let $c_1 \neq c_2$, $t_0 > 1/2$ and $s \in \mathbb{N}$. Let $h_1, H_2, h_{2,1} \in H^s(\mathbb{R}^2)$, $h_{3,1}, h_{3,2} \in H^{s+1}(\mathbb{R}^2)$, and denote $h_2 = h_{2,1} + \partial_x H_2$. Then the unique solution g vanishing at $t = 0$ to*

$$(\partial_t + c_1\partial_x)g = h_1(x - c_1 t, y) + h_2(x - c_2 t, y) + h_{3,1}(x - c_1 t, y)h_{3,2}(x - c_2 t, y)$$

is given by

$$\begin{aligned} g(t,x,y) \quad &= \quad th_1(x - c_1 t, y) + \frac{1}{c_1 - c_2}\int_{x-c_1 t}^{x-c_2 t} h_{2,1}(x', y)dx' \\ &\quad + \frac{1}{c_1 - c_2}\big(H_2(x - c_2 t, y) - H_2(x - c_1 t, y)\big) \\ &\quad + h_{3,1}(x - c_1 t, y)\int_0^t h_{3,2}(x + (c_1 - c_2)s - c_1 t, y)ds. \end{aligned}$$

In particular, $\lim_{t\to\infty} \frac{1}{t}|g(t,\cdot)|_2 = 0$ if and only if $h_1 = 0$, and one then has

$$|g(t,\cdot)|_{H^s} \leq \epsilon(t)t|h_{2,1}|_{H^s} + \frac{2}{|c_1 - c_2|}|H_2|_{H^s} + \mathrm{Cst}\,\frac{\sqrt{t}}{\sqrt{|c_1 - c_2|}}|h_{3,1}|_{H^{s+1}}|h_{3,2}|_{H^{s+1}},$$

where $\epsilon(t) \to 0$ as $t \to \infty$.

REMARK 7.21. When $h_1 = 0$, the contribution of $h_{2,1}$ in the above lemma is the worst one as far as secular growth is concerned since it grows as $o(t)$ while the other terms are $O(1)$ or $O(\sqrt{t})$. This growth rate is sharp in general but can be improved under additional assumptions (see [**216**]).

PROOF. As for the proof of Lemma 7.6, the estimate of the first and third components of g does not cause any difficulty. For the second component $g_{2,1}$ of g (accounting for the contribution of $h_{2,1}$), we get

$$\widehat{g}_{2,1}(t,\xi) = F(t,\xi_1)\widehat{h}_{2,1}(\xi),$$

with $F(t,\xi_1) = e^{-itc_1\xi_1}\int_0^t e^{is(c_1-c_2)\xi_1}ds$. We therefore get

$$\frac{1}{t}|g_{2,1}(t,\cdot)|_2 = \Big(\int_{\mathbb{R}^2}|\frac{1}{t}F(t,\xi_1)|^2|\widehat{h}_{2,1}(\xi)|^2 d\xi\Big)^{1/2}.$$

Since $\frac{1}{t}|F(t,\xi_1)| \leq 1$ and $\frac{1}{t}|F(t,\xi_1)| \to 0$ as $t \to \infty$ for almost all $\xi = (\xi_1, \xi_2) \in \mathbb{R}^2$, we get from Lebesgue's dominated convergence theorem that $\lim_{t\to\infty}\frac{1}{t}|g_{2,1}|_2 = 0$ (the proof for higher Sobolev norm is identical).

For the fourth component g_3 of g, we give just the proof of the L^2 estimate, since the generalization to higher order Sobolev spaces is quite straightforward. We have

$$\begin{aligned} |g_3(t,\cdot)|_2^2 &\leq \frac{t}{|c_1-c_2|}\int_{\mathbb{R}^2}|h_{3,1}(x-c_1t,y)|^2|h_{3,2}(\cdot,y)|_2^2dxdy \\ &= \frac{t}{|c_1-c_2|}\int_{\mathbb{R}}|h_{3,1}(\cdot,y)|_{L^2(\mathbb{R})}^2|h_{3,2}(\cdot,y)|_{L^2(\mathbb{R})}^2dy \\ &\leq \frac{t}{|c_1-c_2|}|h_{3,1}|_{L^2(\mathbb{R}^2)}^2|h_{3,2}|_{L_y^\infty L_x^2}^2, \end{aligned}$$

and the result follows from the continuous embedding $H^1(\mathbb{R}^2)\subset L_y^\infty L_x^2$. $\square$

As previously mentioned, Lemma 7.20 implies that $F_+ + G_+ = F_- - G_- = 0$. From the definition (7.39)–(7.40) of $F_\pm$ and $G_\pm$, these conditions can be written as

$$\begin{aligned} 2\partial_\tau f_+ + \partial_{\tilde{x}}^{-1}\partial_y^2 f_+ + \frac{1}{3}\partial_{\tilde{x}}^3 f_+ + 3f_+\partial_{\tilde{x}} f_+ &= 0, \\ 2\partial_\tau f_- - \partial_{\tilde{x}}^{-1}\partial_y^2 f_- - \frac{1}{3}\partial_{\tilde{x}}^3 f_- - 3f_-\partial_{\tilde{x}} f_- &= 0. \end{aligned}$$

These equations are exactly the KP equations (7.28). Lemma 7.20 then provides us with estimates on u and v or equivalently, through (7.41), on $\zeta_{(1)}$ and $w_{1,(1)}$. The estimate (7.17) is then generalized in the present two-dimensional case into

$$\begin{aligned} |\zeta_{(1)}(t,\cdot,\tau)|_{H^s} + |w_{1,(1)}(t,\cdot,\tau)|_{H^s} &\leq (1+\sqrt{t})\,C(|f_\pm(\cdot,\tau)|_{H^{s+2}}) \\ &\quad + \epsilon(t)tC(|\partial_x^{-1}f_\pm(\cdot,\tau)|_{H^{s+2}}). \end{aligned} \tag{7.43}$$

We now need the following local existence result for the KP equation. The lemma is stated for the right-going KP equation, but the same result holds for the left-going one.

LEMMA 7.22. *Let $s>3$ and $f^0\in H^s\cap\partial_x H^s(\mathbb{R}^2)$. Also let $T>0$.*
(1) *There exists a unique solution $f\in C([0,T];H^s(\mathbb{R}^2))\cap C([0,T],\partial_x H^s(\mathbb{R}^2))$ to the initial value problem*

$$\partial_\tau f + \frac{1}{2}\partial_{\tilde{x}}^{-1}\partial_y^2 f + \frac{1}{6}\partial_{\tilde{x}}^3 f + \frac{3}{2}f\partial_{\tilde{x}} f = 0, \qquad f_{|_{\tau=0}} = f^0,$$

and one has $|f|_{L^\infty([0,T];H^s\cap\partial_x H^s)}\leq C(T,|f^0|_{H^s},|\partial_x f^0|_{H^s})$.
(2) *Moreover, assuming that $f^0\in H^{s+4}(\mathbb{R}^2)$ and $\partial_y^2 f^0\in\partial_x^2 H^{s+2}(\mathbb{R}^2)$, then one has $\partial_y^2 f\in C([0,T];\partial_x^2 H^{s+2}(\mathbb{R}^2))$ and*

$$|\partial_y^2 f|_{L^\infty([0,T];\partial_x^2 H^{s+2})}\leq C(T,|\partial_y^2 f^0|_{H^{s+2}},|f^0|_{H^{s+4}}).$$

PROOF. See [**46**] for the existence and uniqueness of a solution $f\in C([0,T];H^s(\mathbb{R}^2))$ for all $T>0$. The proof of this result is quite involved, and we therefore also refer to [**317**] for the local version of this result (i.e., the same results holds for some $T>0$), which is actually all that we need here. The fact that f also belongs to $C([0,T],\partial_x H^s(\mathbb{R}^d))$ was observed in [**284**] and follows simply from the semi-group formulation

$$f(\tau) = S(\tau)f^0 - \frac{3}{4}\partial_{\tilde{x}}\int_0^t S(\tau-\tau')f(\tau')^2d\tau',$$

where $S(\tau)=\exp(-\tau(\partial_{\tilde{x}}^{-1}\partial_y^2+\frac{1}{6}\partial_{\tilde{x}}^3))$ is clearly unitary on all $H^s(\mathbb{R}^2)$. Since for all τ', $f(\tau')^2\in H^s(\mathbb{R}^2)$ (recall that $s>1$), the second term of the right-hand side is in

$\partial_x H^s(\mathbb{R}^2)$. This is also the case of the first term by assumption on the initial data. Finally, let us note that if

$$\frac{1}{2}\partial_{\tilde{x}}^{-1}\partial_y^2 f = -(\partial_{\tilde{x}}^{-1}\partial_\tau f + \frac{1}{6}\partial_{\tilde{x}}^2 f + \frac{3}{4}f^2),$$

then it is enough to prove that $\partial_\tau f \in C([0,T];\partial_x H^{s+2}(\mathbb{R}))$. Noting that $\tilde{f} = \partial_\tau f$ solves

$$\partial_\tau \tilde{f} + \frac{1}{2}\partial_{\tilde{x}}^{-1}\partial_y^2 \tilde{f} + \frac{1}{6}\partial_{\tilde{x}}^3 \tilde{f} + \frac{3}{2}\partial_{\tilde{x}}(f\tilde{f}) = 0,$$

we get

$$\tilde{f}(\tau) = S(\tau)\tilde{f}^0 - \frac{3}{2}\partial_{\tilde{x}}\int_0^t S(\tau-\tau')f(\tau')\tilde{f}(\tau')d\tau'.$$

From the assumption that $\partial_y^2 f^0 \in \partial_x^2 H^{s+2}(\mathbb{R}^2)$, we get that $f^0 \in \partial_x H^{s+2}$; proceeding as previously, we can deduce from the above formula that $\tilde{f} = \partial_\tau f \in C([0,T];\partial_x H^{s+2}(\mathbb{R}))$. □

Recalling that the initial data for f_+ and f_- are

$$f_+ \,|_{\tau=0} = \frac{\zeta^0 + w_1^0}{2}, \qquad f_- \,|_{\tau=0} = \frac{\zeta^0 - w_1^0}{2},$$

we get from (7.43) and Lemma 7.22 that

$$(7.44) \quad |\zeta_{(1)}(t,\cdot,\tau)|_{H^s} + |w_{1,(1)}(t,\cdot,\tau)|_{H^s} \le \big(1+\sqrt{t}+\epsilon(t)t\big)C(\tau,|(\zeta^0,w_1^0)|_{\partial_x H^{s+3}}).$$

We are now ready to prove that the approximate solution (7.30) is consistent with (7.26) at order $O\big((\mu^{3/2}+\mu^2(1+\sqrt{t}+\epsilon(t)t)\big)$; the secular growth in time therefore is of rate $o(\mu^2 t)$, which is acceptable since $o(\mu^2 t)$ terms remain correctors of the $O(\mu)$ terms over time scales $O(1/\mu)$ — but this is much worse than for the KdV equation for which we had $O(\mu^2\sqrt{t})$, or even $O(\mu)$, instead of $o(\mu^2 t)$.

LEMMA 7.23. *Let $T > 0$ and $s \ge 0$, and $\zeta^0, w_1^0 \in \partial_x H^{s+6}(\mathbb{R}^2)$ be such that $\partial_y^2\zeta^0, \partial_y^2 w_1^0 \in \partial_x^2 H^{s+2}(\mathbb{R}^2)$. Then the approximate solution (ζ_{app}, w_{app}) provided by (7.8), (7.31), and (7.12) solves (7.10) with residuals satisfying, for all $0 \le t \le T/\mu$*

$$R_1^{(1)} = 0, \qquad R_2^{(1/2)} = R_2^{(1)} = 0,$$
$$\mu^{3/2}|R_2^{(3/2)}(t,\cdot,\mu t)|_{H^s} \le \mu^{3/2}C\big(T,|\partial_y^2(\zeta^0,w_1^0)|_{\partial_x^2 H^{s+1}},|(\zeta^0,w_1^0)|_{\partial_x H^{s+6}},|w_2^0|_{H^{s+1}}\big)$$

and, for the terms exhibiting secular growth

$$\mu^2|R_1^\mu(t,\cdot,\mu t)|_{H^s} + \mu^2|R_2^\mu(t,\cdot,\mu t)|_{H^s}$$
$$\le \mu^2 C\big(T,|(\zeta^0,w_1^0)|_{\partial_x H^{s+6}},|\partial_y^2(\zeta^0,w_1^0)|_{\partial_x^2 H^{s+2}},|w_2^0|_{H^{s+3}}\big)(1+\sqrt{t}+\epsilon(t)t\big).$$

PROOF. We have already shown that $R_j^{(1)} = 0$ $(j=1,2)$ and $R_2^{(1/2)} = 0$ in (7.32). We therefore are left to control $R_2^{(3/2)}$ and R_j^μ $(j=1,2)$. Denoting as previously $|f(t,\cdot)|_{H_T^s} = \sup_{0\le\tau\le T}|f(t,\cdot,\tau)|_{H^s}$ and since there are no secularly growing terms in $R_2^{(3/2)}$ we get as in the proof of Lemma 7.9 (with $U_{(j)} = (\zeta_{(j)},w_{(j)})$ $(j=1/2,1,2)$) that (7.33) leads to

$$\begin{aligned}|R_2^{3/2}(t,\cdot,\mu t)|_{H^s} &\le C\big(|\partial_\tau w_{2,(1/2)}(t,\cdot)|_{H_T^s},|w_{2,(1/2)})(t,\cdot)|_{H_T^{s+1}},|U_{(0)}(t,\cdot)|_{H_T^{s+3}}\big)\\ &\le C(|\partial_y^2 f_\pm(t,\cdot)|_{\partial_x^2 H^{s+1}},|f_\pm(t,\cdot)|_{H_T^{s+3}},|w_2^0|_{H^{s+1}}),\end{aligned}$$

where we used (7.37) and the equation (7.28) to get the second inequality. The

estimate on $R_2^{3/2}$ given in the statement of the lemma is then a direct consequence of Lemma 7.22.
We now use the explicit expressions (7.34) and (7.35)–(7.36) to control R_j^μ $(j=1,2)$; for the sake of clarity, we write $|F|_{H^s_T}$ rather than $|F(t,\cdot)|_{H^s_T}$ in the expression below

$$\begin{aligned}|R_j^\mu(t,\cdot,\mu t)|_{H^s} &\leq C\big(|U_{(0)}|_{H_T^{s+1}}\big)\big(1+|\partial_\tau U_{(1)}|_{H_T^s}+|U_{(1)}|_{H_T^{s+3}}+\mu|U_{(1)}|^2_{H_T^{s+1}}\big)\\ &\quad +C\big(|w_{2,(1/2)}|_{H_T^{s+3}}\big)\big(1+|U_{(0)}|_{H_T^{s+1}}+\mu|U_{(1)}|_{H_T^{s+1}}\big).\end{aligned}$$

The only differences between this estimate and the one obtained in the proof of Lemma 7.9 is the third line involving $w_{2,(1/2)}$ and the different growth rate for the secular terms $U_{(1)}$. A straightforward adaptation of the proof of Lemma 7.9 therefore gives the result (we use (7.37) for the terms involving $w_{2,(1/2)}$ and Lemma 7.20 to get the relevant growth rate in the present case). □

7.2.4. Proof of Theorem 7.16. From Lemma 7.23 and Propositions 6.7 and 6.8, we deduce directly the first assertion of the theorem and the estimate, for all $t\in[0,\frac{T}{\mu}]$,

$$\begin{aligned}&|(\zeta,w)_B-(\zeta,w)_{app}|_{L^\infty([0,t];H^{s-1})}\leq \mu^2(t+t^{3/2}+\epsilon(t)t^2)\\ &\times\ C(T,|(\zeta^0,w_1^0)|_{\partial_x H^{s+6}},|\partial_y^2(\zeta^0,w_1^0)|_{\partial_x^2 H^{s+2}},|w_2^0|_{H^{s+3}}).\end{aligned}$$

We also get from (7.30) that

$$\begin{aligned}|(\zeta_{app},w_{app})-(\zeta_{KP},w_{KP})|_{L^\infty([0,t];H^{s-1})} &\leq \mu|(\zeta_{(1)},w_{1,(1)}))|_{L^\infty([0,t]\times[0,\mu t]_\tau;H^{s-1})}\\ &+\sqrt{\mu}|w_{2,(1/2)}))|_{L^\infty([0,t]\times[0,\mu t]_\tau;H^{s-1})},\end{aligned}$$

so that we can use (7.44) and (7.37) to obtain

$$\begin{aligned}&|(\zeta_{app},w_{app})-(\zeta_{KP},w_{KP})|_{L^\infty([0,t];H^{s-1})}\\ &\leq\ (\sqrt{\mu}+\mu\epsilon(t)t)C(T,|(\zeta^0,w_1^0)|_{\partial_x H^{s+2}},|w_2^0|_{H^{s-1}}).\end{aligned}$$

The result stated in Theorem 7.1 follows from a triangular inequality and the Sobolev embedding $H^{t_0}(\mathbb{R}^2)\subset L^\infty(\mathbb{R}^2)$.

7.3. A direct study of unidirectional waves in one dimension

We have shown in Corollary 7.2 that under the KdV regime (7.3), solutions to the water waves equations are approximated with a precision $O(\mu(1+\sqrt{t}))$ by the KdV approximation (7.5): The initial perturbation splits up into two counterpropagating components that evolve according to two uncoupled KdV equations.

It is possible to choose the initial perturbation so that, for instance, the left-going component is zero at leading order (just take $\zeta^0=w^0$ in Corollary 7.2). Physically speaking, this means that we focus our attention on *one* of the two counterpropagating waves after they have split. Investigating this configuration allows one to bring several interesting complements to Corollary 7.2:

- The convergence rate of the KdV approximation is improved to $O(\mu)$ uniformly on $[0,\frac{T}{\mu}]$, because the secular growth effects responsible for the $O(\mu\sqrt{t})$ terms, and that account for coupling effects between the counter-propagating components, are not present anymore.
- The assumption that $\varepsilon=\mu$ specific to the KdV regime (7.3) can be relaxed into the standard assumption $\varepsilon=O(\mu)$ of the long-wave regime.

- It is possible to investigate as in §5.1.2 the regime of *medium amplitude waves*, that is, to assume that $\varepsilon = O(\sqrt{\mu})$ rather than $\varepsilon = O(\mu)$. This regime therefore is more nonlinear than the KdV or long wave regime and surface perturbations are described by equations that can be related to the Camassa-Holm or Degasperis-Procesi equations. In particular, contrary to the KdV/BBM equations, these equations allow for *wave breaking*.

We first investigate the more general regime consisting of shallow water waves ($\mu \ll 1$) of medium amplitude ($\varepsilon = O(\sqrt{\mu})$), in one dimension ($\gamma = 0$) and over flat bottoms ($\beta = 0$). Since equations related to the Camassa-Holm equation can be derived under these assumptions, we call this the *Camassa-Holm regime*

$$\mathcal{A}_{CH} = \{(\varepsilon, \beta, \gamma, \mu), \quad 0 \le \mu \le \mu_0, \quad 0 \le \varepsilon \lesssim \sqrt{\mu}, \quad \beta = \gamma = 0\}. \tag{7.45}$$

The case of long waves ($\varepsilon = O(\mu)$) will then be considered as a particular case allowing for various simplifications. The results below were derived in [**96**].

7.3.1. The Camassa-Holm regime. As shown in §5.1.2.3, the relevant model to describe waves in the CH regime (7.45) is provided by the Green-Naghdi equations (5.22), which can be written in dimension $d = 1$ and over a flat bottom under the form

$$\begin{cases} \zeta_t + \big[(1+\varepsilon\zeta)\overline{v}\big]_x = 0 \\ \overline{v}_t + \zeta_x + \varepsilon\overline{v}\overline{v}_x = \dfrac{\mu}{3}\dfrac{1}{1+\varepsilon\zeta}\big[(1+\varepsilon\zeta)^3(\overline{v}_{xt} + \varepsilon\overline{v}\overline{v}_{xx} - \varepsilon\overline{v}_x^2)\big]_x, \end{cases} \tag{7.46}$$

where, for the sake of clarity, we write $\overline{v} = \overline{V}$ the one-dimensional version of the averaged velocity introduced in (3.31).

There are two different strategies: One can either look for a scalar equation satisfied by the "velocity" (not necessarily $\overline{v}$) and then reconstruct the free surface elevation ζ in terms of this velocity or one can look for an equation satisfied by ζ and reconstruct the velocity in terms of ζ. We will consider the first approach, and then indicate in §7.3.1.2 how to proceed with the second one. We then relate the family of equations derived here to the Camassa-Holm and Degasperis-Procesi equations in §7.3.1.4.

7.3.1.1. *Approximations based on the velocity.* Our goal here is to construct approximate solutions to (7.46) based on solutions to the following generalization of the KdV/BBM equations (see Remark 7.5)

$$u_t + u_x + \frac{3}{2}\varepsilon u u_x + \mu(\mathbf{a}u_{xxx} + \mathbf{b}u_{xxt}) = \varepsilon\mu(\mathbf{c}uu_{xxx} + \mathbf{d}u_x u_{xx}), \tag{7.47}$$

with some conditions on the coefficients **a**, **b**, **c** and **d**. We show that (for well-prepared initial data) the evolution of the averaged velocity $\overline{v}$ is correctly described by an equation of the form (7.47) and construct an approximation for ζ based on this solution. A larger choice for the coefficients **a**, **b**, **c** and **d** is allowed if as in §5.2.1.2, we work with the velocity $v_{\theta,\delta}$ defined as

$$v_{\theta,\delta} = (1 + \mu\theta\mathcal{T})^{-1}(1 + \mu\delta\mathcal{T})\overline{v}, \tag{7.48}$$

with $\theta \ge 0$, $\delta \ge 0$ and $\mathcal{T}$ as in (5.12), that is, in the present case of dimension $d = 1$ and flat bottoms

$$\mathcal{T}\cdot = -\frac{1}{3h}\partial_x(h^3\partial_x\cdot) \qquad (h = 1 + \varepsilon\zeta).$$

Therefore, if $v = v_{\theta,\delta}$ is given by an equation of the form (7.47), the corresponding approximation $\overline{v}_{CH}$ of the average velocity is obtained by inverting (7.48). We therefore get $\overline{v}_{CH} = (1+\mu(\theta-\delta)\mathcal{T})v + O(\mu^2)$, and neglecting the $O(\mu^2)$ terms[8]

$$\overline{v}_{CH} = v + \mu\lambda v_{xx} + 2\mu\varepsilon\lambda v v_{xx} + 3\mu\varepsilon\lambda v_x^2 \qquad (\lambda = -\frac{1}{3}(\theta-\delta)). \tag{7.49}$$

As mentioned above, with the "Camassa-Holm" approximation based on (7.47), we want to approximate unidirectional waves, that is, solutions to (7.46) that essentially go in one direction (e.g., to the right). Such waves are obtained for a particular choice of the surface elevation. For right-going waves, we have seen that this choice imposes $\zeta_{CH} = \overline{v}_{CH} + O(\varepsilon)$. A more precise expression (derived in the computations below) is provided by

$$\zeta_{CH} := \overline{v}_{CH} + \frac{\varepsilon}{4}\overline{v}_{CH}^2 + \mu\frac{1}{6}\partial_{xt}^2\overline{v}_{CH} - \varepsilon\mu\Big[\frac{1}{6}\overline{v}_{CH}\partial_x^2\overline{v}_{CH} + \frac{5}{48}(\partial_x\overline{v}_{CH})^2\Big]. \tag{7.50}$$

Before stating the main result of this section, let us recall that the spaces X^s and Y^s introduced to solve the Green-Naghdi equations (see Proposition 6.4) are defined in the one-dimensional case $d=1$ as

$$Y^s = H^s(\mathbb{R}) \times X^s, \qquad \text{with} \qquad X^s = \{v \in H^s(\mathbb{R}), |v|_{X^s}^2 := |v|_{H^s}^2 + \mu|\partial_x v|_{H^s}^2 < \infty\}.$$

Also note that in the statement below, $\overline{v}_{CH}$ and ζ_{CH} are defined in terms of v through (7.49) and (7.50).

THEOREM 7.24. *Let $s > 3/2$, $D \geq 0$, $\overline{v}^0 \in X^{s+D}$ and $(\varepsilon,\beta,\gamma,\mu) \in \mathcal{A}_{CH}$. Also let $\theta,\delta > 0$, $p \in \mathbb{R}$ and $\lambda = -\frac{1}{3}(\theta-\delta)$, and define*

$$\mathbf{a} = p + \lambda, \quad \mathbf{b} = p - \frac{1}{6} + \lambda, \quad \mathbf{c} = -\frac{3}{2}p - \frac{1}{6} - \frac{3}{2}\lambda, \quad \mathbf{d} = -\frac{9}{2}p - \frac{23}{24} - \frac{3}{2}\lambda.$$

If $\mathbf{b} < 0$, D is large enough, and $1 + \varepsilon\zeta_{CH\,|t=0} \geq h_{min} > 0$, then there exists $T > 0$ such that

(1) *There exists a unique solution $v \in C([0,T/\varepsilon]; X^{s+D})$ to (7.47) with initial data $v^0 = (1+\mu\theta\mathcal{T})^{-1}(1+\mu\delta\mathcal{T})\overline{v}^0$.*
(2) *There exists a unique solution $(\zeta_{GN}, \overline{v}_{GN}) \in C([0,T/\varepsilon]; Y^s)$ to (7.46) with initial condition $(\zeta_{CH\,|t=0}, \overline{v}^0)$.*
(3) *The following estimate holds for all $t \in [0,T/\varepsilon]$*

$$|(\zeta_{GN}, \overline{v}_{GN}) - (\zeta_{CH}, \overline{v}_{CH})|_{L^\infty([0,t]\times\mathbb{R})} \leq \mu^2 t\, C(\mu_0, \frac{1}{h_{min}}, \frac{1}{\mathbf{b}}, \mathbf{a}, \mathbf{b}, \mathbf{c}, \mathbf{d}, |\overline{v}^0|_{X^{s+D}}).$$

REMARK 7.25. Since we know that the Green-Naghdi equations furnish a good approximation to the water waves equations (see Theorem 6.15), a straightforward corollary of the above theorem (in the spirit of Corollary 7.2 above) shows that $(\zeta_{CH}, \overline{v}_{CH})$ is in a $O(\mu^2 t)$-neighborhood of the exact solution to the water waves equations with corresponding initial data.

[8]This is different from the formula used in [**96**] ($\overline{v}_{CH} = v + \mu\lambda v_{xx} + 2\mu\varepsilon\lambda v v_{xx}$), because for the sake of consistency with Chapter 5, we work here with the velocity $v_{\theta,\delta}$ which does not coincide in the CH regime with the velocity at some level line of the fluid domain, the choice used in [**96**]. We refer to §§5.6.2.1–5.6.2.2 for more comments on this point.

7.3.1.2. *Equations on the surface elevation.* Proceeding exactly as in the proof of Theorem 7.24, one can prove that the family of equations

$$\begin{aligned} \zeta_t + \zeta_x + \frac{3}{2}\varepsilon\zeta\zeta_x - \frac{3}{8}\varepsilon^2\zeta^2\zeta_x + \frac{3}{16}\varepsilon^3\zeta^3\zeta_x &+ \mu(\mathbf{a}\zeta_{xxx} + \mathbf{b}\zeta_{xxt}) \\ &= \varepsilon\mu(\mathbf{c}\zeta\zeta_{xxx} + \mathbf{d}\zeta_x\zeta_{xx}) \end{aligned} \tag{7.51}$$

(with some conditions on the coefficients **a**, **b**, **c** and **d**) for the evolution of the surface elevation can be used to construct an approximate solution to the Green-Naghdi equations (and hence to the water waves equations as explained in Remark 7.25).

Contrary to §7.3.1.1, it is ζ that is given by an evolution equation (namely, (7.51)) and an approximation $\overline{v}_{ch}$ of $\overline{v}$ that must be reconstructed in terms of ζ. The formula that stems from the computations is

$$\overline{v}_{ch} := \zeta + \frac{1}{h}\Big(-\frac{\varepsilon}{4}\zeta^2 - \frac{\varepsilon^2}{8}\zeta^3 + \frac{3\varepsilon^3}{64}\zeta^4 - \mu\frac{1}{6}\zeta_{xt} + \varepsilon\mu\big[\frac{1}{6}\zeta\zeta_{xx} + \frac{1}{48}\zeta_x^2\big]\Big).$$

Since the proof of the following result is completely similar to the proof of Theorem 7.24, we omit it.

THEOREM 7.26. *Let $s > 3/2$, $D \geq 0$, $\zeta^0 \in X^{s+D}$ and $(\varepsilon, \beta, \gamma, \mu) \in \mathcal{A}_{CH}$. Also let $q \in \mathbb{R}$ and define*

$$\mathbf{a} = q, \quad \mathbf{b} = q - \frac{1}{6}, \quad \mathbf{c} = -\frac{3}{2}q - \frac{1}{6}, \quad \mathbf{d} = -\frac{9}{2}q - \frac{5}{24}.$$

If $\mathbf{b} < 0$, D is large enough, and $1 + \varepsilon\zeta^0 \geq h_{min} > 0$, then there exists $T > 0$ such that

(1) *There exists a unique solution $\zeta \in C([0, T/\varepsilon]; X^{s+D})$ to (7.47) with initial data ζ^0.*
(2) *There exists a unique solution $(\zeta_{GN}, \overline{v}_{GN}) \in C([0, T/\varepsilon]; Y^s)$ to (7.46) with initial condition $(\zeta^0, \overline{v}_{ch\,|_{t=0}})$.*
(3) *The following estimate holds for all $t \in [0, T/\varepsilon]$*

$$|(\zeta_{GN}, \overline{v}_{GN}) - (\zeta, \overline{v}_{ch})|_{L^\infty([0,t]\times\mathbb{R})} \leq \mu^2 t\, C(\mu_0, \frac{1}{h_{min}}, \frac{1}{\mathbf{b}}, \mathbf{a}, \mathbf{b}, \mathbf{c}, \mathbf{d}, |\zeta^0|_{X^{s+D}}).$$

REMARK 7.27. Choosing $q = 1/12$, equation (7.51) reads

$$\begin{aligned} \zeta_t + \zeta_x + \frac{3}{2}\varepsilon\zeta\zeta_x - \frac{3}{8}\varepsilon^2\zeta^2\zeta_x + \frac{3}{16}\varepsilon^3\zeta^3\zeta_x + \frac{\mu}{12}(\zeta_{xxx} - \zeta_{xxt}) \\ = -\frac{7}{24}\varepsilon\mu(\zeta\zeta_{xxx} + 2\zeta_x\zeta_{xx}). \end{aligned} \tag{7.52}$$

While for any $q \in \mathbb{R}$, (7.51) is an equation for the evolution of the free surface ζ, and all these equations have the same order of accuracy $O(\varepsilon^4, \mu^2)$, it is more advantageous to use (7.52) since it presents better structural properties that we take advantage of for the study of wave breaking. The ratio 2 : 1 between the coefficients of $\zeta_x\zeta_{xx}$ and $\zeta\zeta_{xxx}$ is crucial for such considerations (see [**96**]).

REMARK 7.28. Noting that

$$\frac{3\varepsilon\zeta}{1 + \sqrt{1 + \varepsilon\zeta}} = \frac{3}{2}\varepsilon\zeta - \frac{3}{8}\varepsilon^2\zeta^2 + \frac{3}{16}\varepsilon^3\zeta^3 + O(\varepsilon^4),$$

the result of the theorem is left unchanged if (7.51) is replaced by

$$\zeta_t + \zeta_x + \varepsilon \frac{3\zeta}{1+\sqrt{1+\varepsilon\zeta}}\zeta_x + \mu(\mathbf{a}\zeta_{xxx} + \mathbf{b}\zeta_{xxt}) = \varepsilon\mu(\mathbf{c}\zeta\zeta_{xxx} + \mathbf{d}\zeta_x\zeta_{xx}). \tag{7.53}$$

This equation is more consistent with the simple wave equation (7.64) obtained below in the fully nonlinear regime $\varepsilon = O(1)$.

7.3.1.3. *Proof of Theorem* 7.24. The first point of the theorem follows directly from the well-posedness result given in Section 7.4.2 below; the second one is established in Proposition 6.4. We therefore focus on the proof of the error estimate that first requires some consistency results. In order to simplify the statements, we adopt the following definition.

DEFINITION 7.29. We say that an approximation $(\zeta_{app}, \overline{v}_{app})$ is *consistent* (of order $s \geq 0$ and on $[0, \frac{T}{\varepsilon}]$) with the Green-Naghdi equations (7.46) if (writing $(\zeta, \overline{v})$ instead of $(\zeta_{app}, \overline{v}_{app})$)

$$\begin{cases} \zeta_t + \big[(1+\varepsilon\zeta)\overline{v}\big]_x = \mu^2 r_1 \\ \overline{v}_t + \zeta_x + \varepsilon\overline{v}\,\overline{v}_x = \dfrac{\mu}{3}\dfrac{1}{1+\varepsilon\zeta}\big[(1+\varepsilon\zeta)^3(\overline{v}_{xt} + \varepsilon\overline{v}\,\overline{v}_{xx} - \varepsilon\overline{v}_x^2)\big]_x + \mu^2 r_2; \end{cases}$$

with (r_1, r_2) bounded in $L^\infty([0, \frac{T}{\varepsilon}], H^s(\mathbb{R})^2)$ independently of ε and μ.

The following proposition shows that there is a one-parameter family of equations of the form (7.47) consistent with the Green-Naghdi equations.

PROPOSITION 7.30. *Let $p \in \mathbb{R}$, $s \geq 0$ and assume that*

$$\mathbf{a} = p, \quad \mathbf{b} = p - \frac{1}{6}, \quad \mathbf{c} = -\frac{3}{2}p - \frac{1}{6}, \quad \mathbf{d} = -\frac{9}{2}p - \frac{23}{24}.$$

Then there exists $D > 0$ such that if $\overline{v}_{CH} = v \in C([0, \frac{T}{\varepsilon}]; H^{s+D}(\mathbb{R}))$ solves (7.47) for some $T > 0$, then $(\zeta_{CH}, \overline{v}_{CH})$, with

$$\zeta_{CH} := v + \frac{\varepsilon}{4}v^2 + \mu\frac{1}{6}v_{xt} - \varepsilon\mu\big[\frac{1}{6}vv_{xx} + \frac{5}{48}v_x^2\big],$$

is consistent (of order s and on $[0, \frac{T}{\varepsilon}]$) with the Green-Naghdi equations (7.46).

REMARK 7.31. The proof below shows that the result still holds if v does not solve (7.47) exactly but up to a $O(\mu^2)$ residual.

REMARK 7.32. Johnson [**191**] formally derived equations of the form (7.47) that can of course be understood in the present framework. For instance, one can recover the equations (26a) and (26b) of [**191**] with $p = -\frac{1}{12}$ (and thus $\mathbf{a} = -\frac{1}{12}$, $\mathbf{b} = -\frac{1}{4}$, $\mathbf{c} = -\frac{1}{24}$ and $\mathbf{d} = -\frac{7}{12}$) and $p = \frac{1}{6}$ (and thus $\mathbf{a} = \frac{1}{6}$, $\mathbf{b} = 0$, $\mathbf{c} = -\frac{5}{12}$ and $\mathbf{d} = -\frac{41}{24}$) respectively.

PROOF. **Step 1.** If v solves (7.47), then one also has

$$v_t + v_x + \varepsilon\frac{3}{2}vv_x + \mu a v_{xxt} = \varepsilon\mu\big[bvv_{xx} + cv_x^2\big]_x + O(\mu^2), \tag{7.54}$$

with $a = \mathbf{b} - \mathbf{a}$, $b = \mathbf{c} + \frac{3}{2}\mathbf{a}$ and $c = \frac{1}{2}(\mathbf{d} + 3\mathbf{a} - \mathbf{c})$.
Differentiating (7.47) twice with respect to x, one indeed gets

$$v_{xxx} = -v_{xxt} - \frac{3}{2}\varepsilon\partial_x^2(vv_x) + O(\mu),$$

and we can replace the v_{xxx} term of (7.47) by this expression to get (7.54).

Step 2. We seek w such that if $\zeta = v + \varepsilon w$ and v solves (7.47), then the second equation of (7.46) is satisfied up to a $O(\mu^2)$ term. This is equivalent to checking that

$$\begin{aligned} v_t + [v + \varepsilon w]_x + \varepsilon v v_x - \frac{\mu}{3} v_{xxt} &= \frac{\varepsilon\mu}{3}\big[- v v_{xxt} + [3 v v_{xt} + v v_{xx} - v_x^2]_x\big] + O(\mu^2), \\ &= -\frac{\varepsilon\mu}{3}\big[v v_{xx} + \frac{3}{2} v_x^2\big]_x + O(\mu^2). \end{aligned}$$

The last line is a consequence of the identity $v_t = -v_x + O(\varepsilon, \mu)$ provided by (7.54). The above equation can be recast under the form:

$$\begin{aligned} &\varepsilon w_x + \big[v_t + v_x + \varepsilon\frac{3}{2} v v_x + \mu a v_{xxt} - \varepsilon\mu\big[b v v_{xx} + c v_x^2\big]_x\big] \\ &\quad = \frac{\varepsilon}{2} v v_x + \mu(a + \frac{1}{3}) v_{xxt} - \varepsilon\mu\big[(b + \frac{1}{3}) v v_{xx} + (c + \frac{1}{2}) v_x^2\big]_x + O(\mu^2). \end{aligned}$$

From Step 1, we know that the term in brackets in the left-hand side of this equation is of order $O(\mu^2)$, so that that the second equation of (7.46) is satisfied up to $O(\mu^2)$ terms if

$$\varepsilon w_x = \frac{\varepsilon}{2} v v_x + \mu(a + \frac{1}{3}) v_{xxt} - \varepsilon\mu\big[(b + \frac{1}{3}) v v_{xx} + (c + \frac{1}{2}) v_x^2\big]_x + O(\mu^2),$$

so that we can take

$$\varepsilon w = \frac{\varepsilon}{4} v^2 + \mu(a + \frac{1}{3}) v_{xt} - \varepsilon\mu\big[(b + \frac{1}{3}) v v_{xx} + (c + \frac{1}{2}) v_x^2\big]. \tag{7.55}$$

Step 3. We choose the coefficients a, b and c such that the first equation of (7.46) is also satisfied up to $O(\mu^2)$ terms. This is equivalent to checking that

$$[v + \varepsilon w]_t + [(1 + \varepsilon v) v]_x + \varepsilon^2 [w v]_x = O(\mu^2). \tag{7.56}$$

First note that one infers from (7.55) that

$$\begin{aligned} \varepsilon\partial_t w &= \frac{\varepsilon}{2} v v_t + \mu(a + \frac{1}{3}) v_{xtt} - \varepsilon\mu\big[(b + \frac{1}{3}) v v_{xx} + (c + \frac{1}{2}) v_x^2\big]_t. \\ &= -\frac{\varepsilon}{2} v (v_x + \varepsilon\frac{3}{2} v v_x + \mu a v_{xxt}) - \mu(a + \frac{1}{3}) \partial_{xt}^2 (v_x + \varepsilon\frac{3}{2}(v v_x)) \\ &\quad + \varepsilon\mu\big[(b + \frac{1}{3}) v v_{xx} + (c + \frac{1}{2}) v_x^2\big]_x + O(\mu^2) \\ &= -\varepsilon\frac{1}{2} v v_x - \varepsilon^2\frac{3}{4} v^2 v_x - \mu(a + \frac{1}{3}) v_{xxt} \\ &\quad + \varepsilon\mu\big[(2a + b + \frac{5}{6}) v v_{xx} + (\frac{5}{4} a + c + 1) v_x^2\big]_x + O(\mu^2). \end{aligned}$$

Similarly, one gets

$$\varepsilon^2 [w v]_x = \varepsilon^2\frac{3}{4} v^2 v_x - \varepsilon\mu(a + \frac{1}{3})\big[v v_{xx}\big]_x + O(\mu^2),$$

so that (7.56) is equivalent to

$$v_t + v_x + \frac{3}{2} v v_x - \mu(a + \frac{1}{3}) v_{xxt} = \varepsilon\mu\big[-(a + b + \frac{1}{2}) v v_{xx} - (\frac{5}{4} a + c + 1) v_x^2\big]_x + O(\mu^2).$$

Equating the coefficients of this equation with those of (7.54) shows that the first equation of (7.46) is also satisfied at order $O(\mu^2)$ if the following relations hold:

$$a = -\frac{1}{6}, \qquad b = -\frac{1}{6}, \quad c = -\frac{19}{48},$$

and the conditions given in the statement of the proposition on $\mathbf{a}$, $\mathbf{b}$, $\mathbf{c}$ and $\mathbf{d}$ follow from the expressions of a, b and c given after equation (7.54). □

We will see in §7.3.1.4 below that none of the equations of the one-parameter family considered in Proposition 7.30 is completely integrable. This is why we now derive a wider class of equations of the form (7.47)—and whose solution can still be used as the basis of an approximate solution of the Green-Naghdi equations (7.46) (and thus of the water waves problem). This solution can be achieved by working with the velocity $v_{\theta,\delta}$ introduced in (7.48) rather than $\overline{v}$. The introduction of the parameters $\theta \geq 0$ and $\delta \geq 0$ allows us to derive an approximation consistent with (7.46) built on a two-parameter (p and $\theta - \delta$) family of equations of the form (7.47).

PROPOSITION 7.33. *Let $p \in \mathbb{R}$, $\theta \geq 0$, $\delta \geq 0$, and write $\lambda = -\frac{1}{3}(\theta - \delta)$. Assuming that*

$$\mathbf{a} = p + \lambda, \quad \mathbf{b} = p - \frac{1}{6} + \lambda, \quad \mathbf{c} = -\frac{3}{2}p - \frac{1}{6} - \frac{3}{2}\lambda, \quad \mathbf{d} = -\frac{9}{2}p - \frac{23}{24} - \frac{3}{2}\lambda,$$

there exists $D > 0$ such that if $v = v_{\delta,\theta} \in C([0, \frac{T}{\varepsilon}]; H^{s+D}(\mathbb{R}))$ solves (7.47) for some $T > 0$, then $(\zeta_{app}, \overline{v}_{app})$, with

$$\text{(7.57)} \qquad \overline{v}_{CH} = v + \mu\lambda v_{xx} + 2\mu\varepsilon\lambda v v_{xx} + 3\mu\varepsilon\lambda v_x^2,$$

$$\text{(7.58)} \qquad \zeta_{CH} := \overline{v} + \frac{\varepsilon}{4}\overline{v}^2 + \mu\frac{1}{6}\overline{v}_{xt} - \varepsilon\mu\big[\frac{1}{6}\overline{v}\overline{v}_{xx} + \frac{5}{48}\overline{v}_x^2\big] \qquad (\overline{v} = \overline{v}_{app}),$$

is consistent (of order s and on $[0, \frac{T}{\varepsilon}]$) with the Green-Naghdi equations (7.46).

PROOF. Thanks to Proposition 7.30 and Remark 7.31, it is sufficient to prove that if $\overline{v} = \overline{v}_{app}$ is given by (7.57) then it solves, up to $O(\mu^2)$ terms, an equation of the form (7.47) with coefficients $\mathbf{a}$, $\mathbf{b}$, $\mathbf{c}$ and $\mathbf{d}$ given by the formulas of Proposition 7.30 (i.e., with $\lambda = 0$). This follows by direct computations. □

Deducing the error estimate of Theorem 7.24 from the consistency result of Proposition 7.33 is then a direct consequence of Proposition 6.5.

7.3.1.4. *The Camassa-Holm and Degasperis-Procesi equations.* Among the various types of equations (7.47) with $\mathbf{b} \leq 0$ (a necessary condition for linear well-posedness) there are only two with a bi-Hamiltonian structure: the Camassa-Holm and the Degasperis-Procesi equations [**186**]. Notice that while the KdV equation has a bi-Hamiltonian structure (see [**130**]), this is not the case for the other members of the BBM family of equations (7.7). The importance of a bi-Hamiltonian structure lies in the fact that in general it represents the hallmark of a completely integrable Hamiltonian system whose solitary wave solutions are solitons, that is, localized waves that recover their shape and speed after interacting nonlinearly with another wave of the same type (see [**130, 190**]).

- *Camassa-Holm equations*: The Camassa-Holm (CH) equations are usually written under the form

$$\text{(7.59)} \qquad U_t + \widehat{\kappa} U_x + 3UU_x - U_{txx} = 2U_x U_{xx} + UU_{xxx},$$

with $\widehat{\kappa} \in \mathbb{R}$. A straightforward scaling argument shows that if $\widehat{\kappa} \neq 0$, (7.59) can be written under the form (7.47) by setting $u(t,x) = aU(b(x - vt), ct)$ and $a = \frac{2}{\varepsilon\widehat{\kappa}}$, $b^2 = -\frac{1}{\mathbf{b}\mu}$, $v = \frac{\mathbf{a}}{\mathbf{b}}$, $c = \frac{b}{\widehat{\kappa}}(1 - v)$ (which requires $\mathbf{b} < 0$ and leads to $\mathbf{c} = -\frac{\mathbf{b}}{2}$ and $\mathbf{d} = 2\mathbf{c}$). This motivates the following definition:

DEFINITION 7.34. We say that (7.47) is a Camassa-Holm equation if the following conditions hold:

$$\mathbf{b} < 0, \quad \mathbf{a} \neq \mathbf{b}, \quad \mathbf{b} = -2\mathbf{c}, \quad \mathbf{d} = 2\mathbf{c}.$$

For all $\widehat{\kappa} \neq 0$, the solution u to (7.47) is transformed into a solution U to (7.59) by the transformation

$$U(t,x) = \frac{1}{a}u(\frac{x}{b} + \frac{v}{c}t, \frac{t}{c}),$$

with $a = \frac{2}{\varepsilon\widehat{\kappa}}(1-v)$, $b^2 = -\frac{1}{\mathbf{b}\mu}$, $v = \frac{\mathbf{a}}{\mathbf{b}}$, and $c = \frac{b}{\widehat{\kappa}}(1-v)$.

First derived as a bi-Hamiltonian system by Fokas and Fuchssteiner [**148**], the equation (7.59) gained prominence after Camassa-Holm [**62**] independently re-derived it as an approximation to the Euler equations of hydrodynamics and discovered a number of the intriguing properties of this equation. In [**62**] a Lax pair formulation of (7.59) was found, a fact which lies at the core of showing via direct and inverse scattering [**89, 95**] that (7.59) is a completely integrable Hamiltonian system: for a large class of initial data, solving (7.59) amounts to integrating an infinite number of linear first-order ordinary differential equations which describe the evolution in time of the action-angle variables. The Camassa-Holm equations share with KdV this integrability property as well as the fact that its solitary waves are solitons [**110, 95**]. See [**130**] for a discussion of these properties in the context of the KdV model.

-*Degasperis-Procesi equations*: The Degasperis-Procesi (DP) equations are usually written under the form

$$U_t + \widehat{\kappa}U_x + 4UU_x - U_{txx} = 3U_xU_{xx} + UU_{xxx}, \tag{7.60}$$

with $\widehat{\kappa} \in \mathbb{R}$. The same scaling arguments as for the CH equation motivate the following definition:

DEFINITION 7.35. We say that (7.47) is a Degasperis-Procesi equation if the following conditions hold:

$$\mathbf{b} < 0, \quad \mathbf{a} \neq \mathbf{b}, \quad \mathbf{b} = -\frac{8}{3}\mathbf{c} \quad \mathbf{d} = 3\mathbf{c}.$$

For all $\widehat{\kappa} \neq 0$, the solution u to (7.47) is transformed into a solution U to (7.60) by the transformation

$$U(t,x) = \frac{1}{a}u(\frac{x}{b} + \frac{v}{c}t, \frac{t}{c}),$$

with $a = \frac{8}{3\varepsilon\widehat{\kappa}}(1-v)$, $b^2 = -\frac{1}{\mathbf{b}\mu}$, $v = \frac{\mathbf{a}}{\mathbf{b}}$ and $c = \frac{b}{\widehat{\kappa}}(1-v)$.

Equation (7.60), first derived in [**120**], is also known to have a Lax pair formulation [**118**] and its solitary waves interact like solitons [**255**]. Like the KdV equation (see [**130**]) and the Camassa-Holm equation (see [**228**]), the Degasperis-Procesi equation has infinitely many integrals of motion.

The question now is whether the equations (7.47) with the coefficients allowed by Theorem 7.24 contain CH or DP equations. We worked with the velocity $v_{\theta,\delta}$ defined in (7.48) rather than the averaged velocity $\overline{v}$ because the equations (7.47) found if we only work with $\overline{v}$ (see Proposition 7.30) do not contain CH or DP equations.
Indeed, the one-parameter family of equations (7.47) considered in Proposition 7.30 admits only one for which $\mathbf{d} = 2\mathbf{c}$ (obtained for $p = -\frac{5}{12}$, and thus $\mathbf{c} = \frac{11}{24}$, $\mathbf{d} = \frac{11}{12}$).

However, since $\mathbf{b} = -\frac{7}{12} \neq -2\mathbf{c}$, the corresponding equation is not a Camassa-Holm equation in the sense of Definition 7.34.
Similarly, there is no possible choice of p such that $\mathbf{d} = 3\mathbf{c}$ in Proposition 7.30. Consequently, none of the one-parameter family of equations is a Degasperis-Procesi equation.

It is the extra parameter $\lambda = -\frac{1}{3}(\theta - \delta)$, found by working with $v_{\theta,\delta}$, which allows one to find CH and DP equations in the framework of Theorem 7.24. Indeed, there exists (only) one set of coefficients such that $\mathbf{d} = 2\mathbf{c}$ and $\mathbf{b} = -2\mathbf{c}$ (corresponding to $p = -\frac{1}{3}$ and $\lambda = \frac{1}{12}$, and thus $\mathbf{a} = -\frac{1}{4}$, $\mathbf{b} = -\frac{5}{12} < 0$, $\mathbf{c} = \frac{5}{24}$, $\mathbf{d} = \frac{5}{12}$). The corresponding equation is therefore a Camassa-Holm equation in the sense of Definition 7.34:

$$u_t + u_x + \frac{3}{2}\varepsilon u u_x - \mu(\frac{1}{4}u_{xxx} + \frac{5}{12}u_{xxt}) = \frac{5}{24}\varepsilon\mu(u u_{xxx} + 2u_x u_{xx}). \tag{7.61}$$

There also exists only one set of coefficients such that $\mathbf{d} = 3\mathbf{c}$ and $\mathbf{b} = -\frac{8}{3}\mathbf{c}$ (obtained with $\lambda = \frac{11}{72}$ and $p = -\frac{77}{216}$, and thus $\mathbf{a} = -\frac{11}{54}$, $\mathbf{b} = -\frac{10}{27}$, $\mathbf{c} = \frac{5}{36}$, $\mathbf{d} = \frac{5}{12}$). The corresponding equation is therefore a Degasperis-Procesi equation in the sense of Definition 7.35:

$$u_t + u_x + \frac{3}{2}\varepsilon u u_x - \mu(\frac{11}{54}u_{xxx} + \frac{10}{27}u_{xxt}) = \frac{5}{36}\varepsilon\mu(u u_{xxx} + 3u_x u_{xx}).$$

REMARK 7.36. One of the attractive features of CH and DP equations is that they possess peakons (i.e., peaked solitons) [**62, 118**]. However, these particular solutions can be obtained only in the case $\widehat{\kappa} = 0$ in (7.59) and (7.60). Our analysis shows that the case $\widehat{\kappa} = 0$ is precisely the case that cannot be interpreted as a model for water waves in the present context.

7.3.2. The long-wave regime and the KdV and BBM equations. We are interested here in a subclass of the Camassa-Holm regime $\mathcal{A}_{CH}$ introduced in (7.45) for which surface perturbations are assumed to be of small rather than medium amplitude (i.e., $\varepsilon = O(\mu)$ rather than $\varepsilon = O(\sqrt{\mu})$). This is the *long-wave* regime that we saw in the previous chapter and that is given in the present one-dimensional case ($d = 1$) and flat topography by

$$\mathcal{A}_{LW} = \{(\varepsilon, \beta, \gamma, \mu), \quad 0 \leq \mu \leq \mu_0, \quad 0 \leq \varepsilon \lesssim \mu, \quad \beta = \gamma = 0\}.$$

In this regime, the family of equations (7.47) can be simplified into the following family of KdV/BBM equations

$$u_t + u_x + \frac{3}{2}\varepsilon u u_x + \mu((\mathbf{b} + \frac{1}{6})u_{xxx} + \mathbf{b}u_{xxt}) = 0, \tag{7.62}$$

with $\mathbf{b} \leq 0$ for linear well-posedness. The equations (7.62) can be used to approximate the surface elevation ζ, as shown in the following corollary of Theorem 7.26.

COROLLARY 7.37. *Let* $\mathbf{b} \leq 0$ *and let* $s > 3/2$, $D \geq 0$, $\zeta^0 \in H^{s+D}(\mathbb{R})$ *and* $(\varepsilon, \beta, \gamma, \mu) \in \mathcal{A}_{LW}$. *Also assume that* $1 + \varepsilon\zeta^0 \geq h_{min} > 0$. *Then there exists* $T > 0$ *such that (with* D *large enough)*

(1) *There exists a unique solution* $\zeta_{BBM} \in C([0, T/\varepsilon]; H^{s+D})$ *to (7.62) with initial data* ζ^0.
(2) *There exists a unique solution* $(\zeta_{GN}, \overline{v}_{GN}) \in C([0, T/\varepsilon]; Y^s)$ *to (7.46) with initial condition* $(\zeta^0, \overline{v}^0)$, *where* $\overline{v}^0 = \zeta^0 - \frac{\varepsilon}{4}\zeta^0 - \frac{\mu}{6}\zeta^0_{xt}$.

(3) *The following estimate holds for all* $t \in [0, T/\varepsilon]$

$$|(\zeta_{GN}, \overline{v}_{GN}) - (\zeta_{BBM}, \overline{v}_{BBM})|_{L^\infty([0,t]\times\mathbb{R})} \leq \mu^2 t\, C(\mu_0, \frac{1}{h_{min}}, \mathbf{b}, |\zeta^0|_{H^{s+D}}),$$

where $\overline{v}_{BBM} = \zeta_{BBM} - \frac{\varepsilon}{4}\zeta_{BBM}^2 - \frac{\mu}{6}\partial_{xt}^2\zeta_{BBM}$.

REMARK 7.38. A similar corollary can be derived from Theorem 7.24; in this case, the equation (7.62) is then used to model the evolution of the averaged velocity $\overline{v}$ and an approximation for the surface elevation is given by

$$\zeta = \overline{v} + \frac{\varepsilon}{4}\overline{v}^2 + \mu\overline{v}_{xt},$$

which corresponds to (7.50), where we neglected the terms that are $O(\mu^2)$ under the extra assumption $\varepsilon \leq \mu$ of the long-wave regime considered here.

REMARK 7.39. For all $\mathbf{b} \leq 0$, the equation (7.62) possesses solitary waves[9] of the form

$$\zeta(t, x) = \alpha \mathrm{sech}^2\big[K(x - ct)\big],$$

with

$$c = 1 + \varepsilon\frac{\alpha}{2} \quad \text{and} \quad K = \sqrt{\frac{3}{4}\frac{\varepsilon\alpha}{\mu(1 - 3\mathbf{b}\alpha\varepsilon)}}.$$

(We recall that $\mathbf{b} = 0$ corresponds to the KdV equation, and $\mathbf{b} = -1/6$ to the BBM equation.) These formulas are consistent with the formulas given in Remark 6.19 for the Green-Naghdi equations since they differ at order $O(\epsilon^2)$ for the speed and order $O(\varepsilon)$ for the shape (determined by K). Note that in the particular case $\mathbf{b} = -1/3$, the shape of the solitary wave for (7.62) is exactly the same as for the Green-Naghdi equation (but the speeds still differ at order $O(\varepsilon^2)$).

PROOF. The first point (local existence for the BBM equations (7.62)) is completely classical and goes back to Benjamin, Bona, and Mahony's original paper [**23**] (which also proved global existence).
For the other points of the corollary, one proceeds exactly as in the CH regime, neglecting all the terms that are $O(\mu^2)$ under the small amplitude assumption $\varepsilon \leq \mu$ made here (typically, $O(\varepsilon\mu)$ and $O(\varepsilon^2)$ are terms that have been neglected). □

7.3.3. The fully nonlinear regime. In the fully nonlinear shallow water regime (or briefly, shallow water regime), no smallness assumption is made on ε. The corresponding regime $\mathcal{A}_{SW}$ therefore is the same as the one introduced in §5.1.1, and which reads in dimension $d = 1$ and over flat topographies as

$$\mathcal{A}_{SW} = \{(\varepsilon, \beta, \gamma, \mu), \quad 0 \leq \mu \leq \mu_0, \quad 0 \leq \varepsilon \leq 1, \quad \beta = \gamma = 0\}.$$

We have just seen that for shallow water waves of *small* amplitude (or long waves), the KdV/BBM equations (7.62) furnish an approximation to the Green-Naghdi equations (7.46) of order $O(\mu^2)$; for shallow water waves of *medium* amplitude (the Camassa-Holm regime), the KdV/BBM equations have to be replaced by the more general equations (7.47). It is tempting to think that there exists a generalization to (7.47) that describes the evolution of *large* amplitude unidirectional waves, but such a generalization has not been derived so far.

[9]The stability of these solitary waves was proved very early in the case of the KdV [**22**] and BBM [**33**] equations.

It is, however, possible to derive a scalar approximation to the Green-Naghdi equation of order $O(\mu)$ rather than $O(\mu^2)$. At this level of accuracy, the Green-Naghdi equations can be replaced by the Nonlinear Shallow Water equations (5.7), which in dimension $d = 1$ and for a flat bottom are given by

$$\begin{cases} \partial_t\zeta + \partial_x(h\overline{v}) = 0, \\ \partial_t\overline{v} + \partial_x\zeta + \varepsilon\overline{v}\partial_x\overline{v} = 0, \end{cases} \tag{7.63}$$

with $h = 1 + \varepsilon\zeta$. This hyperbolic system admits *simple waves*, that is, solutions of the form $(\zeta, \overline{v}) = (\zeta, \underline{v}(\zeta))$ or $(\zeta, \overline{v}) = (\underline{\zeta}(\overline{v}), \overline{v})$. The scalar equation that must be satisfied by ζ (in the first case) or $\overline{v}$ (in the second case) and the corresponding mappings $\underline{v}(\cdot)$ or $\underline{\zeta}(\cdot)$ are found by plugging these expressions into (7.63). After simple computations, one finds the equation for ζ is

$$\partial_t\zeta + \partial_x\zeta + \varepsilon\frac{3\zeta}{1+\sqrt{1+\varepsilon\zeta}}\partial_x\zeta = 0, \quad \text{with} \quad \underline{v}(\zeta) = \frac{2}{\varepsilon}\big(\sqrt{1+\varepsilon\zeta} - 1\big). \tag{7.64}$$

For $\overline{v}$, one finds

$$\partial_t\overline{v} + \partial_x\overline{v} + \varepsilon\frac{3}{2}\overline{v}\partial_x\overline{v} = 0, \quad \text{and} \quad \underline{\zeta}(\overline{v}) = \overline{v} + \varepsilon\frac{1}{4}\overline{v}^2. \tag{7.65}$$

Both equations (7.64) and (7.65) are Burgers-type equations that can be solved explicitly by the method of characteristic for times of order $O(1/\varepsilon)$. Since (7.64) and (7.65) are *exact* solutions of the NSW equations (7.63), it is a direct consequence of Theorem 6.10 that (7.64) and (7.65) furnish an approximation of order $O(\mu t)$ over this time scale to the full water waves equations (4.1) with corresponding initial data.

7.4. Supplementary remarks

7.4.1. Historical remarks on the KDV equation. One of the reasons the KdV equation attracted a lot of interest is its ability to model correctly "solitary waves". Such waves were first observed by a naval engineer, John Scott Russell, in 1834:

> I was observing the motion of a boat which was rapidly drawn along a narrow channel by a pair of horses, when the boat suddenly stopped – not so the mass of water in the channel which it had put in motion; it accumulated round the prow of the vessel in a state of violent agitation, then suddenly leaving it behind, rolled forward with great velocity, assuming the form of a large solitary elevation, a rounded, smooth and well-defined heap of water, which continued its course along the channel apparently without change of form or diminution of speed. I followed it on horseback, and overtook it still rolling on at a rate of some eight or nine miles an hour, preserving its original figure some thirty feet long and a foot to a foot and a half in height. Its height gradually diminished, and after a chase of one or two miles I lost it in the windings of the channel. Such, in the month of August 1834, was my first chance interview with that singular and beautiful phenomenon which I have called the *Wave of Translation* [**280**].

> This phenomenon I observed again and again as often as the vessel, after having been put in rapid motion, was suddenly stopped; and the accompanying circumstances of the phenomenon were so uniform, and some consequences of its existence so obvious and important, that I was induced to make *The wave* the subject of numerous experiments [**279**].

In his investigations of what he called the *great primary wave of displacement*, Russell noticed two important features:

(1) The speed of the wave is given by

$$c = \sqrt{g(H_0 + \zeta_{max})},$$

where ζ_{max} is the highest elevation of the wave above the rest height.

(2) They have a definite shape for a given height.

Russell's observations were violently refuted in 1845 by Airy [**8**], who proposed a nonlinear equation for the evolution of the wave which is equivalent to the following Burgers equation[10]

$$\partial_t \zeta + \sqrt{g(H_0 + \frac{3}{2}\zeta)}\partial_x \zeta = 0.$$

From the nonlinear nature of this equation, Airy deduced that it was impossible to have a disturbance propagating without any change of shape. In his opinion, Russell's observations could simply be explained by the fact that the amplitude of the wave is small enough for the linear theory (due to Lagrange) to apply, and he concluded

> We are not disposed to recognize this wave as deserving the epithets "great" or "primary".

Russell argued that his observation that "his" waves have a shape determined by their height is not consistent with Lagrange's linear theory, but Russell was attacked about another aspect by Stokes [**302**] in 1847. Russell had explained the slow decay of the wave by friction effects, but Stokes did not agree:

> Thus the degradation in the height of such waves, which Mr. Russell observed, is not to be attributed wholly, (nor I believe chiefly,) to the imperfect fluidity of the fluid. . .but is an essential characteristic of a solitary wave.

Stokes explained this decay by the phenomenon of dispersion, essentially modeling the evolution of the wave by the linear dispersive equation

$$\partial_t \zeta + \sqrt{gH_0}\partial_x \zeta + \sqrt{gH_0}\frac{H_0^2}{6}\partial_x^3 \zeta = 0$$

(which, quite ironically, is often referred to as the Airy equation); the strong dispersion of this equation indeed induces a decay at rate $O(t^{-1/2})$.

It was Boussinesq who first understood that both nonlinear and dispersive effects should be taken into account to explain Russell's experiment. Looking for

[10] Note that this equation is the first order approximation in ε of (7.64) — and it is written in dimensional form.

solitary waves to the Euler equations, Boussinesq derived in 1871 [**47**] the following formula relating the wave parametrization ζ of the solitary wave to its speed c

$$c^2 = gH_0\big(1+\frac{3}{2}\frac{\zeta}{H_0}+\frac{1}{3}\frac{\zeta''}{\zeta}\big).$$

This formula explains the two crucial observations[11] by Russell mentioned above. In a footnote on page 360 of his 680-page masterpiece *Essai sur la théorie des eaux courantes* [**49**] Boussinesq also derived in 1877 the Korteweg-de Vries equation[12]

$$\partial_t\zeta+\sqrt{gH_0}\partial_x\zeta+\sqrt{gH_0}\frac{H_0^2}{6}\partial_x^3\zeta+\frac{3}{2}\sqrt{gH_0}\frac{\zeta}{H_0}\partial_x\zeta=0,$$

more than twenty years before Korteweg and de Vries rediscovered it [**211**].

Most of the material for these historical remarks has been taken from [**260, 272, 119, 113**], to which we refer readers for more historical aspects related to the Korteweg-de Vries equation.

7.4.2. Large time well-posedness of (7.47) and (7.51). We prove here the well-posedness of the general class of equations

$$u_t+u_x+F(\varepsilon u)u_x+\mu(\mathbf{a}u_{xxx}+\mathbf{b}u_{xxt})=\varepsilon\mu(\mathbf{c}uu_{xxx}+\mathbf{d}u_xu_{xx}), \tag{7.66}$$

with F a smooth mapping defined in a neighborhood $\mathcal{O}$ of the origin, and vanishing at the origin. In particular, (7.66) coincides with (7.51) if one takes $F(x)=\frac{3}{2}x-\frac{3}{8}x^2+\frac{3}{16}x^3$. It coincides with (7.53) if one takes $F(x)=\frac{3x}{1+\sqrt{1+x}}$ and finally, it coincides with (7.47) if $F(x)=\frac{3}{2}x$.

We therefore are interested in solving the initial value problem

$$\begin{cases} u_t+u_x+F(\varepsilon u)u_x+\mu(\mathbf{a}u_{xxx}+\mathbf{b}u_{xxt})=\varepsilon\mu(\mathbf{c}uu_{xxx}+\mathbf{d}u_xu_{xx}),\\ u_{\vert t=0}=u^0\end{cases} \tag{7.67}$$

on a time scale $O(1/\varepsilon)$, and under the condition $\mathbf{b}<0$. We consider here values of ε and μ corresponding to the Camassa-Holm regime

$$\mathcal{A}_{CH}=\{(\varepsilon,\beta,\gamma,\mu),\quad 0\le\mu\le\mu_0,\quad 0\le\varepsilon\le\sqrt{\mu},\quad \beta=\gamma=0\}.$$

In order to state the result, we recall the definition of the spaces X^s

$$\forall s\ge 0,\qquad X^s=H^{s+1}(\mathbb{R})\text{ endowed with the norm }|f|_{X^s}^2=|f|_{H^s}^2+\mu|\partial_x f|_{H^s}^2.$$

We must also assume that the initial condition u^0 takes its values in the domain of definition[13] $\mathcal{O}$ of the nonlinearity F

$$\forall x\in\mathbb{R},\qquad u^0(x)\in\mathcal{O}. \tag{7.68}$$

---

[11] For the first observation, note that after integrating the formula once, one gets

$$\big(\frac{\zeta'}{H_0}\big)^2=3\big(\frac{\zeta}{H_0}\big)^2\big[\big(\frac{c^2}{gH_0}-1\big)-\frac{\zeta}{H_0}\big].$$

At the point where the waves reach its maximal depth ζ_{max}, one has $\zeta'=0$ and therefore $[(\frac{c^2}{gH_0}-1)-\frac{\zeta_{max}}{H_0}]=0$, which yields Russell's formula.

[12] Note that the ODE relating the surface parametrization to the velocity for solitary waves of the KdV equation is

$$c=\sqrt{gH_0}\big(1+\frac{3}{4}\frac{\zeta}{H_0}+\frac{1}{6}\frac{\zeta''}{\zeta}\big),$$

which matches Boussinesq's previous formula up to terms of order $O((\zeta/H_0)^2)$ (or $O(\varepsilon^2)$ in dimensionless version), which as we showed earlier in this chapter, is the size of the residual terms neglected in this approximation.

[13] One can take $\mathcal{O}=\mathbb{R}$ for (7.47) and (7.51); for (7.51), one must take $\mathcal{O}=(-1,+\infty)$.

PROPOSITION 7.40. *Assume that* $\mathbf{b} < 0$ *and let* $s > \frac{3}{2}$ *and* $u^0 \in X^s(\mathbb{R})$, *satisfying (7.68). Then, there exists* $T > 0$ *and a unique solution* $C([0, \frac{T}{\varepsilon}]; X^s(\mathbb{R})) \cap C^1([0, \frac{T}{\varepsilon}]; X^{s-1}(\mathbb{R}))$ *to (7.67). Moreover, for all* $0 \leq t \leq T/\varepsilon$

$$|u(t, \cdot)|_{X^s} \leq C(\mu_0, F, \frac{1}{\mathbf{b}}, \mathbf{c}, \mathbf{d}, |u^0|_{X^s}).$$

PROOF. For all v smooth enough, let us define the "linearized" operator $\mathcal{L}(v, \partial)$ as

$$\begin{aligned}\mathcal{L}(v, \partial) &= (1 + \mu\mathbf{b}\partial_x^2)\partial_t + \partial_x + \mu\mathbf{a}\partial_x^3 + F(\varepsilon v)\partial_x \\ &\quad -\varepsilon\mu\mathbf{c}v\partial_x^3 - \varepsilon\mu\mathbf{d}\big(\frac{1}{2}v_x\partial_x^2 + \frac{1}{2}v_{xx}\partial_x\big).\end{aligned}$$

In order to construct a solution to (7.66) by an iterative scheme, we are led to study the initial value problem

$$\begin{cases} \mathcal{L}(v, \partial)u = \varepsilon f, \\ u_{|t=0} = u^0. \end{cases} \tag{7.69}$$

If v is smooth enough, it is completely standard to check that for all $s \geq 0$, $f \in L^1_{loc}(\mathbb{R}^+_t; H^s(\mathbb{R}_x))$ and $u^0 \in X^s$, there exists a unique solution $u \in C(\mathbb{R}^+; X^s)$ to (7.69) (recall that $\mathbf{b} < 0$). We take for granted the existence of a solution to (7.69) and establish some precise energy estimates on the solution. In order to do so, let us define the "energy" norm

$$\forall s \geq 0, \qquad E^s(u)^2 = |u|^2_{H^s} - \mu\mathbf{b}|\partial_x u|^2_{H^s}.$$

Differentiating $\frac{1}{2}e^{-\varepsilon\lambda t}E^s(u)$ with respect to time, one gets, using the equation (7.69) and integrating by parts

$$\begin{aligned}\frac{1}{2}e^{\varepsilon\lambda t}\partial_t(e^{-\varepsilon\lambda t}E^s(u)^2) &= -\frac{\varepsilon\lambda}{2}E^s(u)^2 + \varepsilon(\Lambda^s f, \Lambda^s u) - \varepsilon(\Lambda^s(V\partial_x u), \Lambda^s u) \\ &\quad +\varepsilon\mu\mathbf{c}(\Lambda^s(v\partial_x^3 u), \Lambda^s u) - \varepsilon\mu\frac{\mathbf{d}}{2}(\Lambda^s(v_x\partial_x u), \Lambda^s\partial_x u),\end{aligned}$$

with $V = \frac{1}{\varepsilon}F(\varepsilon v)$.
Since for all constant coefficient skewsymmetric differential polynomial P (that is, $P^* = -P$), and all h smooth enough, one has

$$(\Lambda^s(hPu), \Lambda^s u) = ([\Lambda^s, h]Pu, \Lambda^s u) - \frac{1}{2}([P, h]\Lambda^s u, \Lambda^s u),$$

we deduce (applying this identity with $P = \partial_x$ and $P = \partial_x^3$)

$$\begin{aligned}\frac{1}{2}e^{\varepsilon\lambda t}\partial_t(e^{-\varepsilon\lambda t}E^s(u)^2) &= -\frac{\varepsilon\lambda}{2}E^s(u)^2 - \varepsilon\big([\Lambda^s, V]\partial_x u, \Lambda^s u\big) + \frac{\varepsilon}{2}((\partial_x V)\Lambda^s u, \Lambda^s u) \\ &-\varepsilon\mu\mathbf{c}\big([\Lambda^s, v]\partial_x^2 u - \frac{3}{2}v_x\Lambda^s\partial_x u - v_{xx}\Lambda^s u, \Lambda^s\partial_x u\big) - \varepsilon\mu\mathbf{c}\big([\Lambda^s, v_x]\partial_x^2 u, \Lambda^s u\big) \\ &-\varepsilon\mu\frac{\mathbf{d}}{2}(\Lambda^s(v_x\partial_x u), \Lambda^s\partial_x u) + \varepsilon(\Lambda^s f, \Lambda^s u).\end{aligned}$$

Here we also used the identities

$$[\Lambda^s, v]\partial_x^3 u = \partial_x\big([\Lambda^s, v]\partial_x^2 u\big) - [\Lambda^s, v_x]\partial_x^2 u$$

and

$$\frac{1}{2}(v_{xxx}\Lambda^s u, \Lambda^s u) = -(v_{xx}\Lambda^s u, \Lambda^s u_x).$$

Since $|u|_{H^s} \leq E^s(u)$ and $\sqrt{\mu}|\partial_x u|_{H^s} \leq \frac{1}{\sqrt{-\mathbf{b}}}E^s(u)$, one gets directly by the Cauchy-Schwartz inequality

$$\begin{aligned} e^{\varepsilon\lambda t}\partial_t(e^{-\varepsilon\lambda t}E^s(u)^2) &\leq \varepsilon C(\mu_0, \frac{1}{\mathbf{b}}, \mathbf{c}, \mathbf{d})(A(u,v)E^s(u) + B(v)E^s(u)^2) \\ &\quad -\varepsilon\lambda E^s(u)^2 + 2|f|_{H^s}E^s(u), \end{aligned}$$

with

$$\begin{aligned} A(u,v) &= |[\Lambda^s, V]\partial_x u|_2 + |[\Lambda^s, v]\partial_x(\sqrt{\mu}\partial_x u)|_2 + |[\Lambda^s, \sqrt{\mu}v_x]\partial_x(\sqrt{\mu}\partial_x u)|_2 \\ &\quad +\sqrt{\mu}|v_x\partial_x u|_{H^s}, \\ B(v) &= |\partial_x V|_\infty + |v_x|_\infty + |\partial_x(\sqrt{\mu}\partial_x v)|_\infty. \end{aligned}$$

Recalling (see point (3) of Proposition B.8 in Appendix B) that for all $s > 3/2$, and all G, U smooth enough, one has

$$|[\Lambda^s, G]U|_2 \leq \mathrm{Cst}\, |\partial_x G|_{H^{s-1}}|U|_{H^{s-1}},$$

and noting that the third point of Proposition B.2 in Appendix B also yields

$$|\partial_x V|_{H^{s-1}} = |F'(\varepsilon v)v_x|_{H^{s-1}} \leq C(F, |v|_{H^s}),$$

it is easy to check that one gets $A(u,v) \leq C(\mu_0, \frac{1}{\mathbf{b}}, F, E^s(v))E^s(u)$ and $B(v) \leq C(\mu_0, \frac{1}{\mathbf{b}}, F, E^s(v))$. Therefore, we obtain

$$e^{\varepsilon\lambda t}\partial_t(e^{-\varepsilon\lambda t}E^s(u)^2) \leq \big(C(\mu_0, M, \frac{1}{\mathbf{b}}, \mathbf{c}, \mathbf{d}, F, E^s(v)) - \lambda\big)E^s(u)^2 + 2\varepsilon E^s(f)E^s(u).$$

Taking $\lambda = \lambda_T$ large enough — how large depends on $C(\mu_0, \frac{1}{\mathbf{b}}, \mathbf{c}, \mathbf{d}, F, E^s(v))$—to have the first term of the right-hand side negative for all $t \in [0, \frac{T}{\varepsilon}]$, one deduces

$$\forall t \in [0, \frac{T}{\varepsilon}], \qquad \partial_t(e^{-\varepsilon\lambda_T t}E^s(u)^2) \leq 2\varepsilon e^{-\varepsilon\lambda_T t}E^s(f)E^s(u).$$

Integrating this differential inequality therefore yields

$$\forall t \in [0, \frac{T}{\varepsilon}], \qquad E^s(u)(t) \leq e^{\varepsilon\lambda_T t}E^0(u^0) + 2\varepsilon\int_0^t e^{\lambda_T(t-t')}E^s(f(t'))dt'.$$

Thanks to this energy estimate, one can conclude classically (see, e.g., [**11**]) to the existence of

$$T = T(\mu_0, |u^0|_{X^s}, \frac{1}{\mathbf{b}}, \mathbf{c}, \mathbf{d}, F) > 0,$$

and of a unique solution $u \in C([0, \frac{T}{\varepsilon}]; X^s)$ to (7.67) as a limit of the iterative scheme

$$u_0 = u^0, \qquad \text{and} \qquad \forall n \in \mathbb{N}, \qquad \begin{cases} \mathcal{L}(u^n, \partial)u^{n+1} = 0, \\ u^{n+1}_{|t=0} = u^0. \end{cases}$$

Since u solves (7.66), we have $\mathcal{L}(u, \partial u)u = 0$ and therefore

$$(\Lambda^{s-1}(1 + \mu\mathbf{b}\partial_x^2)\partial_t u, \Lambda^{s-1}\partial_t u) = -\varepsilon(\Lambda^{s-1}\mathcal{M}(u, \partial)u, \Lambda^{s-1}\partial_t u),$$

with $\mathcal{M}(u, \partial) = \mathcal{L}(u, \partial) - (1 + \mu\mathbf{b}\partial_x^2)\partial_t$. Proceeding as above, one gets

$$E^{s-1}(\partial_t u) \leq C(\mu_0, |u^0|_{X^s}, \frac{1}{\mathbf{b}}, \mathbf{c}, \mathbf{d}, F, E^s(u)),$$

and it follows that the family of solutions is also bounded in $C^1([0, \frac{T}{\varepsilon}]; X^{s-1})$. □

7.4.3. The case of nonflat bottoms. Various generalizations of the scalar equations derived here exist in the case of nonflat bottoms, provided that their variations are sufficiently slow and small. More precisely, let us assume that the bottom is parametrized by

$$z = -1 + \beta b^{(\alpha)}(x), \qquad \text{with} \qquad b^{(\alpha)}(x) = b(\alpha x), \qquad \beta \ll 1, \quad \alpha \ll 1 \tag{7.70}$$

(the smallness of β accounts for the smallness of the bottom variations, and α for their slowness).

7.4.3.1. *Generalization of the KdV equation for nonflat bottoms.* In the long-wave regime $\varepsilon = O(\mu)$ and for flat bottoms, we have seen in §7.3.2 that it is possible to build approximate solutions to the Green-Naghdi equations (and hence to the full water waves equations) based on the KdV[14] equation. More precisely, if ζ solves

$$\zeta_t + \zeta_x + \varepsilon\frac{3}{2}\zeta\zeta_x + \mu\frac{1}{6}\zeta_{xxx} = 0,$$

then $(\zeta, \zeta - \frac{\varepsilon}{4}\zeta^2 - \frac{\mu}{6}\zeta_{xt})$ approximates the solution $(\zeta_{GN}, \overline{v}_{GN})$ of the Green-Naghdi equations at order $O(\mu^2 t)$.
Similarly, if u solves

$$u_t + u_x + \varepsilon\frac{3}{2}uu_x + \mu\frac{1}{6}u_{xxx} = 0,$$

then $(u + \frac{\varepsilon}{4}u^2 + \frac{\mu}{6}u, u)$ approximates the solution $(\zeta_{GN}, \overline{v}_{GN})$ of the Green-Naghdi equations at order $O(\mu^2 t)$.

The generalization of these two KDV equations when the bottom is as in (7.70) and therefore not necessarily flat, is given respectively by

$$\zeta_t + c\zeta_x + \frac{1}{2}c_x\zeta + \varepsilon\frac{3}{2c}\zeta\zeta_x + \mu\frac{1}{6}c^5\zeta_{xxx} = 0,$$

and

$$u_t + cu_x + \varepsilon\frac{3}{2}uu_x + \mu\frac{1}{6}c^5 u_{xxx} + \frac{3}{2}c_x u = 0,$$

where $c(x) = \sqrt{1 - \beta b^{(\alpha)}(x)}$. Note in particular that in variable depth, the KdV equation for the velocity differs from the equation for the free surface elevation. See [**259, 188, 165**] for a formal derivation of these equations, and [**183**] for a rigorous justification under appropriate assumptions on the size of α and β.

7.4.3.2. *Generalization of the CH/DP equations for nonflat bottoms.* The so-called CH regime is more nonlinear than the KdV regime ($\varepsilon = O(\sqrt{\mu})$ versus $\varepsilon = O(\mu)$) and the equations that the surface elevation or the velocity must solve to build an approximation of the Green-Naghdi equations are more complicated than the KdV equation. As shown in §7.3.1, the equations for the velocity u (not necessarily $\overline{v}$) or the surface elevation ζ are (7.47) and (7.51), namely,

$$u_t + u_x + \frac{3}{2}\varepsilon uu_x + \mu(\mathbf{a}u_{xxx} + \mathbf{b}u_{xxt}) = \varepsilon\mu(\mathbf{c}uu_{xxx} + \mathbf{d}u_x u_{xx}),$$

and

$$\begin{aligned}\zeta_t + \zeta_x + \frac{3}{2}\varepsilon\zeta\zeta_x - \frac{3}{8}\varepsilon^2\zeta^2\zeta_x + \frac{3}{16}\varepsilon^3\zeta^3\zeta_x \quad &+ \quad \mu(\mathbf{a}\zeta_{xxx} + \mathbf{b}\zeta_{xxt}) \\ &= \quad \varepsilon\mu(\mathbf{c}\zeta\zeta_{xxx} + \mathbf{d}\zeta_x\zeta_{xx})\end{aligned}$$

with conditions on the coefficients **a**, **b**, **c** and **d** provided by Theorems 7.24 or 7.26 respectively. Various generalizations of these equations are derived and justified in

[14]The results of this section can, of course, be adapted to the BBM equation.

[**183**] under appropriate conditions on α and β. For instance, if $\alpha = \varepsilon$ and $\beta = \mu^{3/2}$, a variable bottom generalization for the surface elevation is given by

$$\begin{aligned}\zeta_t + c\zeta_x + \frac{1}{2}c_x\zeta + \frac{3}{2}\varepsilon\zeta\zeta_x - \frac{3}{8}\varepsilon^2\zeta^2\zeta_x + \frac{3}{16}\varepsilon^3\zeta^3\zeta_x &+ \mu(\mathbf{a}\zeta_{xxx} + \mathbf{b}\zeta_{xxt}) \\ &= \varepsilon\mu(\mathbf{c}\zeta\zeta_{xxx} + \mathbf{d}\zeta_x\zeta_{xx}),\end{aligned}$$

with $c(x) = \sqrt{1 - \beta b^{(\alpha)}(x)}$. Under weaker assumptions on α and β, the dependence on the bottom topography is, of course, stronger (see [**183**]).

7.4.4. Wave breaking. We say that there is wave breaking[15] for a scalar equation as those derived in this chapter, if there exists a time $0 < t^{\varepsilon,\mu} < \infty$ and solutions f to this equation such that

$$f \in L^\infty([0, t^{\varepsilon,\mu}] \times \mathbb{R}) \quad \text{and} \quad \lim_{t \to t^{\varepsilon,\mu}} |\partial_x f(t,\cdot)|_\infty = \infty.$$

Global well-posedness results for the KdV and BBM equations (e.g., [**204, 23**]) show that wave breaking *cannot* occur for these models. This observation led Whitham [**324**] and many others to look for other nonlinear dispersive models that could model breaking waves.

Two particular models, namely, the Camassa-Holm equation [**62**] and the Degasperis-Procesi equation [**120**] raised a lot of interest. They indeed possess smooth solutions that develop singularities according to the wave breaking scenario described above: the solution remains bounded but its slope becomes unbounded. The issue of rigorously justifying these equations as models for the description of water waves has been addressed in §7.3.1.4 above.

It is worth recalling that in the asympotic regime that leads to this model

$$\mathcal{A}_{CH} = \{(\varepsilon, \beta, \gamma, \mu), \quad 0 \le \mu \le \mu_0, \quad 0 \le \varepsilon \lesssim \sqrt{\mu}, \quad \beta = \gamma = 0\},$$

the equations describing the evolution of the velocity unknown and the surface are not the same (although they are identical in the less nonlinear KdV regime): the equations for the velocity are of the form (7.47), while they are of the form (7.51) or (7.53) for the surface elevation.

While wave breaking with the meaning given above can be observed both for the velocity and for the surface elevation, it has been shown in [**96**] that they are of a different nature. For the velocity, one has $\lim_{t \to t^{\varepsilon,\mu}} \partial_x u(t,\cdot) = -\infty$, while for the surface elevation, $\lim_{t \to t^{\varepsilon,\mu}} \partial_x \zeta(t,\cdot) = \infty$. This is one of the reasons why one should be careful when using CH/DP type equations to model wave breaking. More precisely, one should keep in mind that

(1) For standard wave breaking as observed on the beaches, one expects wave breaking on the front part of the wave, and therefore $\lim_{t \to t^{\varepsilon,\mu}} \partial_x \zeta(t,\cdot) = -\infty$ (and not $+\infty$).
(2) Wave breaking on beaches is due to topography, while for the CH/DP type equations, the bottom is assumed to be flat.

[15]This is in some sense a weak definition of wave breaking. An overturning wave breaks according to the definition given here, without necessarily developing any singularity at this point. For a recent example (in infinite depth) of a wave which is initially a graph and which ceases in finite time to be a graph (overturning), see [**66**].

(3) The justification of the CH/DP type equations provided in §7.3.1 holds only under the assumption that the solutions and their derivatives remain bounded, which is obviously not the case near the breaking point. We illustrate this fact in Figure 7.2, which shows that (7.51) and (7.53) provide as expected very similar approximations as long as the wave is not too steep, but that they are dramatically different when the wave steepens.

The fact that CH/DP type equations exhibit wave breaking therefore is important, because this reveals that these equations contain enough nonlinearity to overcome the effects of dispersion (as in real-life wave breaking). We cannot, however, draw definitive practical conclusions from the study of the wave breaking scenarios for these equations.[16]

7.4.5. Full dispersion versions of the scalar shallow water approximations.

7.4.5.1. *One dimensional models.* The KdV/BBM family of equations (7.62) and the CH/DP family (7.47) only differ in their nonlinear terms. Their (linear) dispersion relation is therefore the same; it depends on the parameter $\mathbf{b} \leq 0$ that appears in the linear part of these equations, and is given by

$$\omega_{\mathbf{b}}(k) = k\frac{1-\mu(\mathbf{b}+\frac{1}{6})k^2}{1-\mu\mathbf{b}k^2}, \qquad \mathbf{b} \leq 0$$

(the condition $\mathbf{b} \leq 0$ therefore ensures that the equations (7.62) and (7.47) are linearly well posed). The corresponding celerity is therefore given by

$$c_{\mathbf{b}}(\sqrt{\mu}k) = \frac{1-\mu(\mathbf{b}+\frac{1}{6})k^2}{1-\mu\mathbf{b}k^2},$$

which can be compared to the celerities c_{WW} and c_α of the full water waves equations and of the one parameter family of Green-Naghdi systems (5.36),

$$c_{WW}(\sqrt{\mu}k) = \Big(\frac{\tanh(\sqrt{\mu}k)}{\sqrt{\mu}k}\Big)^{1/2}, \qquad c_\alpha(\sqrt{\mu}k) = \Big(\frac{1+\mu\frac{\alpha-1}{3}k^2}{1+\mu\frac{\alpha}{3}k^2}\Big)^{1/2},$$

as computed in (5.26) and Remark 5.24 respectively (and more generally to the four parameters family of approximations derived in Remark 5.33). The celerity $c_{\mathbf{b}}(\sqrt{\mu}k)$ for the KdV/BBM and CH/DP equations coincides therefore at order $O(\mu^2)$ with $c_{WW}(\sqrt{\mu}k)$ and $c_\alpha(\sqrt{\mu}k)$ and, as in §5.2 of Chapter 5, one can optimize the choice of $\mathbf{b}$ to have the best possible match with the exact linear celerity $c_{WW}(\sqrt{\mu}k)$. One can also propose *full dispersion* variants of the KdV/BBM and CH/DP equations. The full dispersion KdV equation is given by

$$\partial_t u + c_{WW}(\sqrt{\mu}D)u_x + \varepsilon\frac{3}{2}uu_x = 0, \tag{7.71}$$

where $c_{WW}(\sqrt{\mu}D)$ is a Fourier multiplier. Its linear dispersion relation satisfies exactly the linear dispersion relation of the water waves equations. This equation is also known as the Whitham equation after Whitham proposed it in [**322**]; he was expecting that because of its weaker dispersion than the KdV equation, this new equation would allow wave breaking and peaking. It is not difficult to check that

[16] This situation is actually quite similar to the one encountered for the choice of the "best" Green-Naghdi model in §5.2.3. There are certainly some models among the families (7.51) or (7.53) that have a range of validity much wider than expected; as in Figure 5.4, comparisons with experimental data could help us at this point.

Corollary 7.37 still holds for (7.71) since solutions to (7.71) exist and are consistent at order $O(\mu^2)$ with the BBM/KdV equations (7.62) on the relevant time scale.

A full dispersion CH/DP-type equation can be similarly proposed. For instance, for the surface elevation, the equation

$$\partial_t\zeta + c_{WW}(\sqrt{\mu}D)\zeta_x + \varepsilon\frac{3\zeta}{1+\sqrt{1+\varepsilon\zeta}}\zeta_x = \varepsilon\mu(\frac{5}{12}\zeta\zeta_{xxx} + \frac{23}{24}\zeta_x\zeta xx), \tag{7.72}$$

is the full dispersion version of (7.53).

7.4.5.2. *The weakly transverse case.* As in §7.2, we consider here the KP regime (7.25), for which $\gamma = \sqrt{\mu}$ and $\varepsilon = \mu$. The dispersion relation (5.26) for the full linear water waves equations is then given by

$$\begin{aligned}\omega_{WW}(\mathbf{k})^2 &= (\mathbf{k}_1^2+\mu\mathbf{k}_2^2)\frac{\tanh(\sqrt{\mu}\sqrt{\mathbf{k}_1^2+\mu\mathbf{k}_2^2})}{\sqrt{\mu}\sqrt{\mathbf{k}_1^2+\mu\mathbf{k}_2^2}}\\ &= (\mathbf{k}_1^2+\mu\mathbf{k}_2^2)(1-\frac{\mu}{3}\mathbf{k}_1^2)+O(\mu^2)\\ &= \mathbf{k}_1^2\big[1+\mu(\frac{\mathbf{k}_2^2}{\mathbf{k}_1^2}-\frac{1}{3}\mathbf{k}_1^2)\big]+O(\mu^2),\end{aligned}$$

where $\mathbf{k} = (\mathbf{k}_1, \mathbf{k}_2)$. A second order Taylor expansion of the square root of this expression then gives

$$\begin{aligned}\omega_{WW}(\mathbf{k}) &= \pm\mathbf{k}_1\big[1+\frac{\mu}{2}(\frac{\mathbf{k}_2^2}{\mathbf{k}_1^2}-\frac{1}{3}\mathbf{k}_1^2)\big]+O(\mu^2)\\ &= \pm\omega_{KP}(\mathbf{k})+O(\mu^2),\end{aligned}$$

where $\omega_{KP}(\mathbf{k})$ is the dispersion relation associated to the KP equation (7.28),

$$\omega_{KP}(\mathbf{k}) = \mathbf{k}_1 + \mu\frac{1}{2}\frac{\mathbf{k}_2^2}{\mathbf{k}_1} - \mu\frac{1}{6}\mathbf{k}_1^3;$$

the dispersion relation of the KP equation therefores matches at order $O(\mu^2)$ the dispersion relation of the full linear water waves equation in the weakly transverse case $\gamma = \sqrt{\mu}$. More generally, the dispersion relation associated with the KP-BBM family of equations (7.29) indexed by $p \leq 1/6$ is

$$\omega_{KP,p}(\mathbf{k}) = \frac{\mathbf{k}_1 + \mu\frac{1}{2}\frac{\mathbf{k}_2^2}{\mathbf{k}_1} + \mu p\mathbf{k}_1^3}{1+\mu(p-\frac{1}{6})\mathbf{k}_1^2} \qquad (p \leq 1/6),$$

and only differs from the previous one by $O(\mu^2)$ terms. Note finally, that as done above for the KdV/BBM and CH/DP equation, one can write a *full dispersion KP equation*,

$$\partial_t u + c_{WW}(\sqrt{\mu}|D^\gamma|)\Big(1+\mu\frac{D_2^2}{D_1^2}\Big)^{1/2}u_x + \varepsilon\frac{3}{2}uu_x = 0, \tag{7.73}$$

where we recall that $|D^\gamma|$ is the Fourier multiplier $|D^\gamma| = \sqrt{D_1^2+\gamma^2D_2^2}$, with $D_j = -i\partial_j$ $(j = 1, 2)$ and, here, $\gamma = \sqrt{\mu}$.

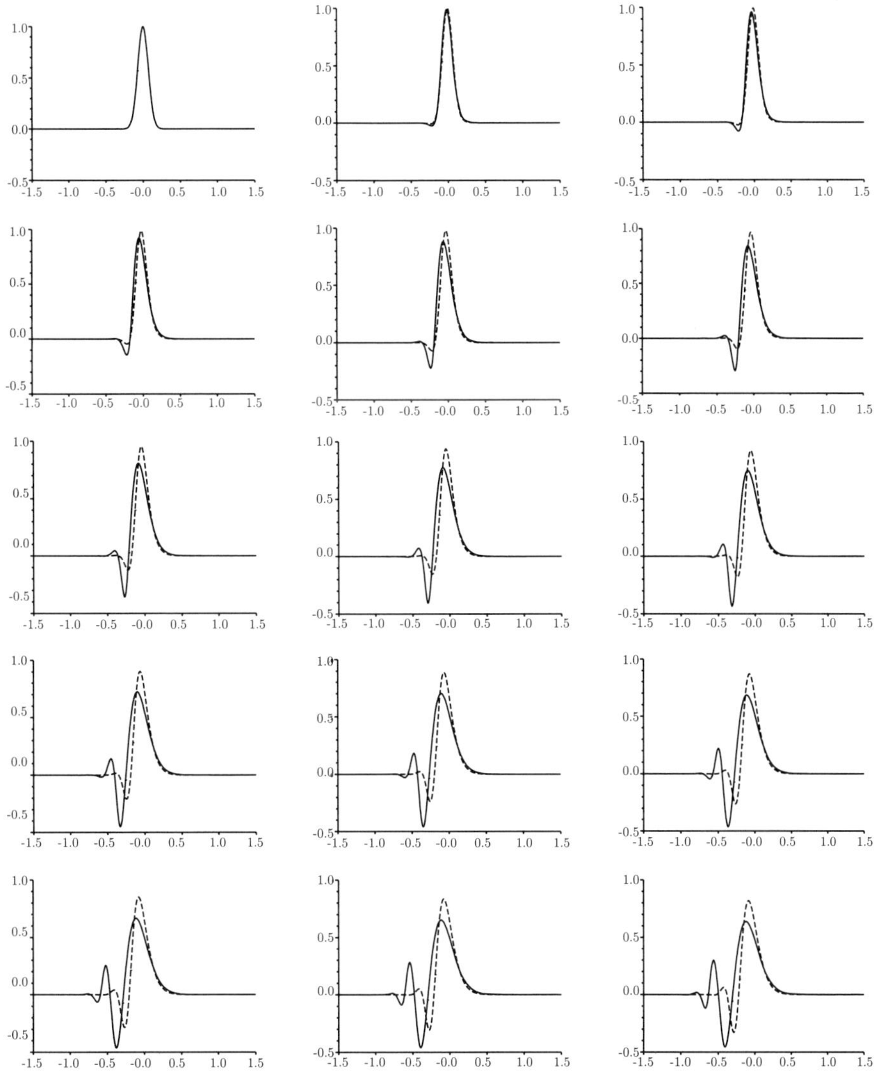

FIGURE 7.2. Left to right, top to bottom: evolution of the surface elevation according to (7.51) and (7.53) — solid and dashed curves respectively.

CHAPTER 8

Deep Water Models and Modulation Equations

In this chapter we discontinue the study of shallow water models: no smallness assumption is made on μ, but the surface elevation must now be small. Asymptotic expansions are then made with respect to the *steepness* $\epsilon = \varepsilon\sqrt{\mu} = a/L_x$ (the ratio of the typical amplitude over the typical horizontal scale) of the waves.

We first consider, in Section 8.1, the so-called "deep water" model formally derived in §8.1.1 from the water waves equations by keeping the leading order terms in the expansion with respect to ϵ. Since the linear properties of the deep water model are the same as those of the water waves equations, it is sometimes called the "full-dispersion" model. We prove, in §8.1.2, that it can be used to construct consistent approximations to the water waves equation. This allows us to bring in §8.1.3 an *almost* full justification of the deep water model (full justification requires a local well-posedness result for these equations that is not known yet). For the sake of completeness, we also give in §8.1.4 the form taken by these equations in infinite depth.

We then investigate, in Section 8.2, the approximation of "wave packets", that is, of fast oscillating waves whose amplitude is slowly varying. In order to describe these wave packets, one has to derive approximations on their slowly varying amplitude. The equations that describe their evolution are called *modulation equations*. They play an important role in mathematical physics since they appear in many different physical contexts (in nonlinear optics, for instance). The relevant *ansatz* is defined in §8.2.1 and a small amplitude expansion of the water waves equations is given in §8.2.2 for wave packets; this allows us to determine the different components of the ansatz in §8.2.3.

Several modulation equations are then derived. We first obtain, in §8.2.4, a Benney-Roskes model with "full-dispersion", from which we derive in §8.2.5 the standard Benney-Roskes model. Focusing on the components of the wave travelling at the group velocity, we obtain, in §8.2.6, the Davey-Stewartson equations in dimension $d = 2$; in dimension $d = 1$, we show, in §8.2.7, that these equations degenerate into a cubic nonlinear equation which is focusing in deep water and defocusing in shallow water.

We then indicate, in Section 8.3, how to adapt the above results to the case of infinite depth. Since the creation of nonoscillating terms by nonlinear interaction of oscillating modes (rectification effects) is smaller in infinite depth than in finite depth, we modify the general *ansatz* in §8.3.1 and show, in §8.3.2, that the cubic nonlinear Schrödinger equation is a relevant model both in dimension $d = 1$ and $d = 2$.

Consistency results are proved for all the models mentioned above, but their full justification remains open in most cases. We briefly comment on this issue in Section 8.4.

Finally, we provide, in Section 8.5, some additional results. We briefly recall in §8.5.1 the historical context of the derivation of the NLS equation in the context of Benjamin-Feir instabilities for periodic wave-trains. In §8.5.2, we propose "full dispersion" versions of the Davey-Stewartson and NLS models, and in §8.5.3 we propose some models with "improved" dispersion relation. Then, in §8.5.4, we mention the Dysthe equation, which is a higher order model taking into account the rectification effects neglected by the standard cubic NLS equation in infinite depth. As mentioned previously, the NLS equation in dimension $d = 1$ is focusing in deep water and defocusing in shallow water; therefore, there is a critical depth at which the nonlinearity vanishes. This situation is addressed in §8.5.5. The inclusion of surface tension is very briefly mentioned in §8.5.6.

Throughout this chapter, for the sake of simplicity, we take $\gamma = 1$ (full transversality) and consider only flat bottoms $b = 0$ (except for the model (8.4)).

8.1. A deep water (or full-dispersion) model

We say that we are in deep water when the shallowness assumption $\mu = \frac{H_0^2}{L_x^2} \ll 1$ made throughout Chapters 5, 6, and 7 is not satisfied. For such regimes, it does not make sense to derive asymptotic models to the water waves equations (5.1) in terms of the parameter μ, and the relevant parameter is the *steepness* ϵ

$$\epsilon = \varepsilon\sqrt{\mu} = \frac{a_{surf}}{L_x}.$$

For the sake of simplicity, we consider only flat bottoms here ($b = 0$) and fully transverse (or isotropic) waves ($\gamma = 1$); the typical "deep water" asymptotic regime investigated in this section therefore is

$$\mathcal{A}_{DW} = \{(\varepsilon, \beta, \gamma, \mu), \quad 0 \leq \mu \leq \mu_{max}, \quad 0 \leq \varepsilon \leq 1, \quad \beta = 0, \quad \gamma = 1, \quad \varepsilon\sqrt{\mu} \leq \epsilon_0\},$$

where $\mu_{max} > 0$ is not necessarily small and ϵ_0 is an upper bound for the steepness of the wave, whose smallness will determine the precision of the model derived in this section. To emphasize the fact that we do not make any shallow water assumption, we do not set the coefficient ν to 1 in the water waves equations (1.28)

$$\text{(8.1)} \qquad \begin{cases} \partial_t\zeta - \dfrac{1}{\mu\nu}\mathcal{G}\psi = 0, \\ \partial_t\psi + \zeta + \dfrac{\varepsilon}{2\nu}|\nabla\psi|^2 - \dfrac{\varepsilon\mu}{\nu}\dfrac{(\frac{1}{\mu}\mathcal{G}\psi + \nabla(\varepsilon\zeta)\cdot\nabla\psi)^2}{2(1+\varepsilon^2\mu|\nabla\zeta|^2)} = 0. \end{cases}$$

REMARK 8.1. We still assume in this section that μ remains bounded ($\mu \leq \mu_{max}$) so that

$$\nu = \frac{\tanh(2\pi\sqrt{\mu})}{2\pi\sqrt{\mu}}$$

remains of order $O(1)$, and it would be equivalent to set it equal to 1 in (8.1). However, the derivation of the "full-dispersion" model below remains valid as $\mu \to \infty$ and therefore $\nu \sim (2\pi\mu)^{-1/2}$. As mentioned in Chapter 1, we have chosen in these notes to assume that μ remains bounded for the sake of simplicity. For a derivation and justification of the "full-dispersion" model in the framework $\mu \to \infty$, see **[12]**. See also §8.1.4 below for the case of infinite depth.

The deep water model is first derived in §8.1.1; a consistency result is then proved in §8.1.2 and the model is then (almost) fully justified in §8.1.3.

8.1.1. Derivation. The "deep water" or "full-dispersion" model below was first derived in [**253, 254, 76**] and couples the evolution of the free surface elevation ζ to the horizontal velocity $\underline{V}$ at the surface

$$\left\{\begin{array}{l}\partial_t\zeta - \frac{1}{\sqrt{\mu}\nu}\mathcal{H}_\mu\underline{V} + \frac{\varepsilon}{\nu}\big(\mathcal{H}_\mu(\zeta\nabla\mathcal{H}_\mu\underline{V}) + \nabla\cdot(\zeta\underline{V})\big) = 0,\\ \partial_t\underline{V} + \nabla\zeta + \frac{\varepsilon}{\nu}\big(\frac{1}{2}\nabla|\underline{V}|^2 - \nu\sqrt{\mu}\nabla\zeta\mathcal{H}_\mu\nabla\zeta\big) = 0,\end{array}\right. \tag{8.2}$$

where $\mathcal{H}_\mu$ is a Fourier multiplier defined as

$$\forall V \in \mathfrak{S}(\mathbb{R}^2)^2, \qquad \widehat{\mathcal{H}_\mu V}(\xi) = -\frac{\tanh(\sqrt{\mu}|\xi|)}{|\xi|}(i\xi)\cdot\widehat{V}(\xi).$$

This deep water model is formally derived by replacing in the water waves equations (5.1) the Dirichlet-Neumann operator $\mathcal{G}\psi$ by its small amplitude expansion given by Proposition 3.44

$$\begin{aligned}\mathcal{G}\psi &= \mathcal{G}_0\psi + \varepsilon\mu\big[-\frac{1}{\mu}\mathcal{G}_0(\zeta(\mathcal{G}_0\psi)) - \nabla\cdot(\zeta\nabla\psi))\big] + O(\varepsilon^2\mu)\\ &= \sqrt{\mu}\mathcal{H}_\mu\nabla\psi + \varepsilon\mu\big[-\mathcal{H}_\mu(\zeta\mathcal{H}_\mu\nabla\psi) - \nabla\cdot(\zeta\nabla\psi)\big] + O(\mu^2),\end{aligned} \tag{8.3}$$

with $\mathcal{G}_0 = \sqrt{\mu}|D|\tanh(\sqrt{\mu}|D|\cdot$. Taking the gradient of the second equation of (8.1), and replacing $\nabla\psi$ by an asymptotic expansion in terms of $\underline{V}$, we get

$$\begin{aligned}\nabla\psi &= \underline{V} + \varepsilon\nabla\zeta\frac{\mathcal{G}\psi + \varepsilon\mu\nabla\zeta\cdot\nabla\psi}{1+\epsilon^2|\nabla\zeta|^2}\\ &= \underline{V} + \epsilon(\mathcal{H}_\mu\underline{V})\nabla\zeta + O(\epsilon^2),\end{aligned}$$

where we used Proposition 3.44 to derive the second identity. A rigorous derivation is provided in the next section.

REMARK 8.2. A generalization to nonflat bottoms has been proposed in [**223**]:

$$\left\{\begin{array}{l}\partial_t\zeta - \frac{1}{\sqrt{\mu}\nu}\mathcal{H}_\mu\underline{V} + \frac{\varepsilon}{\nu}\big(\mathcal{H}_\mu(\zeta\nabla\mathcal{H}_\mu\underline{V}) + \nabla\cdot(\zeta\underline{V})\big) = \frac{\beta}{\nu}\nabla\cdot(B_\mu\underline{V}),\\ \partial_t\underline{V} + \nabla\zeta + \frac{\varepsilon}{\nu}\big(\frac{1}{2}\nabla|\underline{V}|^2 - \nu\sqrt{\mu}\nabla\zeta\mathcal{H}_\mu\nabla\zeta\big) = 0,\end{array}\right. \tag{8.4}$$

with the operator B_μ given by

$$B_\mu = \mathrm{sech}(\sqrt{\mu}|D|)\big[b\,\mathrm{sech}(\sqrt{\mu}|D|)\cdot\big].$$

This system is derived as (8.2) but uses the expansion (3.49) instead of Proposition 3.44.

8.1.2. Consistency of the deep water (or full-dispersion) model. As previously explained, the full justification of an asymptotic model to the water waves equations requires a consistency result that can be of two different types: the water waves equations are consistent with the asymptotic models or the asymptotic model is consistent with the water waves equations. In Chapter 6, we used results of the first type to provide a full justification of the shallow water models derived in Chapter 5.

In order to justify the deep water (or full-dispersion) model (8.2), we will use a consistency result of the second type (mainly because the well-posedness theory for (8.2) is not known yet).

Let us recall that the deep water (or full-dispersion) model is given by

$$(8.5)\qquad \begin{cases} \partial_t \zeta - \frac{1}{\sqrt{\mu}\nu}\mathcal{H}_\mu \underline{V} + \frac{\varepsilon}{\nu}\big(\mathcal{H}_\mu(\zeta\nabla\mathcal{H}_\mu \underline{V}) + \nabla\cdot(\zeta\underline{V})\big) = 0, \\ \partial_t \underline{V} + \nabla\zeta + \frac{\varepsilon}{\nu}\big(\frac{1}{2}\nabla|\underline{V}|^2 - \nu\sqrt{\mu}\nabla\zeta\mathcal{H}_\mu\nabla\zeta\big) = 0, \end{cases}$$

where $\mathcal{H}_\mu$ is a Fourier multiplier defined as

$$\forall V \in \mathfrak{S}(\mathbb{R}^2)^2, \qquad \widehat{\mathcal{H}_\mu V}(\xi) = -\frac{\tanh(\sqrt{\mu}|\xi|)}{|\xi|}(i\xi)\cdot \widehat{V}(\xi).$$

The following proposition shows that this model is consistent with the water waves equations[1]. We recall that $\epsilon = \varepsilon\sqrt{\mu}$.

PROPOSITION 8.3 (Full dispersion model). *Let $N \in \mathbb{N}$, $\psi^0 \in \dot{H}^{N+9/2}(\mathbb{R}^d)$ and $T > 0$. Also let $(\zeta, \underline{V}) \in C([0, \frac{\nu T}{\varepsilon}], H^{N+7/2}(\mathbb{R}^d)^{1+d})$ be a solution of (8.2) with initial condition $(\zeta^0, \nabla\psi^0 - \epsilon(\mathcal{H}_\mu\nabla\psi^0)\nabla\zeta^0)$.*
Then for $0 < \epsilon < \epsilon_0$ with ϵ_0 small enough, there exists $\psi \in C([0, \frac{\nu T}{\varepsilon}]; \dot{H}^{N+2})$ such that

$$\forall 0 \le t \le \frac{\nu T}{\varepsilon}, \qquad |\nabla\psi(t) - \underline{V}(t)|_{H^{N+1}} \le \epsilon t\frac{\varepsilon}{\nu}C(|(\zeta,\underline{V})|_{L^\infty([0,t];H^{N+7/2})})$$

and moreover, (ζ, ψ) is consistent with the water waves equations (8.1) in the following sense

$$\begin{cases} \partial_t\zeta - \frac{1}{\mu\nu}\mathcal{G}\psi = \epsilon\frac{\varepsilon}{\nu}r^1, \\ \partial_t\psi + \zeta + \frac{\varepsilon}{2\nu}|\nabla\psi|^2 - \frac{\varepsilon\mu}{\nu}\frac{(\frac{1}{\mu}\mathcal{G}\psi + \nabla(\varepsilon\zeta)\cdot\nabla\psi)^2}{2(1+\varepsilon^2\mu|\nabla\zeta|^2)} = \epsilon\frac{\varepsilon}{\nu}r^2, \end{cases}$$

with

$$\forall 0 \le t \le \frac{\nu T}{\varepsilon}, \qquad |r^1(t)|_{H^N} + |\mathfrak{P}r^2(t)|_{H^N} \le C(|(\zeta,\underline{V})|_{L^\infty([0,t];H^{N+7/2})}).$$

PROOF. Since $\mathcal{H}_\mu : H^r(\mathbb{R}^2)^2 \mapsto H^r(\mathbb{R}^2)$ is continuous with operator norm bounded from above by 1, the mapping $V \in H^r(\mathbb{R}^2)^2 \mapsto V - \epsilon(\mathcal{H}_\mu V)\nabla\zeta \in H^r(\mathbb{R}^2)^2$ is continuous for all $r \ge 0$ and $\zeta \in H^{r\vee t_0+1}(\mathbb{R}^2)$. Moreover, this mapping is invertible for ϵ small enough, and one can accordingly define $\widetilde{V} := (1-\epsilon\nabla\zeta\mathcal{H}_\mu)^{-1}\underline{V}$, so that $\underline{V} = \widetilde{V} - \epsilon(\mathcal{H}_\mu\widetilde{V})\nabla\zeta$. Replacing $\underline{V}$ by this expression in (8.5) gives

$$(8.6)\qquad \begin{cases} \partial_t\zeta - \frac{1}{\sqrt{\mu}\nu}\mathcal{H}_\mu\widetilde{V} + \frac{\varepsilon}{\nu}\big(\nabla\cdot(\zeta\widetilde{V}) + \mathcal{H}_\mu\nabla(\zeta\mathcal{H}_\mu\widetilde{V})\big) = \epsilon\frac{\varepsilon}{\nu}R^1, \\ \partial_t\widetilde{V} + \nabla\zeta + \frac{1}{2}\frac{\varepsilon}{\nu}\big(\nabla|\widetilde{V}|^2 - \nabla(\mathcal{H}_\mu\widetilde{V})^2\big) = \epsilon\frac{\varepsilon}{\nu}R^2, \end{cases}$$

with

$$(8.7)\qquad \begin{aligned} |R|_{H^{N+3/2}} &\le C(|\zeta|_{H^{N+7/2}}, |\widetilde{V}|_{H^{N+5/2}}) \\ &\le C(|\zeta|_{H^{N+7/2}}, |\underline{V}|_{H^{N+5/2}}). \end{aligned}$$

Now, let $\partial_t + L$ denote the linear part of the above system

$$L := \begin{pmatrix} 0 & -\frac{1}{\sqrt{\mu}\nu}\mathcal{H}_\mu \\ \nabla & 0 \end{pmatrix},$$

[1]As in (8.1) we choose to keep ν in the equations rather than setting it to 1, as we did for shallow water models. See also Remark 8.7.

and $S(t)$ its evolution operator: for all $U = (\zeta, V)$, $S(t)U := u(t)$, where u solves $(\partial_t + L)u = 0$, with initial condition $u_{|_{t=0}} = U$. Since $\mathcal{H}_\mu$ is a Fourier multiplier, one can find an explicit expression for $S(t)$, but we need only the following property: $S(t)$ is unitary on Z^r ($r \in \mathbb{R}$) defined as

$$Z^r := \{U = (\zeta, V) \in H^r(\mathbb{R}^2)^3, |U|_{Z^r} := |\zeta|_{H^r} + \big|\big(\frac{\tanh(\sqrt{\mu}|D|)}{\sqrt{\mu}\nu|D|}\big)^{1/2} V\big|_{H^r} < \infty\}.$$

Writing $\widetilde{u} := (\zeta, \widetilde{V})$, we define $u^\sharp := (\zeta^\sharp, V^\sharp)$ as

$$u^\sharp := \widetilde{u} - \epsilon\frac{\varepsilon}{\nu}\int_0^t S(t-t')R(t')dt' \qquad (R = (R^1, R^2)^T).$$

One immediately checks that $u^\sharp$ solves

$$\begin{cases} \partial_t \zeta^\sharp - \frac{1}{\sqrt{\mu}\nu}\mathcal{H}_\mu V^\sharp + \frac{\varepsilon}{\nu} f^1(\zeta^\sharp, V^\sharp) = \epsilon\frac{\varepsilon}{\nu}k^1, \\ \partial_t V^\sharp + \nabla\zeta^\sharp + \frac{\varepsilon}{\nu}\nabla f^2(\zeta^\sharp, V^\sharp) = \epsilon\frac{\varepsilon}{\nu}\nabla k^2, \end{cases}$$

with initial condition $u^\sharp_{|_{t=0}} = (\zeta^0, \nabla\psi^0)^T$, and where $k^1 := \frac{1}{\epsilon}(f^1(u^\sharp) - f^1(\widetilde{u}))$, $k^2 := \frac{1}{\epsilon}(f^2(u^\sharp) - f^2(\widetilde{u}))$, and

$$f^1(\zeta, V) := \nabla\cdot(\zeta V) + \mathcal{H}_\mu\nabla(\zeta\mathcal{H}_\mu V), \qquad f^2(\zeta, V) := \frac{1}{2}\big(|V|^2 - (\mathcal{H}_\mu V)^2\big).$$

Noting that

$$|V|_{H^r} \lesssim \big|\big(\frac{\tanh(\sqrt{\mu}|D|)}{\sqrt{\mu}\nu|D|}\big)^{1/2} V\big|_{H^{r+1/2}} \lesssim |V|_{H^{r+1/2}},$$

uniformly with respect to μ, and since $S(t)$ is unitary on Z^r, one gets

$$\sup_{[0,\frac{\nu T}{\varepsilon}]} |u^\sharp(t) - \widetilde{u}(t)|_{H^r} \lesssim \epsilon T \sup_{[0,\frac{\nu T}{\varepsilon}]} |R|_{H^{r+1/2}}. \tag{8.8}$$

Together with standard product estimates, this implies that

$$\begin{aligned} |k^1|_{H^N} + |\mathfrak{P}k^2|_{H^N} &\leq TC(T, |\zeta|_{H^{N+1}}, |\widetilde{V}|_{H^{N+1}})|R|_{H^{N+3/2}} \\ &\leq TC(T, |\zeta|_{H^{N+7/2}}, |\underline{V}|_{H^{N+7/2}}), \end{aligned} \tag{8.9}$$

where the last inequality stems from (8.7).
Using the fact that all the terms in the equation on $V^\sharp$ are a gradient of a scalar expression, as well as $V^\sharp_{|_{t=0}}$, it is possible to write $u^\sharp = (\zeta^\sharp, \nabla\psi)^T$, and $(\zeta^\sharp, \psi)$ solves

$$\begin{cases} \partial_t \zeta^\sharp - \frac{1}{\sqrt{\mu}\nu}\mathcal{H}_\mu\nabla\psi + \frac{\varepsilon}{\nu} f^1(\widetilde{\zeta}, \nabla\psi) = \epsilon\frac{\varepsilon}{\nu}k^1, \\ \partial_t\psi + \zeta^\sharp + \frac{\varepsilon}{\nu} f^2(\zeta^\sharp, \nabla\psi) = \epsilon\frac{\varepsilon}{\nu}k^2, \end{cases}$$

with initial condition $(\zeta^\sharp, \psi)_{|_{t=0}} = (\zeta^0, \psi^0)$. Noting now that $\mathcal{G}[0]\psi = \sqrt{\mu}\mathcal{H}_\mu\nabla\psi$, one can check that this system of equations can be written

$$\begin{cases} \partial_t\zeta^\sharp - \frac{1}{\mu\nu}\mathcal{G}^{(1)}\psi = \epsilon\frac{\varepsilon}{\nu}k^1, \\ \partial_t\psi + \zeta + \frac{\varepsilon}{2\nu}|\nabla\psi|^2 - \frac{\varepsilon\mu}{\nu}\frac{\big(\frac{1}{\mu}\mathcal{G}_0\psi + \nabla(\varepsilon\zeta^\sharp)\cdot\nabla\psi\big)^2}{2(1+\varepsilon^2\mu|\nabla\zeta^\sharp|^2)} = \epsilon\frac{\varepsilon}{\nu}k^2, \end{cases}$$

where $\mathcal{G}^{(1)}\psi$ is the first-order expansion of $\mathcal{G}\psi$ provided by Proposition 3.44

$$\mathcal{G}^{(1)}\psi = \mathcal{G}_0\psi - \epsilon\mathcal{G}_0\big(\zeta(\mathcal{G}_0\psi)\big) - \varepsilon\mu\nabla\cdot(\zeta\nabla\psi).$$

One thus gets

$$\begin{cases} \partial_t\zeta^\sharp - \frac{1}{\mu\nu}\mathcal{G}\psi = \epsilon\frac{\varepsilon}{\nu}r^1, \\ \partial_t\psi + \zeta + \frac{\varepsilon}{2\nu}|\nabla\psi|^2 - \frac{\varepsilon\mu}{\nu}\frac{\big(\frac{1}{\mu}\mathcal{G}\psi + \nabla(\varepsilon\zeta^\sharp)\cdot\nabla\psi\big)^2}{2(1+\varepsilon^2\mu|\nabla\zeta^\sharp|^2)} = \epsilon\frac{\varepsilon}{\nu}r^2, \end{cases}$$

with

$$\begin{aligned} r^1 &= k^1 + \frac{1}{\mu\varepsilon\epsilon}\big(\mathcal{G}^{(1)}\psi - \mathcal{G}\psi\big) \\ r^2 &= k^2 + \frac{\big(\frac{1}{\mu}\mathcal{G}\psi + \frac{1}{\mu}\mathcal{G}^{(1)}\psi + 2\varepsilon\nabla\zeta^\sharp\cdot\nabla\psi\big)}{2(1+\varepsilon^2\mu|\nabla\zeta^\sharp|^2)}\frac{(\mathcal{G}_0\psi - \mathcal{G}\psi)}{\epsilon}. \end{aligned}$$

We therefore can deduce from Proposition 3.44 that

$$|r^1|_{H^N} + |\mathfrak{P}r^2|_{H^N} \le |k^1|_{H^N} + |\mathfrak{P}k^2|_{H^N} + M(N+2)|\nabla\psi|_{H^{N+3/2}},$$

where $r = (r^1, r^2)^T$, and the result easily follows from (8.9). □

8.1.3. Almost full justification of the asymptotics. The strategy to justify the deep water model (8.5) is different than the one adopted for the shallow water models. One of the reasons for this is that no local existence/stability result is known for this model. It is, however, possible to prove that any sufficiently regular solution to (8.5), if it exists, is a correct approximation of the solution to the water waves equations (*almost full* justification). To this end, we use the stability Theorem 4.18 associated to the water waves equations. (We had been using in the previous chapters stability properties of the asymptotic models.)
We recall that the parameters $(\varepsilon, \beta, \gamma, \mu)$ belong here to the *deep water* regime

$$\mathcal{A}_{DW} = \{(\varepsilon,\beta,\gamma,\mu),\quad 0\le\mu\le\mu_{max},\quad 0\le\varepsilon\le 1,\quad \beta=0,\quad \gamma=1,\quad \varepsilon\sqrt{\mu}\le\epsilon_0\}.$$

We then have the following result.

THEOREM 8.4. *Let $N > d/2+5$ and $\mathfrak{p} = (\varepsilon,\beta,\gamma,\mu)\in\mathcal{A}_{DW}$. Also let $b=0$ and $U^0 = (\zeta^0,\psi^0)\in E_0^{N+9/2}$ satisfy (4.29).*
If ϵ_0 is small enough, there exists $T = T_{DW} > 0$ that does not depend on $\mathfrak{p}\in\mathcal{A}_{DW}$ and such that
(1) There is a unique solution $U = (\zeta,\psi)\in E^N_{\frac{\nu T}{\varepsilon}}$ to the water waves equations (8.1) with initial data U^0, and we define $\underline{V} = \mathcal{V}[\varepsilon\zeta, 0]\psi$ through (4.13).
(2) If $(\zeta_{DW}, \underline{V}_{DW})\in C([0,\frac{\nu T}{\varepsilon}], H^{N+7/2}(\mathbb{R}^d)^{1+d})$ solves (8.5) with initial condition $(\zeta^0, \nabla\psi^0 - \epsilon(\mathcal{H}_\mu\nabla\psi^0)\nabla\zeta^0)$, then for all $t\in[0,\frac{\nu T}{\varepsilon}]$

$$|\mathfrak{e}_{DW}|_{L^\infty([0,t]\times\mathbb{R}^d)} \le \epsilon t\frac{\varepsilon}{\nu}C\big(\mathcal{E}^{N+9/2}(U^0), \mu_{max}, \epsilon_0, , \frac{1}{h_{min}}, \frac{1}{a_0}\big),$$

where $\mathfrak{e}_B = (\zeta,\underline{V}) - (\zeta_{DW},\underline{V}_{DW})$ and $\mathcal{E}^N(U^0)$ is as in (4.8).

PROOF. The theorem is a straightforward consequence of Proposition 8.3 and Theorem 4.18. □

REMARK 8.5. We emphasize again the fact that Theorem 8.4 brings only an *almost* full justification in the sense that we do not know whether solutions to (8.5) that satisfy the assumptions of (2) exist.

REMARK 8.6. It is possible to derive higher order deep water models by using a higher order expansion of the Dirichlet-Neumann operator (furnished by Proposition 3.44) in (8.3). Such higher order models are implicitly used in the numerical computations of the solutions of the water waves equations in [**103**].

REMARK 8.7. We also recall that μ is assumed to remain bounded throughout these notes ($\mu \leq \mu_{max}$), so that $\nu = \tanh(2\pi\sqrt{\mu})/(2\pi\sqrt{\mu})$ remains of order $O(1)$. We therefore can take $\nu = 1$ in the above theorem and therefore use Theorem 4.18, which has been proved in this setting. We left the dependence on ν explicit because the result remains true as μ grows very large (and therefore $\nu \sim (2\pi\mu)^{-1/2}$), as shown in [**12**]. See also §8.1.4 for the corresponding model in infinite depth.

REMARK 8.8. The deep water model with nonflat bottom (8.4) can be almost fully justified following exactly the same lines.

8.1.4. The case of infinite depth. It can be shown without difficulty that the deep water model takes the following form in infinite depth[2]

$$\begin{cases} \partial_t\zeta - \mathcal{H}\underline{V} + \epsilon\big(\mathcal{H}(\zeta\nabla\mathcal{H}\underline{V}) + \nabla\cdot(\zeta\underline{V})\big) = 0, \\ \partial_t\underline{V} + \nabla\zeta + \epsilon\big(\frac{1}{2}\nabla|\underline{V}|^2 - \nabla\zeta\mathcal{H}\nabla\zeta\big) = 0, \end{cases} \tag{8.10}$$

with $\mathcal{H} = -\frac{1}{|D|}\nabla^T$ and $\epsilon = \varepsilon\sqrt{\mu}$.
With the results of §4.4.3 on the well-posedness of the water waves equations in infinite depth, one can easily adapt the above results to cover the case of infinite depth and establish an almost full justification of (8.10).

8.2. Modulation equations in finite depth

We will now investigate the behavior of *wave packets*, which are slowly modulated oscillating waves. *Modulation equations* describe the time evolution of the amplitude of these oscillations. Such equations play an important role in many physical situations; in the context of water waves, several models have been proposed. Since these models are mostly relevant for deep (or infinite) depth[3], we set $\nu = (2\pi\mu)^{-1/2}$ in the dimensionless water waves equations (1.28) and therefore work with[4]

$$\begin{cases} \partial_t\zeta - \frac{1}{\sqrt{\mu}}\mathcal{G}\psi = 0, \\ \partial_t\psi + \zeta + \frac{\epsilon}{2}|\nabla\psi|^2 - \epsilon\frac{(\frac{1}{\sqrt{\mu}}\mathcal{G}\psi + \epsilon\nabla\zeta\cdot\nabla\psi)^2}{2(1+\epsilon^2|\nabla\zeta|^2)} = 0, \end{cases} \tag{8.11}$$

where we recall (see §1.5) that the parameter ϵ is the *steepness* of the wave

$$\epsilon = \varepsilon\sqrt{\mu} = \frac{a}{L_x} = \frac{\text{typical amplitude}}{\text{typical horizontal scale}}.$$

---

[2]We used the slightly modified dimensionless form of the water waves equations introduced in Footnote 9 in Chapter 4.

[3]It is, however, possible to derive Schrödinger type equations in the shallow water regime. One could perform the computation of this section with (8.1) without setting $\nu = (2\pi\mu)^{-1/2}$ and perform the expansion with respect to ε/ν instead of $\epsilon = \varepsilon\sqrt{\mu}$. One could also directly derive the NLS approximation from a shallow water model (provided it has the required precision). Such derivations have, for instance, been made for the Boussinesq and KdV equations [**1, 306**] (see also [**290**] for a justification).

[4]To avoid dealing with the factor 2π in the equations, we use the same rescaling as in Footnote 9 in Chapter 4.

This is the small parameter we will use to construct (formal) approximation to (8.11) under the form of modulated wave packets.

8.2.1. Defining the ansatz. The linearization around the rest state of (8.11) is given by

$$\partial_t U + \mathcal{A}_0(D)U = 0, \qquad \text{with} \quad \mathcal{A}_0(D) = \begin{pmatrix} 0 & -\frac{1}{\sqrt{\mu}}\mathcal{G}_0(D) \\ 1 & 0 \end{pmatrix}, \tag{8.12}$$

and where $\mathcal{G}_0(D) = \sqrt{\mu}|D|\tanh(\sqrt{\mu}|D|)$ and $U = (\zeta, \psi)^T$. These linearized equations admit real valued *plane wave* solutions under the form[5]

$$U(t,X) = \begin{pmatrix} i\omega\psi_{01} \\ \psi_{01} \end{pmatrix} e^{i\theta} + \text{c.c.}, \tag{8.13}$$

with

$$\theta = \mathbf{k}\cdot X - \omega t, \qquad \omega = \omega(\mathbf{k}) = \big(|\mathbf{k}|\tanh(\sqrt{\mu}|\mathbf{k}|)\big)^{1/2}, \tag{8.14}$$

and where c.c. stands for "complex conjugate". In (8.13), ψ_{01} is a constant; it therefore is natural to look for approximate solutions to (8.11) under the form of *wave packets*

$$U(t,X) = \begin{pmatrix} i\omega\psi_{01}(\epsilon t, \epsilon X) \\ \psi_{01}(\epsilon t, \epsilon X) \end{pmatrix} e^{i\theta} + \text{c.c.}, \tag{8.15}$$

with again $\omega = \big(|\mathbf{k}|\tanh(\sqrt{\mu}|\mathbf{k}|)\big)^{1/2}$. The difference with respect to (8.13) is that the amplitude of the oscillation is now slowly modulated in space and time.
The nonlinearities in (8.11) do not preserve the structure of (8.15) since they create higher order harmonics (i.e., wave packets oscillating with a phase $n\theta$, $n \neq \pm 1$); the approximate solution we want to construct therefore must be of a more general form, namely,

$$U_{app}(t,X) = U_0(t,X) + \epsilon U_1(t,X) + \epsilon^2 U_2(t,X), \tag{8.16}$$

where the leading term U_0 is composed of a wave packet similar to (8.15) and of a nonoscillating term whose presence is made necessary to describe the creation of a mean mode by nonlinear interaction of oscillating modes[6]

$$U_0(t,X) = \begin{pmatrix} i\omega\psi_{01}(\epsilon t, \epsilon X) \\ \psi_{01}(\epsilon t, \epsilon X) \end{pmatrix} e^{i\theta} + \text{c.c.} + \begin{pmatrix} 0 \\ \psi_{00}(\epsilon t, \epsilon X) \end{pmatrix}. \tag{8.17}$$

We also seek the corrector terms U_1, U_2 under the form[7]

$$U_1(t,X) = \begin{pmatrix} \zeta_{11}(\epsilon t, \epsilon X)e^{i\theta} + \zeta_{12}(\epsilon t, \epsilon X)e^{i2\theta} + \text{c.c.} + \zeta_{10}(\epsilon t, \epsilon X) \\ \psi_{11}(\epsilon t, \epsilon x)e^{i\theta} + \psi_{12}(\epsilon t, \epsilon X)e^{i2\theta} + \text{c.c.} \end{pmatrix} \tag{8.18}$$

and

$$U_2(t,X) = \sum_{n=1}^{3} U_{2n}(\epsilon t, \epsilon X)e^{in\theta} + \text{c.c.} + U_{20}(\epsilon t, \epsilon X). \tag{8.19}$$

[5]The relation $\zeta_{01} = i\omega\psi_{01}$ is often called the *polarization condition.*

[6]This phenomenon is called *rectification* in optics [**192, 215**].

[7]In the expression below, we took $\psi_{10} = 0$; it is not obvious *a priori* that it is possible to do so. This follows from the computations below, but we chose to set these coefficients to zero from the beginning in order to simplify the computations.

NOTATION 8.9. We denote by t' and X' the slow variables $t' = \epsilon t$ and $X' = \epsilon X$. The differentiation with respect to t' and X' is written ∂_t', ∇', and we denote $D' = -i\nabla'$, etc., so that, for instance

$$\nabla(\psi_{01}(\epsilon t, \epsilon X)e^{i\theta}) = (i\mathbf{k}\psi_{01} + \epsilon\nabla'\psi_{01})_{|(\epsilon t,\epsilon X)} e^{i\theta}.$$

8.2.2. Small amplitude expansion of (8.11). The strategy consists now of plugging the ansatz (8.16)–(8.19) into (8.11) and choosing the U_{jn} in order to cancel the leading order terms with respect to ϵ. In order to identify these terms, let us write (8.11) under the following abstract form, with $U = (\zeta, \psi)^T$

$$\partial_t U + \mathcal{N}(U) := \mathcal{L}(\partial_t, D)U + \epsilon\mathcal{N}_1(U) = 0, \tag{8.20}$$

where $\mathcal{L}(\partial_t, D)U$ is the "linear part" of the equations and $\epsilon\mathcal{N}_1(U)$ their "nonlinear part"

$$\mathcal{L}(\partial_t, D) = \partial_t + \mathcal{A}_0(D), \qquad \mathcal{N}_1(U) = \begin{pmatrix} -\dfrac{1}{\epsilon\sqrt{\mu}}(\mathcal{G} - \mathcal{G}_0) \\ \dfrac{1}{2}|\nabla\psi|^2 - \dfrac{(\frac{1}{\sqrt{\mu}}\mathcal{G}\psi + \epsilon\nabla\zeta\cdot\nabla\psi)^2}{2(1+\epsilon^2|\nabla\zeta|^2)} \end{pmatrix}.$$

Recalling that $\omega(\cdot)$ is defined in (8.14), for the sake of clarity, we use the notations

$$\mathcal{L}_n = \mathcal{L}(-in\omega, n\mathbf{k}), \qquad \mathcal{L}'_n(D) = \begin{pmatrix} 0 & -2\omega(n\mathbf{k})\nabla\omega(n\mathbf{k})\cdot D \\ 0 & 0 \end{pmatrix}$$

in the following proposition which gives an expansion[8] in terms of ϵ of $\partial_t U_{app} + \mathcal{N}(U_{app})$.

PROPOSITION 8.10. *Let U_{app} be as defined in (8.16)–(8.19). Then one has, with $\theta = \mathbf{k}\cdot X - \omega t$, and using Notation 8.9*

$$\begin{aligned}
\partial_t U_{app} + \mathcal{N}(U_{app}) &= (\epsilon\partial_t' U_{01} + \mathcal{L}(-i\omega, \mathbf{k} + \epsilon D')U_{01})e^{i\theta} + \text{c.c.} + \mathcal{L}(\epsilon\partial_t', \epsilon D')U_{00} \\
&+ \epsilon\sum_{n=1}^{2}(\mathcal{L}_n U_{1n} + \mathcal{N}_{1n})e^{in\theta} + \text{c.c.} + (\mathcal{L}_0 U_{10} + \mathcal{N}_{10}) \\
&+ \epsilon^2\Big[\sum_{n=1}^{3}(\mathcal{L}_n U_{2n} + (\partial_t' + \mathcal{L}'_n(D'))U_{1n} + \mathcal{N}_{2n})e^{in\theta} + \text{c.c.} \\
&\quad (\mathcal{L}_0 U_{20} + \partial_t' U_{10} + \mathcal{N}_{20})\Big] \\
&+ \epsilon^3\sum_{n=-9}^{9} R_{3n}(\epsilon t, \epsilon X)e^{in\theta} + \epsilon^{4-d/2}R_4(t, X),
\end{aligned}$$

with $\mathcal{N}_{11} = 0$ and, writing $\sigma = \tanh(\sqrt{\mu}|\mathbf{k}|)$ and $\sigma_2 = \tanh(2\sqrt{\mu}|\mathbf{k}|)$

$$\mathcal{N}_{10} = \begin{pmatrix} 0 \\ |\mathbf{k}|^2(1-\sigma^2)|\psi_{01}|^2 \end{pmatrix}, \qquad \mathcal{N}_{12} = \begin{pmatrix} -2i\omega|\mathbf{k}|^2(1-\sigma\sigma_2)\psi_{01}^2 \\ -\frac{1}{2}|\mathbf{k}|^2(1+\sigma^2)\psi_{01}^2 \end{pmatrix}.$$

[8]The order of the expansion is chosen in order to include nonlinear and dispersive effects in the modulation equation for ψ_{01}. These effects play a relevant role on a time scale $t' = O(1/\epsilon)$. Over a smaller time scale $t' = O(1)$, the modulation equation is a simple free transport equation at the group velocity.

For the second order nonlinear terms, we have

$$\mathcal{N}_{20} = \begin{pmatrix} 2\omega\mathbf{k}\cdot\nabla'|\psi_{01}|^2 \\ i\mathbf{k}\cdot\nabla'\overline{\psi}_{01}\psi_{01} - i\mathbf{k}\cdot\nabla'\psi_{01}\overline{\psi}_{01} \end{pmatrix},$$

$$\mathcal{N}_{21} = \begin{pmatrix} |\mathbf{k}|^2(1+\sigma^2)\zeta_{12}\overline{\psi}_{01} + 2i\omega|\mathbf{k}|^2(1-\sigma\sigma_2)\overline{\psi}_{01}\psi_{12} - \omega\mathbf{k}\cdot\nabla'\psi_{00}\psi_{01} \\ \quad -|\mathbf{k}|^2(1-\sigma^2)\zeta_{10}\psi_{01} + \omega^2|\mathbf{k}|^3\sigma(1-2\sigma\sigma_2)|\psi_{01}|^2\psi_{01} \\ i\mathbf{k}\cdot\nabla'\psi_{00}\psi_{01} + 2|\mathbf{k}|^2(1-\sigma\sigma_2)\overline{\psi}_{01}\psi_{12} \\ \quad -i\omega|\mathbf{k}|^3\sigma(1-2\sigma\sigma_2)|\psi_{01}|^2\psi_{01} \end{pmatrix},$$

and the $\mathcal{N}_{2n}$ $(n \neq 1)$ can also be explicitly determined.
Moreover, the residual terms satisfy for all $s \geq 0$

$$\sum_{n=-9}^{9} |R_{3n}(\epsilon t,\cdot)|_{H^s} + |R_4(t,\cdot)|_{H^s} \leq C\big(\sum_{ij}|\zeta_{ij}(\epsilon t,\cdot)|_{H^{s+3}}\big)\sum_{ij}|\nabla\psi_{ij}(\epsilon t,\cdot)|_{H^{s+3}}.$$

PROOF. Let us write $U_{app} = (\zeta_{app}, \psi_{app})^T$. In order to prove the proposition, we need to expand $\nabla\psi_{app}$, $\nabla\zeta_{app}$ and $\mathcal{G}_{app}\psi_{app} = \mathcal{G}_{\mu,\gamma}[\epsilon\zeta_{app}, 0]\psi_{app}$ into power of ϵ. We easily get

$$\nabla\psi_{app} = \sum_{j=0}^{2}\epsilon^j\sum_{n\in\mathbb{Z}}(in\mathbf{k}\psi_{jn} + \nabla'\psi_{j-1,n})e^{in\theta}, \tag{8.21}$$

$$\nabla\zeta_{app} = \sum_{j=0}^{2}\epsilon^j\sum_{n\in\mathbb{Z}}(in\mathbf{k}\zeta_{jn} + \nabla'\zeta_{j-1,n})e^{in\theta} \tag{8.22}$$

(recall that ∇' denotes the gradient with respect to ϵX).
The expansion for $\mathcal{G}_{app}\psi_{app}$ is provided by the following lemma. Note that the control of the residual is given in terms of $\epsilon^{d/2}\psi_{app}$ because ψ_{app} is of size $O(\epsilon^{-d/2})$ in $L^2(\mathbb{R}^d)$.

LEMMA 8.11. *Let $s \geq 0$ and $t_0 > d/2$. One has*

$$\begin{aligned}\frac{1}{\sqrt{\mu}}\mathcal{G}_{app}\psi_{app} &= \frac{1}{\sqrt{\mu}}\mathcal{G}_0\psi_{app} + \epsilon\frac{1}{\mu}\mathcal{G}_{1,app}\psi_{app} + \epsilon^2\frac{1}{\mu}\mathcal{G}_{2,app}\psi_{app} \\ &\quad + \epsilon^3\mathcal{G}_{3,app}\psi_{app} + \epsilon^{4-d/2}R_4,\end{aligned}$$

where $\mathcal{G}_{1,app}$ and $\mathcal{G}_{2,app}$ are given by

$$\begin{aligned}\mathcal{G}_{1,app} &= -\mathcal{G}_0(\zeta_{app}(\mathcal{G}_0\cdot)) - \mu\nabla\cdot(\zeta_{app}\nabla\cdot)) \\ \mathcal{G}_{2,app} &= \mathcal{G}_0(\zeta_{app}\mathcal{G}_0(\zeta_{app}\mathcal{G}_0\cdot)) + \mu\frac{1}{2}\Delta(\zeta_{app}^2\mathcal{G}_0\cdot) + \mu\frac{1}{2}\mathcal{G}_0(\zeta_{app}^2\Delta\cdot),\end{aligned}$$

while $\mathcal{G}_{3,app} = \frac{1}{6}d^3\mathcal{G}_{\mu,\gamma}[0,0](\zeta_{app},\zeta_{app},\zeta_{app})\cdot$ and R_4 satisfies

$$|R_4|_{H^s} \leq C(\epsilon|\zeta_{app}|_{H^{s\vee t_0+1}})|\epsilon^{d/2}\nabla\psi_{app}|_{H^s}.$$

PROOF OF THE LEMMA. This is a slight and straightforward extension of Proposition 3.44 to a third-order expansion. The formulas for $\mathcal{G}_{1,app}$ and $\mathcal{G}_{2,app}$ are obtained by replacing ζ by ζ_{app} in (3.40)–(3.41) (an exact formula for $\mathcal{G}_{3,app}$ could be given, but it is not needed). Note that here we use Remark 3.47 since the family $(\zeta_{app})_\epsilon$ is not uniformly bounded in H^s while, in dimension $d = 1, 2$, $(\epsilon\zeta_{app})_\epsilon$ is uniformly bounded. □

In order to put the expansion provided by Lemma 8.11 under a "wave packet" form, we will need the following lemma (recall that $D' = -i\nabla'$, with ∇' the gradient with respect to ϵX). We will use it, together with Lemma 8.11, to derive multiscale expansions of the Dirichlet-Neumann operator.

LEMMA 8.12. *Let $\mathbf{k} \in \mathbb{R}^d$, $s \in \mathbb{R}$, and a smooth enough. Then*

(1) *Denoting $g_0(\xi) := \omega^2(\xi) = |\xi| \tanh(\sqrt{\mu}|\xi|)$, one has*

$$\frac{1}{\sqrt{\mu}}\mathcal{G}_0(a(\epsilon X)e^{i\mathbf{k}\cdot X}) = g_0(\mathbf{k}+\epsilon D')a_{|_{\epsilon X}}e^{i\mathbf{k}\cdot X}.$$

(2) *The above formula can be expanded into powers of ϵ in the following sense*

$$\begin{aligned}\frac{1}{\sqrt{\mu}}\mathcal{G}_0(a(\epsilon X)e^{i\mathbf{k}\cdot X}) &= g_0(\mathbf{k})a(\epsilon X)e^{i\mathbf{k}\cdot X} - i\epsilon\nabla g_0(\mathbf{k})\cdot\nabla' a_{|_{\epsilon X}}e^{i\mathbf{k}\cdot X}\\ &\quad -\epsilon^2\frac{1}{2}\nabla'\cdot\mathcal{H}_{g_0}(\mathbf{k})\nabla' a_{|_{\epsilon X}}e^{i\mathbf{k}\cdot X} + \epsilon^3 R(\epsilon X)e^{i\mathbf{k}\cdot X},\end{aligned}$$

where $\mathcal{H}_{g_0}(\mathbf{k})$ stands for the Hessian matrix of g_0 evaluated at $\mathbf{k}$, and with $|R|_{H^s} \lesssim |a|_{H^{s+3}}$.

PROOF OF THE LEMMA. Since $\frac{1}{\sqrt{\mu}}\mathcal{G}_0 = g_0(D)$, we have

$$\mathcal{F}\big[\frac{1}{\sqrt{\mu}}\mathcal{G}_0(a(\epsilon X)e^{i\mathbf{k}\cdot X})\big](\xi) = g_0(\xi)\frac{1}{\epsilon^d}\widehat{a}(\frac{\xi-\mathbf{k}}{\epsilon})$$

and therefore

$$\begin{aligned}\frac{1}{\sqrt{\mu}}\mathcal{G}_0(a(\epsilon X)e^{i\mathbf{k}\cdot X}) &= \int_{\mathbb{R}^d} e^{iX\cdot\xi}g_0(\xi)\frac{1}{\epsilon^d}\widehat{a}(\frac{\xi-\mathbf{k}}{\epsilon})d\xi\\ &= \int_{\mathbb{R}^d} e^{iX\cdot(\epsilon\xi'+\mathbf{k})}g_0(\epsilon\xi'+\mathbf{k})\widehat{a}(\xi')d\xi, \qquad (8.23)\end{aligned}$$

where we performed the change of variable $\xi' = \frac{\xi-\mathbf{k}}{\epsilon}$ to obtain (8.23). The first point of the lemma follows immediately.
The second point follows easily from a Taylor expansion of $g_0(\epsilon\xi'+\mathbf{k})$ around $\mathbf{k}$ in the right-hand side of (8.23). □

Using this lemma, we get

$$\begin{aligned}\frac{1}{\sqrt{\mu}}\mathcal{G}_0\psi_{app} &= g_0(\mathbf{k}+\epsilon D')\psi_{01\,|_{\epsilon X}}e^{i\theta} + \epsilon g_0(2\mathbf{k})\psi_{12\,|_{\epsilon X}}e^{in\theta} + \text{c.c.}\\ &\quad +\epsilon^2\Big[-\frac{1}{2}\nabla'\cdot\mathcal{H}_{g_0}(0)\nabla'\psi_{00} - i\nabla g_0(2\mathbf{k})\cdot\nabla'\psi_{12\,|_{\epsilon X}}e^{i2\theta} + \text{c.c.}\Big]\\ &\quad +\epsilon^3\big[A_{32}(\epsilon t,\epsilon X)e^{i2\theta} + \text{c.c.} + A_{30}(\epsilon t,\epsilon X)\big], \qquad (8.24)\end{aligned}$$

with $|A_{32}|_{H^s} \lesssim |\psi_{12}|_{H^{s+2}}$ and $|A_{30}|_{H^s} \lesssim |\psi_{00}|_{H^{s+3}}$.
Together with (8.21)–(8.22), Lemma 8.11 also gives

$$\begin{aligned}\epsilon\frac{1}{\mu}\mathcal{G}_{1,app}\psi_{app} &= \epsilon B_{12}(\epsilon t,\epsilon X)e^{2i\theta} + \text{c.c.} + \epsilon^2\sum_{n=1}^{3}B_{2n}(\epsilon t,\epsilon X)e^{in\theta} + \text{c.c.}\\ &\quad +\epsilon^2 B_{20}(\epsilon t,\epsilon X) + \epsilon^3\sum_{n=-4}^{4}B_{3n}(\epsilon t,\epsilon X)e^{in\theta} \qquad (8.25)\end{aligned}$$

with

$$\begin{aligned} B_{12} &= i\omega(2|\mathbf{k}|^2 - g_0(2\mathbf{k})g_0(\mathbf{k}))\psi_{01}^2, \\ B_{20} &= -2\omega\mathbf{k}\cdot\nabla'|\psi_{01}|^2, \\ B_{21} &= -(|\mathbf{k}|^2 + g_0(\mathbf{k})^2)\zeta_{12}\overline{\psi}_{01} - i\omega(2|\mathbf{k}|^2 - g_0(\mathbf{k})g_0(2\mathbf{k}))\overline{\psi}_{01}\psi_{12} \\ &\quad +(|\mathbf{k}|^2 - g_0(\mathbf{k})^2)\zeta_{10}\psi_{01} + \omega\mathbf{k}\cdot\nabla'\psi_{00}\psi_{01}. \end{aligned}$$

The expressions for the other coefficients B_{jn} can easily be determined but do not play any role in the derivation of the modulation equation, so we do not give them. Similarly, we get

$$\epsilon^2\frac{1}{\mu}\mathcal{G}_{2,app}\psi_{app} = \epsilon^2\sum_{n=1}^{3} C_{2n}(\epsilon t,\epsilon X)e^{in\theta} + \text{c.c.} + \epsilon^3\sum_{n=-6}^{6} C_{3n}(\epsilon t,\epsilon X)e^{in\theta} \tag{8.26}$$

with

$$C_{21} = \omega^2 g_0(\mathbf{k})(g_0(\mathbf{k})g_0(2\mathbf{k}) - |\mathbf{k}|^2)|\psi_{01}|^2\psi_{01},$$

while the exact expression of the other coefficients is of no importance.
The proposition then follows from (8.21), (8.22), Lemma 8.11, and (8.24)–(8.26). □

8.2.3. Determination of the ansatz. We show here that it is possible to choose the components of U_{app} in order to cancel all the terms (except the residuals R_{3n} and R_4) in the expansion provided by Proposition 8.10. We first derive the conditions corresponding to the cancellation of the nonoscillating terms and then consider the various harmonics of the oscillating terms.
- *Nonoscillating terms (harmonic zero).* Canceling the nonoscillating terms of the residual up to order $O(\epsilon^3)$ is equivalent to

$$\mathcal{L}(\epsilon\partial_t',\epsilon D')U_{00} + \epsilon\mathcal{L}_0U_{10} + \epsilon\mathcal{N}_{10} + \epsilon^2\big(\mathcal{L}_0U_{20} + \partial_t'U_{10} + \mathcal{N}_{20}\big) = 0$$

or equivalently (recall that $\zeta_{00} = 0$),

$$\begin{cases} \epsilon^2\big[-|D'|\dfrac{\tanh(\epsilon\sqrt{\mu}|D'|)}{\epsilon}\psi_{00} + \partial_t'\zeta_{10} + 2\omega\mathbf{k}\cdot\nabla'|\psi_{01}|^2\big] = 0, \\ \epsilon\big[\partial_t'\psi_{00} + \zeta_{10} + |\mathbf{k}|^2(1-\sigma^2)|\psi_{01}|^2\big] + \epsilon^2\big[\zeta_{20} + \mathcal{N}_{20}^{(2)}\big] = 0, \end{cases} \tag{8.27}$$

where $\mathcal{N}_{20}^{(2)}$ denotes the second component of $\mathcal{N}_{20}$. Cancelling separately the $O(\epsilon)$ and $O(\epsilon^2)$ terms in the second equation provides an explicit expression for ζ_{20}

$$\zeta_{20} = -\mathcal{N}_{20}^{(2)},$$

while the leading order terms ζ_{10} and ψ_{00} for the surface elevation and the velocity potential are given in terms of ψ_{01} by the resolution of the two coupled evolution equations

$$\begin{cases} \partial_t'\zeta_{10} - |D'|\dfrac{\tanh(\epsilon\sqrt{\mu}|D'|)}{\epsilon}\psi_{00} = -2\omega\mathbf{k}\cdot\nabla'|\psi_{01}|^2, \\ \partial_t'\psi_{00} + \zeta_{10} = -|\mathbf{k}|^2(1-\sigma^2)|\psi_{01}|^2. \end{cases} \tag{8.28}$$

- *Second harmonic.* In order to cancel these terms up to order $O(\epsilon^3)$, we can cancel separately the $O(\epsilon)$ and $O(\epsilon^2)$ terms, leading to

$$\mathcal{L}_2U_{12} + \mathcal{N}_{12} = 0 \quad \text{and} \quad \mathcal{L}_2U_{22} + (\partial_t + \mathcal{L}_2'(D'))U_{12} + \mathcal{N}_{22} = 0.$$

Since $\mathcal{L}_2$ is invertible, the first equation gives U_{12} in terms of ψ_{01},

$$\begin{cases} \zeta_{12} = \dfrac{(\sigma^2-3)|\mathbf{k}|^2}{2\sigma^2}\psi_{01}^2, \\ \psi_{12} = i\omega k\dfrac{3+\sigma^4}{4\sigma^3}\psi_{01}^2, \end{cases}$$

while the second equation gives similarly U_{22} in terms of ψ_{01}.

- *Third harmonic.* The third harmonic is present only in the $O(\epsilon^2)$ terms, and since $\mathcal{L}_3$ is invertible, the cancellation of this harmonic determines U_{23}.

- *First harmonic.* Cancelling the first harmonic of the residual gives (after noting that $\mathcal{L}_1 U_{01} = 0$)

$$\begin{aligned} \epsilon\big[\mathcal{L}_1 U_{11} + \partial_t' U_{01} + \frac{\mathcal{L}(-i\omega, \mathbf{k}+\epsilon D') - \mathcal{L}_1}{\epsilon} U_{01}\big] \\ +\epsilon^2\big[\mathcal{L}_1 U_{21} + (\partial_t' + \mathcal{L}_1'(D'))U_{11} + \mathcal{N}_{21}\big] = 0. \end{aligned}$$

The standard way to deal with this condition would be to cancel separately the $O(\epsilon)$ and $O(\epsilon^2)$ terms. In order to derive the "full dispersion" Benney-Roskes model (see §8.2.5), it is more convenient to consider this equation as a whole and to solve it up to $O(\epsilon^3)$ terms that do not affect the residual estimates. We therefore seek to solve

$$\epsilon\mathcal{L}_1(U_{11} + \epsilon U_{21}) = \epsilon(F + \epsilon G) + O(\epsilon^3) \tag{8.29}$$

with

$$\begin{aligned} F &= \begin{pmatrix} -i\omega\partial_t'\psi_{01} + \dfrac{\omega^2(\mathbf{k}+\epsilon D') - \omega^2}{\epsilon}\psi_{01} \\ -\partial_t'\psi_{01} \end{pmatrix} \\ G &= \begin{pmatrix} -\partial_t'\zeta_{11} - 2\omega\nabla\omega(\mathbf{k})\cdot\nabla'\psi_{11} - \mathcal{N}_{21}^{(1)} \\ -\partial_t'\psi_{11} - \mathcal{N}_{21}^{(2)}, \end{pmatrix}. \end{aligned}$$

Since the matrix $\mathcal{L}_1$ is not invertible (recall that we chose $\omega = \omega(\mathbf{k})$), the right-hand side of (8.29) must satisfy a solvability condition to ensure the existence of (a one-dimensional space of) solutions. One readily checks that this solvability condition is $(F_1 + \epsilon G_1) + i\omega(F_2 + \epsilon G_2) = O(\epsilon^2)$, so that (8.29) is equivalent to

$$\begin{cases} -2i\omega\partial_t'\psi_{01} + \dfrac{\omega^2(\mathbf{k}+\epsilon D') - \omega^2}{\epsilon}\psi_{01} - \epsilon\partial_t'(\zeta_{11} + i\omega\psi_{11}) \\ \qquad -2\epsilon\omega c_g\cdot\nabla'\psi_{11} - \epsilon(\mathcal{N}_{21}^{(1)} + i\omega\mathcal{N}_{21}^{(2)}) = O(\epsilon^2), \\ (\zeta_{11} - i\omega\psi_{11}) + \epsilon(\zeta_{21} - i\omega\psi_{21}) = -\partial_t'\psi_{01} - \epsilon(\partial_t'\psi_{11} + \mathcal{N}_{21}^{(2)}) + O(\epsilon^2), \end{cases}$$

where we used the notation

$$c_g = \nabla\omega(\mathbf{k}) \qquad \text{(group velocity)}.$$

We can rewrite these equations under the form

$$\begin{cases} -2i\omega\partial_t'\psi_{01} + \dfrac{\omega^2(\mathbf{k}+\epsilon D') - \omega^2}{\epsilon}\psi_{01} - \epsilon(\partial_t' + c_g\cdot\nabla')(\zeta_{11} + i\omega\psi_{11}) \\ \qquad +\epsilon c_g\cdot\nabla'(\zeta_{11} - i\omega\psi_{11}) - \epsilon(\mathcal{N}_{21}^{(1)} + i\omega\mathcal{N}_{21}^{(2)}) = O(\epsilon^2), \\ (\zeta_{11} - i\omega\psi_{11}) + \epsilon(\zeta_{21} - i\omega\psi_{21}) = -\partial_t'\psi_{01} - \epsilon(\partial_t'\psi_{11} + \mathcal{N}_{21}^{(2)}) + O(\epsilon^2). \end{cases}$$

Note now that the second equation implies that

$$\begin{aligned}\epsilon c_g\cdot\nabla'(\zeta_{11}-i\omega\psi_{11}) &= -\epsilon c_g\cdot\nabla'\partial_t'\psi_{01}+O(\epsilon^2)\\ &= \epsilon(c_g\cdot\nabla')^2\psi_{01}+O(\epsilon^2),\end{aligned}$$

where we used the first equation to get $\partial_t'\psi_{01} = \frac{1}{2i\omega}\frac{\omega^2(\mathbf{k}+\epsilon D')-\omega^2}{\epsilon}\psi_{01}+O(\epsilon) = -c_g\cdot\nabla\psi_{01}+O(\epsilon)$. We therefore take

$$\zeta_{11}-i\omega\psi_{11}=c_g\cdot\nabla'\psi_{01}.$$

Plugging this expression in the previous system then gives

$$\begin{cases} -2i\omega\partial_t'\psi_{01}+\Big(\dfrac{\omega^2(\mathbf{k}+\epsilon D')-\omega^2}{\epsilon}+\epsilon(c_g\cdot\nabla')^2\Big)\psi_{01} \\ \qquad -\epsilon(\partial_t'+c_g\cdot\nabla')(\zeta_{11}+i\omega\psi_{11})-\epsilon(\mathcal{N}_{21}^{(1)}+i\omega\mathcal{N}_{21}^{(2)})=O(\epsilon^2),\\ (\zeta_{21}-i\omega\psi_{21})=-\frac{1}{\epsilon}(\partial_t'-c_g\cdot\nabla')\psi_{01}-(\partial_t'\psi_{11}+\mathcal{N}_{21}^{(2)})+O(\epsilon).\end{cases}$$

At this point, we have determined $\zeta_{11}-i\omega\psi_{11}$, but we still have to choose $\zeta_{11}+i\omega\psi_{11}$, ζ_{21} and ψ_{21}. We proceed as follows:

(1) We determine $\zeta_{11}+i\omega\psi_{11}$ (and therefore ζ_{11} and ψ_{11} since we already know $\zeta_{11}-i\omega\psi_{11}$) by imposing[9]

$$(\partial_t'+c_g\cdot\nabla')(\zeta_{11}+i\omega\psi_{11})=0.$$

(2) We get $\zeta_{21}-i\omega\psi_{21}$ from the second equation (note that $\frac{1}{\epsilon}(\partial_t'-c_g\cdot\nabla')\psi_{01}=O(1)$ since at leading order, ψ_{01} travels at the group velocity).

(3) We can choose $\zeta_{21}+i\omega\psi_{21}$ arbitrarily.

We therefore are left with the following equation on ψ_{01}

$$\partial_t'\psi_{01}+\epsilon\frac{i}{2\omega}\Big[\frac{(\omega^2(\mathbf{k}+\varepsilon D')-\omega^2)}{\epsilon}+\epsilon(\nabla\omega(\mathbf{k})\cdot\nabla')^2\Big]\psi_{01}-\epsilon\frac{i}{2\omega}(\mathcal{N}_{21}^{(1)}+i\omega\mathcal{N}_{21}^{(2)})=O(\epsilon^2).$$

We compute

$$\begin{aligned}\mathcal{N}_{21}^{(1)}+i\omega\mathcal{N}_{21}^{(2)} &= |\mathbf{k}|^4\big(-4-(1-\sigma^2)^2\frac{9}{2\sigma^2}\big)|\psi_{01}|^2\psi_{01}\\ &\quad -2\omega\mathbf{k}\cdot\nabla'\psi_{00}\psi_{01}-|\mathbf{k}|^2(1-\sigma^2)\zeta_{10}\psi_{01}.\end{aligned}$$

Plugging these expressions in the evolution equation on ψ_{01} and dropping $O(\epsilon^2)$ terms, we get

$$\begin{aligned}\partial_t\psi_{01} &+ \frac{i}{2\omega}\Big[\frac{(\omega^2(\mathbf{k}+\epsilon D)-\omega^2)}{\epsilon}+\epsilon(\nabla\omega(\mathbf{k})\cdot\nabla)^2\Big]\psi_{01}\\ &+ \epsilon i\big(\mathbf{k}\cdot\nabla\psi_{00}+\frac{|\mathbf{k}|^2}{2\omega}(1-\sigma^2)\zeta_{10}+2\frac{|\mathbf{k}|^4}{\omega}(1-\alpha)|\psi_{01}|^2\big)\psi_{01}=0,\end{aligned}\tag{8.30}$$

where α is given by

$$\alpha=-\frac{9}{8\sigma^2}(1-\sigma^2)^2.\tag{8.31}$$

[9] This choice is not arbitrary; it is actually the only possible choice to avoid the *secular growth* of the corrector. More precisely, $\zeta_{11}+i\omega\psi_{11}$ grows sublinearly in t' (so that $\epsilon(\zeta_{11}+i\omega\psi_{11})$ remains a $o(1)$ corrector over the relevant time scale $t'=O(1/\epsilon)$), if and only if it solves the transport equation at the group velocity [**215**].

8.2.4. The "full-dispersion" Benney-Roskes model. The full-dispersion Benney-Roskes model is given by the following set of equations, where the leading oscillating term for the velocity potential ψ_{01} is coupled to the leading order nonoscillating terms for the surface elevation and velocity potential ζ_{10} and ψ_{00}

$$(8.32)\quad \begin{cases} \partial_t'\psi_{01} + i\dfrac{\omega(\mathbf{k}+\epsilon D')-\omega}{\epsilon}\psi_{01} \\ \qquad +\epsilon i\big[\mathbf{k}\cdot\nabla'\psi_{00} + \dfrac{|\mathbf{k}|^2}{2\omega}(1-\sigma^2)\zeta_{10} + 2\dfrac{|\mathbf{k}|^4}{\omega}(1-\alpha)|\psi_{01}|^2\big]\psi_{01} = 0, \\ \partial_t'\zeta_{10} - |D'|\dfrac{\tanh(\epsilon\sqrt{\mu}|D'|)}{\epsilon}\psi_{00} = -2\omega\mathbf{k}\cdot\nabla'|\psi_{01}|^2, \\ \partial_t'\psi_{00} + \zeta_{10} = -|\mathbf{k}|^2(1-\sigma^2)|\psi_{01}|^2, \end{cases}$$

where we recall that $\sigma = \tanh(\sqrt{\mu}|\mathbf{k}|)$ while α is given by (8.31) and that the Fourier multiplier $\omega(\mathbf{k}+\epsilon D')$ is given by $\omega(\mathbf{k}+\epsilon D') = (|\mathbf{k}+\epsilon D'|\tanh(\sqrt{\mu}|\mathbf{k}+\epsilon D'|))^{1/2}$. Assuming that $(\psi_{01},\psi_{00},\zeta_{10})$ is provided by the resolution of this system of equations, we can construct the approximate solution U_{app} defined in (8.16) as indicated in §8.2.3. The following proposition shows that this approximation is consistent with the water waves equations.

PROPOSITION 8.13. *Let $s \geq 0$, and $(\psi_{01}, \nabla\psi_{00}, \zeta_{10}) \in C([0,\frac{T}{\epsilon}]; H^{s+5}(\mathbb{R}^d)^{d+2})$ (with $T>0$) be a solution to* (8.32).
Then the approximate solution U_{app} is consistent with the water waves equations (8.20) *in the sense that*

$$\partial_t U_{app} + \mathcal{N}(U_{app}) = \epsilon^3 \sum_{n=-9}^{9} R_{3n}(\epsilon t, \epsilon X)e^{in(\mathbf{k}\cdot X - \omega t)} + \epsilon^{4-d/2}R_4(t,X),$$

with, for all $0 \leq t \leq T/\epsilon^2$,

$$\sum_{n=-9}^{9} |R_{3n}(\epsilon t,\cdot)|_{H^s} + |R_4(t,\cdot)|_{H^s} \leq C(\mu_{max}, |(\psi_{01},\nabla\psi_{00},\zeta_{10})|_{L^\infty([0,\frac{T}{\epsilon}];H^{s+5})}).$$

REMARK 8.14. The proposition assumes that (8.32) is locally well-posed on a time scale $t' = O(1/\epsilon)$ in order for the residual estimate to be of interest. It is also necessary for $(\psi_{01},\psi_{00},\zeta_{10})$ to be uniformly bounded with respect to ϵ on this time scale. Such a result is not known. It is not even known whether the partial results on the standard Benney-Roskes system (see Remark 8.18 below) can be generalized to (8.32) since the dispersion is weaker in the present case.

REMARK 8.15.
(1) The first component of the residual is formally of size $O(\epsilon^3)$, but its L^2 norm is of size $O(\epsilon^{3-d/2})$.
(2) The relevant time scale to observe the dynamics of the full-dispersion Benney-Roskes equation (8.32) is $t' = O(1/\epsilon)$ and therefore $t = O(1/\epsilon^2)$.
(3) It follows that the cumulated approximation error due to the first term of the residual, and over the relevant $O(1/\epsilon^2)$ time scale, is expected to be of size

$O(\epsilon^{1-d/2})$. This is small in dimension $d=1$, but not in dimension $d=2$, where a refined analysis is required[10].

REMARK 8.16. The linear *dispersion relation* associated to (8.32), i.e., the relation defining the set of all $(\omega', \mathbf{k}') \in \mathbb{R}^{1+d}$ such that there exists a plane wave solution of the form[11]

$$\psi_{01} = a e^{i(\mathbf{k}'\cdot X' - \omega' t')}, \qquad \psi_{00} = \zeta_{10} = 0$$

to the linear part of the equations, is given by

$$\omega' = \frac{\omega(\mathbf{k} + \epsilon \mathbf{k}') - \omega}{\epsilon} =: \omega_{FD}(\mathbf{k}').$$

This is exactly the relation we would have found by looking for a solution to the full linear water waves equations under the form (8.15) with ψ_{01} as above. The equations (8.32) therefore provide an *exact* description of the frequency behavior of small perturbations to (8.13).

PROOF. One readily checks that

$$\Big| \frac{\big(\omega^2(\mathbf{k}+\epsilon D') - \omega^2\big) + \epsilon^2 (\nabla\omega(\mathbf{k})\cdot\nabla')^2}{2\omega}\psi_{01} - (\omega(\mathbf{k}+\epsilon D') - \omega)\psi_{01}\Big|_{H^s} \lesssim \epsilon^3 |\psi_{01}|_{H^{s+2}}.$$

It follows that (8.28) and (8.30) are solved up to terms that can be put in the residual without changing the estimates on this latter. The other components of U_{app} are constructed as indicated in §8.2.3 and their norms can be estimated in terms of the norms of $(\psi_{01}, \psi_{00}, \zeta_{10})$. The result then follows from Proposition 8.10, since only the residual terms remain in the expansion of $\partial_t U_{app} + \mathcal{N}(U_{app})$. □

8.2.5. The "standard" Benney-Roskes model. The standard Benney-Roskes model originally derived in [**25**] for the water waves problem (and previously derived by Zakharov and Rubenchik [**334**] for acoustic waves – see also [**335**]) can be deduced from its full dispersion variant by approximating the nonlocal operator $i\omega(\mathbf{k}+\epsilon D')$ by the differential operator $i\omega + \epsilon\nabla\omega(\mathbf{k})\cdot\nabla' - \epsilon^2 \frac{i}{2}\nabla'\mathcal{H}_\omega(\mathbf{k})\nabla'$, which takes a simpler form if we assume (without loss of generality), that $\mathbf{k}$ is oriented along the x coordinate[12] and if we use the following notations

$$\begin{aligned} &\mathbf{k} = |\mathbf{k}|\mathbf{e}_x, \quad \omega(\mathbf{k}) = \underline{\omega}(|\mathbf{k}|), \quad \text{with} \quad \underline{\omega}(r) = \big(r\tanh(\sqrt{\mu}r)\big)^{1/2}, \\ &\omega = \underline{\omega}(|\mathbf{k}|), \quad \omega' = \underline{\omega}'(|\mathbf{k}|), \quad \omega'' = \underline{\omega}''(|\mathbf{k}|). \end{aligned} \tag{8.33}$$

[10]One could, for instance, show that the exact solution U_{ex} of the water waves equation can be written as an infinite sum of wave packets

$$U_{ex}(t,X) = \sum_{n=-\infty}^{\infty} U_{ex,n}(\epsilon t, \epsilon X) e^{in\theta},$$

and then seek approximations for the slowly varying envelopes $U_{ex,n}$; by working out each harmonic separately, this technique allows one to get rid of the fast scale. It has been used widely in nonlinear optics [**128, 192, 215, 220, 85**].

[11]The nonoscillating modes ψ_{00} and ζ_{10} are taken equal to zero since their coupling to ψ_{01} is due only to nonlinear effects.

[12]Recall that we write $X=(x,y)$ when $d=2$ and $X=x$ when $d=1$.

The Benney-Roskes equations can then be written (in dimension $d = 2$, the adaptation to the one-dimensional case is straightforward)

$$(8.34)\quad \begin{cases} \partial'_t\psi_{01} + \omega'\partial'_x\psi_{01} - i\epsilon\frac{1}{2}(\omega''{\partial'_x}^2 + \frac{\omega'}{|\mathbf{k}|}{\partial'_y}^2)\psi_{01} \\ \qquad +\epsilon i\big[|\mathbf{k}|\partial'_x\psi_{00} + \frac{|\mathbf{k}|^2}{2\omega}(1-\sigma^2)\zeta_{10} + 2\frac{|\mathbf{k}|^4}{\omega}(1-\alpha)|\psi_{01}|^2\big]\psi_{01} = 0, \\ \partial'_t\zeta_{10} + \sqrt{\mu}\Delta'\psi_{00} = -2\omega|\mathbf{k}|\partial'_x|\psi_{01}|^2, \\ \partial'_t\psi_{00} + \zeta_{10} = -|\mathbf{k}|^2(1-\sigma^2)|\psi_{01}|^2. \end{cases}$$

A consistency result similar to the one given in Proposition 8.13 holds for (8.34).

PROPOSITION 8.17. *Let $s \geq 0$, and $(\psi_{01}, \nabla\psi_{00}, \zeta_{10}) \in C([0, \frac{T}{\epsilon}]$; then let $H^{s+5}(\mathbb{R}^d)^{d+2})$ (with $T > 0$) be a solution to (8.34).*
Then the approximate solution U_{app} is consistent with the water waves equations (8.20) in the sense that

$$\partial_t U_{app} + \mathcal{N}(U_{app}) = \epsilon^3 \sum_{n=-9}^{9} R_{3n}(\epsilon t, \epsilon X)e^{in(\mathbf{k}\cdot X - \omega t)} + \epsilon^{4-d/2}R_4(t, X),$$

with, for all $0 \leq t \leq T/\epsilon^2$,

$$\sum_{n=-9}^{9} |R_{3n}(\epsilon t, \cdot)|_{H^s} + |R_4(t, \cdot)|_{H^s} \leq C(\mu_{max}, |(\psi_{01}, \nabla\psi_{00}, \zeta_{10})|_{L^\infty([0,\frac{T}{\epsilon}];H^{s+5})}).$$

REMARK 8.18. The proposition assumes that (8.34) is locally well-posed on a time scale $t' = O(1/\epsilon)$. In [**273**], the authors provide a local existence theorem based on smoothing effects due to the dispersion. The existence time thus obtained does not, however, reach the $O(1/\epsilon)$ time scale.

REMARK 8.19.
(1) The dispersion relation (see Remark 8.16) associated to (8.34) is given by

$$\omega_{BR}(\mathbf{k}') = \epsilon\omega' k'_x + \epsilon^2\frac{1}{2}(\omega''(k'_x)^2 + \frac{\omega'}{|\mathbf{k}|}(k'_y)^2),$$

and therefore is a second-order Taylor approximation of the dispersion relation of the full water waves equation in the neighborhood of $(\omega, \mathbf{k})$.
(2) The dispersive properties of (8.34) are worse than those of (8.32), but the standard Benney-Roskes equations contain only differential operators, while the full-dispersion version (8.32) contains nonlocal terms that are more delicate to handle, both numerically and theoretically (see Remark 8.14, for instance).

REMARK 8.20. The residual estimates provided by the proposition are not uniform as $\mu \to \infty$ and cannot be used in infinite depth where a different scaling is required for the ansatz. This will be done in §8.3 below, but we can already guess that the *rectification* effects (i.e., the creation of a nonoscillating mode by nonlinear interaction of oscillating modes) disappear in infinite depth. Indeed, if we formally take $\mu = \infty$ (and therefore $\sigma = 1$) in (8.34), we get $\psi_{00} = 0$, while the last equation gives $\zeta_{10} = 0$.

PROOF. We note that

$$\big|i\omega(\mathbf{k} + \epsilon D')\psi_{01} - \Big[i\omega + \epsilon\nabla\omega(\mathbf{k})\cdot\nabla' - \epsilon^2\frac{i}{2}\nabla'\cdot\mathcal{H}_\omega(\mathbf{k})\nabla'\Big]\psi_{01}\big|_{H^s} \lesssim \epsilon^3|\psi_{01}|_{H^{s+3}},$$

where $\mathcal{H}_\omega(\mathbf{k})$ denotes the Hessian matrix of ω at $\mathbf{k}$, and

$$\big| |D'| \tanh(\epsilon\sqrt{\mu}|D'|)\psi_{00} - (-\epsilon\sqrt{\mu}\Delta')\psi_{00}\big|_{H^s} \lesssim \epsilon^3 |\psi_{00}|_{\nabla H^{s+2}},$$

and moreover that

$$\nabla' \cdot \mathcal{H}_\omega(\mathbf{k})|\nabla|' = \underline{\omega}''(|\mathbf{k}|){\partial'_x}^2 + \frac{\underline{\omega}'(|\mathbf{k}|)}{|\mathbf{k}|}{\partial'_y}^2.$$

Substituting (8.34) with (8.32) brings new terms to the residual, but they do not affect the estimates on the latter. □

8.2.6. The Davey-Stewartson model (dimension $d = 2$). Motivated by the fact that, at leading order, ψ_{01} travels at the group speed $c_g = \nabla\omega(\mathbf{k})$, we look for solutions of the Benney-Roskes model (8.34) under the form

$$\begin{aligned} \psi_{01}(t', X') &= \underline{\psi}_{01}(\epsilon t', X' - c_g t), \\ \psi_{00}(t', X') &= \underline{\psi}_{00}(\epsilon t', X' - c_g t) + \psi^*_{00}(\epsilon t', t', X'), \\ \zeta_{10}(t', X') &= \underline{\zeta}_{10}(\epsilon t', X' - c_g t) + \zeta^*_{10}(\epsilon t', t', X'), \end{aligned}$$

so that $\underline{\psi}_{00}$ and $\underline{\zeta}_{10}$ denote the components of ψ_{00} and ζ_{10} travelling at the group speed. These components can formally be identified by equating all terms travelling (at leading order) at the group velocity in the last two equations of (8.34). Neglecting $O(\epsilon)$ correctors, we can replace ∂'_t by $-c_g \cdot \nabla'$ (we still denote by ∇' the gradient with respect to the variable $X - c_g t$), leading to

$$\begin{cases} -c_g \cdot \nabla' \underline{\zeta}_{10} + \sqrt{\mu}\Delta' \underline{\psi}_{00} = -2\omega \mathbf{k} \cdot \nabla' |\underline{\psi}_{01}|^2, \\ -c_g \cdot \nabla' \underline{\psi}_{00} + \underline{\zeta}_{10} = -|\mathbf{k}|^2 (1 - \sigma^2) |\underline{\psi}_{01}|^2 \end{cases}$$

and therefore

$$\begin{cases} \big[- (c_g \cdot \nabla')^2 + \sqrt{\mu}\Delta' \big] \underline{\psi}_{00} = -(|c_g||\mathbf{k}|(1 - \sigma^2) + 2\omega) \mathbf{k} \cdot \nabla' |\underline{\psi}_{01}|^2 \\ \underline{\zeta}_{10} = c_g \cdot \nabla' \underline{\psi}_{00} - |\mathbf{k}|^2 (1 - \sigma^2) |\underline{\psi}_{01}|^2. \end{cases}$$

The other components ζ^*_{10} and ψ^*_{00} should then be found by solving the homogeneous wave system

$$\begin{cases} \partial'_t \zeta^*_{10} + \sqrt{\mu}\Delta' \psi^*_{00} = 0, \\ \partial'_t \psi^*_{00} + \zeta^*_{10} = 0, \end{cases}$$

with initial conditions

$$\begin{cases} \psi^*_{00|_{t'=0}}(x') &= \psi_{00|_{t'=0}}(x') - \underline{\psi}_{00}\,_{|_{t'=0}}(x'), \\ \zeta^*_{10|_{t'=0}}(x') &= \zeta_{10|_{t'=0}}(x') - \underline{\zeta}_{10}\,_{|_{t'=0}}(x'). \end{cases}$$

In the particular case where $\psi^*_{00|_{t'=0}} = \zeta^*_{10|_{t'=0}} = 0$, the components ζ^*_{10} and ψ^*_{00} remain identically zero for all times. The *Davey-Stewartson approximation* can then be defined as

$$\begin{cases} \psi_{01}(t', x') &= \underline{\psi}_{01}(\epsilon t', x' - c_g t), \\ \psi_{00}(t', x') &= \underline{\psi}_{00}(\epsilon t', x' - c_g, t), \\ \zeta_{10}(t', x') &= \underline{\zeta}_{10}(\epsilon t', x' - c_g t), \end{cases} \tag{8.35}$$

where $\underline{\psi}_{01}(\tau, X')$, $\underline{\psi}_{00}(\tau, X')$ and $\underline{\zeta}_{10}(\tau, X')$ solve the Davey-Stewartson equations, for which the mean field is slaved to the envelope $\underline{\psi}_{01}$ of the wave packet. Using the notations (8.33), these equations are

$$(8.36)\quad \begin{cases} \partial_\tau \underline{\psi}_{01} - i\frac{1}{2}(\omega''\partial_x'^2 + \frac{\omega'}{|\mathbf{k}|}\partial_y'^2)\underline{\psi}_{01} + i(\beta\partial_x'\underline{\psi}_{00} + 2\frac{|\mathbf{k}|^4}{\omega}(1-\tilde{\alpha})|\underline{\psi}_{01}|^2)\underline{\psi}_{01} = 0 \\ [(\sqrt{\mu} - \omega'^2)\partial_x'^2 + \sqrt{\mu}\partial_y'^2]\underline{\psi}_{00} = -2\omega\beta\partial_x'|\underline{\psi}_{01}|^2, \end{cases}$$

with

$$(8.37)\qquad \beta = |\mathbf{k}|(1 + (1-\sigma^2)\frac{\omega'|\mathbf{k}|}{2\omega}), \qquad \tilde{\alpha} = \alpha + \frac{1}{4}(1-\sigma^2)^2,$$

and α as in (8.31), while $\underline{\zeta}_{10}$ is given by

$$(8.38)\qquad \underline{\zeta}_{10} = \underline{\omega}'(|\mathbf{k}|)\partial_x'\underline{\psi}_{00} - |\mathbf{k}|^2(1-\sigma^2)|\underline{\psi}_{01}|^2.$$

REMARK 8.21. Contrary to the Benney-Roskes equations (8.34), the Davey-Stewartson equations are written in a frame moving at the group velocity, and the time derivative is in terms of the variable $\tau = \epsilon^2 t$ rather than $t' = \epsilon t$. One therefore has to use (8.35) to recover an approximation of the envelopes of the leading terms of the wave packet approximation (8.16) to the water waves equations.

The Davey-Stewartson equations have been originally derived by Davey and Stewartson in [**116**] (while Djordjevic and Redekopp [**126**] and Ablowitz and Segur [**2**] generalized them to the case where surface tension is present). The consistency of the Davey-Stewartson approximation was then proved by Craig-Schanz-Sulem [**112**]. Note that this approximation also appears in the context of nonlinear optics where it has been justified in [**86**] (see also [**88**] for the case of nonzero long wave corrections[13]). We give below a result similar to [**112**]. In the statement of the proposition, one first has to define $(\psi_{01}, \psi_{00}, \zeta_{10})$ in terms of $\underline{\psi}_{01}$ through (8.35), before constructing the approximate solution U_{app} as in §8.2.3.

PROPOSITION 8.22. *Let $s \geq 0$, $T > 0$, and $\underline{\psi}_{01} \in C([0,T]; H^{s+5}(\mathbb{R}^2))$ be a solution to (8.36), and let $\underline{\zeta}_{10}$ be given by (8.38).*
Then the approximate solution U_{app} is consistent with the water waves equations (8.20) in the sense that

$$\partial_t U_{app} + \mathcal{N}(U_{app}) = \epsilon^3 \sum_{n=-9}^{9} R_{3n}(\epsilon t, \epsilon X) e^{in(\mathbf{k}\cdot X - \omega t)} + \epsilon^{4-d/2} R_4(t, X),$$

with, for all $0 \leq t \leq T/\epsilon^2$

$$\sum_{n=-9}^{9} |R_{3n}(\epsilon t, \cdot)|_{H^s} + |R_4(t,\cdot)|_{H^s} \leq C(\mu_{max}, |\underline{\psi}_{01}|_{L^\infty([0,T];H^{s+5})}).$$

REMARK 8.23.

(1) The main advantage of the Davey-Stewartson approximation with respect to the Benney-Roskes model is that it allows one to describe the evolution of wave packets over a time scale $t = O(1/\epsilon^2)$ through the resolution of an evolution equation (8.36) on a time scale of order $O(1)$ (recall that $\tau = \epsilon^2 t$) versus $O(1/\epsilon)$ for the Benney-Roskes equation (8.34) (written in terms of $t' = \epsilon t$).

[13]The long wave corrections to the Davey-Stewartson model are, in the context of water waves, the quantity ψ_{00}^* and ζ_{10}^* that have been taken equal to zero here by an adequate choice of initial conditions.

(2) The main drawback of the Davey-Stewartson approximation is that it is less general than the Benney-Roskes model; indeed, the initial conditions have to be well prepared[14] in order for (8.38) and the second equation of (8.36) to hold at $t = 0$. In such a configuration, the nonoscillating modes ψ_{00} and ζ_{10} are "slaved" to the leading oscillating term ψ_{01} through the second equation of (8.36) and (8.38).

REMARK 8.24.

(1) The equations (8.36) belong to the following family of Davey-Stewartson equations

$$\begin{cases} i\partial_\tau \psi + \mathbf{a}\partial_x^2\psi + \mathbf{b}\partial_y^2\psi = (\nu_1|\psi|^2 + \nu_2\partial_x\varphi)\psi, \\ \mathbf{c}\partial_x^2\varphi + \partial_y^2\varphi = -\delta\partial_x|\psi|^2, \end{cases} \tag{8.39}$$

with $\mathbf{b} > 0$ (up to a simple change of unknown if necessary) and $\delta > 0$. Note that these two conditions are obviously satisfied by (8.36). The nature of the Davey-Stewartson equations then depends strongly on the sign of $\mathbf{a}$ and $\mathbf{c}$. See Chapter XII of [**307**] for a detailed discussion. Using the terminology of [**157**], we say that (8.39) is

elliptic-elliptic if $(\operatorname{sgn}\mathbf{a}, \operatorname{sgn}\mathbf{c}) = (+1, +1)$,
hyperbolic-elliptic if $(\operatorname{sgn}\mathbf{a}, \operatorname{sgn}\mathbf{c}) = (-1, +1)$,
elliptic-hyperbolic if $(\operatorname{sgn}\mathbf{a}, \operatorname{sgn}\mathbf{c}) = (+1, -1)$,
hyperbolic-hyperbolic if $(\operatorname{sgn}\mathbf{a}, \operatorname{sgn}\mathbf{c}) = (-1, -1)$.

(2) For the water waves problem, $\mathbf{a} = \frac{1}{2}\omega'' < 0$ since $\omega'' = \underline{\omega}''(|\mathbf{k}|)$, where $\underline{\omega}$ defined in (8.33) is concave[15].
We also note, for the water waves problem, that $\mathbf{c} = \sqrt{\mu} - \omega'^2 > 0$. Writing this condition in variables with dimensions is equivalent to $M < 1$, where $M = C_g/\sqrt{gH_0}$ is the Mach number—with $C_g = \frac{d}{dr}\sqrt{gr\tanh(H_0 r)}_{|r=|k|}$ the group velocity in dimensional form. The fact that $\mathbf{c}$ is positive therefore is a consequence of the fact that the group velocity is always smaller than the phase velocity[16]; it is sometimes said that $\mathbf{c} > 0$ corresponds to a subsonic configuration, while $\mathbf{c} < 0$ is supersonic [**3**].
If follows from the two previous points that the Davey-Stewartson equations (8.36) are of *hyperbolic-elliptic* type. In this situation, (8.39) can be written as a nonlocal nonlinear Schrödinger equation

$$i\partial_\tau\psi + \mathbf{a}\partial_x^2\psi + \mathbf{b}\partial_y^2\psi = (\nu_1|\psi|^2 - \delta\nu_2 E|\psi|^2)\psi,$$

where E is the Fourier multiplier $E = \frac{\partial_x^2}{\mathbf{c}\partial_x^2 + \partial_y^2}$.
Proposition 8.22 assumes that this equation is locally well-posed on a time scale $\tau = O(1)$ for initial data in Sobolev spaces. This has been proved by Ghidaglia

[14] With the notations used above, well-prepared data correspond to $\psi_{00}^* = \zeta_{10}^* = 0$. As shown in [**88**], it is possible to handle general data, but the precision of the approximation is then worse than it is for the Benney-Roskes model (8.34) because the interaction between the Davey-Stewartson approximation and the long wave corrections ψ_{00}^* and ζ_{10}^* are neglected. The Davey-Stewartson model somehow describes the long time asymptotics of the Benney-Roskes equations, when the leading order wave packet and the long wave corrections, which travel at different speeds, do not interact anymore.

[15] This is not necessarily the case in the presence of surface tension, where one can have $\mathbf{a} \geq 0$.

[16] This is not always the case in the presence of surface tension, where one can have $M \geq 1$ and therefore $\mathbf{c} \leq 0$. The case $\mathbf{c} = 0$ is known as the *long wave-short wave resonance* (see [**307**], and [**87**] for an investigation of this resonance in the context of nonlinear optics).

and Saut in [**157**], where situations other than the hyperbolic-elliptic case are also considered.

PROOF. There are some extra residual terms associated with the Davey-Stewartson model, compared to the Benney-Roskes model. These extra terms come from the fact that $\epsilon\partial_\tau \underline{\zeta}_{10}$ and $\epsilon\partial_\tau \underline{\psi}_{00}$ have been discarded in the formal derivation of the equations for $\underline{\zeta}_{10}$ and $\underline{\psi}_{00}$ performed at the beginning of this section. It is easy to check that the contribution of these terms does not affect the residual estimates. □

8.2.7. The nonlinear Schrödinger equation (dimension $d = 1$). The NLS approximation is simply the particular form taken by the Davey-Stewartson equations in dimension $d = 1$. Discarding all the transverse derivatives in (8.36), one easily gets from the second equation that

$$\partial'_x \underline{\psi}_{00} = -\frac{2\omega\beta}{\sqrt{\mu} - \omega'^2} |\underline{\psi}_{01}|^2. \tag{8.40}$$

Replacing $\partial'_x \underline{\psi}_{00}$ by this expression in the first equation yields the following nonlinear Schrödinger equation[17] for $\underline{\psi}_{01}$

$$\partial_\tau \underline{\psi}_{01} - i\frac{1}{2}\omega''\partial_x'^2 \underline{\psi}_{01} + i\delta |\underline{\psi}_{01}|^2 \underline{\psi}_{01} = 0, \tag{8.42}$$

where we used the notation (8.33), and with

$$\delta = 2\frac{|\mathbf{k}|^4}{\omega}(1 - \tilde{\alpha}) - 2\frac{\omega\beta^2}{\sqrt{\mu} - \omega'^2},$$

and β and $\tilde{\alpha}$ as in (8.37), while the nonoscillating surface term $\underline{\zeta}_{10}$ is still given by (8.38).

The NLS equation has been derived for water waves in finite depth by Hasimoto and Ono [**169**] (following the earlier work by Zakharov in infinite depth [**333**]). This approximation also appears in other physical contexts and especially in nonlinear optics, where it is the standard model to describe the evolution of almost monochromatic waves in a weakly nonlinear dispersive medium (see, for instance, [**52, 267**], and [**194, 128, 192, 215, 289**] for a justification). The consistency of the NLS approximation for water waves in finite depth was originally proved in [**111**]. We present a similar result below (this is simply the one-dimensional version of Proposition 8.22).

PROPOSITION 8.25. *Let $s \geq 0$, $T > 0$, and $\underline{\psi}_{01} \in C([0,T]; H^{s+5}(\mathbb{R}))$ be a solution to (8.42), and let $\underline{\zeta}_{10}$ and $\partial'_x \underline{\psi}_{00}$ be given by (8.38) and (8.40). Then the approximate solution U_{app} is consistent with the water waves equations (8.20) in the sense that*

$$\partial_t U_{app} + \mathcal{N}(U_{app}) = \epsilon^3 \sum_{n=-9}^{9} R_{3n}(\epsilon t, \epsilon X) e^{in(\mathbf{k}\cdot X - \omega t)} + \epsilon^{4-d/2} R_4(t, X),$$

[17] The corresponding equation satisfied by $\psi_{01}(t', x') = \underline{\psi}_{01}(\epsilon t', x' - c_g t)$ (with $c_g = \omega'$) is

$$\partial'_t \psi_{01} + \omega' \partial'_x \psi_{01} - i\frac{\epsilon}{2}\partial_x'^2 \omega'' \psi_{01} + i\epsilon\delta |\psi_{01}|^2 \psi_{01} = 0. \tag{8.41}$$

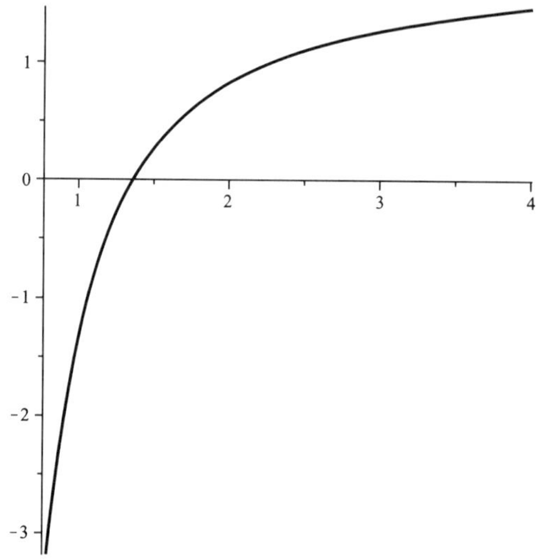

FIGURE 8.1. Dependence on $\mathbf{K}H_0$ of the nonlinearity coefficient δ in (8.42).

with, for all $0 \leq t \leq T/\epsilon^2$,

$$\sum_{n=-9}^{9} |R_{3n}(\epsilon t, \cdot)|_{H^s} + |R_4(t,\cdot)|_{H^s} \leq C(\mu_{max}, |\underline{\psi}_{01}|_{L^\infty([0,T];H^{s+5})}).$$

REMARK 8.26. The NLS approximation presents in dimension $d = 1$ the same advantages and drawbacks as the Davey-Stewartson approximation in dimension $d = 2$ (see Remark 8.23).

REMARK 8.27. The leading oscillating term for the surface elevation is given by $\zeta_{01} = i\omega\psi_{01}$ (see (8.17)). An NLS equation for the envelope $\zeta_{01}(t', x') = \underline{\zeta}_{01}(\epsilon t', x' - c_g t')$ therefore can be deduced from (8.42), namely,

$$\partial_\tau \underline{\zeta}_{01} - i\frac{1}{2}\omega'' {\partial'_x}^2 \underline{\zeta}_{01} + i\frac{\delta}{\omega^2}|\underline{\zeta}_{01}|^2\underline{\zeta}_{01} = 0. \tag{8.43}$$

REMARK 8.28. The sign of the coefficient δ of the nonlinearity depends on $\sqrt{\mu}|\mathbf{k}|$ (or $|\mathbf{K}|H_0$, if $\mathbf{K} = \mathbf{k}/L_x$ is the wave number in dimensional form). There is a critical value $|\mathbf{K}|H_0 \sim 1.363$, above which $\delta > 0$ (focusing case) and below which $\delta < 0$ (defocusing case), as shown in Figure 8.1.

In the neighborhood of this critical value, the nonlinear term vanishes and a new scaling is necessary to observe nonlinear effects (as for the Kawahara equation for long capillary gravity waves, see §9.3.2). This has been done in [**189**]; see §8.5.5.

REMARK 8.29. The proposition assumes the local well-posedness of (8.42) for a time scale $\tau = O(1)$, which is completely standard (since we are in dimension $d = 1$, we also have global well-posedness, regardless of the sign of δ; see [**306**]).

8.3. Modulation equations in infinite depth

We explain here how to formally derive the models studied in §8.2 in the case of finite depth. Using the results of §4.4.3 on the well-posedness of the water waves equations in infinite depth, the same consistency results as in the finite depth case could be derived along the same lines. We recall that the formulation of the water waves equations in infinite depth is provided by (4.65), namely,

$$\begin{cases} \partial_t\zeta - \mathcal{G}[\epsilon\zeta]\psi = 0, \\ \partial_t\psi + \zeta + \frac{\epsilon}{2}|\nabla\psi|^2 - \epsilon\frac{(\mathcal{G}[\epsilon\zeta]\psi + \epsilon\nabla\zeta\cdot\nabla\psi)^2}{2(1+\epsilon^2|\nabla\zeta|^2)} = 0. \end{cases}$$

A small amplitude expansion of these equations can be adapted from §8.2.2.

8.3.1. The ansatz. A quick look at (8.27) shows that the ansatz (8.16)–(8.19) is not valid in infinite depth. Indeed, the term

$$|D'|\frac{\tanh(\epsilon\sqrt{\mu}|D'|)}{\epsilon}\psi_{00},$$

which was treated as a $O(1)$ term in finite depth becomes in infinite depth ($\mu = \infty$), $\frac{1}{\epsilon}|D'|\psi_{00}$, which is obviously not $O(1)$ with respect to ϵ. This means that the nonoscillating component of ψ must be one order smaller in infinite depth (i.e., ψ_{00} must be taken equal to zero, and the leading nonoscillating term for the potential should then be of the form $\epsilon\psi_{10}$). We therefore replace the ansatz (8.16)–(8.19) by

$$U_{app}(t,X) = U_0(t,X) + \epsilon U_1(t,X) + \epsilon^2 U_2(t,X), \tag{8.44}$$

with

$$U_0(t,X) = \begin{pmatrix} i\omega\psi_{01}(\epsilon t,\epsilon X) \\ \psi_{01}(\epsilon t,\epsilon X) \end{pmatrix} e^{i\theta} + \text{c.c.}, \tag{8.45}$$

and with the first corrector U_1 given by [18]

$$U_1(t,X) = \begin{pmatrix} \zeta_{11}(\epsilon t,\epsilon X)e^{i\theta} + \zeta_{12}(\epsilon t,\epsilon X)e^{i2\theta} + \text{c.c.} \\ \psi_{11}(\epsilon t,\epsilon X)e^{i\theta} + \psi_{12}(\epsilon t,\epsilon X)e^{i2\theta} + \text{c.c.} + \psi_{10}(\epsilon t,\epsilon x), \end{pmatrix} \tag{8.46}$$

while the second corrector U_2 keeps the same general form

$$U_2(t,X) = \sum_{n=1}^{3} U_{2n}(\epsilon t,\epsilon X)e^{in\theta} + \text{c.c.} + U_{20}(\epsilon t,\epsilon X). \tag{8.47}$$

8.3.2. The nonlinear Schrödinger equation (dimension $d = 1$ or 2). Proceeding as in §8.2.3, but with the ansatz (8.44)–(8.47)—and using the fact that $\sigma = 1$ in infinite depth—we are led to replace (8.27) by

$$\begin{cases} -|D'|\psi_{10} + 2\omega\mathbf{k}\cdot\nabla'|\psi_{01}|^2 = 0, \\ \zeta_{20} + \mathcal{N}_{20}^{(2)} = 0. \end{cases} \tag{8.48}$$

It is worth pointing out that the leading order nonoscillating term ψ_{10} is determined explicitly in terms of ψ_{01}, while in the case of finite depth, one had to solve the evolution equation (8.27), coupling ψ_{00} to ζ_{10} (which is now taken equal to zero[19]). Proceeding as in the case of finite depth, we can derive the Benney-Roskes equations in infinite depth. It turns out that *in infinite depth, there is no coupling of ψ_{01} with nonoscillating modes*, and that the Benney-Roskes equations coincide with the

[18] As for the case of finite depth, it is difficult to guess *a priori* the particular form of the corrector terms; the expressions given here are the simplest ones consistent with the computations of the next sections.

[19] We therefore assume implicitly that the nonoscillating modes are initially equal to zero.

following nonlinear Schrödinger equation[20] (both in dimension $d = 1$ and $d = 2$), originally derived by Zakharov [**333**]

$$(8.49)\qquad \partial_t'\psi_{01} + \omega'\partial_x'\psi_{01} - i\epsilon\frac{1}{2}\big(\omega''{\partial_x'}^2 + \frac{\omega'}{|\mathbf{k}|}{\partial_y'}^2\big)\psi_{01} + 2\epsilon i\frac{|\mathbf{k}|^4}{\omega}|\psi_{01}|^2\psi_{01} = 0,$$

where we used the notations (8.33) (note that in infinite depth, $\omega = \omega(\mathbf{k}) = |\mathbf{k}|^{1/2}$). Equivalently, writing as in §8.2.6,

$$\psi_{01}(t', X') = \underline{\psi}_{01}(\epsilon t', X' - c_g t),$$

and with $\tau = \epsilon t' = \epsilon^2 t$, we get

$$(8.50)\qquad \partial_\tau\underline{\psi}_{01} - i\frac{1}{2}\big(\omega''{\partial_x'}^2 + \frac{\omega'}{|\mathbf{k}|}{\partial_y'}^2\big)\underline{\psi}_{01} + 2i\frac{|\mathbf{k}|^4}{\omega}|\underline{\psi}_{01}|^2\underline{\psi}_{01} = 0.$$

REMARK 8.30.

(1) Since $\omega'' < 0$ and $\omega' > 0$, the Schrödinger equation (8.42) is said to be *nonelliptic* (or hyperbolic) in dimension $d = 2$. In dimension $d = 1$, since $|\mathbf{k}|^4/\omega > 0$, the one-dimensional NLS equation (8.50) is *focusing.*
(2) The NLS approximation in infinite depth requires the well-posedness of (8.50) for a time scale $\tau = O(1)$, which is again completely standard (but the issue of global well-posedness for the nonelliptic Schrödinger equation in dimension $d = 2$ is much more difficult [**158, 159, 307**]).

REMARK 8.31. Since $\zeta_{01} = i\omega\psi_{01}$, the leading order term for the amplitude solves

$$(8.51)\qquad \partial_t'\zeta_{01} + \omega'\partial_x'\zeta_{01} - i\epsilon\frac{1}{2}\big(\omega''{\partial_x'}^2 + \frac{\omega'}{|\mathbf{k}|}{\partial_y'}^2\big)\zeta_{01} + 2\epsilon i\frac{|\mathbf{k}|^4}{\omega^3}|\zeta_{01}|^2\zeta_{01} = 0,$$

which is exactly the same as (8.51) when $|\mathbf{k}| = 1$ since in infinite depth $\omega^2 = |\mathbf{k}|$.

REMARK 8.32. In [**328**], S. Wu was able to construct a normal form transform that allowed her to prove almost global existence of the water waves equation in dimension $d = 1$, for small data and in infinite depth. The same structure of the equations has been used in [**315**] to fully justify the nonlinear Schrödinger equation in dimension $d = 1$ and in infinite depth.

8.4. Justification of the modulation equations

In Theorem 8.4, we gave the almost full justification[21] of the deep water model (8.2). The strategy was to show that the approximation based on (8.2) was consistent with the water waves equations and to use the stability Theorem 4.18 associated to the water waves equation.

Since we have proved the consistency of all the modulation approximations derived in the previous section, one would expect to proceed as for (8.2) and to obtain the almost full justification of these modulation equations. There are, however, two obstructions:

[20] This equation can be formally derived from the first equation of (8.34) by setting $\zeta_{10} = \psi_{00} = 0$ and $\sigma = 1$.

[21] Recall that an approximation is "almost fully" justified if it provides a good approximation of a solution to the water waves equations *when this approximation is well defined.* This requires a local well-posedness property of the model on the relevant time scale, which is not known for (8.2), (8.32), or (8.34), for instance.

(1) The existence time furnished by Theorem 4.16 for the water waves equation is $t = O(1/\epsilon)$ while, as we have seen, the relevant time scales for the modulation equations derived in this chapter is $t = O(1/\epsilon^2)$.
(2) As noted in Remark 8.15, the residual is of size $O(\epsilon^{3-d/2})$ in Sobolev norm and its cumulated effects on the relevant time scale $t = O(1/\epsilon^2)$ are $O(\epsilon^{1-d/2})$. In dimension $d = 1$, this is small and one can expect to conclude with a stability property similar to the one provided by Theorem 4.18. In dimension $d = 2$, the residual is $O(1)$ and a stability property in Sobolev norm is unlikely to give the result (a refined analysis of the fact that the residual is a sum of wave packets of amplitude $O(\epsilon^3)$ seems to be required).

Of course, the above strategy can be used[22] to justify (or almost justify) the modulation equations derived in this chapter on a time scale[23] $t = O(1/\epsilon)$. However, the typical dispersive and nonlinear effects of these models are not observed at leading order at this time scale for which wave packets basically travel at the group speed without important modifications of the shape of their envelope.

Justifying the modulation equations of this chapter on the relevant time scale therefore requires a better understanding of the water waves equations. The structure of the equations, which allows for the justification of these approximations on the relevant time scale, is certainly related to the structural conditions necessary to obtain global (or almost global) well-posedness results for small data (see the discussion in [**222**]). In infinite depth, almost global well-posedness is known in dimension $d = 1$ [**328**], while global well-posedness holds in dimension $d = 2$ [**329, 156**]. But no such result has been proved yet in finite depth.
Regarding the full justification of modulation equations on the relevant time scale, the only known results are for the NLS approximation and in dimension $d = 1$. In infinite depth, the NLS approximation in dimension $d = 1$ has been justified in [**315**] using the tools developed in [**328**] to obtain almost global existence of the water waves equations in dimension $d = 1$. In finite depth, the result has been established in [**293**] for a toy model sharing many difficulties with the finite depth water waves equations, and very recently in [**134**] for the full water waves equations.

8.5. Supplementary remarks

8.5.1. Benjamin-Feir instability of periodic wave-trains. In 1847, Stokes [**302**] investigated the existence of periodic wave-train solutions to the water waves equations, for which he proposed a formal expansion. The convergence of this expansion for small amplitude waves was proved much later by Nekrasov [**266**] in 1921, Levi-Civita [**231**] in 1925 and, for the case of finite depth, by Struik [**303**] in 1926. It was only in 1978 that the existence of Stokes waves of maximum height on deep water was established by Toland [**314**].

[22]The local well-posedness result of [**273**] is enough to bring a full justification on a time scale $t = O(1/\epsilon)$ (equivalently, $t' = O(1)$). Moreover, the cumulated effects of the residual are then $O(\epsilon^{2-d/2})$ and are therefore small even in dimension $d = 2$, so that the justification holds on $t = O(1/\epsilon)$ regardless of the dimension.

[23]Proceeding as in [**224, 88**] one could expect to extend this justification to $t = O(\frac{|\ln \epsilon|}{\epsilon})$.

However, in 1967 Whitham [**323**], using a general formal variational argument, and Benjamin and Feir, in two famous papers (dealing with theoretical and experimental aspects [**32, 31**]), discovered that periodic wave-trains suffer an instability[24] in deep water (more precisely, when $\mathbf{K}H_0 > 1.363$, $\mathbf{K}$ being the wave number of the wave train). This instability, now called *Benjamin-Feir instability*, is characterized by the appearance of growing modulations. The works of Whitham, and Benjamin and Feir are based on a linear analysis predicting exponential growth for some perturbation frequencies. Because of this exponential growth, linear analysis ceases to be valid, and the modulation analysis leading to the NLS approximation has been developed to describe the evolution of these instabilities. This approach is due to Zakharov [**333**] (1968), for his work in infinite depth, and to Hasimoto and Ono [**169**] (1972), for their work in finite depth.

We provide below the formal argument of [**169**] that allows one to recover the instability threshold $\mathbf{K}H_0 \sim 1.363$ for periodic wave-trains from the NLS approximation of wave packets presented in this chapter. The NLS equation (8.43) admits (nonlinear) plane waves solutions of the form

$$(8.52) \qquad \underline{\zeta}_{01}(\tau, x') = \underline{\zeta}_{01}(\tau) = \frac{a}{2}\exp(-i\Omega\tau), \qquad \text{with} \qquad \Omega = \frac{\delta}{4\omega^2}|a|^2,$$

and where the amplitude $a/2$ is constant. The corresponding approximation for the surface elevation is given by

$$\begin{aligned} \zeta_{app}(t,x) &= \underline{\zeta}_{01}(\epsilon^2 t)\exp\big(i(\mathbf{k}\cdot x - \omega t)\big) + \text{c.c.} \\ &= a\cos\big(\mathbf{k}\cdot x - \omega(1+\epsilon^2\frac{\delta}{4\omega^3}|a|^2)t\big). \end{aligned}$$

Adding the corrector terms of (8.18), this corresponds to the second-order expansion of the Stokes wave train.

Let us now consider a small perturbation of the nonlinear plane wave solution (8.52) under the form

$$\underline{\zeta}(\tau, x') = \frac{1}{2}(a + \dot{a})\exp\big(i(-\Omega\tau + \dot{\theta})\big),$$

where $\dot{a}$ and $\dot{\theta}$ are functions standing for the (small) amplitude and phase perturbations. Plugging this expression into the NLS equation (8.43) and neglecting nonlinear terms gives the following system for $(\dot{a}, \dot{\theta})$

$$\begin{cases} \partial_\tau \dot{a} + \frac{a}{2}\omega''{\partial'_x}^2\dot{\theta} = 0, \\ \partial_\tau\dot{\theta} + \frac{a}{2}\delta\dot{a} - \frac{\omega''}{2a}{\partial'_x}^2\dot{a} = 0. \end{cases}$$

This is a linear system that does not exhibit exponential growth in τ (in other words, the Stokes wave train is stable) if and only if

$$\frac{a}{2}\omega''\xi'^2\big(\frac{a}{2}\delta + \frac{\omega''}{2a}\xi'^2\big) > 0,$$

for all frequencies $\xi' \in \mathbb{R}$. Since $\omega'' < 0$, the Stokes wave train therefore is stable for $\delta < 0$ and unstable for $\delta > 0$. From Remark 8.28, Stokes wave trains are stable in relatively shallow water ($\mathbf{K}H_0 < 1.363$) and unstable in deeper water ($\mathbf{K}H_0 > 1.363$).

[24] Note that this instability may appear for wave-trains far below the maximal height.

One of the main features of the one-dimensional focusing[25] NLS equation is that it possesses solitons. An "experimental confirmation" of the NLS approximation can be found in [**332**], where soliton interactions are observed in wave tanks. The fact that wave trains do not always disintegrate, even in the unstable case $\mathbf{K}H_0 > 1.363$, but sometimes lead to spectacular recurrence effects (i.e., the wave train starts to disintegrate, but then goes back roughly to its initial form) is another experimental observation [**213, 332, 304**] that supports the NLS description of wave trains. A full mathematical justification of the Benjamin-Feir instability can be found in [**58**]; also see the review paper [**123**] for more references on this issue.

8.5.2. Full-dispersion Davey-Stewartson and Schrödinger equations. The Davey-Stewartson equations are actually an order $O(\epsilon)$ approximation of the following *full-dispersion Davey-Stewartson equations*

$$(8.53)\qquad \begin{cases} \partial_\tau \underline{\psi}_{01} + \frac{i}{\epsilon^2}\big(\omega(\mathbf{k}+\epsilon D') - \omega - \epsilon\omega' D'_x\big)\underline{\psi}_{01} \\ \qquad\qquad + i\big(\beta\partial'_x\underline{\psi}_{00} + 2\frac{|\mathbf{k}|^4}{\omega}(1-\tilde{\alpha})|\underline{\psi}_{01}|^2\big)\underline{\psi}_{01} = 0 \\ \big[\frac{|D'|\tanh(\epsilon\sqrt{\mu}|D'|)}{\epsilon} + \omega'\partial_x'^{2}\big]\underline{\psi}_{00} = 2\omega\beta\partial'_x|\underline{\psi}_{01}|^2, \end{cases}$$

where we used the notations (8.33). The dispersion relation of (8.53) is exactly the dispersion relation of the linear water waves equations. This model corresponds to the "Davey-Stewartson version" of the full-dispersion Benney-Roskes model derived in §8.2.4, and a consistency result similar to Proposition 8.22 can be derived also.

REMARK 8.33.

(1) The discussion on the nature of the second equation of the Davey-Stewartson system (8.36) (elliptic vs. nonhyperbolic; see Remark 8.24) is dramatically different from its "full-dispersion" version.

(2) It is, of course, possible to work with the full-dispersion version of only the first or the second of the two equations of the Davey-Stewartson model (8.36). This yields respectively

$$\begin{cases} \partial_\tau \underline{\psi}_{01} + \frac{i}{\epsilon^2}\big(\omega(\mathbf{k}+\epsilon D') - \omega - \epsilon\omega' D'_x\big)\underline{\psi}_{01} \\ \qquad\qquad + i\big(\beta\partial'_x\underline{\psi}_{00} + 2\frac{|\mathbf{k}|^4}{\omega}(1-\tilde{\alpha})|\underline{\psi}_{01}|^2\big)\underline{\psi}_{01} = 0 \\ \big[(\sqrt{\mu} - \omega'^2)\partial_x'^{2} + \sqrt{\mu}\partial_y'^{2}\big]\underline{\psi}_{00} = -2\omega\beta\partial'_x|\underline{\psi}_{01}|^2, \end{cases}$$

and

$$\begin{cases} \partial_\tau \underline{\psi}_{01} - i\frac{1}{2}\big(\omega''\partial_x'^{2} + \frac{\omega'}{|\mathbf{k}|}\partial_y'^{2}\big)\underline{\psi}_{01} + i\big(\beta\partial'_x\underline{\psi}_{00} + 2\frac{|\mathbf{k}|^4}{\omega}(1-\tilde{\alpha})|\underline{\psi}_{01}|^2\big)\underline{\psi}_{01} = 0 \\ \big[\frac{|D'|\tanh(\epsilon\sqrt{\mu}|D'|)}{\epsilon} + \omega'\partial_x'^{2}\big]\underline{\psi}_{00} = 2\omega\beta\partial'_x|\underline{\psi}_{01}|^2. \end{cases}$$

In the one-dimensional case $d = 1$ one can derive the *full-dispersion nonlinear Schrödinger equation* whose dispersive properties exactly match those of the linear

[25]We recall that this corresponds to $\delta > 0$ or equivalently $\mathbf{K}H_0 > 1.363$.

water waves equations (see Remark 8.16). It is given by[26]

$$\partial_\tau \underline{\psi}_{01} + \frac{i}{\epsilon^2}\big(\omega(\mathbf{k}+\epsilon D') - \omega - \epsilon\omega' D'_x\big)\underline{\psi}_{01} + i\delta|\underline{\psi}_{01}|^2\underline{\psi}_{01} = 0, \tag{8.54}$$

with $\omega(\mathbf{k}) = \big(|\mathbf{k}|\tanh(\sqrt{\mu}|\mathbf{k}|)\big)^{1/2}$; the linear part of the standard NLS equation (8.42) is the second-order expansion of the linear part of (8.54).

REMARK 8.34. We have seen in §8.3.2 that in infinite depth, the NLS equation (8.50) is relevant in both dimensions $d = 1$ and $d = 2$. The full-dispersion version of (8.50) is

$$\partial_\tau \underline{\psi}_{01} + \frac{i}{\epsilon^2}\big(\omega(\mathbf{k}+\epsilon D') - \omega - \epsilon\omega' D'_x\big)\underline{\psi}_{01} + 2i\frac{|\mathbf{k}|^4}{\omega}|\underline{\psi}_{01}|^2\underline{\psi}_{01} = 0, \tag{8.55}$$

with $\omega(\mathbf{k}) = |\mathbf{k}|^{1/2}$. Note that while the local well-posedness of (8.55) in $H^s(\mathbb{R}^d)$ with $s > d/2$ does not cause any difficulty, other properties such as the local smoothing for NLS type equations (see [**210**]) cannot be directly applied to (8.55) or other "full dispersion" versions of the classical modulation equations.

8.5.3. The nonlinear Schrödinger approximation with improved dispersion. Proceeding as in Remark 8.16 and using the notations of (8.33), the linear *dispersion relation*[27] associated to the standard NLS equation (8.41) (or (8.49) in infinite depth) is given by

$$\omega' = \omega_{NLS}(\mathbf{k}') \quad \text{with} \quad \omega_{NLS}(\mathbf{k}') = \omega'\mathbf{k}'_1 + \epsilon\Big(\frac{1}{2}\omega''{\mathbf{k}'_1}^2 + \frac{\omega'}{|\mathbf{k}|'}{\mathbf{k}'_2}^2\Big),$$

which is the second-order Taylor expansion at $\epsilon = 0$ of the dispersion relation of the full-dispersion models of §8.5.2, namely,

$$\omega_{FD}(\mathbf{k}') = \frac{\omega(\mathbf{k}+\epsilon\mathbf{k}') - \omega}{\epsilon}$$

(see Remark 8.16). For large wave numbers $\mathbf{k}'$, the dispersive properties of the NLS approximation therefore are a very poor approximation of the dispersive properties of the water waves equations. This problem disappears with the full dispersion models, but their drawback is that they involve nonlocal operators while the NLS equation involves only differential operators. This yields some constraints; for instance, the numerical computations of nonlocal operators is less straightforward than for differential operators, and extensions to initial boundary value problems are quite delicate.

For these reasons, *nonlinear Schrödinger equations with improved dispersion* have been proposed in [**85**] (see also [**220**]) in the context of nonlinear optics where the NLS approximation also plays an important role. In the case of the NLS approximation in infinite depth (8.49), for instance, one obtains the following family of

[26]We recall that $\underline{\psi}_{01}$ is the parametrization of the leading order oscillating mode of the velocity potential in a frame moving at the group velocity and with a rescaling in time, $\psi_{01}(t', x') = \underline{\psi}_{01}(\epsilon t', x' - c_g t')$. We therefore can write (8.54) as the following evolution equation on ψ_{01}

$$\partial_\tau\psi_{01} + \frac{i}{\epsilon}\big(\omega(\mathbf{k}+\epsilon D') - \omega\big)\psi_{01} + i\epsilon\delta|\underline{\psi}_{01}|^2\psi_{01} = 0.$$

[27]We recall that it is the relation defining the set of all $(\omega', \mathbf{k}') \in \mathbb{R}^{1+d}$, such that there exists a plane wave solution of the form $\psi_{01} = ae^{i(\mathbf{k}'\cdot X' - \omega' t')}$, to the linear part of the equation.

equations

$$(1 - i\epsilon\mathbf{b}\cdot\nabla' - \epsilon^2\nabla'\cdot B\nabla')\partial'_t\psi_{01} + \omega'\partial'_x\psi_{01} - \epsilon\frac{i}{2}\big(\omega''{\partial'_x}^2 + \frac{\omega'}{|\mathbf{k}|}{\partial'_y}^2\big)\psi_{01}$$

$$-\epsilon i\omega'\mathbf{b}\cdot\nabla'\partial'_x\psi_{01} + \epsilon^2 C_3(\nabla')\psi_{01} + 2\epsilon i\frac{|\mathbf{k}|^4}{\omega}|\psi_{01}|^2\psi_{01} = 0, \tag{8.56}$$

where $\mathbf{b} \in \mathbb{C}^d$, $B \in \mathcal{M}_{d\times d}(\mathbb{R})$ and $C_3(\nabla)$ is a third-order homogeneous differential operator with real coefficients. We moreover assume that

$$B \text{ is symmetric positive,} \quad \mathbf{b} \in \mathit{Range}(B), \quad \text{and} \quad 4 - \mathbf{b}\cdot(B^{-1}\mathbf{b}) > 0 \tag{8.57}$$

(note that even though $B^{-1}\mathbf{b}$ is not unique when B is not definite, the scalar $\mathbf{b}\cdot(B^{-1}\mathbf{b})$ is uniquely defined). These assumptions ensure that the operator $(1 - i\epsilon\mathbf{b}\cdot\nabla - \epsilon^2\nabla\cdot B\nabla)$ is invertible. All the improved NLS equations (8.56) are asymptotically equivalent to the standard NLS equation (8.49) and satisfy in particular the same consistency properties with the full water waves equations[28].

REMARK 8.35. The standard cubic NLS equation corresponds to the particular case $\mathbf{b} = 0$, $B = 0$ (which satisfies (8.57)).

The dispersion relation associated to (8.56) is given by

$$\omega_{imp}(\mathbf{k}') = \frac{\omega'\mathbf{k}'_1 + \epsilon\Big(\frac{1}{2}\omega''{\mathbf{k}'_1}^2 + \frac{\omega'}{|\mathbf{k}|'}{\mathbf{k}'_2}^2 + 2\mathbf{k}'_1\mathbf{b}\cdot\mathbf{k}'\Big) - \epsilon^2 C_3(\mathbf{k}')}{1 + \epsilon\mathbf{b}\cdot\mathbf{k}' + \mathbf{k}'\cdot B\mathbf{k}'},$$

instead of $\omega_{NLS}(\mathbf{k}')$. With a good choice[29] of $\mathbf{b}$, B and $C_3(\nabla)$, the improved dispersion relation $\omega_{imp}(\mathbf{k}')$ remains very close to the exact dispersion relation $\omega_{FD}(\mathbf{k}')$ even for quite large values of $\mathbf{k}'$, and the dispersive properties of (8.56) therefore are better than for the standard NLS equation (8.49). At the same time, and contrary to the full dispersion NLS equation, (8.56) involves only simple differential operators. See [**85**] for simple numerical computations showing the accuracy of (8.56) in situations where the standard NLS approximations give very poor results.

REMARK 8.36.

(1) Choosing $C_3(\nabla') = -\omega'\nabla'\cdot B\nabla'\partial'_x$, and defining as usual $\underline{\psi}$ by $\psi_{01}(t', X') = \underline{\psi}_{01}(\epsilon t', X' - c_g t')$, we obtain the following "improved" version of Equation (8.50)

$$(1 - i\epsilon\mathbf{b}\cdot\nabla' - \epsilon^2\nabla'\cdot B\nabla')\partial_\tau\underline{\psi}_{01} - i\frac{1}{2}\big(\omega''{\partial'_x}^2 + \frac{\omega'}{|\mathbf{k}|}{\partial'_y}^2\big)\underline{\psi}_{01} + 2i\frac{|\mathbf{k}|^4}{\omega}|\underline{\psi}_{01}|^2\underline{\psi}_{01} = 0.$$

(2) From a physical point of view, the new form of the dispersive terms in (8.56) accounts for the variations of the group velocity when the envelope ψ_{01} of the wave packet involves a large range of frequencies. One could also take into account variations of the polarization of the leading order term (i.e., of the relation $\zeta_{01} = i\omega\psi_{01}$ implicit in (8.17)); this leads to a quasilinear variant of the NLS equation,

[28] This can easily be seen by using a variant of the so-called BBM trick used in §5.2.1 to derive Boussinesq models with improved frequency dispersion. Noting that

$$\partial'_t\psi_{01} + \omega'\partial'_x\psi_{01} = O(\epsilon),$$

one easily gets that the linear terms added in (8.56) to the standard NLS equation satisfy

$$-i\epsilon\mathbf{b}\cdot\nabla(\partial'_t\psi_{01} + \omega'\partial'_x\psi_{01}) - \epsilon^2\nabla\cdot B\nabla\partial'_t\psi_{01} + \epsilon^2 C_3(\nabla')\psi_{01} = O(\epsilon^2),$$

and therefore are of the same size as the terms treated as a residual in the standard NLS approximation.

[29] We refer to §5.2 for similar considerations of the shallow water regime.

as shown in [**135**], in the context of nonlinear optics where such quasilinear terms are often called *self-steepening operators* (see [**28**]).

8.5.4. Higher order approximation: The Dysthe equation. Noting that, when the steepness ϵ is not very small, the NLS approximation does not match very well with some exact computations by Longuet-Higgins [**241, 242**], Dysthe proposed (in the infinite depth case) to include the $O(\epsilon^2)$ terms neglected by the NLS equation (8.49). He obtained the following equation [**140**] (with $\mathbf{k} = \mathbf{e}_x$)

$$
\begin{aligned}
2i\big(\partial_t'\zeta_{01} + \frac{1}{2}\partial_x'\zeta_{01}\big) + \epsilon\big(-\frac{1}{4}{\partial_x'}^2 + \frac{1}{2}{\partial_y'}^2\big)\zeta_{01} - 4\epsilon|\zeta_{01}|^2\zeta_{01}& \\
= \epsilon^2\frac{i}{8}\big({\partial_x'}^3 - 6\partial_x'{\partial_y'}^2\big)\zeta_{01} + \epsilon^2 i\big(6\zeta_{01}^2\partial_x'\overline{\zeta}_{01} - 8|\zeta_{01}|^2\partial_x'\zeta_{01}\big)& \\
+2\epsilon^2\zeta_{01}\big(\partial_x'\psi_{10} - 2i\partial_x'|\zeta_{01}|^2\big);&
\end{aligned}
$$

dropping the $O(\epsilon^2)$ yields the NLS equation (8.51). The main qualitative difference with this new equation is that it contains the first nonoscillating mode ψ_{10} of the velocity potential, which is given by solving (8.48)

$$
\psi_{10} = 2\frac{\partial_x'}{|D'|}|\zeta_{01}|^2.
$$

Plugging this formula in the equation above, we obtain the so-called *Dysthe equation*

$$
\begin{aligned}
2i\big(\partial_t'\zeta_{01} + \frac{1}{2}\partial_x'\zeta_{01}\big) + \epsilon\big(-\frac{1}{4}{\partial_x'}^2 + \frac{1}{2}{\partial_y'}^2\big)\zeta_{01} - 4\epsilon|\zeta_{01}|^2\zeta_{01}& \\
= \epsilon^2\frac{i}{8}\big({\partial_x'}^3 - 6\partial_x'{\partial_y'}^2\big)\zeta_{01} + 2i\epsilon^2\zeta_{01}\big(\zeta_{01}\partial_x'\overline{\zeta}_{01} - \overline{\zeta}_{01}\partial_x'\zeta_{01}\big)& \\
-10\epsilon^2 i|\zeta_{01}|^2\partial_x'\zeta_{01} + 4\epsilon^2\zeta_{01}\frac{{\partial_x'}^2}{|D'|}|\zeta_{01}|^2.&
\end{aligned}
\tag{8.58}
$$

The solutions of the Dysthe equations have been studied by Lo and Mei [**239, 240**] and more recently in [**172**] and show differences with solutions of the NLS equation, and in particular an increase of the group velocity and an asymmetry of the envelope of the surface elevation with respect to the peak of the wave profile. See [**178**] for comparisons with experimental data. Note that a Hamiltonian formulation of these equations has been derived in [**106, 147**]; see [**123**] for a good review of many relevant problems related to the Dysthe equation.

REMARK 8.37. Denoting $\omega(\mathbf{k}) = |\mathbf{k}|^{1/2}$ and noting that

$$
\frac{1}{\epsilon}\big(\omega(\mathbf{e}_x + \epsilon\xi) - \omega(\mathbf{e}_x)\big) = \frac{1}{2}\xi_1 + \epsilon\big(-\frac{1}{8}\xi_2^2 + \frac{1}{4}\xi_2^2\big) + \epsilon^2\big(\frac{1}{16}\xi_1^3 - \frac{3}{8}\xi_1\xi_2^2\big) + O(\epsilon^3),
$$

we deduce that the linear part of the Dysthe equation (8.58) is a third-order expansion of the linear part of the full-dispersion NLS equation (8.55). The Dysthe equation contains both higher order nonlinear and dispersive terms neglected by the standard cubic NLS equation (8.50), while the full-dispersion NLS equation (8.55) contains only higher order linear dispersive terms (at infinite order). A comparison of these models (as well as the models with improved dispersion relation derived in §8.5.3) would bring some insight on the respective importance of dispersive and nonlinear terms in situations where the standard NLS approximation compares poorly with the experiment. One could also derive a full-dispersive Dysthe equation or a Dysthe equation with improved dispersion relation.

8.5.5. The NLS approximation in the neighborhood of $|\mathbf{K}|H_0 = 1.363$. We recall that the NLS equation in finite depth is given by (8.43), namely,

$$i\partial_\tau \underline{\zeta}_{01} + \frac{1}{2}\omega''{\partial'_x}^2 \underline{\zeta}_{01} - \frac{\delta}{\omega^2}|\underline{\zeta}_{01}|^2 \underline{\zeta}_{01} = 0,$$

when written in terms of the surface elevation (and by (8.42) when written in terms of the velocity potential). As noted in Remark 8.28, the coefficient $\delta = \delta(\sqrt{\mu}|\mathbf{k}|)$ of the nonlinearity vanishes for a critical value on $\sqrt{\mu}|\mathbf{k}|$ (or $|\mathbf{K}|H_0$, if $\mathbf{K} = \mathbf{k}/L_x$ is the wave number in dimensional form). This critical value is given by $|\mathbf{K}|H_0 \sim 1.363$; in the neighborhood of this critical value, the equation (8.42) becomes linear and a different scaling is required to observe nonlinear effects. Instead of considering a modulated oscillation given (for the surface elevation and at leading order) by

$$\underline{\zeta}_{01}(\epsilon^2 t, \epsilon(x - \omega' t))e^{i(\mathbf{k}\cdot X - \omega t)} + \text{c.c.},$$

Johnson in [**189**] considered slower modulations in time and space, namely,

$$\underline{\zeta}_{01}(\epsilon^4 t, \epsilon^2(x - \omega' t))e^{i(\mathbf{k}\cdot X - \omega t)} + \text{c.c.}$$

He found that the corresponding equation for $\underline{\zeta}_{01}$ is given by

$$(8.59)\qquad \begin{cases} i\partial_{\tau_1}\underline{\zeta}_{01} + \frac{1}{2}\omega''{\partial''_x}^2\underline{\zeta}_{01} - \frac{\delta}{\omega^2}|\underline{\zeta}_{01}|^2\underline{\zeta}_{01} + a_3|\underline{\zeta}_{01}|^4\underline{\zeta}_{01} \\ \qquad + i\big(a_4|\underline{\zeta}_{01}|^2\partial''_x\underline{\zeta}_{01} - a_5\partial''_x(|\underline{\zeta}_{01}|^2)\underline{\zeta}_{01}\big) - a_6\underline{\zeta}_{01}\partial_\tau\psi = 0, \\ \partial''_x\psi = |\underline{\zeta}_{01}|^2, \end{cases}$$

where τ_1 now stands for $\epsilon^4 t$ and where we denote by ∂''_x the differentiation with respect to $\epsilon^2(x - \omega' t)$.

Johnson used this equation to study the stability of the Stokes wave in the neighborhood of $|\mathbf{K}|H_0 = 1.363$; contrary to what happens with the standard NLS equation, stability now depends on the wave amplitude.

8.5.6. Modulation equations for capillary gravity waves. It is not difficult to include the effects of surface tension in the derivation of all the models considered in this chapter. For instance, Djordjevic and Redekopp [**126**] and Ablowitz and Segur [**2**] derived the Davey-Stewartson equations in the presence of surface tension. Capillary effects can dramatically change the nature of the equations, since the second equation (on the mean mode) can be of hyperbolic type with the terminology of Remark 8.24 (see [**157, 307**]).

See [**173**] for a generalization of the Dysthe equation in the presence of surface tension and [**1, 123**] for further references on this aspect.

CHAPTER 9

Water Waves with Surface Tension

We review in this chapter the changes that have to be carried out if one wants to include capillary effects with the issues addressed in the previous chapters. The full water waves equations with surface tension (or capillary gravity waves equations) are addressed in Section 9.1. The equations are first written down in dimensionless form (§9.1.1) and the physical relevance of capillary effects is briefly discussed in §9.1.2. Sections 9.1.3 to 9.1.7 are then devoted to the proof of a well-posedness theorem that generalizes Theorem 4.16 in the presence of surface tension.

In Section 9.2, we show how to modify the shallow water models derived in Chapter 5 in order to include capillary effects. Both large amplitude models (Green-Naghdi) and small amplitude models (Boussinesq) are addressed.

We then investigate in Section 9.3 the influence of surface tension on the scalar asymptotic models derived in Chapter 7. We show in §9.3.1 that for the KdV equations, this influence induces a change on the value of the third-order dispersive term. There is a critical value for which this coefficient vanishes. We show in §9.3.2 that one has to work in a slightly different regime in order to observe both dispersive and nonlinear effects; in this regime, water waves can be described by the so-called Kawahara equations. We then follow the same approach to analyze, in §§9.3.3 and 9.3.4, the influence of surface tension on the KP approximation.

Finally, the case of deep or infinite water is addressed in Section 9.4 and modulation equations for capillary waves are briefly mentioned in Section 9.5.

9.1. Well-posedness of the water waves equations with surface tension

Various methods exist to handle surface tension in the water waves problem, such as [**331, 29, 14, 15, 295, 99, 297, 262, 4**]; for more methods related to the framework adopted here, see [**277**] and especially [**221**], where the motion of two-fluid interfaces is investigated in a nondimensionalized framework, and surface tension is taken into account. The water waves problem is a particular case of these two-fluid equations, with the density of the upper fluid being set to zero. The local existence Theorem 9.6 below is an adaptation of the main result of [**221**] (with minor modifications allowing for weakly transverse waves).

9.1.1. The equations. Water waves without surface tension were modeled in Chapter 1 by the free surface Euler equations (H1)′–(H9)′ (see §1.1.3). The only change that is required to handle surface tension is to replace (H7)′ by

$$(\mathrm{H7})'_\sigma \qquad P - P_{atm} = \sigma\kappa(\zeta) \quad \text{on} \quad \{z = \zeta(t,X)\},$$

where σ is the surface tension coefficient, and $\kappa(\zeta)$ is the mean curvature of the surface

$$\kappa(\zeta) = -\nabla\cdot\Big(\frac{\nabla\zeta}{\sqrt{1+|\nabla\zeta|^2}}\Big).$$

The Zakharov/Craig-Sulem formulation of the water waves equations has to be consequently modified into

$$(9.1)\qquad \begin{cases} \partial_t\zeta - G[\zeta,b]\psi = 0, \\ \partial_t\psi + g\zeta + \frac{1}{2}|\nabla\psi|^2 - \dfrac{(G[\zeta,b]\psi + \nabla\zeta\cdot\nabla\psi)^2}{2(1+|\nabla\zeta|^2)} = -\dfrac{\sigma}{\rho}\kappa(\zeta). \end{cases}$$

Similarly, the dimensionless version (1.28) of these equations is given by

$$(9.2)\qquad \begin{cases} \partial_t\zeta - \dfrac{1}{\mu\nu}\mathcal{G}\psi = 0, \\ \partial_t\psi + \zeta + \dfrac{\varepsilon}{2\nu}|\nabla^\gamma\psi|^2 - \dfrac{\varepsilon\mu}{\nu}\dfrac{(\frac{1}{\mu}\mathcal{G}\psi + \nabla^\gamma(\varepsilon\zeta)\cdot\nabla^\gamma\psi)^2}{2(1+\varepsilon^2\mu|\nabla^\gamma\zeta|^2)} = -\dfrac{1}{\mathrm{Bo}}\dfrac{\kappa_\gamma(\varepsilon\sqrt{\mu}\zeta)}{\varepsilon\sqrt{\mu}}, \end{cases}$$

where $\mathcal{G} = \mathcal{G}_{\mu,\gamma}[\varepsilon\zeta, \beta b]$, Bo is the*Bond number*, which measures the ratio of gravity forces over capillary forces[1]

$$\mathrm{Bo} = \frac{\rho g L_x^2}{\sigma},$$

where we recall that L_x is the typical horizontal scale in the horizontal direction; finally, $\kappa_\gamma(\zeta)$ is given by

$$\kappa_\gamma(\zeta) = -\nabla^\gamma\cdot\Big(\frac{\nabla^\gamma\zeta}{\sqrt{1+|\nabla^\gamma\zeta|^2}}\Big).$$

9.1.2. Physical relevance. A rough way to evaluate the influence of surface tension on the propagation of water waves consists of discussing its contribution to the linear dispersion relation. Proceeding as in §1.3.2 of Chapter 1, one can check that the dispersion relation (1.22) must be replaced by

$$\begin{aligned} \omega(\xi) &= \Big((g + \frac{\sigma}{\rho}|\xi|^2)|\xi|\tanh(H_0|\xi|)\Big)^{1/2} \\ &= \Big(g(1 + \frac{1}{\mathrm{Bo}}|L_x\xi|^2)|\xi|\tanh(H_0|\xi|)\Big)^{1/2}. \end{aligned}$$

Since by definition, L_x is the characteristic scale, one has $L_x|\xi| \sim 2\pi$ for all the wave numbers ξ where the energy of the wave is concentrated. From the expression of $\omega(\xi)$ above we see that capillary effects are negligible if $1 \ll (2\pi)^{-2}\mathrm{Bo}$, of the same order as the gravity effects if $(2\pi)^{-2}\mathrm{Bo} \sim 1$ and predominant if $(2\pi)^{-2}\mathrm{Bo} \ll 1$.

EXAMPLE 9.1. For the case of an air-water interface, and neglecting the density of the air, the surface tension coefficient is $\sigma \sim 73\times 10^{-3} N\cdot m^{-1}$ (at 20^oC) and the density of sea water $\rho \sim 1030\,\mathrm{kg}\cdot m^{-3}$; the acceleration of gravity is $g = 9.81\, m\cdot s^{-2}$. The capillary effects become of the same order as the gravity effects for $\mathrm{Bo} \sim (2\pi)^2$, i.e., for $L_x \sim 2\pi\big(\frac{\sigma}{\rho g}\big)^{1/2} \sim 1.6\,\mathrm{cm}$. For coastal waves of characteristic length $L_x = 10m$, (linear) capillary effects represent only 0.0003% of the gravity effects.

As shown in Example 9.1, surface tension is in general not relevant to describe wave propagation in coastal oceanography. This is why we have neglected it so far. This does not mean that surface tension is always irrelevant. For smaller scale laboratory experiments, for instance (and possibly with fluids other than water), capillary effects can be important. They are also of fundamental importance for

[1]One should be careful here, because the Bond number is sometimes defined as the inverse of this quantity, especially in the mathematical literature.

the description of wind created ripples, for instance, and can also affect the limiting profile of standing waves [**117, 123**].

The discussion above shows that in the case of an air-water interface, surface tension becomes relevant for very small characteristic scales. It is important to point out that such small characteristic scales can be created by the flow, even if they are not initially present: this corresponds to the appearance of *singularities*. One obvious singularity iswave breaking, and the role of surface tension has indeed been experimentally observed in the late evolution of the crest [**136**].
Another interesting and related example is furnished by the case of two-fluids interfaces (the density of the upper fluid is no longer neglected). A linear analysis similar to the one performed here seems to show that capillary effects are also irrelevant. However, the flow automatically creates singularities (the so-called Kelvin-Helmholtz instabilities) that involve very small characteristic lengths for which surface tension plays a crucial role [**221**].

9.1.3. Linearization around the rest state and energy norm. Following the same strategy as in Chapter 4 for the case of purely gravity waves, we first look at the linearized equations around the rest state $\zeta = 0$, $\psi = 0$

$$\partial_t U + \mathcal{A}_\sigma U = 0, \qquad \text{with} \quad \mathcal{A}_\sigma = \begin{pmatrix} 0 & -\frac{1}{\mu}\mathcal{G}_{\mu,\gamma}[0,\beta b] \\ 1 - \frac{1}{\mathrm{Bo}}\Delta^\gamma & 0 \end{pmatrix}, \tag{9.3}$$

and $U = (\zeta, \psi)^T$.
This system of evolution equations can be made symmetric if we multiply it by the symmetrizer $\mathcal{S}_\sigma$, defined as

$$\mathcal{S}_\sigma = \begin{pmatrix} 1 - \frac{1}{\mathrm{Bo}}\Delta^\gamma & 0 \\ 0 & \frac{1}{\mu}\mathcal{G}_{\mu,\gamma}[0,\beta b] \end{pmatrix}. \tag{9.4}$$

This suggests that a natural energy for the linearized equations (4.6) is given by

$$(\mathcal{S}_\sigma U, U) = |\zeta|_2^2 + \frac{1}{\mathrm{Bo}}|\nabla^\gamma \zeta|^2 + \big(\frac{1}{\mu}\mathcal{G}_{\mu,\gamma}[0,\beta b]\psi, \psi\big).$$

From Proposition 3.12, we know that the second term in the r.h.s. is uniformly equivalent to $|\mathfrak{P}\psi|_2^2$. It therefore is natural to define the energy $\mathcal{E}_\sigma^0(U)$ as

$$\mathcal{E}_\sigma^0(U) = |\zeta|_{H_\sigma^1}^2 + |\mathfrak{P}\psi|_2^2,$$

where for all $r \geq 0$, $H_\sigma^{r+1}(\mathbb{R}^d)$ is the space $H^{r+1}(\mathbb{R}^d)$ endowed with the norm

$$|f|_{H_\sigma^{r+1}}^2 = |f|_{H^r}^2 + \frac{1}{\mathrm{Bo}}|\nabla^\gamma f|_{H^r}^2. \tag{9.5}$$

For the same reasons as in §4.1, we have to define higher energy norms in terms of the "good unknowns"

$$\zeta_{(\alpha)} = \partial^\alpha \zeta, \qquad \psi_{(\alpha)} = \partial^\alpha \psi - \varepsilon \underline{w} \partial^\alpha \zeta. \tag{9.6}$$

There is, however, an important difference with respect to §4.1, namely, that time derivatives are now included in the definition of the energy (hence the summation over $\alpha \in \mathbb{N}^{d+1}$ rather than $\alpha \in \mathbb{N}^d$)

$$\mathcal{E}^N(U) = |\mathfrak{P}\psi|_{H^{t_0+3/2}}^2 + \sum_{\alpha \in \mathbb{N}^{d+1}, |\alpha| \leq N} |\zeta_{(\alpha)}|_{H_\sigma^1}^2 + |\mathfrak{P}\psi_{(\alpha)}|_2^2. \tag{9.7}$$

See Remark 9.9 below for comments on this point.

9.1.4. A linearization formula. Let us first define a natural generalization of the functional space E_T^N introduced in (4.10), namely,

$$E_{\sigma,T}^N = \{U \in C([0,T]; H^{t_0+2} \times \dot{H}^2(\mathbb{R}^d)), \quad \mathcal{E}_\sigma^N(U(\cdot)) \in L^\infty([0,T])\}. \tag{9.8}$$

We also denote by $\mathfrak{m}_\sigma^N(U)$ any constant of the form

$$\mathfrak{m}_\sigma^N(U) = C\big(M, |b|_{H^{N+t_0\vee 1+1}}, \mathcal{E}_\sigma^N(U), \frac{1}{\mathrm{Bo}}\big), \tag{9.9}$$

with M as in (4.4).

REMARK 9.2. Since the Bond number Bo is generally very large, we make some effort to get controls in terms of $\mathfrak{m}_\sigma^N(U)$, which depends[2] on $1/\mathrm{Bo}$, not on Bo. The situations where $\mathrm{Bo} \ll 1$ did not seem relevant enough in the present framework (see the discussion in §9.1.2) to justify a specific treatment.

The following proposition is a generalization of Proposition 4.5 in the presence of surface tension. The main difference is that the residual R_α is now controlled in H_σ^1-norm rather than L^2, which forces us to include the subprincipal term in the expansion of $\partial^\alpha \mathcal{G}\psi$. Before stating the proposition, we need the following notations

$$\zeta_{\langle\check{\alpha}\rangle} = (\zeta_{(\check{\alpha}^1)}, \dots, \zeta_{(\check{\alpha}^{d+1})}), \qquad \psi_{\langle\check{\alpha}\rangle} = (\psi_{(\check{\alpha}^1)}, \dots, \psi_{(\check{\alpha}^{d+1})}), \tag{9.10}$$

where $\check{\alpha}^j = \alpha - \mathbf{e}_j$. The operator $\mathcal{G}_{(\alpha)}\psi_{\langle\check{\alpha}\rangle}$ that arises in the description of the subprincipal part of $\partial^\alpha(\mathcal{G}\psi)$ is then given by

$$\mathcal{G}_{(\alpha)}\psi_{\langle\check{\alpha}\rangle} = \sum_{j=1}^{d+1} \alpha_j d\mathcal{G}(\partial_j\zeta)\psi_{(\check{\alpha}^j)}, \tag{9.11}$$

where we used Notation 3.20 for the shape derivative of $\mathcal{G}$.

PROPOSITION 9.3. *Let $t_0 > d/2$ and $N \in \mathbb{N}$. Moreover, assume that $N \geq t_0 + t_0 \vee 2 + 3/2$, $b \in H^{N+1\vee t_0+1}(\mathbb{R}^d)$ and $U = (\zeta, \psi) \in E_{\sigma,T}^N$ satisfy (4.3). Then, for all $\alpha \in \mathbb{N}^{d+1}$ with $1 \leq |\alpha| \leq N$, one has*

$$\frac{1}{\mu}\partial^\alpha(\mathcal{G}\psi) = \frac{1}{\mu}\mathcal{G}\psi_{(\alpha)} + \max\{\varepsilon, \beta\}R_\alpha, \qquad (|\alpha| \leq N{-}1)$$

$$\frac{1}{\mu}\partial^\alpha(\mathcal{G}\psi) = \frac{1}{\mu}\mathcal{G}\psi_{(\alpha)} - \varepsilon\nabla^\gamma \cdot (\zeta_{(\alpha)}\underline{V}) + \frac{1}{\mu}\mathcal{G}_{(\alpha)}\psi_{\langle\check{\alpha}\rangle} + \max\{\varepsilon, \beta\}R_\alpha, \qquad (|\alpha| = N)$$

with $(\zeta_{(\alpha)}, \psi_{(\alpha)})$ as in (4.9) and $\underline{V}$ as in Theorem 3.21, while R_α satisfies

$$|R_\alpha|_{H_\sigma^1} \leq \mathfrak{m}_\sigma^N(U).$$

PROOF. As in the proof of Proposition 4.5, we focus on the most difficult case, namely $|\alpha| = N$, and we deal only with the case of flat bottoms (the generalization to nonflat bottoms follows the same lines as for Proposition 4.5). The proof requires the following lemma, which is a straightforward generalization of Lemma 4.6 and whose proof is omitted.

LEMMA 9.4. *Under the assumptions of the proposition, and with $r \geq 0$ and $\delta \in \mathbb{N}^d$ such that $r + |\delta| \leq N - 1$, one has*

$$|\mathfrak{P}\partial^\delta\psi|_{H_\sigma^{r+1}} \leq \mathfrak{m}_\sigma^N(U).$$

[2]We recall that when the notation $C(\cdot, \dots)$ is used, it is always assumed that C is a nondecreasing function of its arguments.

Denoting by $\{\partial^\alpha, \mathcal{G}\}\psi$ the Poisson bracket (with $\check{\alpha}^j$ as in (9.11))

$$\{\partial^\alpha, \mathcal{G}\}\psi = \sum_{j=1}^{d+1} \alpha_j d\mathcal{G}(\partial_j\zeta)\partial^{\check{\alpha}^j}\psi,$$

we can write the following generalization of (4.14)

$$\partial^\alpha \mathcal{G}\psi = \mathcal{G}\partial^\alpha\psi + d\mathcal{G}(\partial^\alpha\zeta)\psi + \mathcal{G}_{(\alpha)}\psi_{\langle\check{\alpha}\rangle} + \varepsilon\mu R_\alpha, \tag{9.12}$$

with $\varepsilon\mu R_\alpha = (\{\partial^\alpha, \mathcal{G}\}\psi - \mathcal{G}_{(\alpha)}\psi_{\langle\check{\alpha}\rangle}) + \varepsilon\mu R'_\alpha$ and where $\varepsilon\mu R'_\alpha$ is a sum of terms of the form

$$d^j\mathcal{G}(\partial^{\iota^1}\zeta, \dots, \partial^{\iota^j}\zeta)\partial^\delta\psi =: A_{j,\delta,\iota}, \tag{9.13}$$

where $j \in \mathbb{N}$, $\iota = (\iota^1, \dots, \iota^j) \in \mathbb{N}^{dj}$ and $\delta \in \mathbb{N}^d$ satisfy

$$\sum_{i=1}^{j} |\iota^i| + |\delta| = N, \qquad 0 \leq |\delta| \leq N-2 \quad \text{and} \quad |\iota^i| < N. \tag{9.14}$$

(Because of the additional term in (9.12), one has $|\delta| \leq N-2$ here, while we had $|\delta| \leq N-1$ in (4.16).)
As in the proof of Proposition 4.5, we have to control the $A_{j,\delta,\iota}$; the difference is that these terms must now be controlled in H^1_σ and not only in L^2. Still denoting by l an integer such that $|\iota^l| = \max_{1\leq m\leq j}|\iota^m|$, we distinguish the same cases as in the proof of Proposition 4.5:
(i) The case $|\delta| = j = 1$ and $\iota = \iota^1$ (and thus $|\iota| = N-1$). Proceeding exactly as in the proof of Proposition 4.5, one readily checks that (4.17) can be generalized into

$$\frac{1}{\mu}|A_{j,\delta,\iota}|_{H^1_\sigma} \leq \varepsilon M\Big(\sum_{\alpha\in\mathbb{N}^{d+1}, |\alpha|\leq N} |\partial^\alpha\zeta|_{H^1_\sigma}\Big)|\mathfrak{P}\partial^\delta\psi|_{H^{t_0+5/2}_\sigma}.$$

Using Lemma 9.4 (recall that $t_0 + 3 \leq N$) we therefore get

$$|A_{j,\delta,\iota}|_{H^1_\sigma} \leq \varepsilon\mu\mathfrak{m}^N_\sigma(U). \tag{9.15}$$

(ii) The case $|\delta| \leq N-1 \vee t_0 - 3/2$ and $|\iota^l| < N-1$. Here again, a direct adaptation of the proof of Proposition 4.5 shows that (9.15) holds.
(iii) The case $N-1 \vee t_0 - 1/2 \leq |\delta| \leq N-2$. Using the second estimate of Proposition 3.28, we now get

$$|A_{j,\delta,\iota}|_{H^1_\sigma} \leq M\varepsilon^j\mu \prod_{m=1}^{j} |\partial^{\iota^m}\zeta|_{H^{1\vee t_0+2}_\sigma}|\mathfrak{P}\partial^\delta\psi|_{H^2_\sigma},$$

and one concludes as in Proposition 4.5.
(iv) The case $|\delta| = N-1$. This case will not be considered here, since (9.14) implies that $|\delta| \leq N-2$.
(v) The case $|\delta| = 0$ and $|\iota^l| = N-1$. This case can be straightforwardly adapted from the proof of Proposition 4.5.

We therefore have proved that the residual R'_α satisfies the estimate stated in the proposition. Note now that $\frac{1}{\mu}(\{\partial^\alpha, \mathcal{G}\}\psi - \mathcal{G}_{(\alpha)}\psi_{\langle\check{\alpha}\rangle})$ is a sum of terms of the form

$$\frac{1}{\sqrt{\mu}}d\mathcal{G}(\partial^\iota\zeta)(\frac{1}{\sqrt{\mu}}\underline{w}\partial^\delta\zeta) \qquad (|\iota| = 1, \quad \iota + \delta = \alpha)$$

that are bounded in H^1_σ-norm by $\varepsilon\mathfrak{m}^N_\sigma(U)$, owing to Theorem 3.21 and Proposition 3.28. We deduce that R_α satisfies the estimate of the proposition. □

9.1.5. The quasilinear system. As in Chapter 4, we want to quasilinearize the equations. We still denote by $\underline{\mathfrak{a}}$ and $\mathcal{B}[U]$ the quantities defined in (4.20) and (4.21) respectively, but the presence of surface tension forces us to replace the operator $\mathcal{A}[U]$ by $\mathcal{A}_\sigma[U]$ defined as

$$\mathcal{A}_\sigma[U] = \begin{pmatrix} 0 & -\frac{1}{\mu}\mathcal{G} \\ \underline{\mathfrak{a}} - \frac{1}{\mathrm{Bo}}\nabla^\gamma \cdot \mathcal{K}(\varepsilon\sqrt{\mu}\nabla^\gamma\zeta)\nabla^\gamma & 0 \end{pmatrix}, \tag{9.16}$$

where the term $\frac{1}{\mathrm{Bo}}\nabla^\gamma \cdot \mathcal{K}(\varepsilon\sqrt{\mu}\nabla^\gamma\zeta)\nabla^\gamma$ comes from the linearization of the surface tension term

$$\frac{1}{\varepsilon\sqrt{\mu}}\partial^\alpha\kappa_\gamma(\varepsilon\sqrt{\mu}\zeta) = -\nabla^\gamma \cdot \mathcal{K}(\varepsilon\sqrt{\mu}\nabla^\gamma\zeta)\nabla^\gamma\partial^\alpha\zeta + \mathcal{K}_{(\alpha)}[\varepsilon\sqrt{\mu}\nabla\zeta]\zeta_{\langle\check{\alpha}\rangle} + \dots, \tag{9.17}$$

where the dots stand for derivatives of ζ of order lower or equal to $|\alpha|$, and $\zeta_{\langle\check{\alpha}\rangle}$ is as in (9.10). The positive definite $d \times d$ matrix $\mathcal{K}[\nabla^\gamma\zeta]$ is defined as

$$\mathcal{K}(\nabla^\gamma\zeta) = \frac{(1+|\nabla^\gamma\zeta|^2)\mathrm{Id} - \nabla^\gamma\zeta \otimes \nabla^\gamma\zeta}{(1+|\nabla^\gamma\zeta|^2)^{3/2}}, \tag{9.18}$$

while the second-order operator $\mathcal{K}_{(\alpha)}[\nabla\zeta]$ is given for all $F = (f_1, \dots, f_{d+1})^T$ by

$$\mathcal{K}_{(\alpha)}[\nabla\zeta]F = -\nabla^\gamma \cdot \Big[\sum_{j=1}^{d+1}\big(d\mathcal{K}(\nabla^\gamma\partial_j\zeta)\nabla^\gamma f_j + d\mathcal{K}(\nabla^\gamma f_j)\nabla^\gamma\partial_j\zeta\big)\Big]. \tag{9.19}$$

We can now define the antidiagonal operator $\mathcal{C}_\alpha[U]$

$$\mathcal{C}_\alpha[U] = \begin{pmatrix} 0 & -\frac{1}{\mu}\mathcal{G}_{(\alpha)} \\ \frac{1}{\mathrm{Bo}}\mathcal{K}_{(\alpha)}[\varepsilon\sqrt{\mu}\nabla\zeta] & \end{pmatrix} \tag{9.20}$$

and state the main result of this section.

PROPOSITION 9.5. *Let $T > 0$, $t_0 > d/2$ and N be as in Proposition 4.5. If $U = (\zeta, \psi) \in E^N_{\sigma,T}$ and $b \in H^{N+1\vee t_0+1}(\mathbb{R}^d)$ satisfy (4.3) uniformly on $[0,T]$ and solve (9.2), then for all $\alpha \in \mathbb{N}^{d+1}$ with $1 \le |\alpha| \le N$, the couple $U_{(\alpha)} = (\zeta_{(\alpha)}, \psi_{(\alpha)})$ solves*

$$\partial_t U_{(\alpha)} + \mathcal{A}_\sigma[U]U_{(\alpha)} = \max\{\varepsilon,\beta\}(R_\alpha, S_\alpha)^T, \qquad (|\alpha| < N),$$
$$\partial_t U_{(\alpha)} + \mathcal{A}_\sigma[U]U_{(\alpha)} + \mathcal{B}[U]U_{(\alpha)} + \mathcal{C}_\alpha[U]U_{\langle\check{\alpha}\rangle} = \max\{\varepsilon,\beta\}(R_\alpha, S_\alpha)^T, \qquad (|\alpha| = N),$$

where $U_{\langle\check{\alpha}\rangle} = (\zeta_{\langle\check{\alpha}\rangle}, \psi_{\langle\check{\alpha}\rangle})^T$ and the residuals R_α and S_α satisfy the estimates

$$|R_\alpha|_{H^1_\sigma} + |\mathfrak{P}S_\alpha|_2 \lesssim \mathfrak{m}^N_\sigma(U)$$

uniformly on $[0,T]$.

PROOF. The first equation is obtained directly by applying ∂^α to the first equation of (9.2) and using Proposition 9.3.
For the second equation, the only difference with the zero surface tension case (Proposition 9.5) is the treatment of the surface tension term whose linearization is directly given by (9.17). □

9.1.6. Initial condition. As for the case without surface tension, we want to prove that the energy $\mathcal{E}_\sigma^N(U)$ defined in (9.7), namely,

$$\mathcal{E}_\sigma^N(U) = |\nabla\psi|_{H^{t_0+2}}^2 + \sum_{\alpha\in\mathbb{N}^{d+1}, |\alpha|\leq N} |\zeta_{(\alpha)}|_{H^1_\sigma}^2 + |\mathfrak{P}\psi_{(\alpha)}|_2^2,$$

is controlled for positive times by the energy at $t = 0$. The difference is that the summation now involves space-time derivatives and not only space derivatives. We therefore have to specify which sense we give to the initial energy $\mathcal{E}_\sigma^N(U^0)$.
If we denote, for all $\alpha \in \mathbb{N}^{d+1}$, $\alpha = (\alpha_0, \alpha_1, \dots, \alpha_d)$ so that $\partial_t^{\alpha_0}$ corresponds to the time derivatives of ∂^α, the problem is to choose initial values $U^0_{(\alpha)}$ for $(U_{(\alpha)})_{|t=0}$ (with $U_{(\alpha)} = (\zeta_{(\alpha)}, \psi_{(\alpha)})$) when $\alpha_0 > 0$, in terms of U^0 and its spacial derivatives. This can be done by a finite induction. When $\alpha_0 = 0$ (no time derivative), we take

$$U^0_{(\alpha)} = \big(\partial^\alpha\zeta^0, \partial^\alpha\psi^0 - \varepsilon\underline{w}^0\partial^\alpha\zeta^0\big)^T \quad \text{with} \quad \underline{w}^0 = w[\varepsilon\zeta^0, \beta b]\psi^0.$$

Now, let $1 \leq n \leq N$ and assume that $U_{(\beta)\,|t=0} = U^0_{(\beta)}$ has been chosen for all $\beta \in \mathbb{N}^{d+1}$ with $\beta_0 < n$. We note that for all α with $\alpha_0 = n$ we have

$$U_{(\alpha)\,|t=0} = \big(\partial_t\zeta_{(\alpha')}, \partial_t\psi_{(\alpha')} + \varepsilon\partial_t\underline{w}\partial^{\alpha'}\zeta\big)_{|t=0},$$

with $\alpha' = (\alpha_0 - 1, \alpha_1, \dots, \alpha_d)$, and we therefore are led to set initial conditions for $\partial_t U_{(\alpha')}$, which is achieved by using Proposition 9.5.

The initial energy, which we denote slightly abusively by $\mathcal{E}_\sigma^N(U^0)$, therefore is defined as

$$\mathcal{E}_\sigma^N(U^0) = |\nabla\psi^0|_{H^{t_0+2}}^2 + \sum_{\alpha\in\mathbb{N}^{d+1}, |\alpha|\leq N} |\zeta^0_{(\alpha)}|_{H^1_\sigma}^2 + |\mathfrak{P}\psi^0_{(\alpha)}|_2^2, \tag{9.21}$$

with $U^0_{(\alpha)}$ as constructed above.

9.1.7. Well-posedness of the water waves equations with surface tension. We prove here the following generalization of Theorem 4.16; we recall that $\mathcal{E}_\sigma^N(U^0)$ is defined in (9.21).

THEOREM 9.6. *Let $t_0 > d/2$, $N \geq t_0 + t_0\vee 2 + 3/2$, and $(\varepsilon, \mu, \beta, \gamma)$ satisfy (4.28). Also let $b \in H^{N+1\vee t_0+1}(\mathbb{R}^d)$, and U^0 be such that $\mathcal{E}_\sigma^N(U^0) < \infty$, and satisfying*

$$\exists h_{min} > 0, \quad \exists a_0 > 0, \qquad 1 + \varepsilon\zeta^0 - \beta b \geq h_{min} \quad \text{and} \quad \mathfrak{a}(U^0) \geq a_0, \tag{9.22}$$

with $\mathfrak{a}(U^0)$ as defined in (4.27).
Then there exists $T > 0$ and a unique solution $U \in E^N_{\sigma, T/\varepsilon\vee\beta}$ to (9.2) with initial data U^0. Moreover,

$$\frac{1}{T} = c^1_{WW,\sigma} \quad \text{and} \quad \sup_{t\in[0, \frac{T}{\varepsilon\vee\beta}]} \mathcal{E}_\sigma^N(U(t)) = c^2_{WW,\sigma},$$

with $c^j_{WW,\sigma} = C(\mathcal{E}_\sigma^N(U^0), \mu_{max}, \frac{1}{h_{min}}, \frac{1}{a_0}, |b|_{H^{N+1\vee t_0+1}}, \frac{1}{\mathrm{Bo}})$ *for* $j = 1, 2$.

REMARK 9.7. In the presence of surface tension, the assumption $\mathfrak{a}(U^0) \geq a_0 > 0$ (for the Rayleigh-Taylor criterion, see §4.3.5) is not necessary to obtain well-posedness [**29, 14, 15, 295, 179, 99, 297, 262, 277, 4**]. The consequence is an existence time $\tilde{T}$ such that $1/\tilde{T}$ depends on the Bond number Bo (not only on $1/\mathrm{Bo}$). This existence time therefore is much too small for most applications to oceanography, as explained in §9.1.2. Imposing the assumption $\mathfrak{a}(U^0) \geq a_0 > 0$

allows us to recover an existence time of the same order as in Theorem 4.16. (In particular, it is straightforward to deduce from Theorem 4.16 that, as $\sigma \to 0$, solutions to (9.2) converge towards solutions to (1.28).)

REMARK 9.8. As in Chapter 4, it is not difficult to derive a stability property associated to Theorem 9.6, similar to Theorem 4.18 in the absence of surface tension. The extension to the case of nonasymptotically flat bottoms and infinite depth can also be derived as in §4.4.1 and §4.4.3 respectively.

PROOF. Here we emphasize the difference with respect to the proof of Theorem 4.16, in particular on the construction of the symmetrizers and on the derivation of the energy estimates. The other points are as in Theorem 4.16 and are omitted here. Here again, we write $\epsilon = \max\{\varepsilon, \beta\}$ throughout the proof.

A symmetrizer for the "quasilinear" system

$$\partial_t U_{(\alpha)} + \big(\mathcal{A}_\sigma[U] + \mathcal{B}[U]U_{(\alpha)} + \mathcal{C}_\alpha[U]U_{\langle\check{\alpha}\rangle}\big) = \epsilon \begin{pmatrix} R_\alpha \\ S_\alpha \end{pmatrix} \tag{9.23}$$

must also symmetrize the subprincipal terms included in $\mathcal{C}_\alpha[U]$ (see (9.20)), which is not the case with the symmetrizer used in the proof of Theorem 4.16. We therefore construct a new symmetrizer $\mathcal{S}_\sigma[U]$ as

$$\mathcal{S}_\sigma[U] = \mathcal{S}^1[U] + \mathcal{S}^2_\alpha[U],$$

where $\mathcal{S}^1[U]$ and $\mathcal{S}^2_\alpha[U]$ are diagonal operators defined as

$$\begin{aligned} \mathcal{S}^1[U] &= \operatorname{diag}(\mathfrak{a} - \frac{1}{\mathrm{Bo}}\nabla^\gamma \cdot \mathcal{K}(\varepsilon\sqrt{\mu}\nabla^\gamma\zeta)\nabla^\gamma, \frac{1}{\mu}\mathcal{G}), \\ \mathcal{S}^2_\alpha[U] &= \operatorname{diag}(\frac{1}{\mathrm{Bo}}\mathcal{K}_{(\alpha)}[\varepsilon\sqrt{\mu}\nabla\zeta], \frac{1}{\mu}\mathcal{G}_{(\alpha)}), \quad (\text{or } \mathcal{S}^2_\alpha[U] = 0 \quad \text{if} \quad |\alpha| < N). \end{aligned}$$

This leads us to replace the energies $\mathcal{F}^{[l]}(U)$ defined in (9.24) by

$$\mathcal{F}^{[l]}_\sigma(U) = \sum_{0\leq|\alpha|\leq l} \mathcal{F}^\alpha_\sigma(U), \tag{9.24}$$

for all $1 \leq l \leq N$, and with

$$\begin{aligned} \mathcal{F}^\alpha_\sigma(U) &= \frac{1}{2}([\mathcal{S}^1[U], U_{(\alpha)}) & (\alpha \neq 0), \\ \mathcal{F}^0_\sigma(U) &= |\zeta|^2_{H^1_\sigma} + \frac{1}{\mu}(\psi, \mathcal{G}_\mu[0]\psi) & (\alpha = 0), \end{aligned} \tag{9.25}$$

where $U_{(\alpha)} = (\zeta_{(\alpha)}, \psi_{(\alpha)})$.

REMARK 9.9.
(1) Multiplying (9.23) by $\mathcal{S}^1[U]$ symmetrizes the highest order terms, but since $\mathcal{S}^1[U]$ is a second order operator, the subprincipal terms need to be symmetrized also (otherwise the commutator term is not controlled by the energy). This explains the presence of $\mathcal{S}^2_\alpha[U]$ in the symmetrizer.
(2) The component $\mathcal{S}^2_\alpha[U]$ does not appear here; this is because $(\mathcal{S}^2_\alpha[U]\partial_t U_{(\alpha)}, U_{(\alpha)})$ can be controlled by the energy without using the equation to replace the time derivative by space derivatives. This is made possible by the fact that the energy controls both space *and* time derivatives. This method was introduced to handle surface tension in [**277**] (see also [**29**]).

The inequalities of Lemma 4.27 can then be generalized into

$$\mathcal{E}_\sigma^j(U) \le (M + \frac{4}{a_0})\mathcal{F}_\sigma^{[j]}(U) \quad \text{and} \quad \mathcal{F}^{[j]}(U) \le \mathfrak{m}_\sigma^N(U)\mathcal{E}_\sigma^j(U), \tag{9.26}$$

for all $0 \le j \le N$, and where $\mathcal{E}_\sigma^N(U)$ and $\mathfrak{m}_\sigma^N(U)$ are defined in (9.7) and (9.9).

The main difference in the derivation of the energy estimate is that (4.39) must now be replaced by

$$\frac{d}{dt}\Big(\mathcal{F}_\sigma^\alpha(U) + (U_{(\alpha)}, \mathcal{S}_\alpha^2 U_{\langle\check{\alpha}\rangle})\Big) = \sum_{j=1}^4 A_j + B_1 + B_2, \tag{9.27}$$

with

$$A_1 = \frac{1}{2}\big([\partial_t, \mathcal{S}^1]U_{(\alpha)}, U_{(\alpha)}\big), \qquad A_2 = -\big((\mathcal{S}^1)\mathcal{B}U_{(\alpha)}, U_{(\alpha)}\big),$$
$$A_3 = \big(U_{(\alpha)}, \partial_t(\mathcal{S}_\alpha^2 U_{\langle\check{\alpha}\rangle})\big), \qquad A_4 = -\big(\mathcal{B}U_{(\alpha)}, \mathcal{S}_\alpha^2 U_{\langle\check{\alpha}\rangle}\big),$$

and

$$B_1 = \epsilon\big((R_\alpha, S_\alpha)^T, (\mathcal{S}^1)U_{(\alpha)} + \mathcal{S}_\alpha^2 U_{\langle\check{\alpha}\rangle}\big).$$

As in the proof of Theorem 4.16, we now provide some control on these terms. We use the same notations as in the corresponding part of the proof of Theorem 4.16.
- *Control of* A_1. The only difference with the zero surface tension case is that one must add $\frac{1}{\mathrm{Bo}}\big((\partial_t \mathcal{K}(\varepsilon\sqrt{\mu}\nabla^\gamma\zeta))\nabla\zeta_{(\alpha)}, \nabla\zeta_{(\alpha)}\big)$ to A_{11}. This new term is obviously bounded from above by $\varepsilon\mathfrak{m}_\sigma^N(U)$.
- *Control of* A_2. As for A_1 an extra term due to the capillary term in $\mathcal{S}^2$ has to be taken into account; it is also easily controlled by $\varepsilon\mathfrak{m}_\sigma^N(U)$.
- *Control of* A_3. From the definition of $\mathcal{S}_\alpha^2$ and the fact that time derivatives are allowed in the energy (9.7), it is easy to get that $|A_3| \le \varepsilon\mathfrak{m}_\sigma^N(U)$.
- *Control of* A_4. One can write

$$A_4 = \varepsilon\big(\nabla^\gamma\cdot(\underline{V}\zeta_{(\alpha)}), \frac{1}{\mathrm{Bo}}\mathcal{K}_{(\alpha)}\zeta_{\langle\check{\alpha}\rangle}\big) - \varepsilon\big(\underline{V}\cdot\nabla^\gamma\psi_{(\alpha)}, \frac{1}{\mu}\mathcal{G}_{(\alpha)}\psi_{\langle\check{\alpha}\rangle}\big).$$

The first term can be directly controlled by the Cauchy-Schwartz inequality and standard product estimates; for the second one, we write

$$\varepsilon\big(\underline{V}\cdot\nabla^\gamma\psi_{(\alpha)}, \frac{1}{\mu}\mathcal{G}_{(\alpha)}\psi_{\langle\check{\alpha}\rangle}\big) = \varepsilon\Big(\frac{\nabla^\gamma}{(1+\sqrt{\mu}|D^\gamma|)}\psi_{(\alpha)}, (1+\sqrt{\mu}|D^\gamma|)^{1/2}(\underline{V}\frac{1}{\mu}\mathcal{G}_{(\alpha)}\psi_{\langle\check{\alpha}\rangle})\Big).$$

We deduce from the Cauchy-Schwarz inequality that this quantity is bounded from above in absolute value by

$$\varepsilon|\underline{V}|_{H^{t_0}}|\mathfrak{P}\psi_{(\alpha)}|_2\big(|\frac{1}{\mu}\mathcal{G}_{(\alpha)}\psi_{(\alpha)}|_2 + \mu^{1/4}|\frac{1}{\mu}\mathcal{G}_{(\alpha)}\psi_{(\alpha)}|_{H^{1/4}}\big).$$

The fact that this quantity is itself controlled by $\varepsilon\mathfrak{m}_\sigma^N(U)$ is then a direct consequence of the first two points of Proposition 3.28.
- *Control of* B_1. Using the bounds on R_α and S_α established in Proposition 9.5, one readily gets $|B_1| \le \epsilon\mathfrak{m}_\sigma^N(U)$.

Gathering all the information coming from the above estimates, we deduce from (9.27) that for all $1 \le |\alpha| \le N$

$$\frac{d}{dt}\Big(\mathcal{F}_\sigma^\alpha(U) + (U_{(\alpha)}, \mathcal{S}_\alpha^2 U_{\langle\check{\alpha}\rangle})\Big) \le \epsilon\mathfrak{m}^N(U). \tag{9.28}$$

As in the proof of Theorem 4.16, we also obtain that

$$\frac{d}{dt}\Big(\mathcal{F}^0_\sigma(U)\Big) \leq \epsilon \mathfrak{m}^N(U). \tag{9.29}$$

Summing (9.28) and (9.29) over all $\alpha \in \mathbb{N}^{d+1}$, first with $|\alpha| \leq N-1$ and then with $|\alpha| \leq N$ (and recalling that $\mathcal{S}^2_\alpha = 0$ if $|\alpha| < N$), we deduce that

$$\begin{aligned} \frac{d}{dt}\Big(\mathcal{F}^{N-1}_\sigma(U)\Big) &\leq \epsilon \mathfrak{m}^N_\sigma(U), \\ \frac{d}{dt}\Big(\mathcal{F}^N_\sigma(U) + \sum_{|\alpha|=N}(U_{(\alpha)}, \mathcal{S}^2_\alpha U_{\langle \check{\alpha}\rangle})\Big) &\leq \epsilon \mathfrak{m}^N_\sigma(U), \end{aligned}$$

and therefore, for any constant $\mathfrak{C}$,

$$\frac{d}{dt}\widetilde{\mathcal{F}}^N_\sigma(U) \leq \epsilon \mathfrak{m}^N(U), \tag{9.30}$$

with $\widetilde{\mathcal{F}}^N_\sigma(U) = \mathcal{F}^N_\sigma(U) + \mathfrak{C}\mathcal{F}^{N-1}_\sigma(U) + \sum_{|\alpha|=N}(U_{(\alpha)}, \mathcal{S}^2_\alpha U_{\langle\check{\alpha}\rangle})$.
Now, it follows from the definition of $\mathcal{S}^2_\alpha$ that there exists a constant $c_3(U) \leq M$ such that

$$\sum_{|\alpha|=N} |(U_{(\alpha)}, \mathcal{S}^2_\alpha U_{\langle\check{\alpha}\rangle})| \leq \frac{1}{2}\mathcal{F}^N_\sigma(U) + c_3(U)\mathcal{F}^{N-1}_\sigma(U).$$

As long as $\mathfrak{C} \geq c_3(U)$ (and $\mathfrak{C} \leq M$), one therefore has

$$\frac{1}{2}\mathcal{F}^N_\sigma(U) \leq \widetilde{\mathcal{F}}^N_\sigma(U) \leq M\mathcal{F}^N_\sigma(U).$$

With the help of (9.26), this in turn implies that

$$\mathcal{E}^N_\sigma(U) \leq 2(M + \frac{4}{a_0})\widetilde{\mathcal{F}}^N_\sigma(U). \tag{9.31}$$

It therefore follows from (9.30) that

$$\frac{d}{dt}\widetilde{\mathcal{F}}^N_\sigma(U^\iota) \leq \epsilon \widetilde{\mathfrak{m}}^N_\sigma(U), \tag{9.32}$$

where $\widetilde{\mathfrak{m}}^N_\sigma(U) = C\big(M, \widetilde{\mathcal{F}}^N_\sigma(U), \frac{1}{\mathrm{Bo}}, \frac{1}{a_0}\big)$. This gives us a bound on $\widetilde{\mathfrak{m}}^N_\sigma(U)$ on a time interval of the form $[0, T/\epsilon]$ with T as in the statement of the theorem and, using (9.31), we get the desired bound on $\mathcal{E}^N_\sigma(U)$.
The uniqueness part of the proof follows the same lines as the proof of Theorem 4.16 and is omitted here. □

9.2. Shallow water models (systems) with surface tension

We recall that the shallow water models derived in §5.1 and coupling the surface elevation ζ to the average velocity $\overline{V}$ were obtained by taking the gradient of the second equation of the water waves system (5.1) and replacing $\nabla\psi$ by an asymptotic expansion in terms of ζ and $\overline{V}$. In the presence of surface tension, one has to replace (5.1) by (9.2), but the general strategy is the same. The only difference is that capillary terms must be added to the right-hand side of the equation on $\overline{V}$; this contribution is given by the gradient of the right-hand side of (9.2), namely,

$$\frac{1}{\mathrm{Bo}}\nabla^\gamma\nabla^\gamma\cdot\big(\frac{\nabla^\gamma\zeta}{\sqrt{1+\varepsilon^2\mu|\nabla^\gamma\zeta|^2}}\big).$$

Let us define the *rescaled Bond number* bo as

$$\mathrm{bo} = \frac{\rho g H_0^2}{\sigma},$$

where H_0 is the characteristic water depth, so that $\frac{1}{\mathrm{Bo}} = \mu\frac{1}{\mathrm{bo}}$ and the capillary term above can be written as

$$\frac{1}{\mathrm{Bo}}\nabla^\gamma\nabla^\gamma\cdot\Big(\frac{\nabla^\gamma\zeta}{\sqrt{1+\varepsilon^2\mu|\nabla^\gamma\zeta|^2}}\Big) = \frac{\mu}{\mathrm{bo}}\nabla^\gamma\Delta^\gamma\zeta + O(\varepsilon^2\mu^2). \tag{9.33}$$

REMARK 9.10. It is assumed in the asymptotic expansion (9.33) that bo is not too small, so that $\frac{\mu}{\mathrm{bo}} = O(\mu)$. This assumption seems quite reasonable since, for water, one can compute with the numerical values given in §9.1.2 that in order to have $\mathrm{bo}_{water} \leq 1$, the still water depth must satisfy $H_0 \leq (\frac{\sigma}{\rho g})^{1/2} \sim 2.7\mathrm{mm}$.

9.2.1. Large amplitude models. Let us first consider the most general case of the shallow water asymptotic regime

$$\mathcal{A}_{SW} = \{(\varepsilon,\beta,\gamma,\mu),\quad 0\leq\mu\leq\mu_0,\quad 0\leq\varepsilon\leq 1,\quad 0\leq\beta\leq 1,\quad 0\leq\gamma\leq 1\},$$

for some $0 < \mu_0 \leq \mu_{max}$; we recall that the important point about this regime is that no smallness assumption is made on the surface elevation or bottom variations.

The capillary terms (9.33) are of order $O(\mu)$, the size of the error term for the Nonlinear Shallow Water equations (5.7). The conclusion is that, as an asymptotic model, *the Nonlinear Shallow Water equations (*5.7*) are unaffected by the presence, or lack thereof, of surface tension.*

For the Green-Naghdi equations (5.11) or equivalently (5.15), the situation is different; indeed, the precision of this model is $O(\mu^2)$ and the $O(\mu)$ capillary effects must be added to the equations to keep the same accuracy. In the presence of surface tension, the Green-Naghdi equations (5.15) therefore become

$$\begin{cases} \partial_t\zeta + \nabla^\gamma\cdot(h\overline{V}) = 0,\\ [I+\mu\mathcal{T}]\big(\partial_t\overline{V} + \varepsilon(\overline{V}\cdot\nabla^\gamma)\overline{V}\big) + (1-\frac{\mu}{\mathrm{bo}}\Delta^\gamma)\nabla^\gamma\zeta + \mu\varepsilon\mathcal{Q}_1(\overline{V}) = 0, \end{cases} \tag{9.34}$$

where $\mathcal{Q}_1(\overline{V})$ is as in (5.16).
More generally, the Green-Naghdi equations with improved frequency dispersion (9.35) become

$$\begin{cases} \partial_t\zeta + \nabla^\gamma\cdot(h\overline{V}) = 0,\\ [1+\mu\alpha\mathcal{T}]\Big(\partial_t\overline{V} + \varepsilon(\overline{V}\cdot\nabla^\gamma)\overline{V} + \dfrac{\alpha-1}{\alpha}(1-\dfrac{\mu}{\mathrm{bo}}\Delta^\gamma)\nabla^\gamma\zeta\Big)\\ \qquad\qquad + \dfrac{1}{\alpha}(1-\dfrac{\mu}{\mathrm{bo}}\Delta^\gamma)\nabla^\gamma\zeta + \mu\varepsilon\mathcal{Q}_1(\overline{V}) = 0. \end{cases} \tag{9.35}$$

Using Theorem 9.6, a local well-posedness result for these systems similar to those of Propositions 6.4 and 6.5 would automatically guarantee the *full justification* of these models along the same lines as Theorem 6.15.

9.2.2. Small amplitude models. We now consider the case of small amplitude models ($\varepsilon = O(\mu)$, long wave regime); these models correspond to the various Boussinesq systems derived in §5.1.3. Since these models approximate the water waves equations up to $O(\mu^2)$ terms, the $O(\mu)$ capillary effects must be included to keep the same precision in the presence of surface tension.

The Boussinesq-Peregrine model with large ($\beta = O(1)$) topography variations (5.23) therefore becomes

$$(9.36) \qquad \begin{cases} \partial_t \zeta + \nabla^\gamma \cdot (h\overline{V}) = 0, \\ (I + \mu \mathcal{T}_b)\partial_t \overline{V} + (1 - \frac{\mu}{\text{bo}}\Delta^\gamma)\nabla^\gamma \zeta + \varepsilon(\overline{V} \cdot \nabla^\gamma)\overline{V} = 0, \end{cases}$$

where $h_b := 1 - \beta b$ and $\mathcal{T}_b$ are as in (5.24). As for the Green-Naghdi equations above, similar generalizations exist for the corresponding Boussinesq models with improved frequency dispersion, (5.28) and (5.31).

When the bottom is almost flat ($\beta = O(\mu)$), we have derived in Chapter 5 the four parameters family of Boussinesq systems (5.34), which, in the presence of surface tension becomes

$$(9.37) \qquad \begin{cases} (1 - \mu \mathbf{b}\Delta^\gamma)\partial_t \zeta + \nabla^\gamma \cdot (hV_{\theta,\delta}) + \mu \mathbf{a}\Delta^\gamma \nabla^\gamma \cdot V_{\theta,\delta} = 0, \\ (1 - \mu \mathbf{d}\Delta^\gamma)\partial_t V_{\theta,\delta} + \nabla^\gamma \zeta + \varepsilon(V_{\theta,\delta} \cdot \nabla^\gamma)V_{\theta,\delta} + \mu(\mathbf{c} - \frac{1}{\text{bo}})\Delta^\gamma \nabla^\gamma \zeta = 0, \end{cases}$$

where we recall that $V_{\theta,\delta}$ is given in terms of $\overline{V}$ by (5.30), while $\mathbf{a}$, $\mathbf{b}$, $\mathbf{c}$ and $\mathbf{d}$ are given by (5.35).
Let us consider a particular system belonging to this family: the one for which all the dispersive effects are carried by the term $\Delta^\gamma \nabla^\gamma \zeta$, that is, we take $\mathbf{a} = \mathbf{b} = \mathbf{d} = 0$ (and therefore $\mathbf{c} = 1/3$, owing to (5.35)). The capillary effects therefore are of the same nature, but the opposite sign, as the dispersive terms of the standard (zero surface tension) Boussinesq systems. We therefore can draw a qualitative conclusion from this analogy, namely, that in the Boussinesq regime with almost flat bottoms, *dispersive effects act as a negative surface tension* of value

$$(9.38) \qquad \sigma_{virtual} = -\frac{1}{3}\rho g H_0^2.$$

When the surface tension coefficient σ satisfies $\sigma = -\sigma_{virtual}$ (or equivalently, when $\text{bo} = 3$), then these effects are in perfect balance and disappear. The system (9.37) does not contain any dispersive term and now coincides[3] with the Nonlinear Shallow Water equations (5.7).

REMARK 9.11. The same generalization also holds for the four-parameter family of systems with symmetric nonlinearity (5.40): one just has to replace the coefficient $\mathbf{c}$ by $\mathbf{c} - \frac{1}{\text{bo}}$. In the presence of surface tension, the conditions stated in Definition 5.38 to define the subclass Σ' of fully symmetric systems must be replaced by $\mathbf{b} \geq 0$, $\mathbf{d} \geq 0$ and $\mathbf{a} = \mathbf{c} - \frac{1}{\text{bo}}$. It is important to note that even with surface tension, this class remains nonempty. For instance, one can take $\mathbf{b} = \mathbf{d} = 0$ and $\mathbf{a} = \mathbf{c} = \frac{1}{6} - \frac{1}{2\text{bo}}$ (or, using (5.35), $\frac{\theta - \delta}{3} = \frac{1}{2\text{bo}} - \frac{1}{6}$, $\lambda = -\delta$ and $\alpha = -\theta$). The justification of these models as an approximation to the water waves equation with surface tension therefore follows from Theorem 9.6 with exactly the same arguments as in the zero surface tension case (see Theorem 6.20).

[3]Though the asymptotic model is the same as (5.7), the situation is slightly different here. We have made a small amplitude assumption for the surface and bottom variations, $\varepsilon = O(\mu)$, $\beta = O(\mu)$. On the other hand, the accuracy of the model is $O(\mu^2)$ here rather than $O(\mu)$ for (5.7).

9.3. Asymptotic models: Scalar equations

9.3.1. Capillary effects and the KdV approximation. We justified in Theorem 7.1 the KdV approximation in the so-called KdV regime

$$\mathcal{A}_{KdV} = \{(\varepsilon,\beta,\gamma,\mu),\quad 0\le\mu\le\mu_0,\quad \varepsilon=\mu,\quad \beta=\gamma=0\}. \tag{9.39}$$

We started from one of the Boussinesq systems with symmetric nonlinearity (5.40), since the latter had previously been justified in Theorem 6.20. In the same regime, we can apply exactly the same procedure in the presence of surface tension, since one can also derive and justify fully symmetric Boussinesq systems (see Remark 9.11). We can (see also Remark 9.11) start from

$$\begin{cases} \partial_t\zeta + \partial_x((1+\frac{\mu}{2}\zeta)w) + \mu(\frac{1}{6}-\frac{1}{2\mathrm{bo}})\partial_x^3 w = 0,\\ \partial_t w + (1+\frac{\mu}{2}\zeta)\partial_x\zeta + \mu(\frac{1}{6}-\frac{1}{2\mathrm{bo}})\partial_x^3\zeta + \mu\frac{3}{2}w\partial_x w = 0,\end{cases} \tag{9.40}$$

which has the same "KdV-KdV" structure as (7.4).
Proceeding exactly as in Chapter 7, one easily checks that the KdV approximation (7.5)–(7.6) must be replaced by the following generalization (referred to as the KdV_σ approximation)

$$\begin{array}{rcl} \zeta_{KdV,\sigma}(t,x) &=& f_+(x-t,\mu t) + f_-(x+t,\mu t),\\ w_{KdV,\sigma}(t,x) &=& f_+(x-t,\mu t) - f_-(x+t,\mu t),\end{array} \tag{9.41}$$

where f_+ and f_- solve

$$\begin{array}{l} \partial_\tau f_+ + (\frac{1}{6}-\frac{1}{2\mathrm{bo}})\partial_{\tilde{x}}^3 f_+ + \frac{3}{2}f_+\partial_{\tilde{x}} f_+ = 0,\\ \partial_\tau f_- - (\frac{1}{6}-\frac{1}{2\mathrm{bo}})\partial_{\tilde{x}}^3 f_- - \frac{3}{2}f_-\partial_{\tilde{x}} f_- = 0.\end{array} \tag{9.42}$$

The following results are then straightforward adaptations of the results proved in §7.1.1, in the absence of surface tension:

(1) Theorem 7.1 and the improvement given in Corollary 7.12 remain true in the presence of surface tension: solutions to the Boussinesq system (9.40) are well approximated on a time scale of order $O(1/\varepsilon)$ by the KdV_σ approximation (9.41)–(9.42) with the same error estimates as in the zero surface tension case.
(2) Corollary 7.2 remains true in the presence of surface tension: the KdV_σ approximation (9.41)–(9.42) also furnishes a good approximation to the full water waves equations.

As previously explained in §9.2.2, dispersive effects act in the present regime as a negative surface tension $\sigma_{virtual}<0$ given by (9.38). When the real surface tension is small, its contribution therefore is to reduce the coefficient of the dispersive term in the KdV approximation. When it is very big, the sign of this coefficient is changed. This has some important qualitative consequences (e.g., solitary waves are then of depression type[4]), but the most drastic qualitative difference with the zero surface tension case is obtained for the critical value of the surface tension that *cancels* the dispersive term in the KdV equation. This is achieved when

$$\mathrm{bo}=3 \quad \text{or equivalently} \quad \sigma=\frac{1}{3}\rho g H_0^2.$$

[4]This had already been observed by Korteweg and de Vries in their paper [**211**].

This configuration will be specifically addressed in the next section.

EXAMPLE 9.12.
(1) Using the numerical values given in §9.1.2, one readily checks that for water, this critical regime is obtained when the depth at rest is $H_0 \sim 4.7$mm.
(2) It seems that the only observation of depression solitary waves in a fluid, predicted when bo < 3, is the one, reported in [**145**], in a thin layer of mercury for which bo $= 1.23 < 3$ ($H_0 = 2.12$mm, $\rho = 13.5 \times 10^3$kg/m^3, $\sigma = 0.483N/m$). Note that for this experimental setting, dynamic viscosity should be taken into account.

9.3.2. The Kawahara approximation. As shown in Example 9.12, the physical relevance of the critical regime bo $= 3$ is quite restricted, but, for the sake of completeness, let us briefly discuss this situation, which is quite puzzling since dispersive effects have altogether disappeared from the KdV$_\sigma$ equation (which therefore has degenerated into a Burgers equation).

The KdV regime (9.39) assumed that $\varepsilon = \mu$ in order for the nonlinear effects and the dispersive effects to be of the same order; indeed, ε measures the nonlinear effects and μ the dispersive effects. (This is obvious in the formulation (7.62) of the KdV/BBM equations.) When bo $= 3$, the vanishing of dispersive term in the KdV$_\sigma$ equation means that *dispersive effects are of smaller order than nonlinear effects.*

If one is interested in obtaining a scalar asymptotic equation with both nonlinear and dispersive terms, as for the KdV equation, one therefore has to work in a less nonlinear scaling than the KdV regime (9.39). This regime, which we will call the *Kawahara regime*, is

$$\mathcal{A}_{Kawa} = \{(\varepsilon, \beta, \gamma, \mu), \quad 0 \leq \mu \leq \mu_0, \quad \varepsilon = \mu^2, \quad \beta = \gamma = 0\}. \tag{9.43}$$

The difference with the KdV regime (9.39) is that $\varepsilon = \mu^2$ rather than $\varepsilon = \mu$; nonlinear effects therefore are smaller and, as we will see, of the same order as dispersive effects. We assume that bo is very close to its critical value 3 in the following sense

$$\frac{1}{\text{bo}} = \frac{1}{3} + \mu\kappa, \tag{9.44}$$

for some $\kappa \in \mathbb{R}$.

In the Kawahara regime (9.43) and under the assumption (9.44), it is possible to derive and justify the so-called Kawahara approximation, which described the surface elevation as follows

$$\zeta_{Kawa}(t, x) = f_+(x - t, \mu^2 t) + f_-(x + t, \mu^2 t) \tag{9.45}$$

(with, of course, a similar formula for the velocity), where f_+ and f_- solve

$$\begin{aligned} \partial_\tau f_+ - \frac{1}{2}\kappa\partial_{\tilde{x}}^3 f_+ + \frac{1}{90}\partial_{\tilde{x}}^5 f_+ + \frac{3}{2} f_+ \partial_{\tilde{x}} f_+ = 0, \\ \partial_\tau f_- + \frac{1}{2}\kappa\partial_{\tilde{x}}^3 f_- - \frac{1}{90}\partial_{\tilde{x}}^5 f_- - \frac{3}{2} f_- \partial_{\tilde{x}} f_- = 0. \end{aligned} \tag{9.46}$$

(1) Theorem 9.6 shows that solutions to the water waves equations exist on a time scale $O(1/\varepsilon) = O(1/\mu^2)$.

(2) One proves with the method developed in §3.6.1 that in the Kawahara regime, the averaged velocity $\overline{V}$ is related to $\nabla^\gamma\psi$ by the formula

$$\overline{V} = \nabla^\gamma\psi + \frac{\mu}{3}\Delta^\gamma\nabla^\gamma\psi + \mu^2\frac{2}{15}(\Delta^\gamma)^2\nabla^\gamma\psi + O(\mu^3)$$

and therefore

$$\nabla^\gamma\psi = \overline{V} - \mu\frac{1}{3}\Delta^\gamma\overline{V} - \mu^2\frac{1}{45}(\Delta^\gamma)^2\overline{V} + O(\mu^3)$$

(only the one-dimensional version of this formula is needed here).

(3) One deduces as in §5.1 that the solutions to the water waves equations are consistent at order $O(\mu^3)$ with the following higher order Boussinesq system (in which we performed the substitution $\varepsilon = \mu^2$)

$$(9.47)\qquad \begin{cases} \partial_t\zeta + \nabla^\gamma\cdot((1+\mu^2\zeta)\overline{V}) = 0, \\ \partial_t\overline{V} + \nabla^\gamma\zeta + \mu^2(\overline{V}\cdot\nabla^\gamma)\overline{V} - \mu^2\kappa\Delta^\gamma\nabla^\gamma\zeta + \mu^2\frac{1}{45}(\Delta^\gamma)^2\nabla^\gamma\zeta = 0. \end{cases}$$

(Here again, only the one-dimensional version of this system is needed — just replace ∇^γ with ∂_x. The $2d$ version will be used in §9.3.4 below.)

(4) This higher order Boussinesq system (or any "improved" version in the sense of §§5.2.1 and 5.3) is justified as in §6.2.3.

(5) The Kawahara approximation can be defined and justified exactly as the KdV approximation (see §7.1.1).

We will not provide details on these different points, since they should not cause any major new difficulties. Note that the Kawahara approximation was first derived by Kakutani and Ono [**195**] to describe hydromagnetic waves in plasmas, then by Hasimoto for water waves [**168**], and later by Kawahara [**203**]. The mathematical justification of this approximation for water waves was done by Schneider and Wayne [**292**] and improved by Iguchi [**180**] with different methods. (See [**133**] for another proof.)

9.3.3. Capillary effects and the KP approximation. The justification of the KP approximation has been done in Theorem 7.16, in the so-called KP regime

$$(9.48)\qquad \mathcal{A}_{KP} = \{(\varepsilon,\beta,\gamma,\mu),\quad 0\le\mu\le\mu_{max},\quad \varepsilon=\mu,\quad \gamma=\sqrt{\varepsilon},\quad \beta=0\}.$$

As for the KdV approximation, we started from a Boussinesq systems with symmetric nonlinearity (7.26), which, in the presence of surface tension, can be generalized into

$$(9.49)\qquad \begin{cases} \partial_t\zeta + \nabla^\gamma\cdot((1+\frac{\varepsilon}{2}\zeta)W) + \mu(\frac{1}{6}-\frac{1}{2\mathrm{bo}})\Delta^\gamma\nabla^\gamma\cdot W = 0, \\ \partial_t W + (1+\frac{\varepsilon}{2}\zeta)\nabla^\gamma\zeta + \mu(\frac{1}{6}-\frac{1}{2\mathrm{bo}})\Delta^\gamma\nabla^\gamma\zeta \\ \qquad + \frac{\varepsilon}{2}\big[(W\cdot\nabla^\gamma)W + W\nabla^\gamma\cdot W + \frac{1}{2}\nabla^\gamma|W|^2\big] = 0. \end{cases}$$

As for the KdV approximation, the presence of surface tension does not significantly modify the analysis performed in §7.2 and one readily checks that the KP approximation (7.27)–(7.28) must be replaced by the following generalization (referred to as the KP$_\sigma$ approximation)

$$(9.50)\qquad \begin{aligned} \zeta_{KP}(t,x,y) &= f_+(x-t,y,\mu t) + f_-(x+t,y,\mu t), \\ w_{1,KP}(t,x,y) &= f_+(x-t,y,\mu t) - f_-(x+t,y,\mu t), \end{aligned}$$

where $f_+ = f(\tilde{x}, y, \tau)$ and $f_- = f(\tilde{x}, y, \tau)$ solve the KP equations

$$(9.51)\qquad \begin{aligned} \partial_\tau f_+ + \frac{1}{2}\partial_{\tilde{x}}^{-1}\partial_y^2 f_+ + (\frac{1}{6} - \frac{1}{2\mathrm{bo}})\partial_{\tilde{x}}^3 f_+ + \frac{3}{2} f_+ \partial_{\tilde{x}} f_+ = 0, \\ \partial_\tau f_- - \frac{1}{2}\partial_{\tilde{x}}^{-1}\partial_y^2 f_- - (\frac{1}{6} - \frac{1}{2\mathrm{bo}})\partial_{\tilde{x}}^3 f_- - \frac{3}{2} f_- \partial_{\tilde{x}} f_- = 0. \end{aligned}$$

As for the KdV_σ approximation, one easily checks that the results proved in §7.2 for the KP approximation in the absence of surface tension remain true when capillary effects are included.

REMARK 9.13. As in the KdV case, there is a critical value for the surface tension, corresponding to $\mathrm{bo} = 3$. When $\mathrm{bo} > 3$, then the KP_σ equations (9.51) are of the same nature as the KP equations (7.28) derived in the absence of surface tension. Such equations are commonly referred to as KP2 equations in the literature. When $\mathrm{bo} < 3$ (strong surface tension), the dispersion coefficient changes sign. The corresponding equation is then called the KP1 equation. The mathematical behavior of the KP1 and KP2 equations are quite different in many aspects, but this does not affect the present analysis, where we essentially need local well-posedness results for these equations (see Lemma 7.22). See [**208**] for a recent review of KP1/KP2 equations.

9.3.4. The weakly transverse Kawahara approximation. When the surface tension is close to its critical value, $\mathrm{bo} \sim 3$, then the third-order dispersive term disappears from the KP_σ equation, thus reproducing the same phenomenon observed for the KdV_σ approximation.

Inspired by §9.3.2, we consider a less nonlinear regime in order to be able to obtain dispersive and nonlinear effects of the same order. This regime is a natural weakly transverse variation of the *Kawahara regime* (9.43), namely,

$$(9.52)\qquad \mathcal{A}_{Kawa,2d} = \{(\varepsilon, \beta, \gamma, \mu), \quad 0 \leq \mu \leq \mu_0, \quad \varepsilon = \mu^2, \quad \gamma = \sqrt{\varepsilon}, \quad \gamma = 0\}.$$

Assuming that $\mathrm{bo} \sim 3$ in the following sense

$$\frac{1}{\mathrm{bo}} = \frac{1}{3} + \mu\kappa,$$

for some $\kappa \in \mathbb{R}$, the Kawahara approximation (9.45)–(9.46) can be generalized as follows in the weakly transverse regime (9.52)

$$(9.53)\qquad \zeta_{Kawa}(t, x) = f_+(x - t, y, \mu^2 t) + f_-(x + t, y, \mu^2 t)$$

(with a similar formula for the velocity), where f_+ and f_- solve

$$(9.54)\qquad \begin{aligned} \partial_\tau f_+ + \frac{1}{2}\partial_{\tilde{x}}^{-1}\partial_y^2 f_+ - \frac{1}{2}\kappa\partial_{\tilde{x}}^3 f_+ + \frac{1}{90}\partial_{\tilde{x}}^5 f_+ + \frac{3}{2} f_+ \partial_{\tilde{x}} f_+ = 0, \\ \partial_\tau f_- - \frac{1}{2}\partial_{\tilde{x}}^{-1}\partial_y^2 f_- - \frac{1}{2}\kappa\partial_{\tilde{x}}^3 f_- - \frac{1}{90}\partial_{\tilde{x}}^5 f_- - \frac{3}{2} f_- \partial_{\tilde{x}} f_- = 0. \end{aligned}$$

The procedure to justify this approximation is exactly the same as the procedure in §9.3.2 (this is why we had given the $2d$ version of (9.47)). This weakly transverse generalization of the Kawahara equation has been derived and justified in [**263**]; also note that these equations are known to be globally well-posed [**286**].

9.4. Asymptotic models: Deep and infinite water

When μ is not small, we have seen that it is possible to obtain asymptotic models through small amplitude expansions in the deep water regime

$$\mathcal{A}_{DW} = \{(\varepsilon, \beta, \gamma, \mu), \quad 0 \leq \mu \leq \mu_{max}, \quad 0 \leq \varepsilon\sqrt{\mu} \leq \epsilon_0, \quad \beta = 0, \quad \gamma = 1\},$$

with $\epsilon_0 \ll 1$. This allowed us to derive the full dispersion model (8.2). In the presence of surface tension, this model reads

$$\left\{ \begin{array}{l} \partial_t\zeta - \frac{1}{\sqrt{\mu}\nu}\mathcal{H}_\mu \underline{V} + \frac{\varepsilon}{\nu}\big(\mathcal{H}_\mu(\zeta\nabla\mathcal{H}_\mu\underline{V}) + \nabla\cdot(\zeta\underline{V})\big) = 0, \\ \partial_t\underline{V} + (1 - \frac{1}{\widetilde{\mathrm{bo}}}\Delta)\nabla\zeta + \frac{\varepsilon}{\nu}\big(\frac{1}{2}\nabla|\underline{V}|^2 - \nu\sqrt{\mu}\nabla\zeta\mathcal{H}_\mu\nabla\zeta\big) = 0, \end{array} \right. \tag{9.55}$$

where we introduced $\widetilde{\mathrm{bo}}$ such that $\mu/\mathrm{bo} = 1/\widetilde{\mathrm{bo}}$, i.e.,

$$\widetilde{\mathrm{bo}} = \frac{\rho g L_x^2}{\sigma}.$$

(This quantity does not depend on the depth and therefore is the relevant one for large or infinite depth.) The same kind of justification that was made in Theorem 8.4, without surface tension, can be made here.

REMARK 9.14. The infinite depth version of this system can be derived as in Section 8.1.4, namely,

$$\left\{ \begin{array}{l} \partial_t\zeta - \mathcal{H}\underline{V} + \epsilon\big(\mathcal{H}(\zeta\nabla\mathcal{H}\underline{V}) + \nabla\cdot(\zeta\underline{V})\big) = 0, \\ \partial_t\underline{V} + (1 - \frac{1}{\widetilde{\mathrm{bo}}}\Delta)\nabla\zeta + \epsilon\big(\frac{1}{2}\nabla|\underline{V}|^2 - \nabla\zeta\mathcal{H}\nabla\zeta\big) = 0. \end{array} \right.$$

9.5. Modulation equations

See Section 8.5.6 for comments and references on modulation equations for capillary gravity waves.

APPENDIX A

More on the Dirichlet-Neumann Operator

We present in this section some additional results on the Dirichlet-Neumann operator. These results are not necessary for the study of the water waves developed in these notes but are of general interest.

We first prove in §A.1 the *shape analyticity* of the Dirichlet-Neumann operator, i.e., we show that it depends analytically on the surface and the bottom parametrizations. This result holds both in finite and infinite depth.

We then prove in §A.2 that the Dirichlet-Neumann is *essentially self-adjoint*; the fact that it it symmetric is quite straightforward and has been proved in Chapter 3. Proving that it is self-adjoint is a bit more complicated, and we provide here an elementary proof.

We also briefly address in §A.3 the *invertibility* properties of the Dirichlet-Neumann operators. Because of singularities arising at low frequencies, the inverse of a Dirichlet-Neumann operator is not always defined; however, we exhibit a general configuration for which this inverse is well defined.

Finally, we give in §A.4 a few comments on the *symbolic analysis* of the Dirichlet-Neumann operator.

A.1. Shape analyticity of the Dirichlet-Neumann operator

A.1.1. Shape analyticity of the velocity potential. Let $0 \leq s \leq t_0 + 1/2$ and $\psi \in \dot{H}^{s+1/2}(\mathbb{R}^d)$. We have seen that to every $\Gamma = (\zeta, b) \in H^{t_0+1}(\mathbb{R}^d)^2$ ($t_0 > d/2$) satisfying (2.6) one can associate an admissible diffeomorphism Σ_Γ—and therefore a matrix $P(\Sigma_\Gamma)$ given by formula (2.13)—and a variational solution ϕ_Γ to (2.12). We are interested here in the regularity of the mapping $\Gamma \mapsto \phi_\Gamma$; since Γ determines the shape of the fluid domain, this regularity is called *shape regularity*.

Let us first introduce the notion of analytic mappings on Banach spaces (see [**60**] for more details).

DEFINITION A.1 (Analyticity in Banach spaces). Let X and Y be Banach spaces and $U \subset X$ be an open subset of X. A mapping $F : U \to Y$ is *analytic* if for all $x_0 \in U$ one can write, in a neighborhood of x_0

$$F(x) = \sum_{k=0}^{\infty} F_k(x - x_0)^k,$$

where the $F_k : X^k \to Y$ are symmetric, bounded, k-linear operators, such that

$$\exists N > 0, \quad \exists r > 0, \quad \forall k \in \mathbb{N}, \qquad \|F^k\|_{X^k \to Y} \leq N r^{-k}.$$

REMARK A.2. With the same notations as in Definition A.1, and denoting by $d^j F(x_0)(\mathbf{h})$ the j-th derivative ($j \in \mathbb{N}$) of F at x_0 and in the direction $\mathbf{h} \in X^j$, it is classical to check that

$$d^j F(x_0)(\mathbf{h}) = j!\, F_j(\mathbf{h}).$$

We will require throughout this section that the mapping $\Gamma \mapsto \Sigma_\Gamma$ is *shape analytic* in the following sense:

DEFINITION A.3. Let $t_0 > d/2$.

(1) We denote by $\mathbf{\Gamma} \subset H^{t_0+1}(\mathbb{R}^d)^2$ the (open) set
$$\mathbf{\Gamma} = \{\Gamma = (\zeta, b) \in H^{t_0+1}(\mathbb{R}^d)^2, \quad (2.6) \text{ is satisfied}\}.$$
(2) For all $\Gamma = (\zeta, b) \in \mathbf{\Gamma}$, we write Ω_Γ the fluid domain
$$\Omega_\Gamma = \{(X, z), -1 + \beta b(X) < z < \varepsilon\zeta(X)\}.$$
(3) We say that a mapping $\Gamma \in \mathbf{\Gamma} \mapsto \Sigma_\Gamma$ is *shape analytic* if:
   a- For all $\Gamma \in \mathbf{\Gamma}$, $\Sigma_\Gamma : \Omega_\Gamma \to \mathcal{S}$ is an admissible diffeomorphism.
   b- The mapping
$$\mathfrak{Q} : \begin{array}{ccl} \mathbf{\Gamma} & \to & (L^\infty H^{t_0})^{(d+1)\times(d+1)} \\ \Gamma & \mapsto & \mathfrak{Q}(\Gamma) = Q(\Sigma_\Gamma) = P(\Sigma_\Gamma) - I, \end{array}$$
   is analytic in the sense of Definition A.1.

(4) We say that it is in addition *regularizing* if one can replace $L^\infty H^{t_0}$ by $H^{t_0+1/2,2}$ in (3)-b above[1].

EXAMPLE A.4. Examples 2.14 and 2.15 provide two shape analytic mappings (this is an easy consequence of the explicit expression of $Q(\Sigma)$ given in Example 2.21). Example 2.15 is in addition regularizing (this is proved as in Proposition 2.18 using the product estimates of Corollaries B.5 and B.6).

It is a direct consequence of Proposition 2.36 that given any regularizing, shape analytic, mapping $\Gamma \mapsto \Sigma_\Gamma$, one can define, for all $0 \le s \le t_0 + 1/2$ and $\psi \in \dot{H}^{s+1/2}$, a mapping $\mathfrak{A}_\psi$ as

$$\mathfrak{A}_\psi : \begin{array}{ccl} \mathbf{\Gamma} & \to & \dot{H}^{s+1}(\mathcal{S}) \\ \Gamma & \mapsto & \phi_\Gamma, \end{array} \tag{A.1}$$

where ϕ_Γ is the unique variational solution to (2.12). The following proposition shows that this mapping is analytic in the sense of Definition A.1.

PROPOSITION A.5. *Let $0 \le s \le t_0 + 1/2$ and $\psi \in \dot{H}^{s+1/2}(\mathbb{R}^d)$ and assume that $\Gamma \in \mathbf{\Gamma} \mapsto \Sigma_\Gamma$ is shape analytic and regularizing. Then the mapping $\mathfrak{A}_\psi$ is analytic.*

PROOF. By assumption, we know that in a small enough neighborhood of $\Gamma_0 \in \mathbf{\Gamma}$, one can write

$$\mathfrak{Q}(\Gamma) = \sum_{k=0}^{\infty} \mathfrak{Q}_k (\Gamma - \Gamma_0)^k,$$

where the $\mathfrak{Q}_k : X^k \to Y$ (with $X = H^{t_0+1}(\mathbb{R}^d)^2$, $Y = (H^{t_0+1/2,2})^{(d+1)\times(d+1)}$) are bounded, symmetric, and k-linear, and moreover satisfy

$$\forall k \in \mathbb{N}, \qquad \|\mathfrak{Q}_k\|_{X^k \to Y} \le N r^{-k}, \tag{A.2}$$

[1] Owing to Proposition 2.12, this is a stronger condition than (3)-b.

for some $r > 0$ and $N > 0$. We want to prove that in a possibly smaller neighborhood of Γ_0, one can write

$$\mathfrak{A}_\psi(\Gamma) = \sum_{k=0}^{\infty} \mathfrak{A}_k(\Gamma - \Gamma_0)^k,$$

where the $\mathfrak{A}_k : X^k \to Z$ (with $Z = \dot{H}^{s+1}(\mathcal{S})$) are bounded, symmetric, and k-linear, and moreover satisfy

$$\forall k \in \mathbb{N}, \qquad \|\mathfrak{A}_k\|_{X^k \to Z} \leq \underline{N}\,\underline{r}^{-k}, \tag{A.3}$$

for some $\underline{r} > 0$ and $\underline{N} > 0$.
Since by definition $\mathfrak{A}_\psi(\Gamma) = \phi_\Gamma$, a straightforward identification of the terms of the same homogeneity in $(\Gamma - \Gamma_0)$, shows that if they exist, the $a_k := \mathfrak{A}_\psi(\Gamma - \Gamma_0)^k$ must necessarily solve the following induction formula:

$$\left\{ \begin{array}{l} \nabla^{\mu,\gamma} \cdot (I + Q_0)\nabla^{\mu,\gamma} a_0 = 0, \\ a_{0\,|z=0} = \psi, \qquad \mathbf{e}_z \cdot (I + Q_0)\nabla_{X,z} a_0 = 0, \end{array} \right. \tag{A.4}$$

and, for all $k \geq 1$

$$\left\{ \begin{array}{l} \nabla^{\mu,\gamma} \cdot (I + Q_0)\nabla^{\mu,\gamma} a_k = -\nabla^{\mu,\gamma} \cdot \sum_{j=1}^k Q_j \nabla^{\mu,\gamma} a_{k-j}, \\ a_{k\,|z=0} = 0, \quad \mathbf{e}_z \cdot (I + Q_0)\nabla^{\mu,\gamma} a_k = \mathbf{e}_z \cdot \sum_{j=1}^k Q_j \nabla^{\mu,\gamma} a_{k-j\,|z=-1}, \end{array} \right. \tag{A.5}$$

where we used the condensed notation $Q_k = \mathfrak{Q}_k(\Gamma - \Gamma_0)^k$. The following lemma shows that this inductive system of equations is well-posed.

LEMMA A.6. *Let the assumptions of Proposition* A.5 *be satisfied. Then there exists a unique solution* $(a_k)_{k \in \mathbb{N}}$ *to* (A.4)*-*(A.5) *with* $a_k \in \dot{H}^{s+1}(\mathcal{S})$. *Moreover, there exists* $\underline{r} > 0$ *such that*

$$\forall k \in \mathbb{N}, \qquad \|\Lambda^s \nabla^{\mu,\gamma} a_k\|_2 \leq \underline{N}\,\underline{r}^{-k} |\Gamma - \Gamma_0|^k_{H^{t_0+1}},$$

with $\underline{N} = \sqrt{\mu} M_0 |\mathfrak{P}\psi|_{H^s}$.
If $s = t_0 + 1/2$, *then the same estimate holds on* $\|\nabla^{\mu,\gamma} a_k\|_{H^{t_0+1/2,1}}$.

PROOF. Noting that $Q_0 = Q(\Sigma_{\Gamma_0})$, we can use Corollary 2.40 to see that there exists a unique variational solution to (A.4) and that

$$\|\Lambda^s \nabla^{\mu,\gamma} a_0\|_2 \leq \underline{N},$$

with $\underline{N} = \sqrt{\mu} M_0 |\mathfrak{P}\psi|_{H^s}$ (and M_0 as in (2.10)). For all $k \geq 1$, let us construct a_k assuming the existence of $a_0, \dots a_{k_1}$ satisfying (for some $\underline{r} > 0$)

$$\forall j = 0, \dots k-1, \qquad \|\Lambda^s \nabla^{\mu,\gamma} a_j\|_2 \leq \underline{N}\,\underline{r}^{-j} |\Gamma - \Gamma_0|^j_{H^{t_0+1}}, \tag{A.6}$$

and such that the same bound holds on $\|\nabla^{\mu,\gamma} a_j\|_{H^{s,1}}$ if $s = t_0 + 1/2$.
Let us first consider the case $0 \leq s \leq t_0$. Defining $\mathbf{g}_k = \sum_{j=1}^k Q_j \nabla^{\mu,\gamma} a_{k-j}$, we get by Corollary B.5 that

$$\forall 0 \leq s \leq t_0, \qquad \|\Lambda^s \mathbf{g}_k\|_2 \leq \sum_{j=1}^k \|Q_j\|_{L^\infty H^{t_0}} \|\Lambda^s \nabla^{\mu,\gamma} a_{k-j}\|_2.$$

We then deduce from (A.2) the continuous embedding $H^{t_0+1/2,1} \subset L^\infty H^{t_0}$ and the induction assumption that

$$\forall 0 \leq s \leq t_0, \qquad \|\Lambda^s \mathbf{g}_k\|_2 \leq \Big(\sum_{j=1}^k N r^{-j} \underline{N}\,\underline{r}^{-(k-j)}\Big) |\Gamma - \Gamma_0|^k_{H^{t_0+1}}.$$

It therefore is a direct consequence of Lemma 2.38 (with $\mathbf{g} = \mathbf{g}_k$) that there exists a unique a_k solving (A.5) and that

$$\forall 0 \leq s \leq t_0, \qquad \|\Lambda^s \nabla^{\mu,\gamma} a_k\|_2 \leq \big(M_0 \sum_{j=1}^{k} N r^{-j} \underline{N}\, \underline{r}^{-(k-j)}\big)|\Gamma - \Gamma_0|^k_{H^{t_0+1}}.$$

Choosing $\underline{r}$ small enough to have $NM_0 \sum_{j=1}^{\infty}(\frac{\underline{r}}{r})^j \leq 1$, one deduces that (A.6) holds also for $j = k$, and the next step of the induction therefore is established in the case $0 \leq s \leq t_0$.
For $s = t_0 + 1/2$, Corollary B.5(2) and the continuous embedding $H^{t_0+1/2,1} \subset L^\infty H^{t_0}$ show that

$$\|\mathbf{g}_k\|_{H^{t_0+1/2,1}} \lesssim \sum_{j=1}^{k} \|Q_j\|_{H^{t_0+1/2,1}} \|\nabla^{\mu,\gamma} a_{k-j}\|_{H^{t_0+1/2,1}},$$

and one can conclude the next step of the induction as above.
Finally, for $t_0 \leq s \leq t_0 + 1/2$, the result is obtained by interpolation. □

Quite obviously, the dependence of a_k on $(\Gamma - \Gamma_0)$ is homogeneous of order k and we can thus deduce from the lemma that one can find a symmetric k-linear mapping $\mathfrak{A}_k$ such that $\mathfrak{A}_k(\Gamma - \Gamma_0)^k = a_k$ and satisfying (A.3). In particular, the series $b = \sum_{k \geq 1} a_k$ converges absolutely in $H^1_{0,surf}(\mathcal{S})$ for Γ close enough to Γ_0. We now have to show that $a_0 + b = \phi_\Gamma$. Setting

$$\mathbf{b}_{jk} = (1 + Q_j)\nabla^{\mu,\gamma} a_k,$$

a direct consequence of the lemma and of (A.2) is that $\sum_{j,k \in \mathbb{N}} \|\mathbf{b}_{jk}\|_2 < \infty$ for Γ close enough to Γ_0. It follows that for all $\varphi \in H^1_{0,surf}(\mathcal{S})$

$$\int_{\mathcal{S}} \nabla^{\mu,\gamma} b \cdot P(\Sigma)\nabla^{\mu,\gamma}\varphi = \sum_{k=0}^{\infty} \int_{\mathcal{S}} \sum_{j=0}^{k-1} \mathbf{b}_{j,k-j} \cdot \nabla^{\mu,\gamma}\varphi.$$

Now, by definition of the $\mathbf{b}_{jk}$ and because $(a_k)_{k \in \mathbb{N}}$ solves (A.4)–(A.5), one has

$$\forall k \geq 0, \qquad \int_{\mathcal{S}} \sum_{j=0}^{k} \mathbf{b}_{j,k-j} \cdot \nabla^{\mu,\gamma}\varphi = 0.$$

It follows that

$$\begin{aligned} \int_{\mathcal{S}} \nabla^{\mu,\gamma} b \cdot P(\Sigma)\nabla^{\mu,\gamma}\varphi &= -\sum_{k=0}^{\infty} \int_{\mathcal{S}} \mathbf{b}_{k,0} \cdot \nabla^{\mu,\gamma}\varphi \\ &= -\int P(\Sigma)\nabla^{\mu,\gamma} a_0 \cdot \nabla^{\mu,\gamma}\varphi \end{aligned}$$

and therefore that $a_0 + b$ is a variational solution to (2.12). By the uniqueness of the variational solution, we deduce that $a_0 + b = \phi_\Gamma$, and the result follows. □

As noted in Remark A.2, Proposition A.5 provides some information on the derivatives of the mapping $\mathfrak{A}_\psi$. The following proposition gives more precise information in the case when Σ is regularizing. In the statement below, we use Notation 2.10.

PROPOSITION A.7. *Let* $t_0 > d/2$ *and* $0 \leq s \leq t_0 + 1/2$, $\psi \in \dot{H}^{s+1/2}(\mathbb{R}^d)$, *and* $\Gamma = (\zeta, b) \in \mathbf{\Gamma}$. *Also let* Σ *be as shape analytic and regularizing. For all* $j \in \mathbb{N}$ *and* $(\mathbf{h}, \mathbf{k}) = (h_1, \dots h_j, k_1, \dots, k_j) \in H^{t_0+1}(\mathbb{R}^d)^{2j}$, *one has*

$$\|\Lambda^s \nabla^{\mu,\gamma} d^j \mathfrak{A}_\psi(\Gamma)(\mathbf{h}, \mathbf{k})\|_2 \leq \sqrt{\mu} M_0 \prod_{m=1}^{j} |(\varepsilon h_m, \beta k_m)|_{H^{t_0+1}} |\mathfrak{P}\psi|_{H^s}.$$

If $s = t_0 + 1/2$, *then the same estimate holds on* $\|\nabla^{\mu,\gamma} d^j \mathfrak{A}_\psi(\Gamma)(\mathbf{h}, \mathbf{k})\|_{H^{t_0+1/2,1}}$.

PROOF. For the sake of simplicity, we write $d^j \mathfrak{A}(\mathbf{h}, \mathbf{k}) = d^j \mathfrak{A}_\psi(\Gamma)(\mathbf{h}, \mathbf{k})$. Differentiating (2.12) with respect to Γ, we get that $d^j \mathfrak{A}_\psi(\Gamma)(\mathbf{h}, \mathbf{k}) = \psi^\mathfrak{h}$ for $j = 0$ and that for $j \geq 1$

$$\text{(A.7)} \qquad \begin{cases} \nabla^{\mu,\gamma} \cdot P \nabla^{\mu,\gamma} d^j \mathfrak{A}(\mathbf{h}, \mathbf{k}) = -\nabla^{\mu,\gamma} \cdot \mathbf{g}, \\ d^j \mathfrak{A}(\mathbf{h}, \mathbf{k})_{|_{z=0}} = 0, \quad \mathbf{e}_z \cdot P(\Sigma) \nabla^{\mu,\gamma} d^j \mathfrak{A}(\mathbf{h}, \mathbf{k}) = -\mathbf{e}_z \cdot \mathbf{g}_{|_{z=-1}} \end{cases}$$

with $P = P(\Sigma)$ and

$$\mathbf{g} = \sum d^{j_1} P(\mathbf{h}, \mathbf{k})_I \nabla^{\mu,\gamma} d^{j_2} \mathfrak{A}(\mathbf{h}, \mathbf{k})_{II},$$

the summation being taken over all integers j_1 and j_2 such that $j_1 + j_2 = j$ and $j_2 < j$, and on all the $2j_1$ and $2j_2$ uplets $(\mathbf{h}, \mathbf{k})_I$ and $(\mathbf{h}, \mathbf{k})_{II}$ whose coordinates form a permutation of the coordinates of $(\mathbf{h}, \mathbf{k})$. We therefore can invoke Lemma 2.38 to get

$$\text{(A.8)} \qquad \forall 0 \leq s \leq t_0 + 1/2, \qquad \|\Lambda^s \nabla^{\mu,\gamma} d^j \mathfrak{A}(\mathbf{h}, \mathbf{k})\|_2 \leq M_0 \|\Lambda^s \mathbf{g}\|_2.$$

Let us first consider the case $0 \leq s \leq t_0$. Using the first product estimate of Corollary B.5, we can bound $\|\Lambda^s \mathbf{g}\|_2$ from above by a sum of terms of the form

$$\begin{aligned} &\|d^{j_1} P(\mathbf{h}, \mathbf{k})_I\|_{L^\infty H^{t_0}} \|\Lambda^s \nabla^{\mu,\gamma} d^{j_2} \mathfrak{A}(\mathbf{h}, \mathbf{k})_{II}\|_2 \\ \leq\ & M_0 \prod_{j \in I} |(\varepsilon h_j, \beta k_j)|_{H^{t_0+1/2}} \|\Lambda^s \nabla^{\mu,\gamma} d^{j_2} \mathfrak{A}(\varepsilon \mathbf{h}, \beta \mathbf{k})_{II}\|_2. \end{aligned}$$

Since $j_2 < j$, the result therefore follows by induction for $0 \leq s \leq t_0$.
For $t_0 \leq s \leq t_0 + 1/2$, the modifications to be made are exactly the same as those made in the proof of Lemma A.6. □

In Proposition A.7, the control requires *at least* the $H^{t_0+1/2}$-norm of the components of $(\mathbf{h}, \mathbf{k})$, even if $s = 0$. In the following proposition, we show that it is possible to relax this constraint when Σ is as in Example 2.15 (the consequence is that the control now requires at least the $H^{t_0+1/2}$-norm of $\mathfrak{P}\psi$).

PROPOSITION A.8. *Let* $t_0 > d/2$, $\Gamma = (\zeta, b) \in \mathbf{\Gamma}$, $0 \leq s \leq t_0$, *and* $\psi \in \dot{H}^{t_0+1/2}(\mathbb{R}^d)$. *Also let* Σ *be as in Example* 2.15.
For all $j \in \mathbb{N}$ *and* $(\mathbf{h}, \mathbf{k}) = (h_1, \dots h_j, k_1, \dots, k_j) \in H^{t_0+1}(\mathbb{R}^d)^{2j}$, *one has*

$$\begin{aligned} \|\Lambda^s \nabla^{\mu,\gamma} d^j \mathfrak{A}_\psi(\Gamma)(\mathbf{h}, \mathbf{k})\|_2 \ &\leq\ \sqrt{\mu} M_0 |(\varepsilon h_1, \beta k_1)|_{H^{s+1/2}} \\ &\times \prod_{m>1} |(\varepsilon h_m, \beta k_m)|_{H^{t_0+1}} |\mathfrak{P}\psi|_{H^{t_0+1/2}}. \end{aligned}$$

REMARK A.9. In the statement of the proposition, the weak control is on the first coordinate of $(\mathbf{h}, \mathbf{k})$, but since the $d^j \mathfrak{A}_\psi(\Gamma)(\mathbf{h}, \mathbf{k})$ is symmetric in the (h_m, k_m), one could replace it by (h_m, k_m) for any $1 \leq m \leq j$.

PROOF. We use the same notations and approach as in the proof of Proposition A.7. The only difference is in the control of $\mathbf{g}$ in (A.8). We distinguish two cases:
- *If* (h_1, k_1) *belongs to* $(\mathbf{h}, \mathbf{k})_{II}$, then we proceed exactly as for Proposition A.7 and get the result by an induction argument.
- *If* (h_1, k_1) *belongs to* $(\mathbf{h}, \mathbf{k})_I$, then we use Corollary B.5 to control $\mathbf{g}$ as follows

$$\|\Lambda^s \mathbf{g}\|_2 \leq \sum \|\Lambda^s d^{j_1} P(\mathbf{h}, \mathbf{k})_I\|_2 \|\nabla^{\mu,\gamma} d^{j_2} \mathfrak{A}(\mathbf{h}, \mathbf{k})_{II}\|_{L^\infty H^{t_0}}.$$

Since $\|\cdot\|_{L^\infty H^{t_0}} \lesssim \|\cdot\|_{H^{t_0+1/2,1}}$, we can deduce from Proposition A.7 that

$$\|\Lambda^s \mathbf{g}\|_2 \leq \sqrt{\mu} M_0 \sum \|\Lambda^s d^{j_1} P(\mathbf{h}, \mathbf{k})_I\|_2 |(\varepsilon \mathbf{h}, \beta \mathbf{k})_{II}|_{H^{t_0+1}} |\mathfrak{P}\psi|_{H^{t_0+1/2}},$$

for all $0 \leq s \leq t_0$. We thus need the following lemma.

LEMMA A.10. *Let* $j \in \mathbb{N}^*$, $0 \leq s \leq t_0$, *and* $(\mathbf{h}, \mathbf{k}) \in H^{t_0+1}(\mathbb{R}^d)^{2j}$; *one has*

$$\|\Lambda^s d^j Q(\mathbf{h}, \mathbf{k})\|_2 \leq M_0 |(\varepsilon h_1, \beta k_1)|_{H^{s+1/2}} \prod_{m \geq 2} |(\varepsilon h_m, \beta k_m)|_{H^{t_0+1}}.$$

PROOF OF LEMMA A.10. Recall that an explicit expression of $Q(\Sigma)$ is given in Example 2.21. Since $\zeta \mapsto \sigma$ is linear (with $\Sigma(X, z) = (X, z + \sigma(X, z))$), it is easy to deduce the result from Corollaries B.5 and B.6 and the regularizing properties of σ. □

Thanks to this lemma and the above estimate on $\|\Lambda^s \mathbf{g}\|_2$, the proposition is proved. □

A.1.2. Shape analyticity of the Dirichlet-Neumann operator. The analyticity of the Dirichlet-Neumann operator is proved with full details in the case of finite depth. We then indicate how to adapt this result to the case of infinite depth.

A.1.2.1. *The case of finite depth.* This section is devoted to the proof of the first point of Theorem 3.21 that we recall here for convenience. As in Definition A.3, we denote by $\mathbf{\Gamma}$ the set of all $(\zeta, b) \in H^{t_0+1}(\mathbb{R}^d)$ such that (3.2) is satisfied. Let $0 \leq s \leq t_0 + 1/2$ ($t_0 > d/2$). Given $\psi \in \dot{H}^{s+1/2}(\mathbb{R}^d)$, we are interested here in the mapping

$$\mathcal{G}_{\mu,\gamma}[\varepsilon\cdot, \beta\cdot] : \begin{array}{ccc} \mathbf{\Gamma} \subset H^{t_0+1}(\mathbb{R}^d)^2 & \to & H^{s-1/2}(\mathbb{R}^d), \\ \Gamma = (\zeta, b) & \mapsto & \mathcal{G}_{\mu,\gamma}[\varepsilon\zeta, \beta b]\psi. \end{array} \tag{A.9}$$

As in Notation 3.20, we denote by $d^j \mathcal{G}_{\mu,\gamma}[\varepsilon\zeta, \beta b](\mathbf{h}, \mathbf{k})\psi$, or simply $d^j \mathcal{G}(\mathbf{h}, \mathbf{k})\psi$ the j-th order derivative of (A.9) at (ζ, b) in the direction $(\mathbf{h}, \mathbf{k})$ and if the bottom is kept fixed, we denote by $d^j \mathcal{G}_{\mu,\gamma}[\varepsilon\zeta, \beta b](\mathbf{h})\psi$, or simply $d^j \mathcal{G}(\mathbf{h})\psi$ the j-th order partial derivative of (A.9) at ζ in the direction $\mathbf{h}$.

The shape analyticity property we prove here is the following.

THEOREM A.11. *Let* $t_0 > d/2$. *Then for all* $0 \leq s \leq t_0 + 1/2$ *and* $\psi \in \dot{H}^{s+1/2}(\mathbb{R}^d)$, *the mapping* (A.9) *is analytic in the sense of Definition* A.1.

REMARK A.12. Theorem A.11 shows analyticity in $\mathbf{\Gamma}$ and not only for small perturbations of the flat surface (and of a flat bottom). Calderon [**61**] and Coifmann-Meyer [**83**] proved that when $d = 1$, the Dirichlet-Neumann is shape analytic as a mapping $H^1(\mathbb{R}) \to L^2(\mathbb{R})$ for small Lipschitz perturbations of the flat surface; this result has been extended by Craig-Schanz-Sulem [**112**] (see also [**109**]) for $d = 2$ under the slightly stronger assumption of small C^1 perturbations of the surface;

Hu and Nicholls gave in [**176**] a simpler proof of the analyticity of the Dirichlet-Neumann as an operator mapping $C^{1+\alpha}$ into C^α for small $C^{1+\alpha}$ perturbations of the flat surface (see also [**177**] for improvements of this result).

PROOF. *Throughout this proof, Σ is assumed to be the regularizing diffeomorphism of Example* 2.15.
For all $\Gamma = (\zeta, b) \in \mathbf{\Gamma}$, let us write $\mathfrak{G}(\Gamma) = \mathcal{G}_{\mu,\gamma}[\varepsilon\zeta, \beta b]\psi$. We recall that in order to prove analyticity of (A.9), we have to check that for all $\Gamma_0 \in \mathbf{\Gamma}$, one can write for all Γ in a neighborhood of Γ_0

$$\mathfrak{G}(\Gamma) = \sum_{k=0}^{\infty} \mathfrak{G}_k(\Gamma - \Gamma_0),$$

where the mappings $\mathfrak{G}_k : X^k \to Y$ (with $X = H^{t_0+1}(\mathbb{R}^d)^2$, $Y = H^{s+1/2}(\mathbb{R}^d)$) are bounded, symmetric, and k-linear and moreover satisfy

$$\forall k \in \mathbb{N}, \qquad \|\mathfrak{G}_k\|_{X^k \to Y} \leq N r^{-k}, \tag{A.10}$$

for some $r > 0$ and $N > 0$.
If such a result is true then, by identifying the terms of same degree in $(\Gamma - \Gamma_0)$, one deduces from (3.12) that

$$\big(\Lambda^{s-1/2} g_k, \varphi\big) = \sum_{j=0}^{k} \int_{\mathcal{S}} \Lambda^s \big(P_j \nabla^{\mu,\gamma} a_{k-j}\big) \Lambda^{-1/2} \nabla^{\mu,\gamma} \varphi^\dagger, \tag{A.11}$$

where we used the notations (with $\mathfrak{Q}$ and $\mathfrak{A}_\psi$ as in Definition A.3 and (A.1)),

$$g_j = \mathfrak{G}_j(\Gamma - \Gamma_0)^j, \quad Q_j = \mathfrak{Q}_j(\Gamma - \Gamma_0)^j, \qquad a_j = \mathfrak{A}_\psi(\Gamma - \Gamma_0)^j,$$

and $P_j = Q_j$ if $j \geq 1$, $P_0 = I + Q_0$. Conversely, (A.11) can be used to define $\mathfrak{G}_k$. We are thus left to prove (A.10), or equivalently

$$\forall k \in \mathbb{N}, \qquad |g_k|_{H^{s-1/2}} \leq N r^{-k} |\Gamma - \Gamma_0|_{H^{t_0+1}}^k. \tag{A.12}$$

Proceeding with the same duality argument as in the proof of Theorem 3.15, we get from (A.11) that for all $0 \leq s \leq t_0 + 1/2$

$$|g_k|_{H^{s-1/2}} \lesssim \mu^{1/4} \sum_{j=0}^{k} \|Q_j \nabla^{\mu,\gamma} a_{k-j}\|_{H^{s,0}}.$$

We therefore deduce that

$$\forall 0 \leq s \leq t_0, \qquad |g_k|_{H^{s-1/2}} \lesssim \mu^{1/4} \sum_{j=0}^{k} \|Q_j\|_{L^\infty H^{t_0}} \|\Lambda^s \nabla^{\mu,\gamma} a_{k-j}\|_2.$$

Since $\mathfrak{Q}$ is shape analytic and regularizing in the sense of Definition A.3, and owing to Lemma A.6, we know that there exist two constants $N_1 > 0$ and $r_1 > 0$ such that

$$\forall j \in \mathbb{N}, \qquad \|Q_j\|_{H^{t_0+1/2,1}} + \|\nabla^{\mu,\gamma} a_j\|_{H^{s,0}} \leq N_1 r_1^{-j} |\Gamma - \Gamma_0|_{H^{t_0+1}}^j.$$

Therefore, one has $|g_k|_{H^{s-1/2}} \lesssim \mu^{1/4} N_1^2 k r_1^{-k} |\Gamma - \Gamma_0|_{H^{t_0+1}}^k$, which implies (A.12) for some $0 < r < r_1$. The result is therefore proved for $0 \leq s \leq t_0$. To extend it to $t_0 \leq s \leq t_0 + 1/2$, we proceed exactly as in the proof of Lemma A.6. $\square$

A.1.2.2. *The case of infinite depth.* In infinite depth, the mapping (A.9) must be replaced, for all $0 \leq s \leq t_0 + 1/2$ ($t_0 > d/2$) and $\psi \in H^{s+1/2}(\mathbb{R}^d)$, by

$$\text{(A.13)} \qquad \mathcal{G}[\epsilon\cdot] : \begin{array}{ccc} H^{t_0+1}(\mathbb{R}^d)^2 & \to & H^{s-1/2}(\mathbb{R}^d), \\ \zeta & \mapsto & \mathcal{G}[\epsilon\zeta]\psi, \end{array}$$

where the Dirichlet-Neumann operator in infinite depth $\mathcal{G}[\epsilon\zeta]$ is as defined in §3.7.3. We denote by $d^j\mathcal{G}[\epsilon\zeta](\mathbf{h})\psi$, or simply $d^j\mathcal{G}(\mathbf{h})\psi$ the j-th order derivative of (A.13) at ζ in the direction $\mathbf{h} \in H^{t_0+1}(\mathbb{R}^d)^j$.
Theorem A.11 can then be adapted as follows.

THEOREM A.13. *Let $t_0 > d/2$. Then for all $0 \leq s \leq t_0 + 1/2$ and $\psi \in H^{s+1/2}(\mathbb{R}^d)$, the mapping (*A.13*) is analytic in the sense of Definition* A.1.

PROOF. The proof follows exactly the same lines as the proof of Theorem A.11. Indeed, Lemma A.6 is the key element in that proof, and it can be easily adapted to the infinite depth case by using Corollary 2.49 instead of Corollary 2.40 for the elliptic regularity estimates. □

A.2. Self-Adjointness of the Dirichlet-Neumann operator

An interesting corollary to Proposition 3.32 is that the Dirichlet-Neumann operator is not only symmetric for the L^2-scalar product, but also essentially self-adjoint. This result can be found in [**277**] with a proof using symbolic analysis of the Dirichlet-Neumann operator (see §3.7 in Chapter 3); we give an elementary proof that relies on Proposition 3.32.

PROPOSITION A.14. *Let $t_0 > d/2$ and $\zeta, b \in H^{t_0+1}(\mathbb{R}^d)$ be such that (*3.2*) is satisfied. Then $\mathcal{G}$ has a self-adjoint realization on $L^2(\mathbb{R}^d)$ with domain $H^1(\mathbb{R}^d)$.*

PROOF. We already know from Proposition 3.9 and Remark 3.10 that $\mathcal{G}$ is symmetric on $H^{3/2}$ for the L^2-scalar product. In order to prove the proposition, we need to consider the closure $\overline{\mathcal{G}}$ of $\mathcal{G}$ with domain

$$D(\overline{\mathcal{G}}) = \{\psi \in L^2, \exists(\psi_n)_n, \psi_n \in H^\infty, \psi_n \to \psi \in L^2, \mathcal{G}\psi_n \text{ converges in } L^2\}.$$

What we need to prove is that $\overline{\mathcal{G}}$ is self-adjoint, and that $D(\overline{\mathcal{G}}) = H^1$. Recalling that the adjoint $\overline{\mathcal{G}}^*$ has domain

$$D(\overline{\mathcal{G}}^*) = \{\psi \in L^2, \quad \exists C > 0, \quad \forall \psi' \in D(\overline{\mathcal{G}}), \quad |(\psi, \mathcal{G}\psi')| \leq C|\psi'|_2\},$$

the result follows from the following series of inclusions

$$H^1(\mathbb{R}^d) \subset D(\overline{\mathcal{G}}) \subset D(\overline{\mathcal{G}}^*) \subset H^1(\mathbb{R}^d),$$

which we will now prove. The first inclusion is a direct consequence of the density of H^∞ in H^1 and of the continuity of $\mathcal{G}$ as a mapping from H^1 to L^2 (see Theorem 3.15). The second inclusion is a consequence of the aforementioned symmetry of $\mathcal{G}$. For the third inclusion, we need the following elliptic regularity result.

LEMMA A.15. *Let $t_0 > d/2$ and $\zeta, b \in H^{t_0+1}(\mathbb{R}^d)$ be such that (*3.2*) is satisfied. Also let $\psi \in L^2(\mathbb{R}^d)$ such that $\mathcal{G}\psi \in L^2(\mathbb{R}^d)$. Then $\psi \in H^1(\mathbb{R}^d)$.*

PROOF OF THE LEMMA. A direct consequence of Proposition 3.12 is that under the assumptions of the lemma, one has $|\mathfrak{P}\psi|_2 < \infty$. From the definition of $\mathfrak{P}$, this implies that ψ, which is assumed to be in $L^2(\mathbb{R}^d)$, also satisfies $\nabla\psi \in H^{-1/2}(\mathbb{R}^d)^d$. This is enough to conclude that $\psi \in H^{1/2}(\mathbb{R}^d)$.

In order to prove that it belongs to $H^1(\mathbb{R}^d)$, let $(u_n)_{n\in\mathbb{N}}$ be a sequence of functions in $H^\infty(\mathbb{R}^d)$ converging to u in $H^{1/2}(\mathbb{R}^d)$, and with $(\mathcal{G}u_n)_n$ bounded in $L^2(\mathbb{R}^d)$. Now, we can use Proposition 3.12 to get

$$\begin{aligned} |\mathfrak{P}\Lambda^{1/2}u_n|_2^2 &\leq M(\Lambda^{1/2}u_n, \mathcal{G}\Lambda^{1/2}u_n) \\ &= M\Big((\Lambda u_n, \mathcal{G}u_n) + (\Lambda^{1/2}u_n, [\Lambda^{1/2}, \mathcal{G}]u)\Big). \end{aligned}$$

Using the Cauchy-Schwarz inequality for the first component of the right-hand side, and Proposition 3.32 for the second one, we get

$$|\mathfrak{P}u_n|_{H^{1/2}}^2 \leq M\big(|u_n|_{H^1}|\mathcal{G}u_n|_2 + |\Lambda^{1/2}u_n|_2|\mathfrak{P}u_n|_{H^{1/2}}\big).$$

Since $|u_n|_{H^1} \leq |u_n|_2 + |\mathfrak{P}u_n|_{H^{1/2}}$, we deduce from the fact that $(\Lambda^{1/2}u_n)_n$ and $(\mathcal{G}u_n)_n$ are both bounded in L^2 that there exist two constants A and B independent of n such that $|\mathfrak{P}u_n|_{H^{1/2}}^2 \leq A|\mathfrak{P}u_n|_{H^{1/2}} + B$. This readily implies that $(\mathfrak{P}u_n)_n$ remains bounded in $H^{1/2}(\mathbb{R}^d)$ and therefore that $\mathfrak{P}u \in H^{1/2}(\mathbb{R}^d)$. Since we already know that $u \in H^{1/2}$, we conclude that $u \in H^1$. □

Using this lemma, the proof is now complete. □

A.3. Invertibility of the Dirichlet-Neumann operator

It follows from Proposition 3.12 that $\mathcal{G} : \dot{H}^{1/2} \to H^{-1/2}$ is one-to-one. It is however obviously not onto. This can for instance be observed in the case of a flat bottom and surface, where we know that $\mathcal{G} = \sqrt{\mu}|D^\gamma|\tanh(\sqrt{\mu}|D^\gamma|)$.
If $\varphi \in \mathcal{S}(\mathbb{R}^d)$ belongs to the range of $\mathcal{G}$, then

$$\Lambda^{-1/2}\frac{1}{\sqrt{\mu}|D^\gamma|\tanh(\sqrt{\mu}|D^\gamma|)}\nabla^\gamma\varphi \in L^2(\mathbb{R}^d),$$

and therefore

$$\int_{\mathbb{R}^d}\frac{1}{1+|\xi|^2}\frac{1}{\mu\tanh(\sqrt{\mu}|\xi|)^2}|\widehat{\varphi}(\xi)|^2 d\xi < \infty.$$

Quite obviously, the singularity at the origin makes this condition impossible to hold if $\widehat{\varphi}$ does not vanish at the origin.

What we show here is that the range of a linear operator $A : X \mapsto H^{s-1/2}(\mathbb{R}^d)$ (X a topological vector space) belongs to the range of $\mathcal{G}$ if A satisfies the following property

$$\text{(A.14)} \qquad \forall 0 \leq s \leq t_0 + 1/2, \qquad (\Lambda^s A\psi_1, \Lambda^s\psi_2) \leq \mu M_A(\psi_1)|\mathfrak{P}\psi_2|_{H^s},$$

for all $\psi_1 \in X$, $\psi_2 \in \dot{H}^{s+1/2}(\mathbb{R}^d)$ and for some positive constant $M_A(\psi_1)$.

EXAMPLE A.16.
(1) The Dirichlet-Neumann operator $\mathcal{G}$ satisfies this property with $X = \dot{H}^{s+1/2}(\mathbb{R}^d)$ and $M_A(\psi) = |\mathfrak{P}\psi|_2$.
(2) The Fourier multiplier $|D^\gamma|$ also satisfies this property with $X = H^{s+1/2}(\mathbb{R}^d)$ and $M_{|D^\gamma|}(\psi) = \frac{1}{\mu}|(1+\sqrt{\mu}D^\gamma)^{1/2}\psi|_2$. The partial derivative operators ∂_x and $\gamma\partial_y$ also satisfy the property with the same constant.

PROPOSITION A.17. *Let $t_0 > d/2$ and $\zeta \in H^{t_0+1}(\mathbb{R}^d)$ be such that (2.6) is satisfied. Also let A be a linear operator satisfying (A.14).*
Then all $0 \leq s \leq t_0 + 1/2$, the mapping

$$\mathcal{G}^{-1}\circ A : \begin{array}{rcl} X & \to & \dot{H}^{s+1/2}(\mathbb{R}^d) \\ \psi & \mapsto & \mathcal{G}^{-1}(A\psi) \end{array}$$

is well defined and one has, for all $\psi \in X$,

$$|\mathfrak{P}\mathcal{G}^{-1} \circ A\psi|_{H^s} \leq M_0 M_A(\psi).$$

REMARK A.18. We deduce from Example A.16 and the proposition that $\mathcal{G}^{-1} \circ |D^\gamma|$ is well defined on $H^{s+1/2}(\mathbb{R}^d)$ with values in $\dot{H}^{s+1/2}(\mathbb{R}^d)$ and that

$$\forall 0 \leq s \leq t_0 + 1/2, \qquad |\mathfrak{P}\mathcal{G}^{-1} \circ |D^\gamma|\psi|_{H^s} \leq \frac{1}{\mu} M_0 |(1+\sqrt{\mu}|D|)^{1/2}\psi|_{H^s}.$$

PROOF. Let us prove that $(\mathcal{G})^{-1}A\psi$ is well defined in $\dot{H}^{1/2}(\mathbb{R}^d)$. We first prove that there exists a unique variational solution $\phi \in \dot{H}^1(\mathcal{S})$ to

$$\text{(A.15)} \qquad \begin{cases} \nabla^{\mu,\gamma} \cdot P(\Sigma)\nabla^{\mu,\gamma}\phi = 0 \quad \text{in} \quad \mathcal{S}, \\ \partial_{\mathbf{n}}\phi_{|_{z=0}} = A\psi, \qquad \partial_{\mathbf{n}}\phi^-{}_{|_{z=-1}} = 0, \end{cases}$$

where Σ is a regularizing diffeomorphism in the sense of Definition 2.17.
For all $\varphi \in C^\infty(\overline{\mathcal{S}^-}) \cap \dot{H}^1(\mathcal{S}^-)$, one deduces from (A.14) that

$$\begin{aligned} (A\psi, \varphi_{|_{z=0}}) &\leq \mu M_A(\psi)|\mathfrak{P}\varphi_{|_{z=0}}|_2 \\ &\leq \mu M_0 M_A(\psi)\|\nabla^{\mu,\gamma}\varphi\|_2, \end{aligned}$$

where we used Remark 3.13 to derive the second inequality. The linear form $\varphi \mapsto (\mathcal{G}\psi, \varphi_{|_{z=0}})$ is thus continuous on $\dot{H}^1(\mathcal{S})$; moreover, $\nabla^{\mu,\gamma} \cdot P(\Sigma)\nabla^{\mu,\gamma}$ is obviously coercive on this space, so that the existence/uniqueness of a variational solution to (A.15) follows classically from the Lax-Milgram theorem.
We can therefore define $\widetilde{\psi} = \phi_{|_{z=0}}$ so that one obviously has $\mathcal{G}\widetilde{\psi} = A\psi$ and $\widetilde{\psi} = \mathcal{G}^{-1} \circ A\psi$. Since such a solution is obviously unique, the operator $\mathcal{G}^{-1} \circ A$ is well defined.

We will now prove the estimates given in the statement of the proposition. Noting that $|\mathfrak{P}\mathcal{G}^{-1} \circ A\psi|_2 = |\mathfrak{P}\widetilde{\psi}|_2$, with $\widetilde{\psi} = \phi_{|_{z=0}}$ as constructed above, we deduce from Proposition 3.19 that

$$\text{(A.16)} \qquad \sqrt{\mu}|\mathfrak{P}\mathcal{G}^{-1} \circ A\psi|_2 \leq M_0 \|\Lambda^s \nabla^{\mu,\gamma}\phi\|_2.$$

In order to get an estimate on the r.h.s. of this inequality, let us apply Λ^s to the l.h.s. of (A.15), multiply it by $\Lambda^s\phi$ and integrate by parts, to get

$$\int_{\mathcal{S}} P\Lambda^s\nabla^{\mu,\gamma}\phi \cdot \Lambda^s\nabla^{\mu,\gamma}\phi = -\int_{\mathbb{R}^d} \Lambda^s(A\psi)\Lambda^s\widetilde{\psi} + \int_{\mathcal{S}} [\Lambda^s, Q]\nabla^{\mu,\gamma}\phi \cdot \nabla^{\mu,\gamma}\Lambda^s\phi.$$

We can therefore deduce from the coercivity of $P(\Sigma)$, (A.14) and the Cauchy-Schwarz inequality that

$$k(\Sigma)\|\Lambda^s\nabla^{\mu,\gamma}\phi\|_2^2 \leq \mu M_A(\psi)|\mathfrak{P}\widetilde{\psi}|_{H^s} + \|[\Lambda^s, Q]\nabla^{\mu,\gamma}\phi\|_2\|\Lambda^s\nabla^{\mu,\gamma}\phi\|_2.$$

Since $\sqrt{\mu}|\mathfrak{P}\widetilde{\psi}|_{H^s} \leq M_0\|\Lambda^s\nabla^{\mu,\gamma}\phi\|_2$ by Proposition 3.19, one can deduce that

$$\|\Lambda^s\nabla^{\mu,\gamma}\phi\|_2 \leq \sqrt{\mu}M_0 M_A(\psi) + M_0\|[\Lambda^s, P(\Sigma)]\nabla^{\mu,\gamma}\phi\|_2.$$

Proceeding as for (2.17), we get

$$\forall 0 \leq s \leq t_0 + 1/2, \qquad \|\Lambda^s\nabla^{\mu,\gamma}\phi\|_2 \leq \sqrt{\mu}M_0 M_A(\psi),$$

from which the result follows. □

A.4. Remarks on the symbolic analysis of the Dirichlet-Neumann operator

An aspect that we do not treat in these notes because we do not need it is the *symbolic analysis* of the Dirichlet-Neumann operator. It is known [**310, 316, 218, 297**] that the Dirichlet-Neumann operator (in its version with dimensions (1.2)) is related to the square root of the Laplace Beltrami operator associated to the surface through the formula

$$G[\zeta, b] - g_1(X, D) = R_0, \tag{A.17}$$

where R_0 is an operator of order zero and

$$g_1(x,\xi) = \sqrt{|\xi|^2 + (|\nabla\zeta|^2|\xi|^2 - (\nabla\zeta\cdot\xi)^2)},$$

and where the pseudo-differential operator $\sigma(x, D)$ associated to a symbol $\sigma(x,\xi)$ is defined as

$$\sigma(x,D)u = (2\pi)^{-d}\int_{\mathbb{R}^d} e^{ix\cdot\xi}\sigma(x,\xi)\widehat{u}(\xi)d\xi.$$

A refinement of this identity is a paralinearization formula derived in [**9, 4**], which may also be used to take into account the subprincipal terms in the symbolic analysis of the Dirichlet-Neumann operator. Let us for instance state this formula, including the first subprincipal term (this is used in [**4**] to handle surface tension). Defining

$$g_0(x,\xi) = \frac{1+|\nabla\zeta|^2}{2g_1}\big\{\nabla\cdot(\alpha_1\nabla\zeta) - i\partial_\xi g_1\cdot\nabla\alpha_1\big\},$$

with

$$\alpha_1 = \frac{1}{1+|\nabla\zeta|^2}(g_1 + i\nabla\zeta\cdot\xi),$$

the paralinearization formula reads

$$G[\zeta,b]\psi = T_{g_1+g_0}(\psi - T_{\underline{w}}\zeta) - T_{\underline{V}}\cdot\nabla\zeta + f(\zeta,\psi),$$

where T_σ denotes the paradifferential operator of symbol σ, $\underline{w}$ and $\underline{V}$ are as in Theorem 3.21 (with ε, μ and γ set to one), and $f(\zeta,\psi)$ satisfies the estimate

$$|f(\zeta,\psi)|_{H^{s+1/2}} \le C(|\zeta|_{H^{s+1/2}})|\nabla\psi|_{H^{s-1}}.$$

It is worth emphasizing the fact that neither g_0 nor g_1 involves the bottom parametrization b. This is because the contribution of the bottom to the Dirichlet-Neumann operator is analytic by standard elliptic theory and therefore transparent to any homogeneous symbolic expansion (at any order). In the shallow water limit, where the role of the bottom topography is crucial, this is problematic. In dimensionless variables, this difficulty appears as a singularity as $\mu \to 0$. For instance, formula (A.17) can be written in dimensionless variables as

$$\frac{1}{\sqrt{\mu}}\mathcal{G} - g_1^\mu(x,D) = \frac{1}{\sqrt{\mu}}\widetilde{R}_0^\mu, \tag{A.18}$$

where

$$g_1^\mu(x,\xi) = \sqrt{|\xi|^2 + \varepsilon^2\mu(|\nabla\zeta|^2|\xi|^2 - (\nabla\zeta\cdot\xi)^2)},$$

where $\widetilde{R}_0^\mu$ is an operator of order zero, with operator norm uniformly bounded with respect to $\mu \in (0,1)$. Using (homogeneous) symbolic analysis to describe the Dirichlet-Neumann operator therefore induces a singularity of order $O(1/\sqrt{\mu})$ as $\mu \to 0$, and this is why we avoided it in these notes.

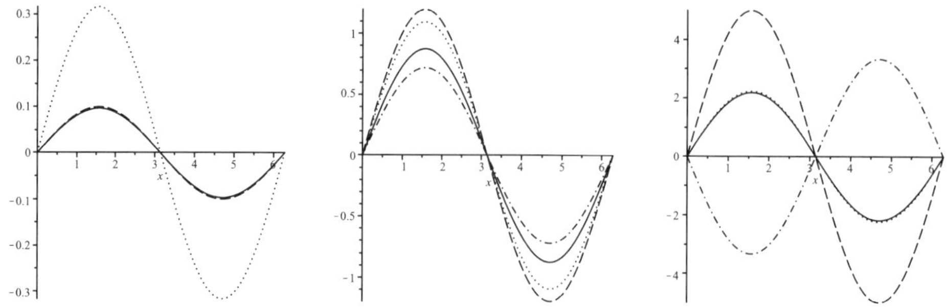

FIGURE A.1. Comparison of $\mathcal{G}_{\mu,\gamma}[0,0]\psi$ with $\psi = \sin x$ (solid line) with various approximations: first- and second- order shallow water approximations (dash and dashdot respectively), and symbolic approximation (dots) for $\mu = 0.1$, $\mu = 1.2$ and $\mu = 5$. See Example A.19.

EXAMPLE A.19. One can compute explicitly $\mathcal{G}_{\mu,\gamma}[0,0]\sin = \sqrt{\mu}\tanh(\sqrt{\mu})\sin$; the symbolic approximation (A.18) and the shallow water expansions of order 1 and 2 (see Remark 3.39) yield respectively

$$\mathcal{G}_{\mu,\gamma}[0,0]\sin \sim \sqrt{\mu}\sin, \qquad \mathcal{G}_{\mu,\gamma}[0,0]\sin \sim \mu\sin; \qquad \mathcal{G}_{\mu,\gamma}[0,0]\sin \sim (\mu - \frac{\mu^2}{3})\sin.$$

The accuracy of these approximations depends strongly on μ (and therefore on the depth). (See Figure A.1.)

Another alternative would have been to use a nonhomogeneous symbolic analysis of the Dirichlet-Neumann operator, including the infinitely smoothing contribution of the bottom

$$\frac{1}{\sqrt{\mu}}\mathcal{G} - g^\mu(x,D) = \varepsilon\sqrt{\mu}\widetilde{R}^\mu, \tag{A.19}$$

where the symbol $g^\mu(x,\xi)$ is given by

$$g^\mu(x,\xi) = g_1^\mu \tanh\Big(\sqrt{\mu}(1+\varepsilon\zeta)\int_{-1}^0 \frac{|\xi|^2 + \varepsilon^2\mu(z+1)^2(|\nabla\zeta|^2|\xi|^2 - (\nabla\zeta\cdot\xi)^2)}{1+\varepsilon^2\mu(z+1)^2|\nabla\zeta|^2}dz\Big).$$

This formula has been derived in [**221**] to handle the shallow-water limit in the two-fluid problem. Contrary to the standard formula (A.18), the symbolic approximation furnished by (A.19) is not singular as $\mu \to 0$. However, its use induces technical difficulties, and this is why we preferred the much simpler approach used here and based on [**12**] (see also [**181**]), which does not require symbolic analysis.

A.5. Related operators

A.5.1. The Laplace equation with nonhomogeneous Neumann condition at the bottom. Let Φ solve the potential equation with *nonhomogeneous* Neumann condition at the bottom

$$\begin{cases} \Delta^{\mu,\gamma}\Phi = 0, & -1+\beta b < z < \varepsilon\zeta, \\ \Phi_{|_{z=\varepsilon\zeta}} = \psi, & \sqrt{1+\beta^2|\nabla b|^2}\partial_{\mathbf{n}}\Phi_{|_{z=-1+\beta b}} = B. \end{cases} \tag{A.20}$$

(We have seen in §1.6 of Chapter 1 that this problem is relevant for the analysis of the water waves problem with moving bottom.) As in Chapter 2, it is possible to use an admissible diffeomorphism Σ from the flat strip $\mathcal{S}$ onto Ω to transform (A.20) into the following elliptic boundary value problem on $\mathcal{S}$, with $\phi = \Phi \circ \Sigma$ and $P(\Sigma)$ as in (2.13)

$$\left\{ \begin{array}{ll} \nabla^{\mu,\gamma} \cdot P(\Sigma)\nabla^{\mu,\gamma}\phi = 0, & -1 < z < 0, \\ \phi_{|_{z=0}} = \psi, \qquad \partial_{\mathbf{n}}\phi_{|_{z=-1}} = B, & \end{array} \right. \tag{A.21}$$

and where the conormal derivative at 0 or -1 is now $\partial_{\mathbf{n}} = \mathbf{e}_z \cdot P(\Sigma)\nabla^{\mu,\gamma}\phi_{|_{z=0,-1}}$.

We can generalize Definition 2.29 of variational solutions to the present case of a nonhomogeneous Neumann condition at the bottom. We thus say that ϕ is a variational solution of (A.21) if there exists $\widetilde{\phi} \in H^1_{0,surf}(\mathcal{S})$ such that $\phi = \widetilde{\phi} + \psi^\dagger$ and

$$\int_{\mathcal{S}} \nabla^{\mu,\gamma}\widetilde{\phi} \cdot P(\Sigma)\nabla^{\mu,\gamma}\varphi = -\int_{\mathcal{S}} \nabla^{\mu,\gamma}\psi^\dagger \cdot P(\Sigma)\nabla^{\mu,\gamma}\varphi - \int_{\mathbb{R}^d} B \cdot \varphi_{|_{z=-1}},$$

for all $\varphi \in H^1_{0,surf}(\mathcal{S})$. Proposition 2.31 can then be generalized in the following way.

PROPOSITION A.20. *Let Σ be an admissible diffeomorphism, $\psi \in \dot{H}^{1/2}(\mathbb{R}^d)$ and $B \in H^{-1/2}(\mathbb{R}^d)$. Then there exists a unique variational solution ϕ to* (A.21)*. Moreover, if*

$$\frac{1}{k(\sigma)} + \|P(\Sigma)\| \leq M_0,$$

with $k(\Sigma)$ and M_0 as in Lemma 2.26 *and* (2.10)*, then*

$$\|\nabla^{\mu,\gamma}\phi\|_2 \leq M_0\big(\sqrt{\mu}|\mathfrak{P}\psi|_2 + |\frac{1}{(1+\sqrt{\mu}|D^\gamma|)^{1/2}}B|_2\big).$$

PROOF. As for Proposition 2.31, the result follows from the Lax-Milgram theorem. The only difference is that we need to show that the linear form $\varphi \in H^1_{0,surf}(\mathcal{S}) \mapsto \int_{\mathbb{R}^d} B \cdot \varphi_{|_{z=-1}}$ is continuous. This is a consequence of the following inequality, which, together with Proposition 2.31 and Lemma 2.34, also implies the estimate of the statement

$$\Big|\int_{\mathbb{R}^d} B \cdot \varphi_{|_{z=-1}}\Big| \lesssim |\frac{1}{(1+\sqrt{\mu}|D^\gamma|)^{1/2}}B|_2 \|\nabla^{\mu,\gamma}\varphi\|_2.$$

In order to prove this inequality, we just have to note that

$$\Big|\int_{\mathbb{R}^d} B \cdot \varphi_{|_{z=-1}}\Big| \lesssim |\frac{1}{(1+\sqrt{\mu}|D^\gamma|)^{1/2}}B|_2 |(1+\sqrt{\mu}|D^\gamma|)^{1/2}\varphi_{|_{z=-1}}|_2$$

and the trace formula

$$\forall \varphi \in H^1_{0,surf}, \qquad |(1+\sqrt{\mu}|D^\gamma|)^{1/2}\varphi_{|_{z=-1}}|_2 \lesssim \|\nabla^{\mu,\gamma}\varphi\|_2,$$

which can be derived with slight modifications to the proof of Proposition 2.12. □

REMARK A.21. The higher order estimates of Proposition 2.36 can easily be generalized to the present case.

A.5.2. The Neumann-Neumann, Dirichlet-Dirichlet, and Neumann-Dirichlet operators. Following Iguchi [**182**], we define the *Dirichlet-Neumann*, *Neumann-Neumann*, *Dirichlet-Dirichlet*, and *Neumann-Dirichlet* operators (denoted respectively by $\mathcal{G} = \mathcal{G}_{\mu,\gamma}[\varepsilon\zeta, \beta b]$, $\mathcal{G}^{NN} = \mathcal{G}^{NN}_{\mu,\gamma}[\varepsilon\zeta, \beta b]$, $\mathcal{G}^{DD} = \mathcal{G}^{DD}_{\mu,\gamma}[\varepsilon\zeta, \beta b]$, $\mathcal{G}^{ND} = \mathcal{G}^{ND}_{\mu,\gamma}[\varepsilon\zeta, \beta b]$) by the relations

$$\begin{aligned} \mathcal{G}\psi + \mathcal{G}^{NN} B &= \sqrt{1+\varepsilon^2|\nabla\zeta|^2}\partial_{\mathbf{n}}\Phi_{|_{z=\varepsilon\zeta}}, \\ \mathcal{G}^{DD}\psi + \mathcal{G}^{ND} B &= \psi. \end{aligned}$$

(We recall that $\partial_{\mathbf{n}}$ stands for the *upward* conormal derivative, $\partial_{\mathbf{n}} = \partial_z - \varepsilon\mu\nabla^\gamma\zeta\cdot\nabla^\gamma$ at the surface, and $\partial_{\mathbf{n}} = \partial_z - \beta\mu\nabla^\gamma b\cdot\nabla^\gamma$ at the bottom.)

REMARK A.22. These definitions of course coincide with the definitions of the Dirichlet-Neumann and Neumann operators given in (1.29) and (1.42) respectively.

REMARK A.23. The operators $\mathcal{G}$, $\mathcal{G}^{NN}$, $\mathcal{G}^{DD}$, $\mathcal{G}^{ND}$ can be alternatively defined as

$$\begin{aligned} \mathcal{G}\psi + \mathcal{G}^{NN} B &= \mathbf{e}_z\cdot P(\Sigma)\nabla^{\mu,\gamma}\phi_{|_{z=0}}, \\ \mathcal{G}^{DD}\psi + \mathcal{G}^{ND} B &= \psi. \end{aligned}$$

Most of the properties of the Dirichlet-Neumann operator $\mathcal{G}$ proved in Chapter 3 can be adapted to $\mathcal{G}^{NN}$, $\mathcal{G}^{DD}$ and $\mathcal{G}^{ND}$. For instance, one easily checks that $\mathcal{G}^{ND}$ is symmetric for the L^2 scalar product, while the adjoint of $\mathcal{G}^{NN}$ is $-\mathcal{G}^{DD}$

$$\begin{aligned} \forall\psi_1,\psi_2\in\mathfrak{S}(\mathbb{R}^d), &\qquad (\mathcal{G}^{ND}\psi_1,\psi_2) = (\psi_1, \mathcal{G}^{DN}\psi_2), \\ \forall B,\psi\in\mathfrak{S}(\mathbb{R}^d), &\qquad (\mathcal{G}^{NN}B,\psi) = -(B,\mathcal{G}^{DD}\psi), \end{aligned}$$

(see Proposition 3.9 for an equivalent property on $\mathcal{G}$).
We show here how the shape derivative formula of Theorem 3.21 can be generalized and provide a shallow water asymptotic expansion of $\mathcal{G}^{NN}$ that can be used to derive shallow water models when the bottom is moving.

A.5.3. A generalized shape derivative formula. Adapting the proof of Theorem 3.21 given in Chapter 3, T. Iguchi gave in [**182**] a general shape derivative formula for the Dirichlet-Neumann operator $\mathcal{G}$ with a *nonhomogeneous* Neumann condition B at the bottom. In agreement with Notation 3.20, we denote by $d\mathcal{G}_{\mu,\gamma}[\varepsilon\zeta,\beta b](h,k)\psi$ (or simply $d\mathcal{G}(h,k)\psi$) the derivative of the mapping

$$\mathcal{G}_{\mu,\gamma}[\varepsilon\cdot,\beta\cdot]: \begin{array}{ccc} \mathbf{\Gamma}\subset H^{t_0+1}(\mathbb{R}^d)^2 & \to & H^{s-1/2}(\mathbb{R}^d), \\ \Gamma = (\zeta,b) & \mapsto & \mathcal{G}_{\mu,\gamma}[\varepsilon\zeta,\beta b]\psi \end{array}$$

at (ζ, b) and in the direction (h,k). We use similar definitions for $d\mathcal{G}^{NN}(h,k)B$, $d\mathcal{G}^{ND}(h,k)B$ and $d\mathcal{G}^{DD}(h,k)B$.

With these notations, the generalization of the shape derivative formula of Theorem 3.21 is given by [**182**]

$$d\mathcal{G}(h,0)\psi + d\mathcal{G}^{NN}(h,0)B = -\varepsilon\mathcal{G}(h\underline{w}) - \varepsilon\mu\nabla^\gamma\cdot(h\underline{V}), \tag{A.22}$$

with

$$\underline{w} = \frac{\mathcal{G}\psi + \mathcal{G}^{NN}B + \varepsilon\mu\nabla^\gamma\zeta\cdot\nabla^\gamma\psi}{1+\varepsilon^2\mu|\nabla^\gamma\zeta|^2} \quad\text{and}\quad \underline{V} = \nabla^\gamma\psi - \varepsilon\underline{w}\nabla^\gamma\zeta.$$

There also exist formulas for the shape derivatives with respect to bottom variations

$$d\mathcal{G}(0,k)\psi + d\mathcal{G}^{NN}(0,k)B = -\beta\mu\mathcal{G}^{NN}\nabla^\gamma\cdot(k\tilde{V}), \tag{A.23}$$

with

$$\tilde{w} = \frac{B + \beta\mu\nabla^\gamma b\cdot\nabla^\gamma(\mathcal{G}^{DD}\psi + \mathcal{G}^{ND}B)}{1+\beta^2\mu|\nabla^\gamma b|^2} \quad\text{and}\quad \tilde{V} = \nabla^\gamma(\mathcal{G}^{DD}\psi + \mathcal{G}^{ND}B) - \beta\tilde{w}\nabla^\gamma b.$$

(See [**182**] for a proof.)

A.5.4. Asymptotic expansion of the Neumann-Neumann operator. According to Remark A.23, the Neumann-Neumann operator is defined as

$$\mathcal{G}^{NN}B = \mathbf{e}_z\cdot P(\Sigma)\nabla^{\mu,\gamma}\phi_{|_{z=0}}, \tag{A.24}$$

with

$$\begin{cases} \nabla^{\mu,\gamma}\cdot P(\Sigma)\nabla^{\mu,\gamma}\phi = 0,\\ \phi_{|_{z=0}} = 0, \qquad \partial_{\mathbf{n}}\phi_{|_{z=0}} = B. \end{cases} \tag{A.25}$$

As in §3.6.1, we look for an approximate solution ϕ_{app} to (A.25) under the form

$$\phi_{app}(X,z) = \sum_{j=0}\mu^j\phi_j(X,z).$$

The method is exactly the same as for Lemma 3.42, but the expansion differs due to the difference on the boundary conditions. With the notations of §3.6.1, the ϕ_j is determined by the induction relation

$$\frac{1}{h}\partial_z^2\phi_0 = 0, \qquad \phi_{0\,|_{z=0}} = 0, \qquad \frac{1}{h}\partial_z\phi_{0\,|_{z=0}} = B$$

and

$$\begin{cases} \frac{1}{h}\partial_z^2\phi_j = -A(\nabla^\gamma,\partial_z)\phi_{j-1},\\ \phi_{j\,|_{z=0}} = 0, \qquad \frac{1}{h}\partial_z\phi_{j\,|_{z=-1}} = -\mathbf{b}(\nabla^\gamma,\partial_z)\phi_{j-1\,|_{z=-1}}, \end{cases} \qquad (1\le j\le n).$$

so that they can all be computed explicitly; for ϕ_0, one gets

$$\phi_0 = zhB \quad\text{and therefore}\quad \phi = zhB + O(\mu).$$

Plugging this into (A.24) yields

$$\mathcal{G}^{NN}B = B + O(\mu).$$

The approximations of $\mathcal{G}^{NN}B$ of order $O(\mu^n)$ are obtained in the same way but require the computations of the ϕ_j $(0\le j\le n-1)$.

A.5.5. Asymptotic expansion of the averaged velocity. Let us give here an asymptotic expansion of the asymptotic velocity that generalizes Proposition 3.37 to the case of a moving bottom. With the notations from the introduction, the velocity potential Φ must solve, when the bottom is moving, the following boundary value problem (with ν set to 1)

$$\begin{cases} \Delta^{\mu,\gamma}\Phi = 0, \qquad -1+\beta b\le z\le\varepsilon\zeta,\\ \Phi_{|_{z=\varepsilon\zeta}} = \psi, \qquad \sqrt{1+\beta^2|\nabla b|^2}\partial_{\mathbf{n}}\Phi_{|_{z=-1+\beta b}} = \nu\frac{\beta}{\varepsilon\delta}\partial_\tau b, \end{cases}$$

or equivalently, with Σ an admissible diffeomorphism and $\Phi = \phi\circ\Sigma$,

$$\begin{cases} \nabla^{\mu,\gamma}\cdot P(\Sigma)\nabla^{\mu,\gamma}\phi = 0, & -1<z<0,\\ \phi_{\vert z=0} = \psi, \qquad \partial_{\mathbf{n}}\phi_{\vert z=-1} = \mu\frac{\beta}{\varepsilon\delta}\partial_\tau b, \end{cases}$$

which is a particular case of (A.21) with $B = \mu\frac{\beta}{\varepsilon\delta}\partial_\tau b$. We moreover assume that

$$\beta \le \varepsilon,$$

so that β/ε can be treated as $O(1)$ terms in the asymptotic expansions below.

Combining Lemma 3.42 and the results of §A.5.4, it is straightforward to get

$$\phi = \psi + \mu\big(\phi_1 + z\frac{\beta}{\varepsilon\delta}h\partial_\tau b\big) + O(\mu^2), \tag{A.26}$$

with ϕ_1 as in Lemma 3.42.

If the averaged velocity is defined as in (3.31)

$$\overline{V} = \overline{V}_{\mu,\gamma}[\varepsilon\zeta,\beta b]\psi = \frac{1}{h}\int_{-1+\beta b}^{\varepsilon\zeta}\nabla^\gamma\Phi(\cdot,z)dz \qquad (h = 1+\varepsilon\zeta-\beta b),$$

then we deduce as in Proposition 3.37 an asymptotic expansion of $\overline{V}$ through the formula (3.35)

$$\overline{V} = \frac{1}{h}\int_{-1}^{0}\big[h\nabla^\gamma\phi - (z\nabla^\gamma h + \varepsilon\nabla^\gamma\zeta)\partial_z\phi\big]dz,$$

by replacing ϕ by its asymptotic expansion (A.26). We thus find

$$\overline{V} = \nabla^\gamma\psi + \mu\Big[\overline{V}_1 - \frac{\beta}{\varepsilon\delta}\big(\frac{1}{2}h\nabla^\gamma\partial_\tau b + \varepsilon\nabla^\gamma\zeta\partial_\tau b\big)\Big] + O(\mu^2), \tag{A.27}$$

where $\overline{V}_1$ is as in Proposition 3.37.

APPENDIX B

Product and Commutator Estimates

We give here some product and commutator estimates used in various parts of this book. All the results of this section hold for any nonzero $d \in \mathbb{N}$ (and not only $d = 1, 2$).
It will be convenient to use the following notation:

NOTATION B.1. For all $r \in \mathbb{R}$, we write

$$A_s + \langle B_s \rangle_{s>r} = \begin{cases} A_s & \text{if} \quad s \leq r, \\ A_s + B_s & \text{if} \quad s > r. \end{cases}$$

B.1. Product estimates

B.1.1. Product estimates for functions defined on $\mathbb{R}^d$. We first give two standard product estimates for scalar functions defined on $\mathbb{R}^d$ and a classical Moser estimate for nonlinearities [**11, 174**].

PROPOSITION B.2. *Let $t_0 > d/2$.*
(1) *Let $s \geq -t_0$ and $f \in H^s \cap H^{t_0}(\mathbb{R}^d)$, $g \in H^s(\mathbb{R}^d)$. Then $fg \in H^s(\mathbb{R}^d)$ and*

$$|fg|_{H^s} \lesssim |f|_{H^{t_0}} |g|_{H^s} + \langle |f|_{H^s} |g|_{H^{t_0}} \rangle_{s>t_0}.$$

(2) *Let $s_1, s_2 \in \mathbb{R}$ be such that $s_1 + s_2 \geq 0$. Then for all $s \leq s_j$ ($j = 1, 2$) and $s < s_1 + s_2 - d/2$, and all $f \in H^{s_1}(\mathbb{R}^d)$, $g \in H^{s_2}(\mathbb{R}^d)$, one has $fg \in H^s(\mathbb{R}^d)$ and*

$$|fg|_{H^s} \lesssim |f|_{H^{s_1}} |g|_{H^{s_2}}.$$

(3) *Let $F : \mathbb{R} \mapsto \mathbb{R}$ be a smooth function such that $F(0) = 0$. Also let $s \geq 0$ and $f \in H^s \cap L^\infty(\mathbb{R}^d)$. Then $F(f) \in H^s(\mathbb{R}^d)$ and*

$$|F(f)|_{H^s} \leq C(|f|_\infty) |f|_{H^s}.$$

REMARK B.3. When $s \geq 0$, an alternative to the first estimate is

$$|fg|_{H^s} \lesssim |f|_\infty |g|_{H^s} + |f|_{H^s} |g|_\infty.$$

The next proposition gives estimates on $f/(1+g)$ when $1+g$ does not vanish.

PROPOSITION B.4. *Let $t_0 > d/2$, $s \geq -t_0$ and $c_0 > 0$. Also let $f \in H^s(\mathbb{R}^d)$ and $g \in H^s \cap H^{t_0}(\mathbb{R}^d)$ be such that for all $X \in \mathbb{R}^d$, one has $1 + g(X) \geq c_0$. Then $\frac{f}{1+g}$ belongs to $H^s(\mathbb{R}^d)$ and*

$$\Big|\frac{f}{1+g}\Big|_{H^s} \leq C\Big(\frac{1}{c_0}, |g|_{H^{t_0}}\Big)\Big(|f|_{H^s} + \big\langle |f|_{H^{t_0}} |g|_{H^s} \big\rangle_{s>t_0}\Big).$$

PROOF. Noting that

$$\frac{f}{1+g} = f - f\frac{g}{1+g},$$

one can use Proposition B.2 to get

$$(B.1)\qquad |\frac{f}{1+g}|_{H^s} \lesssim |f|_{H^s}(1+|\frac{g}{1+g}|_{H^{t_0}}) + \big\langle |f|_{H^{t_0}}|\frac{g}{1+g}|_{H^s}\big\rangle_{s>t_0}.$$

We are thus led to control $\frac{g}{1+g}$ in H^s for $s \geq 0$ ($s \geq t_0$ would be enough). Let us first consider the case when s is an integer. For all $\alpha \in \mathbb{N}^d$, with $|\alpha| = k$, one can check by induction that $\partial^\alpha\big(\frac{g}{1+g}\big)$ is a sum of terms of the form

$$I = \frac{1}{(1+g)^k}\partial^\beta g \prod_{j=0}^{k-|\beta|} \prod_{\gamma_j\in\mathbb{N}^d, |\gamma_j|=j} (\partial^{\gamma_j} g)^{r_{\gamma_j}},$$

where $\beta \in \mathbb{N}^d$ and the $r_{\gamma_j} \in \mathbb{N}$ satisfy the relation

$$|\beta| + \sum_{j=0}^{k-|\beta|} j \sum_{\gamma_j\in\mathbb{N}^d, |\gamma_j|=j} r_{\gamma_j} = k.$$

It is straightforward to get

$$(B.2)\qquad |I|_2 \leq \frac{1}{c_0^k}|\partial^\beta g \prod_{j=0}^{k-|\beta|} \prod_{\gamma_j\in\mathbb{N}^d, |\gamma_j|=j} (\partial^{\gamma_j} g)^{r_{\gamma_j}}|_2.$$

In order to control the L^2-norm that appears in the r.h.s., note that

$$\frac{1}{2k/|\beta|} + \sum_{j=1}^{k-|\beta|} \sum_{\gamma_j\in\mathbb{N}^d, |\gamma_j|=j} \frac{r_{\gamma_j}}{2k/j} = \frac{1}{2}.$$

Using Young's inequality, we can thus deduce from (B.2) that

$$|I|_2 \leq \frac{1}{c_0^k}|\partial^\beta g|_{2k/|\beta|} \prod_{j=0}^{k-|\beta|} \prod_{\gamma_j\in\mathbb{N}^d, |\gamma_j|=j} |\partial^{\gamma_j} g|_{2k/j}^{r_{\gamma_j}}.$$

Since by Gagliardo-Nirenberg's inequality, one has, for all $\gamma \in \mathbb{N}^d$

$$|\partial^\gamma g|_{2k/|\gamma|} \lesssim |g|_\infty^{1-|\gamma|/k} |g|_{H^k}^{|\gamma|/k},$$

we deduce that

$$|I|_2 \leq \frac{1}{c_0^k} C(|g|_\infty)|g|_{H^k}.$$

This provides us an L^2 estimate on $\partial^\alpha\big(\frac{g}{1+g}\big)$ for all $|\alpha| = k$ (and of course $|\alpha| \leq k$); we thus get

$$\forall s \in \mathbb{N}, \qquad |\frac{g}{1+g}|_{H^s} \leq C(\frac{1}{c_0}, |g|_\infty)|g|_{H^s}.$$

This estimate remains valid for noninteger values of s by interpolation and the result therefore follows from (B.1). $\square$

B.1.2. Product estimates for functions defined on the flat strip $\mathcal{S}$. We give here two corollaries of Propositions B.2 and B.4 respectively for functions defined on the strip $\mathcal{S}$. We recall that the spaces $L^\infty H^s$ and $H^{s,k}$ were introduced in Definition 2.11 and that the notation $a \vee b$ stands for $\max\{a,b\}$.

COROLLARY B.5. *Let $t_0 > d/2$.*
(1) *If $s \geq -t_0$, $f \in L^\infty H^{s\vee t_0}$ and $g \in H^{s,0}$, one has $fg \in H^{s,0}$ and*

$$\|fg\|_{H^{s,0}} \lesssim \|f\|_{L^\infty H^{t_0}} \|g\|_{H^{s,0}} + \big\langle \|f\|_{L^\infty H^s} \|g\|_{H^{t_0,0}} \big\rangle_{s>t_0}.$$

(2) *If $s \geq -t_0$, $f, g \in L^\infty H^{t_0} \cap H^{s,0}$ (if $s \leq t_0$, $g \in H^{s,0}$ is enough), one has $fg \in H^{s,0}$ and*

$$\|fg\|_{H^{s,0}} \lesssim \|f\|_{L^\infty H^{t_0}} \|g\|_{H^{s,0}} + \big\langle \|f\|_{H^{s,0}} \|g\|_{L^\infty H^{t_0}} \big\rangle_{s>t_0}.$$

(3) *Let $s_1, s_2 \in \mathbb{R}$ be such that $s_1 + s_2 \geq 0$. Then for all $s \leq s_j$ $(j = 1,2)$ and $s < s_1 + s_2 - d/2$, and all $f \in L^\infty H^{s_1}$, $g \in H^{s_2,0}$, one has $fg \in H^{s,0}$ and*

$$\|fg\|_{H^{s,0}} \lesssim \|f\|_{L^\infty H^{s_1}} \|g\|_{H^{s_2,0}}.$$

PROOF. In order to prove the first point, write

$$\begin{aligned}
\|fg\|^2_{H^{s,0}} &= \int_{-1}^0 |f(\cdot,z)g(\cdot,z)|^2_{H^s} dz \\
&\lesssim \int_{-1}^0 \big(|f(\cdot,z)|_{H^{t_0}} |g(\cdot,z)|_{H^s} + \langle |f(\cdot,z)|_{H^s} |g(\cdot,z)|_{H^{t_0}} \rangle_{s>t_0}\big)^2 dz,
\end{aligned}$$

where we used the second estimate of Proposition B.2 to derive the last inequality. We thus have

$$\begin{aligned}
\|fg\|^2_{H^{s,0}} &\lesssim \int_{-1}^0 \big(|f(\cdot,z)|^2_{H^{t_0}} |g(\cdot,z)|^2_{H^s} + \langle |f(\cdot,z)|^2_{H^s} |g(\cdot,z)|^2_{H^{t_0}} \rangle_{s>t_0}\big) dz \\
&\lesssim \|f\|^2_{L^\infty H^{t_0}} \int_{-1}^0 |g(\cdot,z)|^2_{H^s} dz + \Big\langle \|f\|^2_{L^\infty H^s} \int_{-1}^0 |g(\cdot,z)|^2_{H^{t_0}} dz \Big\rangle_{s>t_0}. \qquad \text{(B.3)}
\end{aligned}$$

Since by definition one has $\int_{-1}^0 |g(\cdot,z)|^2_{H^s} dz = \|g\|_{H^{s,0}}$, the first part of the corollary follows easily. To derive the second estimate, note that one can replace (B.3) by

$$\|fg\|^2_{H^{s,0}} \lesssim \|f\|^2_{L^\infty H^{t_0}} \int_{-1}^0 |g(\cdot,z)|^2_{H^s} dz + \Big\langle \|g\|^2_{L^\infty H^{t_0}} \int_{-1}^0 |f(\cdot,z)|^2_{H^s} dz \Big\rangle_{s>t_0},$$

and proceed as above. The third estimate is deduced along the same lines from the second estimate of Proposition B.2. □

COROLLARY B.6. *Let $t_0 > d/2$, $s > -t_0$ and $c_0 > 0$. Let $f \in H^{s,0}$ and $g \in L^\infty H^{s\vee t_0}$ be such that for all $(X,z) \in \mathcal{S}$, $1 + g(X,z) \geq c_0$.*
Then $\frac{f}{1+g}$ belongs to $H^{s,0}$ and

$$\Big\|\frac{f}{1+g}\Big\|_{H^{s,0}} \leq C\Big(\frac{1}{c_0}, \|g\|_{L^\infty H^{t_0}}\Big)\Big(\|f\|_{H^{s,0}} + \big\langle \|\Lambda^{t_0} f\|_2 \|g\|_{L^\infty H^s} \big\rangle_{s>t_0}\Big).$$

PROOF. This result can be deduced from Proposition B.4 as in the proof of Corollary B.5. □

B.2. Commutator estimates

B.2.1. Commutator estimates for functions defined in $\mathbb{R}^d$. We need only commutator estimates between a Fourier multiplier and a function. Before stating these estimates, let us define the class $\mathcal{S}^s$ of Fourier multipliers of order s (the notation $\lceil d/2 \rceil$ used in the definition of the seminorm refers to the lowest integer larger or equal to $d/2$.)

DEFINITION B.7. We say that a Fourier multiplier $\sigma(D)$ is of order s ($s \in \mathbb{R}$) and write $\sigma \in \mathcal{S}^s$ if $\xi \in \mathbb{R}^d \mapsto \sigma(\xi) \in \mathbb{C}$ is smooth and satisfies

$$\forall \xi \in \mathbb{R}^d, \forall \beta \in \mathbb{N}^d, \qquad \sup_{\xi \in \mathbb{R}^d} \langle \xi \rangle^{|\beta| - s} |\partial^\beta \sigma(\xi)| < \infty.$$

We also introduce the seminorm

$$\mathcal{N}^s(\sigma) = \sup_{\beta \in \mathbb{N}^d, |\beta| \le 2 + d + \lceil \frac{d}{2} \rceil} \sup_{\xi \in \mathbb{R}^d} \langle \xi \rangle^{|\beta| - s} |\partial^\beta \sigma(\xi)|.$$

The main commutator estimates we will need are gathered in the following proposition; the precise dependence of these estimates on σ (through $\mathcal{N}^s(\sigma)$) will be used in Corollary B.15 below. Refined commutator estimates are also given in Proposition B.10 below.

PROPOSITION B.8. *Let $t_0 > d/2$, $s \ge 0$ and $\sigma \in \mathcal{S}^s$.*

(1) *If $f \in H^s \cap H^{t_0+1}$ then, for all $g \in H^{s-1} \cap H^{t_0}$, one has*

$$|[\sigma(D), f]g|_2 \lesssim \mathcal{N}^s(\sigma)\big(|\nabla f|_\infty |g|_{H^{s-1}} + |\nabla f|_{H^{s-1}} |g|_\infty\big).$$

(2) *If $0 \le s \le t_0 + 1$ and $f \in H^{t_0+1}$ then, for all $g \in H^{s-1}$, one has*

$$|[\sigma(D), f]g|_2 \lesssim \mathcal{N}^s(\sigma) |\nabla f|_{H^{t_0}} |g|_{H^{s-1}}.$$

(3) *If $-t_0 < r \le t_0 + 1 - s$ and $f \in H^{t_0+1}$ then, for all $g \in H^{r+s-1}$, one has*

$$\big|[\sigma(D), f]g\big|_{H^r} \lesssim \mathcal{N}^s(\sigma) |\nabla f|_{H^{t_0}} |g|_{H^{r+s-1}}.$$

PROOF. The first point is a generalization of the well-known Kato-Ponce estimate [**202**] involving only the gradient of f (one has $|f|_{H^s}$ in the standard Kato-Ponce estimate). Their proof can be found in [**219**] (Theorems 3.ii and 5.ii).
The second point of the proof is a Calderon-Coifmann-Meyer type estimate which also depends only on ∇f. (The standard estimate—see Prop. 4.2 of [**309**]—depends on $|f|_{H^{t_0+1}}$.) We refer to Theorem 6.ii of [**219**] for a proof.
The third point, which is a generalization of the second one, can also be found in [**219**]. □

Using Notation B.1, a direct corollary of Proposition B.8 is given below.

COROLLARY B.9. *Let $t_0 > d/2$, $s \ge 0$ and $\sigma \in \mathcal{S}^s$. If $f \in H^s \cap H^{t_0+1}$ then, for all $g \in H^{s-1}$,*

$$|[\sigma(D), f]g|_2 \lesssim \mathcal{N}^s(\sigma)\big(|\nabla f|_{H^{t_0}} |g|_{H^{s-1}} + \big\langle |\nabla f|_{H^{s-1}} |g|_{H^{t_0}} \big\rangle_{s > t_0+1}\big).$$

PROOF. One has to use the first point of Proposition B.8 (together with the continuous Sobolev embedding $H^{t_0} \subset L^\infty$) when $s > t_0 + 1$ and the second point when $0 \le s \le t_0 + 1$ (or the third one to obtain the second estimate of the corollary). □

Various generalizations of these commutator estimates can be found in [**219**]: commutator estimates involving pseudo-differential operators with rough symbols, negative order operators, higher order commutator estimates, etc. We give in the proposition below another kind of generalization of the Calderon-Coifmann-Meyer estimate (the third point in Proposition B.8); the basic idea is that the commutator $[\Lambda^s, f]g$ can cost less than t_0+1 derivative on f if one accepts to gain less than one derivative on g.

PROPOSITION B.10. *Let $t_0 > d/2$. The following commutator estimates hold.*
(1) *Let $\delta \in (0,1)$ and $-d/2 < s < d/2+\delta$. Then for all $f \in H^{d/2+\delta}(\mathbb{R}^d)$ and $g \in H^{s-\delta}(\mathbb{R}^d)$, one has*

$$|[\Lambda^s, f]g|_2 \lesssim |f|_{H^{d/2+\delta}} |g|_{H^{s-\delta}}.$$

(2) *Let $\delta \in [0,1]$ and $-d/2 < s \le t_0+\delta$. Then for all $f \in H^{t_0+\delta}(\mathbb{R}^d)$ and $g \in H^{s-\delta}(\mathbb{R}^d)$, one has*

$$|[\Lambda^s, f]g|_2 \lesssim |f|_{H^{t_0+\delta}} |g|_{H^{s-\delta}}.$$

(3) *Then for all $-d/2 < s \le t_0+1$ and all $0 < \underline{\delta} \le 1$ such that $\underline{\delta} < t_0 - d/2$, we have*

$$|[\Lambda^s, f]g|_2 \lesssim |f|_{H^{\max\{s,t_0\}}} |g|_{H^{\min\{t_0, s-\underline{\delta}\}}}.$$

REMARK B.11. In the statement of the proposition, one can replace $\Lambda^s = (1-\Delta)^{s/2}$ by $\sigma(D)$, with $\sigma \in \mathcal{S}^s$.

REMARK B.12. The estimates of the proposition can be seen as improvements of the Calderon-Coifmann-Meyer estimate (the third point in Proposition B.8), which is a particular case ($\delta = 1$) of the second point.

PROOF. **Step 1.** *A brief reminder about paraproducts.* We recall here basic notations and results concerning paraproducts. Throughout this proof, $\psi \in C_0^\infty(\mathbb{R}^d)$ denotes a smooth bump function such that

$$\psi(\xi) = 1 \quad \text{if} \quad |\xi| \le 1/2 \quad \text{and} \quad \psi(\xi) = 0 \quad \text{if} \quad |\xi| \ge 1, \tag{B.4}$$

and we define $\varphi \in C_0^\infty(\mathbb{R}^d)$ as

$$\varphi(\xi) = \psi(\xi/2) - \psi(\xi), \qquad \forall \xi \in \mathbb{R}^d, \tag{B.5}$$

so that φ is supported in the annulus $1/2 \le |\xi| \le 2$, and one has

$$1 = \psi(\xi) + \sum_{p\ge 0} \varphi(2^{-p}\xi), \qquad \forall \xi \in \mathbb{R}^d. \tag{B.6}$$

For all $p \in \mathbb{Z}$, we introduce the functions φ_p, supported in $2^{p-1} \le |\xi| \le 2^{p+1}$, and defined as

$$\varphi_p = 0 \quad \text{if} \quad p < -1, \qquad \varphi_{-1} = \psi, \qquad \varphi_p(\cdot) = \varphi(2^{-p}\cdot) \quad \text{if} \quad p \ge 0. \tag{B.7}$$

This allows us to give the classical definition of Zygmund spaces (see [**311**]): for all $r \in \mathbb{R}$, $C_*^r(\mathbb{R}^d)$ is the set of all $u \in \mathcal{S}'(\mathbb{R}^d)$ such that

$$|u|_{C_*^r} := \sup_{p\ge -1} 2^{pr} |\varphi_p(D)u|_\infty < \infty. \tag{B.8}$$

We recall the continuous embeddings $H^s(\mathbb{R}^d) \subset C_*^{s-d/2}(\mathbb{R}^d)$, for all $s \in \mathbb{R}$, and $L^\infty(\mathbb{R}^d) \subset C_*^0(\mathbb{R}^d)$. We also recall that

$$\forall \delta \in (0,1), \quad \forall f \in C_*^\delta(\mathbb{R}^d), \quad \forall x,y \in \mathbb{R}^d, \qquad |f(x+y) - f(x)| \lesssim |f|_{C_*^\delta} |y|^\delta. \tag{B.9}$$

We can now define the paraproduct operator T_f (see [**42, 258, 71**]) as

$$T_f = \sum_{p\geq -1} \psi_{p-N}(D)f\,\varphi_p(D),$$

where $\psi_j = \psi(2^{-j}\cdot)$; every product of function can therefore be decomposed as

$$fg = T_f g + T_g f + R(f,g), \tag{B.10}$$

with

$$R(f,g) = \sum_{p\geq -1}\sum_{|p-q|\leq N} \varphi_q(D)f\varphi_p(D)g.$$

The properties we need in this proof of the paraproduct and residual operators T_f, $R(\cdot,\cdot)$ are the following

$$\forall r>0, \forall s\in\mathbb{R}, \qquad |T_f g|_{H^s} \lesssim |f|_{C_*^{-r}}|g|_{H^{s+r}}, \tag{B.11}$$

$$\forall s\in\mathbb{R}, \qquad |T_f g|_{H^s} \lesssim |f|_\infty |g|_{H^s}, \tag{B.12}$$

$$\forall s+r>0, \qquad |R(f,g)|_{H^{s+r-d/2}} \lesssim |f|_{H^s}|g|_{H^r}. \tag{B.13}$$

Finally, let us recall the classical characterization of Sobolev spaces (see Theorem 2.2.1 of [**71**] or Lemma 9.4 of [**311**]).

LEMMA B.13. *Let $(u_p)_{p\geq -1}$ be a sequence of $\mathcal{S}'(\mathbb{R}^d)$ such that for all $p\geq 0$, $\widehat{u_p}$ is supported in $A2^{p-1}\leq |\xi|\leq B2^{p+1}$, for some $A,B>0$, and such that $\widehat{u_{-1}}$ is compactly supported.*
If, for some $s\in\mathbb{R}$, $\sum_{p\geq -1} 2^{2ps}|u_p|_2^2<\infty$, then

$$\sum_{p\geq -1} u_p =: u \in H^s(\mathbb{R}^d) \quad \textit{and} \quad |u|_{H^s}^2 \leq \mathrm{Cst}\sum_{p\geq -1} 2^{2ps}|u_p|_2^2.$$

Conversely, if $u\in H^s(\mathbb{R}^d)$, then

$$\sum_{p\geq -1} 2^{2ps}|\varphi_p(D)u|_2^2 \leq \mathrm{Cst}\,|u|_{H^s}^2.$$

Step 2. *Paralinearization of the commutator.* Using the paraproduct decomposition (B.10), it is straightforward to check that

$$[\Lambda^s, f]g = [\Lambda^s, T_f]g + \Lambda^s T_g f - T_{\Lambda^s g}f + \Lambda^s R(f,g) + R(f,\Lambda^s g). \tag{B.14}$$

The proof of the proposition consists of providing estimates for all the components of this decomposition.
Step 3. *A commutator estimate for the paraproduct operator T_f.* We prove here that[1]

$$\forall\delta\in(0,1), \qquad \forall s\in\mathbb{R}, \qquad |[\Lambda^s, T_f]g|_2 \lesssim |f|_{C_*^\delta}|g|_{H^{s-\delta}}. \tag{B.15}$$

Let us first write

$$[\Lambda^s, T_f]g = \sum_{p\geq -1}[\Lambda^s, \psi_{p-N}(D)f]\varphi_p(D)g.$$

[1]Straightforward modifications of the proof show that this estimate can be generalized to $\delta=0$ and $\delta=1$, provided that $|f|_{C_*^\delta}$ is replaced by $|f|_\infty$ and $|\nabla f|_\infty$ respectively.

Since for all $p \geq 0$, $[\Lambda^s, \psi_{p-N}(D)f]\varphi_p(D)g$ is supported in an annulus of the form $\{A2^{p-1} \leq |\xi| \leq B2^{p+1}\}$, the characterization of Sobolev spaces given at the end of Step 1 allows us to write

$$|[\Lambda^s, T_f]g|_2^2 \lesssim \sum_{p\geq -1} 2^{2ps}|[\Lambda^s, \psi_{p-N}(D)f]\varphi_p(D)g|_2^2, \tag{B.16}$$

and we therefore are led to control $|[\Lambda^s, \psi_{p-N}(D)f]\varphi_p(D)g|_2^2$. Denoting $g_p = \varphi_p(D)g$ and noting that there exists $\widetilde{\varphi} \in C_0(\mathbb{R}^d\backslash\{0\})$ such that

$$[\Lambda^s, \psi_{p-N}(D)f]g_p = [\Lambda^s\widetilde{\varphi}_p(D), \psi_{p-N}(D)f]g_p,$$

we can write

$$[\Lambda^s, \psi_{p-N}(D)f]g_p = \int_{\mathbb{R}^d} 2^{pd}H_s(2^p(x-y))[\psi_{p-N}(D)f(x) - \psi_{p-N}(D)f(y)]g_p(y)dy,$$

where H_s is the inverse Fourier transform of $\xi \mapsto \langle 2^p\xi\rangle^s\widetilde{\varphi}(\xi)$. A standard convolution estimate therefore gives

$$\begin{aligned}|[\Lambda^s, \psi_{p-N}(D)f]g_p|_2 &\lesssim 2^{-p\delta}\big|H_s(x)|x|^\delta\big|_{L^1_x}\Big|\frac{[\psi_{p-N}(D)f(x) - \psi_{p-N}(D)f(\cdot)]}{|x-\cdot|^\delta}g_p\Big|_2 \\ &\lesssim 2^{p(s-\delta)}\Big|\frac{[\psi_{p-N}(D)f(x) - \psi_{p-N}(D)f(\cdot)]}{|x-\cdot|^\delta}g_p\Big|_2,\end{aligned}$$

where we used the fact that $\big|H_s(x)|x|^\delta\big|_{L^1_x} \lesssim 2^{ps}$, which stems from the fact that $\widetilde{\varphi} \in C_0(\mathbb{R}^d\backslash\{0\})$. We can now use (B.9) to get

$$|[\Lambda^s, \psi_{p-N}(D)f]g_p|_2 \lesssim 2^{p(s-\delta)}|f|_{C^\delta_*}|g_p|_2.$$

Plugging this into (B.16) and once again using (for g) the characterization of Sobolev spaces given at the end of Step 1, we obtain (B.15).

Step 4. One has

$$\forall\delta \in \mathbb{R}, \quad \forall s < d/2 + \delta, \qquad |T_g f|_{H^s} \lesssim |f|_{H^{d/2+\delta}}|g|_{H^{s-\delta}}. \tag{B.17}$$

The result follows directly from (B.11), with $r = d/2+\delta-s > 0$ and the continuous embedding $C_*^{-r}(\mathbb{R}^d) \subset H^{-r+d/2}(\mathbb{R}^d)$.

Step 5. One has

$$\forall\delta > -d/2, \quad \forall s \in \mathbb{R}, \qquad |T_{\Lambda^s g}f|_2 \lesssim |f|_{H^{d/2+\delta}}|g|_{H^{s-\delta}}. \tag{B.18}$$

This is again a direct consequence of (B.11), with $r = d/2 + \delta > 0$.

Step 6. For the first residual term, we get

$$\forall\delta \in \mathbb{R}, \quad \forall s > -d/2, \qquad |R(f,g)|_{H^s} \lesssim |f|_{H^{d/2+\delta}}|g|_{H^{s-\delta}} \tag{B.19}$$

as a direct consequence of (B.13) with $\widetilde{s} = d/2 + \delta$ and $\widetilde{r} = s - \delta$ (where $\widetilde{r}$ and $\widetilde{s}$ correspond to the indexes r and s in (B.13)).

Step 7. For the second residual term, we obtain

$$\forall\delta \in \mathbb{R}, \quad \forall s \in \mathbb{R}, \qquad |R(f,\Lambda^s g)|_2 \lesssim |f|_{H^{d/2+\delta}}|g|_{H^{s-\delta}}, \tag{B.20}$$

which is also a direct consequence of (B.13) with $\widetilde{s} = d/2 + \delta$ and $\widetilde{r} = -\delta$.

Step 8. The first point of the proposition follows directly from Steps 2 to 7. For the second estimate of the proposition, the proof is the same, with only small modifications in Steps 3 and 4. To allow $\delta = 0$ and $\delta = 1$ in Step 3, we use Footnote1 and the continuous embedding $W^{k,\infty}(\mathbb{R}^d) \subset H^{t_0+k}(\mathbb{R}^d)$ ($k \in \mathbb{N}$). In Step 4, we have to take $r = t_0 + \delta - s$, which yields the result for $s < t_0 + \delta$. For $s = t_0 + \delta$, the

result follows from (B.12) and the continuous embedding $H^{t_0}(\mathbb{R}^d) \subset L^\infty(\mathbb{R}^d)$.
Let us now prove the third estimate of the proposition. Let $\underline{\delta} \in (0,1]$ be such that $\underline{\delta} < t_0 - d/2$ and $\widetilde{t_0} = t_0 - \underline{\delta} > d/2$. Also let $k_0 \in \mathbb{N}$ be the smallest integer such that $k_0(1-\underline{\delta}) \geq \underline{\delta}$. We have to consider three cases:
- *Case 1:* $-d/2 \leq s \leq t_0 = \widetilde{t_0} + \underline{\delta}$. Using the second estimate of the proposition with $\delta = \underline{\delta}$, we get

$$\begin{aligned} |[\Lambda^s, f]g|_2 &\lesssim |f|_{H^{\widetilde{t_0}+\underline{\delta}}} |g|_{H^{s-\underline{\delta}}} \\ &\lesssim |f|_{H^{t_0}} |g|_{H^{\min\{s-\underline{\delta}, t_0\}}}, \end{aligned}$$

since $\widetilde{t_0} + \underline{\delta} = t_0$ and $s - \underline{\delta} \leq \widetilde{t_0} \leq t_0$.
- *Case 2:* $t_0 \leq s \leq t_0 + 1 - \underline{\delta}$. The second point of the proposition, with $\delta = s - \widetilde{t_0} \in [\underline{\delta}, 1]$, yields

$$\begin{aligned} |[\Lambda^s, f]g|_2 &\lesssim |f|_{H^{\widetilde{t_0}+\delta}} |g|_{H^{s-\delta}} \\ &\lesssim |f|_{H^s} |g|_{H^{\min\{s-\underline{\delta}, t_0\}}}, \end{aligned}$$

since $s - \delta = \widetilde{t_0} \leq t_0$.
- *Case 3: If* $k_0 > 1$, $1 \leq k < k_0$ *and* $t_0 + k(1-\underline{\delta}) \leq s \leq t_0 + (k+1)(1-\underline{\delta})$. Let us define $\underline{t_0} = t_0 + k(1-\underline{\delta}) - \underline{\delta}$; from the definition of $\underline{\delta}$ and k_0, we get that $d/2 < \underline{t_0} < t_0$. Also let $\delta = s - \underline{t_0} \in [\underline{\delta}, 1]$. Using the second estimate of the proposition, we get

$$\begin{aligned} |[\Lambda^s, f]g|_2 &\lesssim |f|_{H^{\underline{t_0}+\delta}} |g|_{H^{s-\delta}} \\ &\lesssim |f|_{H^s} |g|_{H^{\min\{s-\underline{\delta}, t_0\}}}, \end{aligned}$$

since $\delta \geq \underline{\delta}$ and $s - \delta = \underline{t_0} < t_0$.
- *Case 4:* $t_0 + k_0(1-\underline{\delta}) \leq s \leq t_0 + 1$. We once again use the second estimate of the proposition with $\delta = s - t_0 \in [k_0(1-\underline{\delta}), 1]$ to get

$$\begin{aligned} |[\Lambda^s, f]g|_2 &\lesssim |f|_{H^{t_0+\delta}} |g|_{H^{s-\delta}} \\ &\lesssim |f|_{H^s} |g|_{H^{\min\{s-\underline{\delta}, t_0\}}}, \end{aligned}$$

since $s - \delta = t_0$ and $\delta \geq k_0(1-\underline{\delta}) \geq \underline{\delta}$ by definition of k_0. □

In the last result of this section, we give a uniform estimate for commutators involving "smoothed" symbols (the uniformity is with respect to the smoothing parameter). Such "smoothed" symbols are introduced in the following definition:

DEFINITION B.14. Let $s \geq 0$, $\sigma \in \mathcal{S}^s$ and $1 > \delta > 0$. We define the "smoothed" Fourier multiplier $\sigma_\delta(D)$ as

$$\sigma_\delta(D) = \sigma(D)\chi(\delta\Lambda), \qquad (\Lambda = (1+|D|^2)^{1/2}),$$

where χ is any real valued, smooth, and compactly supported function equal to 1 in a neighborhood of the origin.

The following corollary is then a direct consequence of Corollary B.9, since $\mathcal{N}^s(\sigma_\delta) \lesssim \mathcal{N}^s(\sigma)$. Similar adaptations can be brought to Proposition B.10.

COROLLARY B.15. *Let* $t_0 > d/2$, $s \geq 0$ *and* $\sigma \in \mathcal{S}^s$. *If* $f \in H^s \cap H^{t_0+1}$ *then, for all* $g \in H^{s-1}$ *and all* $\delta > 0$,

$$|[\sigma_\delta(D), f]g|_2 \lesssim \mathcal{N}^s(\sigma)\big(|\nabla f|_{H^{t_0}} |g|_{H^{s-1}} + \big\langle |\nabla g|_{H^{s-1}} |g|_{H^{t_0}} \big\rangle_{s>t_0+1}\big)$$

(in particular, the r.h.s. is independent of δ*).*

B.2.2. Commutator estimates for functions defined on $\mathcal{S}$. From the commutator estimates given in the previous section, one can easily deduce some commutator estimates for functions defined on the flat strip $\mathcal{S} = \mathbb{R}^d \times (-1, 0)$. For instance, we get the following corollary (with the convention that $\sigma_\delta(D) = \sigma(D)$ when $\delta = 0$ and recalling that the spaces $L^\infty H^s$ and $H^{s,0}$ are defined in Definition 2.11). It is important to note that the right-hand side of the three inequalities stated below is independent of δ.

COROLLARY B.16. *Let $t_0 > d/2$, $s \geq 0$ and $\sigma \in \mathcal{S}^s$.*

(1) *If $u \in L^\infty H^s \cap L^\infty H^{t_0+1}$ then, for all $v \in H^{s-1,0}$ and all $\delta \geq 0$,*

$$\|[\sigma_\delta(D), u]v\|_2 \lesssim \mathcal{N}^s(\sigma)\Big(\|\nabla u\|_{L^\infty H^{t_0}} \|v\|_{H^{s-1,0}} + \big\langle \|\nabla u\|_{L^\infty H^{s-1}} \|v\|_{H^{t_0,0}} \big\rangle_{s>t_0+1}\Big).$$

(2) *If $u \in H^{s,0} \cap L^\infty H^{t_0+1}$ then, for all $v \in H^{s-1,0} \cap L^\infty H^{t_0}$ and all $\delta \geq 0$,*

$$\|[\sigma_\delta(D), u]v\|_2 \lesssim \mathcal{N}^s(\sigma)\big(\|\nabla u\|_{L^\infty H^{t_0}} \|v\|_{H^{s-1,0}} + \big\langle \|\nabla u\|_{H^{s-1,0}} \|v\|_{L^\infty H^{t_0}} \big\rangle_{s>t_0+1}\big).$$

(3) *If $u \in H^{s\vee t_0+1,0}$ then, for all $v \in L^\infty H^{s-1}$ and all $\delta \geq 0$,*

$$\|[\sigma_\delta(D), u]v\|_2 \lesssim \mathcal{N}^s(\sigma)\big(\|\nabla u\|_{H^{t_0,0}} \|v\|_{L^\infty H^{s-1}} + \big\langle \|\nabla u\|_{H^{s-1,0}} \|v\|_{L^\infty H^{t_0}} \big\rangle_{s>t_0+1}\big).$$

PROOF. We prove only the first estimate; the other estimate requires only small modifications to the proof

$$\|[\sigma_\delta(D), u]v\|_2^2 = \int_{-1}^0 |[\sigma_\delta(D), u(\cdot, z)]v(\cdot, z)|_2^2 dz,$$

so that we deduce from Corollaries B.9 ($\delta = 0$) and B.15 ($\delta > 0$) that

$$\begin{aligned} \|[\sigma_\delta(D), u]v\|_2^2 \ \lesssim\ & \mathcal{N}^s(\sigma)^2 \int_{-1}^0 \Big[|\nabla u(\cdot, z)|_{H^{t_0}}^2 |v(\cdot, z)|_{H^{s-1}}^2 \\ & + \big\langle |\nabla u(\cdot, z)|_{H^{s-1}}^2 |v(\cdot, z)|_{H^{t_0}}^2 \big\rangle_{s>t_0+1}\Big] dz, \end{aligned} \tag{B.21}$$

and the result follows easily. □

Finally, we give a last corollary that stems from Proposition B.10. (We omit the proof, since it follows the same lines as the proof of Corollary B.16.) As above, these estimates can be generalized to "smoothed" versions of Λ^s, namely, $(\Lambda^s)_\delta$.

COROLLARY B.17. *Let $t_0 > d/2$. Also let $-d/2 < s \leq t_0 + 1$ and $0 < \underline{\delta} \leq 1$ such that $\underline{\delta} < t_0 - d/2$. Then*

(1) *For all $u \in L^\infty H^{\max\{s,t_0\}}$ and $v \in H^{\min\{t_0, s-\underline{\delta}\},0}$, we have*

$$\|[\Lambda^s, u]v\|_2 \lesssim \|u\|_{L^\infty H^{\max\{s,t_0\}}} \|\Lambda^{\min\{t_0, s-\underline{\delta}\}} v\|_2.$$

(2) *For all $u \in H^{\max\{s,t_0\},0}$ and $v \in L^\infty H^{\min\{t_0, s-\underline{\delta}\}}$, we have*

$$\|[\Lambda^s, u]v\|_2 \lesssim \|\Lambda^{\max\{s,t_0\}} u\|_2 \|v\|_{L^\infty H^{\min\{t_0, s-\underline{\delta}\}}}.$$

B.3. Product and commutator with C^k functions

In order to handle bottoms that are not asymptotically flat (see, for instance, §2.5.2), one must be able to handle product and commutators with functions that are not of Sobolev type, but rather of class $W^{k,\infty}(\mathbb{R}^d)$. It is therefore natural[2] to

[2]We recall that we have the following continuous injections

$$\forall k \in \mathbb{N}, \qquad W^{k,\infty}(\mathbb{R}^d) \subset C_*^k(\mathbb{R}^d), \qquad C_*^{k+r}(\mathbb{R}^d) \subset W^{k,\infty}(\mathbb{R}^d) \quad (\forall r > 0).$$

provide product and commutator estimates with functions in the Zygmund class $C^s_*(\mathbb{R}^d)$ defined in (B.8). We prove here the Zygmund version of the standard product and commutator estimates of Propositions B.2 and B.8; the other estimates from the previous sections can be adapted in the same way.

PROPOSITION B.18. *The following estimates hold.*

(1) *Let* $s \in \mathbb{R}$, $r > 0$ *and* $f \in C^{|s|+r}_*(\mathbb{R}^d)$, $g \in H^s(\mathbb{R}^d)$. *Then* $fg \in H^s(\mathbb{R}^d)$ *and*

$$|fg|_{H^s} \lesssim |f|_{C_*^{|s^-|+r}} |g|_{H^s} + \big\langle |f|_{C^{s+r}_*} |g|_2 \big\rangle_{s>0},$$

where $s^- = \min\{s, 0\}$.

(2) *For all* $s \geq 0$, $r > 0$ *and* $f \in C^{1+r}_*(\mathbb{R}^d) \cap C^{s+r}_*(\mathbb{R}^d)$, *and all* $g \in H^{s-1}(\mathbb{R}^d)$, *one has*

$$\big|[\Lambda^s, f]g\big|_2 \lesssim |f|_{C^{1+r}_*} |g|_{H^{s-1}} + \big\langle |f|_{C^{s+r}_*} |g|_2 \big\rangle_{s>1}.$$

REMARK B.19.

(1) When $s = k$ is an integer, the product estimate of the proposition can be improved with standard tools. For instance

$$\forall k \in \mathbb{Z}, \qquad |fg|_{H^k} \lesssim |f|_{W^{|k|,\infty}} |g|_{H^k}.$$

Similar improvements hold for the commutator estimate when s is an even integer (or if Λ^s, with s an integer, is replaced by a differential operator).

(2) Further extensions to product and commutator estimates on the strip $\mathcal{S}$ can be obtained as in §B.1.2 and §B.2.2 and are left to the reader.

PROOF. The product estimate follows easily from the paraproduct identity (B.10), the estimates (B.11)–(B.13), as well as the following ones that we prove here

$$\text{(B.22)} \qquad \forall t \geq 0, \forall r > t, \forall s \in \mathbb{R}, \qquad |T_f g|_{H^s} \lesssim |f|_{H^{-t}} |g|_{C^{s+r}_*},$$

$$\text{(B.23)} \qquad \forall s + r > 0, \qquad |R(f, g)|_{H^{s+r}} \lesssim |f|_{H^s} |g|_{C^r_*}.$$

Recall that

$$T_f g = \sum_{p \geq -1} \psi_{p-N}(D) f \varphi_p(D) g,$$

and note that

$$\begin{aligned} 2^{2ps} |\psi_{p-N}(D) f \varphi_p(D) g|_2^2 &\leq 2^{2ps} |\Lambda^t \psi_{p-N}(D) \Lambda^{-t} f|_2^2 \, |\varphi_p(D) g|_\infty^2 \\ &\lesssim 2^{2ps}\, 2^{2pt} |f|_{H^{-t}}^2 \, 2^{-2p(s+r)} |g|^2_{C^{s+r}_*}. \end{aligned}$$

It follows that

$$\sum_{p \geq -1} 2^{2ps} |\psi_{p-N}(D) f \varphi_p(D) g|_2^2 \lesssim \Big(\sum_{p \geq -1} 2^{2p(t-r)} \Big) |f|^2_{H^{-t}} \, |g|^2_{C^{s+r}_*}.$$

Since the summation converges for $r > t$, (B.22) follows from Lemma B.13. For (B.23), which is classical, we refer, for instance, to Theorem 2.4.1 of [**71**] (see also Proposition 23 of [**219**]).

For the commutator estimate, we use the decomposition (B.14), namely,

$$[\Lambda^s, f]g = [\Lambda^s, T_f]g + \Lambda^s T_g f - T_{\Lambda^s g} f + \Lambda^s R(f, g) + R(f, \Lambda^s g).$$

For the first component of the right-hand side, we get (see Footnote 1)

$$\forall s \in \mathbb{R}, \qquad |[\Lambda^s, T_f]g|_2 \lesssim |f|_{W^{1,\infty}} |g|_{H^{s-1}}.$$

For the second and third ones, we use (B.22) to get

$$\forall r>0, \forall s\geq 0, \qquad |T_g f|_{H^s} + |T_{\Lambda^s g} f|_2 \lesssim |f|_{C_*^{1+r}} |g|_{H^{s-1}} + \big\langle |f|_{C_*^{s+r}} |g|_2 \big\rangle_{s>1}.$$

Finally, the last two components can be controlled using (B.23)

$$\forall s\geq 0, \quad \forall r>0, \qquad |R(f,g)|_{H^s} + |R(f,\Lambda^s g)|_2 \lesssim |f|_{C_*^{1+r}} |g|_{H^{s-1}}.$$

Bringing all these estimates together yields the commutator estimate stated in the proposition. □

B.4. Product and commutator in uniformly local Sobolev spaces

B.4.1. Uniformly local Sobolev spaces. Uniformly local Sobolev spaces have been introduced by Kato [**200**]; they have the some local regularity properties as the standard Sobolev spaces but contain functions that do not vanish at infinity (such as smooth step functions or periodic functions). They are defined as follows.

DEFINITION B.20. Let $\phi \in \mathcal{D}(\mathbb{R}^d)$ (not identically zero) and $s\in\mathbb{R}$. We denote by $H^s_{ul}(\mathbb{R}^d)$ the set of all measurable functions f on $\mathbb{R}^d$ such that $|f|_{H^s_{ul}} < \infty$, where

$$|f|_{H^s_{ul}} = \sup_{x\in\mathbb{R}^d} |\phi(\cdot - x) f|_{H^s}.$$

REMARK B.21. The choice of the smooth and compactly supported function ϕ in the above definition is not important, since a different choice leads to an equivalent norm. Similarly, one defines an equivalent norm by taking

$$|f|_{H^s_{ul}} = \sup_{q\in\mathbb{Z}^d} |\Phi(\cdot - q) f|_{H^s},$$

where $\Phi\in\mathcal{D}(\mathbb{R}^d)$ with support in the ball $B(0,1)$, equal to 1 in $B(0,1/2)$, and such that

$$\forall x\in\mathbb{R}^d, \qquad \sum_{q\in\mathbb{Z}^d} \Phi(x-q) = 1.$$

While standard Sobolev spaces are obviously continuously embedded in their uniformly local version

$$\forall s\in\mathbb{R}, \quad \forall f\in H^s(\mathbb{R}^d), \qquad |f|_{H^s_{ul}} \lesssim |f|_{H^s},$$

the norms $|\cdot|_{H^s_{ul}}$ and $|\cdot|_{H^s}$ are not equivalent on $H^s(\mathbb{R}^d)$, i.e., the spaces H^s and H^s_{ul} cannot be permuted in the above inequality. One has, however, the following property, where we recall the notation $\langle x\rangle = (1+|x|^2)^{1/2}$. This property ensures in particular that uniformly local Sobolev spaces are included in $\mathcal{S}'(\mathbb{R}^d)$.

PROPOSITION B.22. *Let $s\in\mathbb{R}$ and $N\geq d+1$. Then one has*

$$\sup_{x\in\mathbb{R}^d} |\langle\cdot - x\rangle^{-N} f|_{H^s} \lesssim |f|_{H^s_{ul}},$$

for all $f\in H^s_{ul}(\mathbb{R}^d)$.

PROOF. We reproduce here the proof of [**7**]. Owing to Remark B.21, one has

$$|\langle\cdot - x\rangle^{-N} f|_{H^s} \leq \sum_{q\in\mathbb{Z}^d} |\langle\cdot - x\rangle^{-N}\Phi(\cdot - q) f|_{H^s}.$$

Note now that

$$\langle y-x\rangle^{-N}\Phi(y-q)f(y) = \frac{1}{\langle x-q\rangle^N}\frac{\langle x-q\rangle^N}{\langle x-y\rangle^N}\tilde\Phi(y-q)\Phi(y-q)f(y),$$

where $\tilde{\Phi} \in \mathcal{D}(\Omega)$ is equal to 1 on the support of Φ, and that the mapping

$$y \mapsto \frac{\langle x-q\rangle^N}{\langle x-y\rangle^N}\tilde{\Phi}(y-q)$$

is bounded together with all its derivatives and with semi-norms uniformly bounded with respect to x and q. We therefore obtain

$$\sum_{q\in\mathbb{Z}^d} |\langle \cdot - x\rangle^{-N}\Phi(\cdot-q)f|_{H^s} \lesssim \sum_{q\in\mathbb{Z}^d}\frac{1}{\langle x-q\rangle^N}|f|_{H^s_{ul}} \lesssim |u|_{H^s_{ul}}.$$

The last inequality is a consequence of the assumption $N \geq d+1$. Taking the supremum over all $x \in \mathbb{R}^d$ then yields the result. □

B.4.2. Product estimates. As for standard Sobolev spaces, one has the continuous embedding

$$\text{(B.24)} \qquad \forall s > d/2, \qquad H^s_{ul}(\mathbb{R}^d) \subset L^\infty(\mathbb{R}^d)\cap C(\mathbb{R}^d).$$

The following straightforward generalizations of Proposition B.2 also hold.

PROPOSITION B.23. *Let $t_0 > d/2$.*

(1) *Let $s \geq -t_0$ and $f \in H^s \cap H^{t_0}(\mathbb{R}^d)$, $g \in H^s_{ul}(\mathbb{R}^d)$. Then $fg \in H^s(\mathbb{R}^d)$ and*

$$|fg|_{H^s} \lesssim |f|_{H^{t_0}}|g|_{H^s_{ul}} + \big\langle |f|_{H^s}|g|_{H^{t_0}_{ul}}\big\rangle_{s>t_0}.$$

(1′) *Let $s \geq -t_0$ and $f \in H^s_{ul} \cap H^{t_0}_{ul}(\mathbb{R}^d)$, $g \in H^s_{ul}(\mathbb{R}^d)$. Then $fg \in H^s_{ul}(\mathbb{R}^d)$ and*

$$|fg|_{H^s_{ul}} \lesssim |f|_{H^{t_0}_{ul}}|g|_{H^s_{ul}} + \big\langle |f|_{H^s_{ul}}|g|_{H^{t_0}_{ul}}\big\rangle_{s>t_0}.$$

(2) *Let $s_1, s_2 \in \mathbb{R}$ be such that $s_1+s_2 \geq 0$. Then for all $s \leq s_j$ ($j=1,2$) and $s < s_1+s_2-d/2$, and all $f \in H^{s_1}(\mathbb{R}^d)$, $g \in H^{s_2}_{ul}(\mathbb{R}^d)$, one has $fg \in H^s(\mathbb{R}^d)$ and*

$$|fg|_{H^s} \lesssim |f|_{H^{s_1}}|g|_{H^{s_2}_{ul}}.$$

(2′) *Let $s_1, s_2 \in \mathbb{R}$ be such that $s_1+s_2 \geq 0$. Then for all $s \leq s_j$ ($j=1,2$) and $s < s_1+s_2-d/2$, and all $f \in H^{s_1}_{ul}(\mathbb{R}^d)$, $g \in H^{s_2}_{ul}(\mathbb{R}^d)$, one has $fg \in H^s_{ul}(\mathbb{R}^d)$ and*

$$|fg|_{H^s_{ul}} \lesssim |f|_{H^{s_1}_{ul}}|g|_{H^{s_2}_{ul}}.$$

(3) *Let $F: \mathbb{R} \mapsto \mathbb{R}$ be a smooth function such that $F(0)=0$. Also let $s \geq 0$ and $f \in H^s_{ul} \cap L^\infty(\mathbb{R}^d)$. Then $F(f) \in H^s_{ul}(\mathbb{R}^d)$ and*

$$|F(f)|_{H^s_{ul}} \leq C(|f|_\infty)|f|_{H^s_{ul}}.$$

PROOF. Let us prove, for instance, the third point. Choosing $\tilde{\phi} \in \mathcal{D}(\mathbb{R}^d)$ such that $\tilde{\phi}$ is equal to 1 on the support of ϕ, one has for all $x \in \mathbb{R}$, $\phi(\cdot - x)F(f) = \phi(\cdot-x)F(\tilde{\phi}(\cdot-x)f)$, and therefore, using the third point of Proposition B.2

$$\begin{aligned} |\phi(\cdot-x)F(f)|_{H^s} &\leq C(|\tilde{\phi}(\cdot-x)f|_\infty)|\tilde{\phi}(\cdot-x)f|_{H^s} \\ &\leq C(|f|_\infty)|\tilde{\phi}(\cdot-x)f|_{H^s}, \end{aligned}$$

and taking the sup over all $x \in \mathbb{R}^d$ yields the result. □

The next proposition is a similar generalization of Proposition B.4.

PROPOSITION B.24. *Let $t_0 > d/2$, $s \geq -t_0$ and $c_0 > 0$. Also let $f \in H^s(\mathbb{R}^d)$ and $g \in H^s_{ul} \cap H^{t_0}(\mathbb{R}^d)$ be such that for all $X \in \mathbb{R}^d$, one has $1 + g(X) \geq c_0$. Then $\frac{f}{1+g}$ belongs to $H^s(\mathbb{R}^d)$ and*

$$|\frac{f}{1+g}|_{H^s} \leq C(\frac{1}{c_0}, |g|_{H^{t_0}_{ul}})\big(|f|_{H^s} + \big\langle |f|_{H^{t_0}}|g|_{H^s_{ul}}\big\rangle_{s>t_0}\big).$$

The extensions to product estimates on the flat strip similar to those of §B.1.2 can easily be deduced from the above propositions; they are omitted here.

B.4.3. Commutator estimates. We give here only a partial generalization of Corollary B.9 to uniformly local Sobolev spaces. Other generalizations are left to the reader. We consider here only a commutator with $\Lambda^s = (1-\Delta)^s$ when s is an even integer. The case of s an odd integer can be deduced quite easily (see [**200**]), but the commutator estimate is more delicate to derive for noninteger values of s, and a fortiori for a general Fourier multiplier of order s, $\sigma \in \mathcal{S}^s$.

COROLLARY B.25. *Let $t_0 > d/2$ and $k \in \mathbb{N}$, $k \neq 0$. If $f \in H^{2k}_{ul} \cap H^{t_0+1}_{ul}$ then, for all $g \in H^{2k-1}$,*

$$|[\Lambda^{2k}, f]g|_2 \lesssim \mathcal{N}^s(\sigma)\big(|\nabla f|_{H^{t_0}_{ul}}|g|_{H^{s-1}} + \big\langle |\nabla f|_{H^{s-1}_{ul}}|g|_{H^{t_0}}\big\rangle_{s>t_0+1}\big).$$

PROOF. Since Λ^{2k} is a differential operator of order $2k$, we can write

$$[\Lambda^{2k}, f]g = \sum_{\alpha,\beta\in\mathbb{N}^d, 0<|\alpha|\leq 2k, |\alpha|+|\beta|=2k} *_{\alpha,\beta}\partial^\alpha f \partial^\beta g,$$

where the $*_{\alpha,\beta}$ are numerical constants of no importance. Owing to the assumptions made on f and g, we have

$$\partial^\alpha f \in H^{2k-|\alpha|}_{ul} \cap H^{t_0+1-|\alpha|}_{ul}, \qquad \partial^\beta g \in H^{|\alpha|-1},$$

and the result therefore follows from the product estimates of Proposition B.23. □

APPENDIX C

Asymptotic Models: A Reader's Digest

We present here a brief review of the various asymptotic models derived in these notes. We refer to Figure C.1 for a graphical presentation of the shallow water models and to the following comments for a reminder of the elements that can be found in these notes concerning the range of validity, the derivation, and the justification of these asymptotic models.

C.1. What is a *fully justified* asymptotic model?

Let us first recall that the dimensionless parameters ε, μ, β and γ (respectively called nonlinearity, shallowness, topography, and transversality parameters) are defined in (1.17). The physical configurations considered in most of these notes range from shallow to moderately deep water, which amounts to assuming that

$$0 < \mu < \mu_{max},$$

for some $\mu_{max} > 0$ not necessarily small, but finite. The dimensionless water waves equations are then given by

$$\text{(C.1)} \qquad \begin{cases} \partial_t \zeta - \frac{1}{\mu}\mathcal{G}\psi = 0, \\ \partial_t \psi + \zeta + \frac{\varepsilon}{2}|\nabla^\gamma \psi|^2 - \frac{\varepsilon}{\mu}\frac{(\mathcal{G}\psi + \varepsilon\mu\nabla^\gamma\zeta\cdot\nabla^\gamma\psi)^2}{2(1+\varepsilon^2\mu|\nabla^\gamma\zeta|^2)} = 0. \end{cases}$$

See also (1.28) for a more general nondimensionlazation allowing very deep water configurations, and (4.65) in infinite depth.

Further assumptions on the size of ε, μ, β and γ correspond to asymptotic regimes, to which one can associate one or several asymptotic models. Given a particular asymptotic regime, the main issue regarding the validity of an asymptotic model is then the following:

(1) Do the solutions of the water waves equations exist on the relevant time scale?
(2) Do the solutions of the asymptotic model exist on the same time scale?
(3) Are these solutions close to the solution of the water waves equations with corresponding initial data? And how close?

Full justification. When the answer to these three questions is positive, we say that the asymptotic model is *fully justified*. Theorem 4.16 provides an existence theorem[1] for the water waves equations (C.1) that basically always brings a positive answer to the first question. Answering the second question requires a specific

[1] See Theorem 4.36 for infinite depth and Theorem 9.6 in the presence of surface tension.

analysis[2] of the asymptotic model under consideration. It would have required a lot of place and energy to perform this analysis for all the models derived here, and we therefore focused our attention on a few models that cover the most important physical configurations. It is also this analysis of the asymptotic model that allows us to give a positive answer to the third question and therefore to give a full justification of these models. **Almost full justification**. For the asymptotic models for which we did *not* provide local existence and stability results, we cannot provide a full justification. However, the strategy proposed here is quite robust and a full justification would follow directly from the proof of such well-posedness results (which are sometimes straightforward and sometimes more involved). Indeed, for these models, our analysis allows us to say that solutions to these models, if they exist, provide good approximations to the water waves equations[3]. We then say that these models are *almost fully justified* (see §6.2.4 for more details).

Below is a list of the models derived in these notes, with a reminder of their range of validity and their justification status (i.e., fully or almost fully justified).

C.2. Shallow water models

C.2.1. Low precision models. When μ is small but without further assumption on the parameters ε, β and γ, the roughest model, obtained by dropping all $O(\mu)$ terms in (C.1), is the **Nonlinear Shallow Water** (or Saint-Venant) equations (5.7). This model is *fully justified* in Theorem 6.10: for times $t \in [0, \frac{T}{\varepsilon \vee \beta}]$ (with $T > 0$ independent of μ), its precision is $O(\mu t)$. In particular, denoting by (ζ, ψ) the solution to (C.1) and by $(\zeta_{SW}, \overline{V}_{SW})$ the solution to (5.7) with corresponding initial data, one has

$$\forall t \in [0, \frac{t}{\varepsilon \vee \beta}], \qquad \big|(\zeta, \nabla\psi) - (\zeta_{SW}, \overline{V}_{SW})\big|_{L^\infty([0,t]\times\mathbb{R}^d)} \leq C\mu t.$$

C.2.2. High precision models. A better precision is obtained when the $O(\mu)$ terms are kept in the equations and only $O(\mu^2)$ terms are dropped. The relevant time scale is still $t \in [0, \frac{T}{\varepsilon \vee \beta}]$, but the precision of the models is improved into $O(\mu^2 t)$. It is possible to work in full generality (no assumption on ε, β and γ) or to make some extra physical assumptions (e.g., small topography variations $\beta = O(\mu)$). This leads to many different models[4] (we refer to Figure C.1 for a graphical presentation):

♯1 *Large surface and topography variations:* $\varepsilon = O(1)$, $\beta = O(1)$. The **Green-Naghdi** equations (also called Serre or fully nonlinear Boussinesq equations) are the most general (but most complicated) of the models presented here. The

---

[2] More precisely, a local existence theorem on the relevant time scale, and a stability property with respect to perturbations (see, for instance, Propositions 6.1 and 6.3 for the particular case of the nonlinear shallow water equations).

[3] One has to prove only that the solutions to these models are consistent with other models that are fully justified (e.g., solutions to the Green-Naghdi equations with improved frequency dispersion are consistent at order $O(\mu^2)$ with the standard Green-Naghdi equations, or with the water waves equations themselves (as for the deep water model (8.2)). The result then follows by the stability property already proved for the latter equations (e.g., Proposition 6.5 for the GN equations, Theorem 4.18 for the water waves equations).

[4] These models are different because they are derived under different assumptions, but the relevant time scale and the precision of the approximation is the same for all of them.

standard form of these equations is given in (5.11), or equivalently (5.15). Variants with improved frequency dispersion are given in (5.36) and §5.2.2.2. The *full* justification of the standard Green-Naghdi equations (5.11) is given in Theorem 6.15. Its variants are *almost fully* justified (with the terminology introduced above).

♯2 *Small surface and large topography variations:* $\varepsilon = O(\mu)$, $\beta = O(1)$. To this regime corresponds the **Boussinesq-Peregrine** model (5.23), which requires a small amplitude assumption (namely, $\varepsilon = O(\mu)$) that is not required in ♯1. Variants with improved frequency dispersion are (5.28) and (5.31). All these models are *almost fully* justified.

♯3 *Medium surface and topography variations:* $\varepsilon = O(\sqrt{\mu})$, $\beta = O(\sqrt{\mu})$. In this regime, the Green-Naghdi equations (5.11) can be simplified into the **medium amplitude Green-Naghdi** equations (5.21). This model is *almost fully* justified.

♯4 *Large surface and small topography variations:* $\varepsilon = O(1)$, $\beta = O(\mu)$. Somehow symmetric to the Boussinesq-Peregrine model, the **Green-Naghdi equations with almost flat bottom** (5.18) allow for large amplitude waves, but over small amplitude topography variations. This model is *almost fully* justified[5].

♯5 *Medium surface and small topography variations:* $\varepsilon = O(\sqrt{\mu})$, $\beta = O(\mu)$. The Green-Naghdi equations can then be simplified into (5.22); this model is *almost fully* justified.

♯6 *Small surface and topography variations:* $\varepsilon = O(\mu)$, $\beta = O(\mu)$. This is the well-known long waves regime for which the **Boussinesq** system (5.25) can be derived. Many variants with different frequency dispersion are given in (5.34), and variants with symmetric nonlinearity are also given in (5.40) and (5.41). The Boussinesq systems with symmetric dispersion and nonlinearity are *fully justified* in Theorem 6.20; all the other Boussinesq systems are *almost fully justified.*

C.2.3. Approximation by scalar equations. When the surface elevation is of medium or small amplitude ($\varepsilon = O(\sqrt{\mu})$ or $\varepsilon = O(\mu)$) and for flat bottoms[6] ($\beta = 0$), it is possible in some cases to approximate the water waves equations (C.1) by scalar equations rather than systems. The relevant time scale is $O(1/\varepsilon)$ for all the models described below, but the precision may vary.

♯7 *Medium amplitude wave in* 1d: $\varepsilon = O(\sqrt{\mu})$, $d = 1$. For well-prepared initial data, the **Camassa-Holm type equations** (7.47) (velocity based approximation) and (7.51) or (7.53) (surface elevation based approximation) are *fully* justified in Theorems 7.24 and 7.26 respectively. The precision of the approximation is $O(\mu^2 t)$. See also (7.72) for a version of these equations with full dispersion.

♯8 *Small amplitude waves in* 1d: $\varepsilon = O(\mu)$, $d = 1$. When $\varepsilon = \mu$ (KdV regime) then the **KdV/BBM equations** (7.7) are *fully* justified and furnish an approximation of precision $O(\mu\sqrt{t})$ in general and $O(\mu)$ under a decay assumption on the initial data (see Corollaries 7.2 and 7.12). Under the weaker assumption $\varepsilon = O(\mu)$, but for well-prepared initial data only, the precision is $O(\mu^2 t)$ (see Corollary 7.37). See also (7.71) for the KdV equation with full dispersion.

---

[5]In the case of flat bottoms, this model coincides, however, with the standard Green-Naghdi equations (5.11) and is therefore *fully* justified.

[6]We have given in Chapter 7 some references where this assumption is weakened.

♯9 *Small amplitude, weakly transverse 2d waves:* $\varepsilon = \mu$, $d = 2$ *and* $\gamma = \sqrt{\varepsilon}$. The **KP/KP-BBM equations** (7.29) are *fully* justified in Theorem 7.16 under strong assumptions on the initial data and with a $o(1)$ precision only. See also (7.73) for the KP equation with full dispersion.

C.3. Deep water and infinite depth models

Shallow water expansions are irrelevant when μ is not small, but small amplitude expansions are possible. More precisely, asymptotic models can be derived with respect to the *steepness* of the wave

$$\epsilon = \varepsilon\sqrt{\mu} = \frac{a}{L_x},$$

where we recall (see §1.3.1) that a is the typical amplitude and L_x the typical horizontal length of the wave.
With a formal $O(\epsilon^2)$ precision, we derived in this context the so-called **full dispersion model** (8.2) in finite depth (see also (8.4) for nonflat bottoms and (8.10) for infinite depth). This model is *almost fully*[7] justified in Theorem 8.4.

C.4. Modulation equations

See Figure C.2 for a graphical presentation of the comments presented below and devoted to modulation equations in finite and infinite depth.

C.4.1. Modulation equations in finite depth. Modulation equations are generally used in rather deep water. We therefore use the "deep water" nondimensionalization of the water waves equations

$$\text{(C.2)} \qquad \begin{cases} \partial_t\zeta - \dfrac{1}{\sqrt{\mu}}\mathcal{G}\psi = 0, \\ \partial_t\psi + \zeta + \dfrac{\epsilon}{2}|\nabla\psi|^2 - \epsilon\dfrac{(\frac{1}{\sqrt{\mu}}\mathcal{G}\psi + \epsilon\nabla\zeta\cdot\nabla\psi)^2}{2(1+\epsilon^2|\nabla\zeta|^2)} = 0. \end{cases}$$

The so-called *modulation equations* are evolution equations of slowly modulated *wave packets* of the form

$$\begin{pmatrix} \zeta(t,X) \\ \psi(t,X) \end{pmatrix} = \begin{pmatrix} i\omega\psi_{01}(\epsilon t,\epsilon X) \\ \psi_{01}(\epsilon t,\epsilon X) \end{pmatrix} e^{i(\mathbf{k}\cdot X-\omega t)} + \text{c.c.} + O(\epsilon),$$

with $\omega = \omega(\mathbf{k}) = \big(|\mathbf{k}|\tanh(\sqrt{\mu}|\mathbf{k}|)\big)^{1/2}$.
The evolution of such wave packets on a time scale $t = O(1/\epsilon)$ is governed by a transport equation at the group velocity $c_g = \nabla\omega(\mathbf{k})$. Over the larger time scale $t = O(1/\epsilon^2)$, more complex nonlinear and dispersive effects must be taken into account. The equations below have a formal $O(\epsilon^3)$ precision.

♯10 *Full dispersion and standard Benney-Roskes equations.* These equations couple the evolution of the amplitude ψ_{01} of the leading order term of the wave packet to a nonoscillating mode created by the nonlinearities. A version with full dispersion[8] is proposed in (8.32). The classical version of the Benney-Roskes equation is then recovered by a simple Taylor expansion of the dispersive terms

[7] The obstruction to a full justification is that there is no local well-posedness result for these deep water models.

[8] This means that its linear dispersive properties are the same as those of the water waves equations.

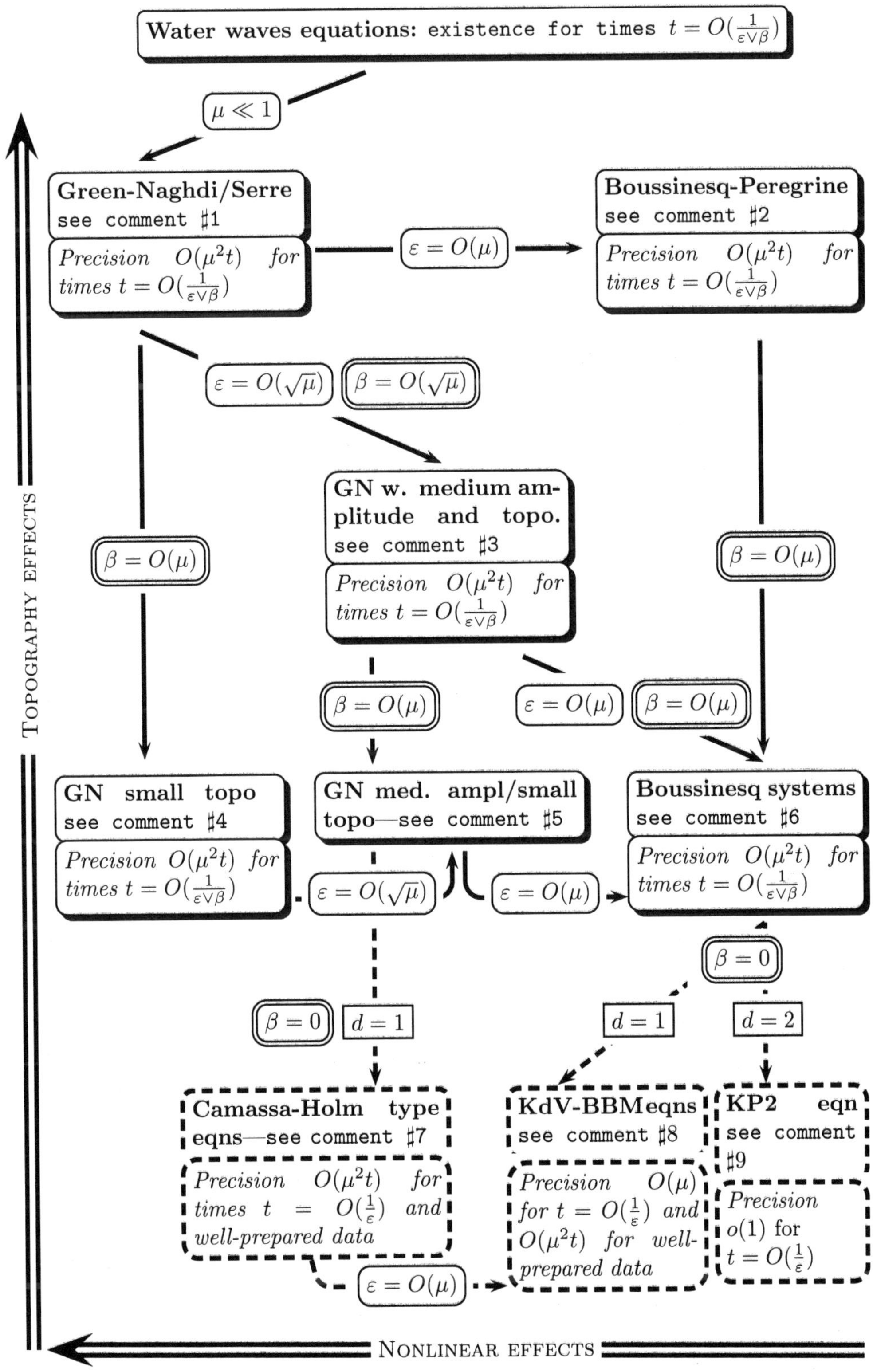

FIGURE C.1. Shallow water models. Approximations by systems correspond to solid boxes; approximations by scalar equations correspond to dashed boxes. See §C.2.2 and §C.2.3 for comments.

in (8.34). Assuming[9] that these equations are well-posed on the relevant time scale, a consistency result is given in Propositions 8.13 and 8.17 respectively.

♯11 *Davey-Stewartson, cubic Schrödinger equations, and variants.* For well-prepared initial data, one can write the Benney-Roskes equations in a frame moving at the group velocity and approximate them by the Davey-Stewartson equations (8.36) in dimension $d = 2$. In dimension $d = 1$, these equations degenerate into a cubic nonlinear Schrödinger equation (8.25). For both models, the mean mode is now slaved to the oscillating mode[10]. Consistency results[11] for the approximations based on these equations are given in Propositions 8.22 and 8.25 respectively. Variants of these equations with full or improved dispersion are given in §8.5.2 and §8.5.3 respectively.

C.4.2. Modulation equations in infinite depth. In infinite depth, the water waves equations are given by (4.65)

$$\begin{cases} \partial_t\zeta - \mathcal{G}[\epsilon\zeta]\psi = 0, \\ \partial_t\psi + \zeta + \dfrac{\epsilon}{2}|\nabla\psi|^2 - \epsilon\dfrac{(\mathcal{G}[\epsilon\zeta]\psi + \epsilon\nabla\zeta\cdot\nabla\psi)^2}{2(1+\epsilon^2|\nabla\zeta|^2)} = 0. \end{cases}$$

In infinite depth, the creation of a nonoscillating mode by nonlinear interaction of the oscillating modes (rectification) is less important.

♯12 *Nonlinear Schrödinger equations and variants.* Since the nonoscillating modes are of lower order in infinite depth, the Benney-Roskes equation degenerate into the cubic nonlinear Schrödinger equation (8.49) in both dimensions $d = 1$ and $d = 2$. Variants of these equations with full or improved dispersion are given in §8.5.2 and §8.5.3 respectively. The same kind of consistency results as in the finite depth case can be established[12].

♯13 *Higher order model.* A higher order model with formal $O(\epsilon^4)$ precision is provided by the Dysthe equation (8.58).

C.5. Influence of surface tension

C.5.1. On shallow water models. See §9.2, §9.3, and §9.4 for the modifications that need to be made to the shallow water models in the presence of surface tension. These modifications are essentially changes in the third order dispersive terms of the models. Strong qualitative differences are observed in three situations (of limited physical interest):

(1) *Critical Bond number, 1d waves:* $\frac{1}{\mathrm{Bo}} = \mu\frac{1}{\mathrm{bo}}$, *with* $\mathrm{bo} \sim 3$, *and* $d = 1$. Dispersive effects disappear from the KdV equation and can be observed only in a less nonlinear regime where a good approximation is provided by the **Kawahara** equation (9.46).

(2) *Critical Bond number, weakly transverse 2d waves:* $\frac{1}{\mathrm{Bo}} = \mu\frac{1}{\mathrm{bo}}$, *with* $\mathrm{bo} \sim 3$, *and* $d = 2$, $\gamma = \sqrt{\varepsilon}$. The natural extension of the previous situation is provided by the **weakly transverse Kawahara equation** (9.54).

[9] Such a property is not known. There is a local well-posedness result for (8.34) in [**273**], but not on the relevant time scale.

[10] It is given by an evolution equation coupled to the equation on the oscillating mode in the Benney-Roskes model.

[11] For the NLS equation, a full justification has recently been established in [**134**].

[12] In the one-dimensional case $d = 1$, the NLS approximation has been fully justified in [**315**].

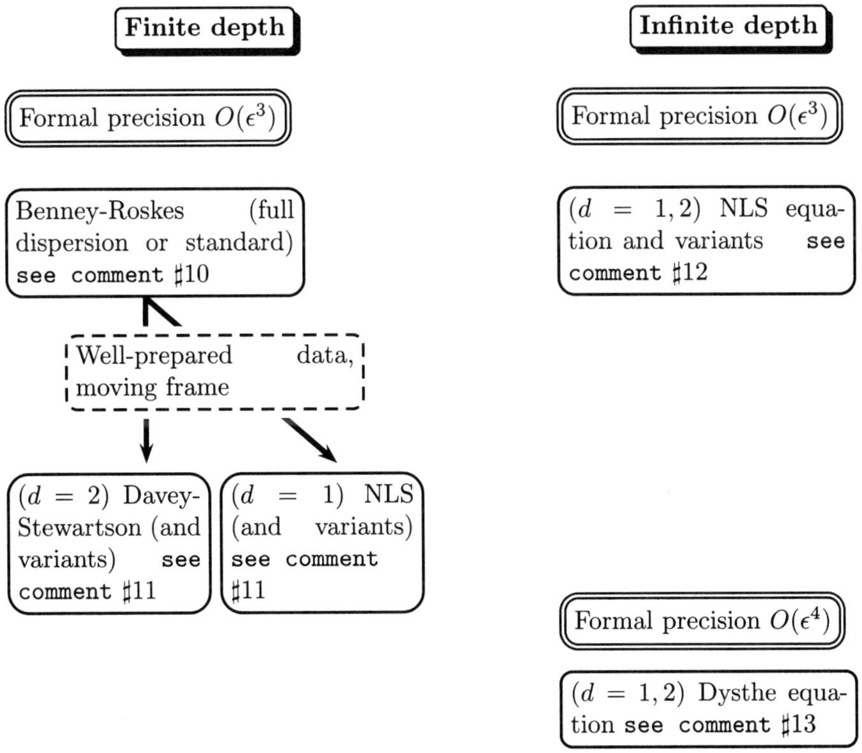

FIGURE C.2. Modulation equations in finite and infinite depth.

(3) *Strong surface tension, weakly transverse 2d waves:* bo < 3, $d = 2$, $\gamma = \sqrt{\varepsilon}$. The sign of the third-order dispersive terms in the KP equations (7.29) changes: it is now a **KP1 equation** (as opposed to the KP2 equation [**208**]).

C.5.2. On deep water models and modulation equations. Deep water models with surface tension are derived in §9.4, and capillary effects on modulation equations are mentioned in §8.5.6.

Bibliography

[1] M. J. ABLOWITZ, *Nonlinear dispersive waves.* Asymptotic analysis and solitons. Cambridge Texts in Applied Mathematics. Cambridge University Press, New York, 2011. xiv+348 pp.

[2] M. J. ABLOWITZ, H. SEGUR, *On the evolution of packets of water waves*, J. Fluid Mech. **92** (1979), 691-715.

[3] M. J. ABLOWITZ, H. SEGUR, *Solitons and the inverse scattering transform*, SIAM Studies in Applied Mathematics (Philadelphia), 1981.

[4] T. ALAZARD, N. BURQ, C. ZUILY, *On the Cauchy problem for the water waves with surface tension*, Duke Math. J. **158** (2011), 413-499.

[5] T. ALAZARD, N. BURQ, C. ZUILY, *Strichartz estimates for water waves*, Ann. Sci. École Norm. Sup., to appear.

[6] T. ALAZARD, N. BURQ, C. ZUILY, *On the Cauchy problem for gravity water waves*, submitted.

[7] T. ALAZARD, N. BURQ, C. ZUILY, *Cauchy theory for the water waves system in the uniformly local Sobolev spaces*, submitted.

[8] G. B. AIRY, *Tides and waves*, Encyclopedia Metropolitana, vol. 5 (1845), 291-396.

[9] T. ALAZARD, G. MÉTIVIER, *Paralinearization of the Dirichlet to Neumann operator, and regularity of three-dimensional water waves*, Comm. Partial Differential Equations **34** (2009), 1632-1704.

[10] S. ALINHAC, *Existence d'ondes de rarfaction pour des systmes quasi-linaires hyperboliques multidimensionnels* (French) [Existence of rarefaction waves for multidimensional hyperbolic quasilinear systems] Comm. Partial Differential Equations **14** (1989), 173-230.

[11] S. ALINHAC, P. GÉRARD, *Opérateurs pseudo-différentiels et théorème de Nash-Moser*, Savoirs Actuels. [Current Scholarship]. InterEditions, Paris, 1991.

[12] B. ALVAREZ-SAMANIEGO, D. LANNES, *Large time existence for 3d water-waves and asymptotics*, Invent. Math. **171** (2008), 485-541.

[13] B. ALVAREZ-SAMANIEGO, D. LANNES, *A Nash-Moser theorem for singular evolution equations. Application to the Serre and Green-Naghdi equations*, Indiana Univ. Math. J. **57** (2008), 97-131.

[14] D. M. AMBROSE, N. MASMOUDI, *The zero surface tension limit of two-dimensional water waves*, Comm. Pure Appl. Math. **58** (2005), 1287-1315.

[15] D. M. AMBROSE, N. MASMOUDI, *The zero surface tension limit of three-dimensional water waves*, Indiana Univ. Math. J. **58** (2009), 479-521.

[16] V. I. ARNOLD, *Sur la géométrie differentielle des groups de Lie de dimension infinie et ses applications à l'hydrodynamique des fluides parfaits*, Ann. Inst. Fourier (Grenoble) **16** (1966), 319-361.

[17] W. ARTILES, A. NACHBIN, *Asymptotic nonlinear wave modeling through the Dirichlet-to-Neumann operator*, Methods Appl. Anal. **11** (2004), 475-492.

[18] C. BARDOS, D. LANNES, *Mathematics for 2d interfaces*, in Panorama et Synthèses, to appear.

[19] E. BARTHELEMY, *Nonlinear shallow water theories for coastal waves*, Surveys in Geophysics **25** (2004), 315-337.

[20] T. BEALE, T. HOU, J. LOWENGRUB, *Growth rates for the linearized motion of fluid interfaces away from equilibrium*, Comm. Pure Appl. Math. **46** (1993), 1269-1301.

[21] S. BEJI, J. A. BATTJES, *Experimental investigation of wave propagation over a bar*, Coastal Engineering **19** (1993), 151-162.

[22] T. B. BENJAMIN, *The stability of solitary waves*, Proc. Roy. Soc. London A **328** (1972), 153-183.

[23] T. B. BENJAMIN, J. L. BONA, J. J. MAHONY, *Model equations for long waves in nonlinear dispersive systems*, Phil. Trans. Roy. Soc. London A **272** (1972), 47-78.

[24] T. BENJAMIN, P. OLVER, *Hamiltonian structure, symmetries and conservations laws for water waves*, J. Fluid Mech. **125** (1982), 137-185.

[25] D. J. BENNEY, G. J. ROSKES, *Wave instabilities*, Stud. Appl. Math. **48** (1969), 377-385.

[26] W. BEN YOUSSEF, T. COLIN, *Rigorous derivation of Korteweg-de Vries-type systems from a general class of nonlinear hyperbolic systems*, M2AN Math. Model. Numer. Anal. **34** (2000), 873-911.

[27] W. BEN YOUSSEF, D. LANNES, *The long wave limit for a general class of 2D quasilinear hyperbolic problems*, Commun. Partial Differ. Equations **27** (2002), 979-1020.

[28] L. BERGÉ, S. SKUPIN, *Modeling ultrashort filaments of light*, DCDS **23** (4) 2009, 1099-1139.

[29] K. BEYER, M. GÜNTHER, *On the Cauchy problem for a capillary drop. I. Irrotational motion*, Math. Methods Appl. Sci. **21** (1998), 1149-1183.

[30] K. BEYER, M. GÜNTHER, *The Jacobi equation for irrotational free boundary flows*, Analysis (Munich) **20** (2000), 237-254.

[31] T. B. BENJAMIN, *Instability of periodic wavetrains in nonlinear dispersive systems*, Proc. Roy. Soc. A **299** (1967), 59-75.

[32] T. B. BENJAMIN, J. E. FEIR, *The disintegration of wave trains on deep water. Part 1. Theory*, J. Fluid Mech. **27** (1967), 417-430.

[33] J. L. BONA, *On the stability of solitary waves*, Proc. Roy. Soc. London A **344** (1975), 363-374.

[34] J. L. BONA, M. CHEN, J.-C. SAUT, *Boussinesq equations and other systems for small-amplitude long waves in nonlinear dispersive media. I. Derivation and linear theory*, J. Nonlinear Sci. **12** (2002), 283-318.

[35] J. L. BONA, M. CHEN, J.-C. SAUT, *Boussinesq equations and other systems for small amplitude long waves in nonlinear dispersive media: II. The nonlinear theory*, Nonlinearity **17** (2004), 925-952.

[36] J. L. BONA, T. COLIN, D. LANNES, *Long wave approximations for water waves*, Arch. Ration. Mech. Anal. **178** (2005), 373-410.

[37] J. L. BONA, D. LANNES, J.-C. SAUT, *Asymptotic models for internal waves*, J. Math. Pures Appl., **89** (2008), 538-566.

[38] J. L. BONA, Y. LIU, M. M. TOM, *The Cauchy problem and stability of solitary-wave solutions for RLW-KP-type equations*, J. Differential Equations **185** (2002), 437-482.

[39] J. L. BONA, R. SMITH, *A model for the two-way propagation of water waves in a channel*, Math. Proc. Cambridge Phil. Soc. **79** (1976), 167-182.

[40] P. BONNETON, E. BARTHELEMY, F. CHAZEL, R. CIENFUEGOS, D. LANNES, F. MARCHE, M. TISSIER, *Fully nonlinear weakly dispersive modelling of wave transformation, breaking and runup*, Eur. J. of Mech. B/Fluids, to appear.

[41] P. BONNETON, F. CHAZEL, D. LANNES, F. MARCHE, M. TISSIER, *A splitting approach for the fully nonlinear and weakly dispersive Green-Naghdi model*, J. Comput. Phys **230** (2011), 1479-1498.

[42] J.-M. BONY, *Calcul symbolique et propagation des singularités pour les équations aux dérivés partielles non linéaires*, Ann. Sci. École. Norm. Sup. (4) **14** (1981), no. 2, 209–246.

[43] A. DE BOUARD, W. CRAIG, O. DIAZ-ESPINOSA, P. GUYENNE, C. SULEM, *Long wave expansions for water waves over random topography*, Nonlinearity **21** (2008), 2143-2178.

[44] F. BOUCHUT, A. MANGENEY-CASTELNAU, B. PERTHAME, J. P. VILOTTE, *A new model of Saint Venant and Savage-Hutter type for gravity driven shallow water flows*, C. R. Math. Acad. Sci. Paris **336** (2003), 531-536.

[45] F. BOUCHUT, M. WESTDICKENBERG, *Gravity driven shallow water models for arbitrary topography*, Comm. in Math. Sci. **2** (2004) 359-389.

[46] J. BOURGAIN *On the Cauchy problem for the Kadomtsev-Petviashvili equation*, Geom. Funct. Anal. **3** (1993), 315-341.

[47] J. BOUSSINESQ, *Théorie générale des mouvements qui sont propagés dans un canal rectangulaire horizontal*, C. R. Acad. Sci. Paris **73** (1871), 256-260.

[48] J. BOUSSINESQ, *Théorie des ondes et des remous qui se propagent le long d'un canal rectangulaire horizontal, en communiquant au liquide contenu dans ce canal des vitesses sensiblement pareilles de la surface au fond*, J. Math. Pures Appl. **17** (1872), 55-108.

[49] J. Boussinesq, *Essai sur la theorie des eaux courantes*, Mémoires présentés par divers savants à l'Acad. des Sci. Inst. Nat. France, XXIII (1877), 1-680.

[50] J. Boussinesq, *Lois de l'extinction de la houle en haute mer*, C. R. Acad. Sci. Paris **121** (1895), 15-20.

[51] M. Boutounet, L. Chupin, P. Noble, J.-P. Vila, *Shallow waters viscous flows for arbitrary topography*, Commun. Math. Sci. **6** (2008), 29-55.

[52] R. W. Boyd. Nonlinear optics (Academic Press, 1992).

[53] Y. Brenier, *Minimal geodesics on groups of volume-preserving maps and generalized solutions of the Euler equations*, Commun. Pure Appl. Math. **52** (1999), 411-452.

[54] D. Bresch, *Shallow-Water Equations and Related Topics*, Handbook of Differential Equations: Evolutionary Equations **5** (2009), 1-104.

[55] D. Bresch, G. Métivier, *Global existence and uniqueness for the lake equations with vanishing topography: Elliptic estimates for degenerate equations*, Nonlinearity **19** (2006), 591-610.

[56] D. Bresch, G. Métivier, *Anelastic limits for Euler-type systems*, Appl. Math. Res. Express. AMRX **2** (2010), 119-141.

[57] D. Bresch, P. Noble, *Mathematical Justification of a Shallow Water Model*, Methods Appl. Anal. **14** (2007), 87-117.

[58] T. J. Bridges, A. Mielke, *A proof of the Benjamin-Feir instability*, Arch. Rational Mech. Anal. **133** (1995), 145-198.

[59] B. Buffoni, M. D. Grovesy, S. M. Sung, E. Wahlen, *Existence and conditional energetic stability of three-dimensional fully localised solitary gravity-capillary water waves*, submitted.

[60] B. Buffoni, J. Toland, *Analytic Theory of Global Bifurcation*, Princeton University Press (2003).

[61] A. P. Calderon, *Cauchy integrals on Lipschitz curves and related operators*, Proc. Natl. Acad. Sci. USA **75** (1977), 1324-1327.

[62] R. Camassa, D. Holm, *An integrable shallow water equation with peaked solitons, Phys. Rev. Lett.* **71** (1993), 1661-1664.

[63] R. Camassa, D. D. Holm, C. D. Levermore, *Long-time effects of bottom topography in shallow water*, Phys. D **98** (1996), 258-286.

[64] R. Camassa, D. D. Holm, C. D. Levermore, *Long-time shallow-water equations with a varying bottom*, J. Fluid Mech. **349** (1997), 173-189.

[65] A. Castro, D. Cordoba, C. Fefferman, F. Gancedo, M. Gómez-Serrano, *Splash singularities for water waves*, Proc. Natl. Acad. Sci. **109** (2012), 733-738.

[66] A. Castro, D. Cordoba, C. Fefferman, F. Gancedo, M. Lopez-Fernandez, *Rayleigh-Taylor breakdown for the Muskat problem with applications to water waves*, Annals of Math. **175** (2012), 909-948.

[67] M. Cathala, The nonlinear shallow water equations with nonsmooth topographies: The case of a step, in preparation.

[68] F. Chazel, *Influence of bottom topography on long water waves*, M2AN Math. Model. Numer. Anal. **41** (2007), 771-799.

[69] F. Chazel, M. Benoit, A. Ern, S. Piperno, *A double-layer Boussinesq-type model for highly nonlinear and dispersive waves*, Proc. R. Soc. Lond. A **465** (2009), 2319-2346.

[70] F. Chazel, D. Lannes, F. Marche, *Numerical simulation of strongly nonlinear and dispersive waves using a Green-Naghdi model*, J. Sci. Comput. **48** (2011), 105-116.

[71] J.-Y. Chemin, *Fluides parfaits incompressibles*, Astérisque No. 230 (1995), 177 pp.

[72] J.-Y. Chemin, B. Desjardins, I. Gallagher, E. Grenier, *Basics of Mathematical Geophysics, An introduction to rotating uids and the Navier- Stokes equations*, Oxford Lecture Ser. Math. Appl. **32** (2006), University Press.

[73] M. Chen, *Equations for bi-directional waves over an uneven bottom*, Mathematical and Computers in Simulation **62** (2003), 3-9.

[74] Q. Chen, J. T. Kirby, R. A. Dalrymple, F. Shi, E. B. Thornton, *Boussinesq modeling of longshore currents*, Journal of Geophysical Research **108** (2003).

[75] C. H. Cheng, D. Coutand, S. Shkoller, *On the motion of vortex sheets with surface tension in the 3D Euler equations with vorticity*, Commun. Pure Appl. Math. **61** (2008), 1715-1752.

[76] W. Choi, *Nonlinear evolution equations for two-dimensional surface waves in a fluid of finite depth*, J. Fluid Mech. **295** (1995), 381-394.

[77] H. Christianson, V. Hur, G. Staffilani, *Strichartz estimates for the water-wave problem with surface tension*, Comm. Partial Differential Equations **35** (2010), 2195-2252.

[78] D. Christodoulou, H. Lindblad, *On the motion of the free surface of a liquid*, Comm. Pure Appl. Math. **53** (2000), 1536-1602.

[79] L. Chupin, *Roughness effect on the Neumann boundary condition*, Asymptotic Anal. **78** (2012), 85-121.

[80] R. Cienfuegos, E. Barthélemy, P. Bonneton, *A fourth-order compact nite volume scheme for fully nonlinear and weakly dispersive boussinesq-type equations. Part I: Model development and analysis*, Int. J. Numer. Meth. Fluids **56** (2006), 1217-1253.

[81] D. Clamond, *Note on the velocity and related fields of steady irrotational two-dimensional surface gravity waves*, Phil. Trans. Roy. Soc. A **370**, 1572-1586.

[82] D. Clamond, D. Dutykh, *Practical use of variational principles for modeling water waves*, submitted, http://arxiv.org/abs/1002.3019/, 2011.

[83] R. Coifman, Y. Meyer, *Nonlinear harmonic analysis and analytic dependence*, Pseudo-differential Operators and Applications (Notre Dame, IN, 1984), AMS, Providence, RI, 1985, pp. 71-78.

[84] J. Colliander, M. Keel, G. Staffilani, H. Takaoka, T. Tao, *Sharp global well-posedness results for KdV and modified KdV on* $\mathbb{R}$ *and* $\mathcal{T}$, J. Amer. Math. Soc. **16** (2003), 705-749.

[85] M. Colin, D. Lannes, *Short pulses approximations in dispersive media*, SIAM J. Math. Anal. **41** (2009), 708-732.

[86] T. Colin, *Rigorous derivation of the nonlinear Schrödinger equation and Davey-Stewartson systems from quadratic hyperbolic systems*, Asymptotic Analysis **31** (2002), 69-91.

[87] T. Colin, D. Lannes, *Long-wave short-wave resonance for nonlinear geometric optics*, Duke Math. J. **107** (2001), 351-419.

[88] T. Colin, D. Lannes, *Justification of and long-wave correction to Davey-Stewartson systems from quadratic hyperbolic systems*, Discrete and Continuous Dynamical Systems **11** (2004), 83-100.

[89] A. Constantin, *On the scattering problem for the Camassa-Holm equation, Proc. Roy. Soc. London A* **457** (2001), 953-970.

[90] A. Constantin, *A Hamiltonian formulation for free surface water waves with nonvanishing vorticity*, J. Nonl. Math. Phys. **12** (2005), 202-211.

[91] A. Constantin, *Nonlinear water waves with applications to wave-current interactions and tsunamis*, CBMS-NSF Regional Conference Series in Applied Mathematics **81** (2011).

[92] A. Constantin, *On the particle paths in solitary water waves*, Q. Appl. Math. **68** (2010), 81-90.

[93] A. Constantin, *On the recovery of solitary wave profiles from pressure measurements*, J. Fluid. Mech. **699** (2012), 376-384.

[94] A. Constantin, J. Escher, *Wave breaking for nonlinear nonlocal shallow water equations, Acta Mathematica* **181** (1998), 229-243.

[95] A. Constantin, V. S. Gerdjikov, R. I. Ivanov, *Inverse scattering transform for the Camassa-Holm equation, Inverse Problems* **22** (2006), 2197-2207.

[96] A. Constantin, D. Lannes, *The hydrodynamical relevance of the Camassa-Holm and Degasperis- Procesi equations*, Arch. Ration. Mech. Anal. **192** (2009), 165-186.

[97] A. Córdoba, D. Córdoba, F. Gancedo, *Interface evolution: Water waves in 2-D. Adv. Math.* **223** (2010), 120-173.

[98] A. Córdoba, D. Céordoba, F. Gancedo, *Interface evolution: The Hele-Shaw and Muskat problems*, Annals of Math. **173** (2011), 477-544.

[99] D. Coutand, S. Shkoller, *Well-posedness of the free-surface incompressible Euler equations with or without surface tension*, J. Amer. Math. Soc. **20** (2007), 829-930.

[100] D. Coutand, S. Shkoller, *On the finite-time splash and splat singularities for the 3-D free-surface Euler equations* (2012), preprint.

[101] W. Craig, *An existence theory for water waves and the Boussinesq and Korteweg-de Vries scaling limits*, Comm. Partial Differential Equations **10** (1985), 787-1003.

[102] W. Craig, M. D. Groves, *Hamiltonian long-wave approximations to the water-wave problem. Wave Motion* **19** (1994), 367-389.

[103] W. Craig, P. Guyenne, J. Hammack, D. Henderson, C. Sulem, *Solitary water wave interactions*, Phys. Fluids **18** (2006), 057106.

[104] W. Craig, P. Guyenne, H. Kalisch, *Hamiltonian long-wave expansions for free surfaces and interfaces*, Communication on Pure and Applied Mathematics **58** (2005), 1587-1641.

[105] W. Craig, P. Guyenne, D. Nicholls, C. Sulem, *Hamiltonian long-wave expansions for water waves over a rough bottom*, Proc. Royal Society A **461** (2005), 1-35.

[106] W. Craig, P. Guyenne, D. Nicholls, C. Sulem, *Hamiltonian Modulation Theory for Water Waves on Arbitrary Depth*, Proceedings of the Twenty-first (2011) International Offshore and Polar Engineering Conference Maui, Hawaii, USA, June 19-24, 2011.

[107] W. Craig, D. Lannes, C. Sulem, *Water waves over a rough bottom in the shallow water regime*, submitted.

[108] W. Craig, A.-M. Matei, *On the regularity of the Neumann problem for free surfaces with surface tension*, Proc. Amer. Math. Soc. **135** (2007), 2497-2504.

[109] W. Craig, D. P. Nicholls, *Traveling two and three dimensional capillary gravity water waves*, SIAM J. Math. Anal. **32** (2000), 323-359.

[110] W. Craig, C. Sulem, *Numerical simulation of gravity waves*, J. Comput. Phys **108** (1993), 73-83.

[111] W. Craig, C. Sulem, P.-L. Sulem, *Nonlinear modulation of gravity waves: A rigorous approach*, Nonlinearity **5** (1992), 497-522.

[112] W. Craig, U. Schanz, C. Sulem, *The modulational regime of three-dimensional water waves and the Davey-Stewartson system*, Ann. Inst. Henri Poincaré Anal. Non Linéaire **14** (1997), 615-667.

[113] A. D. D. Craik, *The origins of water wave theory*, Annu. Rev. Fluid Mech. **36** (2004), 1-28.

[114] C. M. Dafermos, *Hyperbolic conservation laws in continuum physics*, volume 325 of Grundlehren der Mathematischen Wissenschaften. Springer-Verlag, Berlin, third edition, 2010.

[115] R. A. Dalrymple, S. T. Grilli, J. T. Kirby, *Tsunamis and challenges for accurate modeling*, Oceanography **19** (2006), 142-151.

[116] A. Davey, K. Stewartson, *On three-dimensional packets of surface waves*, Proc. Roy. Soc. Lond. A **338** (1974), 101-110.

[117] M. Debiane and C. Kharif, *A new limiting form for steady periodic gravity waves with surface tension on deep water*, Phys. Fluids **8** (1996), 2780-82.

[118] A. Degasperis, D. Holm, A. Hone, *A new integrable equation with peakon solutions, Theor. Math. Phys.* **133** (2002), 1461-1472.

[119] O. Darrigol, *The spirited horse, the engineer, and the mathematician: Water waves in nineteenth-century hydrodynamics*, Arch. Hist. Exact Sci. **58** (2003), 21-95.

[120] A. Degasperis, M. Procesi, Asymptotic integrability, in *Symmetry and perturbation theory* (A. Degasperis & G. Gaeta, eds.), pp 23–37, World Scientific, Singapore, 1999.

[121] J. Deny, J.-L. Lions, *Les espaces de Beppo Levi*, Ann. Inst. Fourier Grenoble **5** (1953-54), 497-522.

[122] F. Dias, A. I. Dyachenko, V. E. Zakharove, *Theory of weakly damped free-surface flows: A new formulation based on potential flow solutions*, Physics Letters A **372** (2008), 1297-1302.

[123] F. Dias, C. Kharif, *Nonlinear gravity and capillary-gravity waves*, Annu. Rev. Fluid Mech. **31**, 301-346. Annual Reviews, Palo Alto, CA, 1999.

[124] M. W. Dingemans, *Comparison of computations with Boussinesq-like models and laboratory measurements*, Report H-1684.12, **32** Delft Hydraulics.

[125] M. W. Dingemans, *Water wave propagation over uneven bottoms. Part 1: Linear wave propagation; Part 2: Non-linear wave propagation*, World Scientific (1997).

[126] V. D. Djordjevic, L. G. Redekopp, *On two-dimensional packets of capillary-gravity waves*, J. Fluid Mech. **79** (1977), 703-714.

[127] A. R. van Dongeren, J. A. Battjes, T. T. Janssen, J. van Noorloos, K. Steenhauer, G. Steenbergen, A. J. H. M. Reniers, *Shoaling and shoreline dissipation of low frequency waves*, Journal of Geophysical Research **112** (2007).

[128] P. Donnat, J.-L. Joly, G. Métivier, J. Rauch, *Diffractive nonlinear geometric optics*, Sémin. Equ. Dériv., Partielles XVII, Ecole Polytechnique, Palaiseau, 1996, pp. 1-23.

[129] V. Dougalis, D. Mitsotakis, J.-C. Saut, *On some Boussinesq systems in two space dimensions: Theory and numerical analysis*, Math. Model. Num. Anal. **41** (2007), 825-854.

[130] P. G. Drazin, R. S. Johnson, *Solitons: An introduction*, Cambridge University Press, Cambridge, 1992.

[131] V. Duchêne, *Asymptotic shallow water models for internal waves in a two-fluid system with a free surface*, SIAM J. Math. Anal. **42** (2010), 2229-2260.

[132] V. Duchêne, *Boussinesq/Boussinesq systems for internal waves with a free surface, and the KdV approximation*, M2AN Math. Model. Numer. Anal. **46** (2011), 145-185.

[133] W.-P. Düll, *Validity of the Korteweg-de Vries approximation for the two-dimensional water wave problem in the arc length formulation*, Comm. Pure Appl. Math. **65** (2012), 381-429.

[134] W.-P. Düll, G. Schneider, C. E. Wayne, *Justification of the Nonlinear Schrödinger equation for the evolution of gravity driven 2D surface water waves in a canal of finite depth*, preprint.

[135] E. Dumas, D. Lannes, J. Szeftel, *Some variants of the focusing NLS equations. Derivation, justification and open problems*, in preparation.

[136] J. H. Duncan, H. Qiao, V. Philomin, A. Wenz, *Gentle spilling breakers: Crest profile evolution*, J. Fluid Mech. **379** (1999), 191-222.

[137] A. Dutrifoy, A. Majda, *The dynamics of equatorial long waves: A singular limit with fast variable coeffcients*, Commun. Math. Sci. **4** (2006), pp. 375-397.

[138] D. Dutykh, *Modélisation mathématique des tsunamis*, Thèse de doctorat de l'Ecole Normale Supérieure de Cachan (2007).

[139] D. Dutykh, F. Dias, *Dissipative Boussinesq equations*, C. R. Mécanique **335** (2007), 559-583.

[140] K. B. Dysthe, *Note on a modification to the nonlinear Schrödinger equation for application to deep water waves*, Proc. Royal Society A **369** (1979), 105-114.

[141] D. G. Ebin, *The equations of motion of a perfect fluid with free boundary are not well posed*, Comm. Partial Differential Equations **12** (1987), 1175-1201.

[142] G. Ebin, J. Marsden, *Groups of diffeomorphisms and the motion of an incompressible fluid*, Ann. Math. **92** (1970), 102-163.

[143] C. Eckart, *Variation principles of hydrodynamics*, Phys. Fluids **3** (1960), 421-427.

[144] J. Escher, T. Schlurmann, *On the recovery of the free surface from the pressure within periodic traveling water waves*, J. Nonlin. Math. Phys. **15** (2008), suppl. 2, 50-57.

[145] E. Falcon, C. Laroche, S. Fauve, *Observation of depression solitary waves on a thin fluid layer*, Phys. Rev. Lett. **89** (2002), 204501.

[146] J. Fan and T. Ozawa, *Regularity criterion for a Bona-Colin-Lannes system*, Nonlinear Anal. **71** (2009), 2634-2639.

[147] F. Fedele, D. Dutykh, *Hamiltonian form and solitary waves of the spatial Dysthe equations*, JETP Letters, **94** (2011), 840-844.

[148] A. S. Fokas, B. Fuchssteiner, *Symplectic structures, their Bäcklund transformation and hereditary symmetries*, *Physica D* **4** (1981), 821-831.

[149] I. Gallagher, L. Saint-Raymond, *On the influence of the Earth's rotation on geophysical flows*, Handbook of Mathematical Fluid Dynamics, Elsevier (S. Friedlander & D. Serre, eds.), 2006.

[150] I. Gallagher, L. Saint-Raymond, *Mathematical study of the betaplane model: Equatorial waves and convergence results*, Mém. Soc. Math. Fr. (2007).

[151] T. Gallay, G. Schneider, *KP description of unidirectional long waves. The model case.*, Proc. R. Soc. Edinb., Sect. A, Math. **131** (2001), 885-898.

[152] J. Garnier, R. Kraenkel, A. Nachbin, *Optimal Boussinesq model for shallow-water waves interacting with a microstructure*, Phys. Rev. E **76** (2007), 046311.

[153] J. Garnier, J. C. Muñoz Grajales, A.Nachbin, *Effective behavior of solitary waves over random topography*, Multiscale Model. Simul., **6** (2007) 995-1025.

[154] D. Gérard-Varet, E. Dormy, *On the ill-posedness of the Prandtl equation*, J. Amer. Math. Soc. **23** (2010), 591-609.

[155] J. F. GERBEAU, B. PERTHAME, *Derivation of Viscous Saint-Venant System for Laminar Shallow Water; Numerical Validation*, Discrete and Continuous Dynamical Systems, Ser. B **1** (2001), 89-102.
[156] P. GERMAIN, N. MASMOUDI, J. SHATAH, *Global solutions for the gravity water waves equation in dimension 3*, Ann. of Math. **175** (2012), 691-754.
[157] J.-M. GHIDAGLIA, J.-C. SAUT, *On the initial value problem for the Davey-Stewartson systems*, Nonlinearity **3** (1990), 475-506.
[158] J.-M. GHIDAGLIA, J.-C. SAUT, *Nonelliptic Schrödinger Equations*, J. Nonlinear Sci. **3** (1993), 169-195.
[159] J.-M. GHIDAGLIA, J.-C. SAUT, *Nonexistence of travelling wave solutions to nonelliptic nonlinear Schrödinger equations*, J. Nonlinear Sci. **6** (1996), 139-145.
[160] A. E. GILL, *Atmosphere-Ocean Dynamics*, volume 30 of International Geophysics Series. Academic Press, 1982.
[161] M. F. GOBBI, J. T. KIRBY, G. WEI, *A fully nonlinear Boussinesq model for surface waves. II. Extension to $O(kh)^4$*, J. Fluid Mech. **405** (2000), 181-210.
[162] E. GREEN, P. M. NAGHDI, *A derivation of equations for wave propagation in water of variable depth*, J. Fluid Mech. **78** (1976), 237.
[163] H. P. GREENSPAN, *The Theory of Rotating Fluids*, Cambridge University Press, London, 1968.
[164] E. GRENIER, *On the nonlinear instability of Euler and Prandtl equations*, Comm. Pure Appl. Math. **53** (2000), 1067-1091.
[165] E. VAN GROESEN, S. R. PUDJAPRASETYA, *Uni-directional waves over slowly varying bottom. I. Derivation of a KdV-type of equation*, Wave Motion **18** (1993), 345-370.
[166] Y. GUO, T. NGUYEN, *A note on the Prandtl boundary layers*, arXiv:1011.0130.
[167] J. HAMILTON, *Differential equations for long-period gravity waves on fluid of rapidly varying depth*, J. Fluid Mech. **83** (1977), 289-310.
[168] H. HASIMOTO, *Water waves*, Kagaku **40** (1970), 401-408 [Japanese].
[169] H. HASIMOTO, H. ONO, *Nonlinear modulation of gravity waves*, J. Phys. Soc. Japan **33** (1972), 805-811.
[170] K. R. HELFRICH, W. K. MELVILLE, *Long nonlinear internal waves*, Annual Review of Fluid Mechanics **38** (2006), 395-425.
[171] J. W. HERIVEL, *The derivation of the equations of motion of an ideal fluid by Hamilton's principle*, Proc. Cambridge Phil. Soc. **51** (1955), 344-349.
[172] K. HENDERSON, D. H. PEREGRINE, J. W. DOLD, *Unsteady water wave modulations: Fully nonlinear solutions and comparison with the nonlinear Schrödinger equation*, Wave Motion **29** (1999), 341-361.
[173] S. J. HOGAN, *The fourth-order evolution equation for deep-water gravity-capillary waves*, Proc. Royal Society A **402** (1985), 359-372.
[174] L. HÖRMANDER, *Lectures on nonlinear hyperbolic differential equations*, Mathématiques & Applications (Berlin) [Mathematics & Applications], **26**. Springer-Verlag, Berlin, 1997.
[175] T. HOU, J. S. LOWENGRUB, M.J. SHELLEY, *Removing the Stiffness from Interfacial Flows with Surface Tension*, J. Comput. Phys. **114** (1994), 312-338.
[176] B. HU, D. P. NICHOLLS, *Analyticity of Dirichlet-Neumann operators on Holder and Lipschitz domains*, SIAM J. Math. Anal. **37** (2005), 302-320.
[177] B. HU, D. P. NICHOLLS, *The Domain of Analyticity of Dirichlet-Neumann Operators*, Proceedings of the Royal Society of Edinburgh A **140** (2010), 367-389.
[178] H.-H. HWUNG, W.-S. CHIANG, S.-C. HSIAO, *Observations on the evolution of wave modulation*, Proc. Royal Society A **463** (2007), 85-112.
[179] T. IGUCHI, *A long wave approximation for capillary-gravity waves and an effect of the bottom*, Comm. Partial Differential Equations **32** (2007), 37-85.
[180] T. IGUCHI, *A long wave approximation for capillary-gravity waves and the Kawahara equation*, Bulletin of the Institute of Mathematics Academia Sinica **2** (2008), 179-220.
[181] T. IGUCHI, *A shallow water approximation for water waves*, J. Math. Kyoto Univ. **49** (2009), 13-55.
[182] T. IGUCHI, *A mathematical analysis of tsunami generation in shallow water due to seabed deformation*, submitted.
[183] S. ISRAWI, *Variable depth KDV equations and generalizations to more nonlinear regimes*, M2AN **44** (2010), 347-370.

[184] S. ISRAWI, *Large Time existence for 1D Green-Naghdi equations*, Nonlinear Analysis: Theory, Methods & Applications **74** (2011), 81-93.

[185] S. ISRAWI, *Derivation and analysis of a new 2D Green-Naghdi system*, Nonlinearity **23** (2010), 2889-2904.

[186] R. I. IVANOV, *On the integrability of a class of nonlinear dispersive wave equations, J. Nonlinear Math. Phys.* **12** (2005), 462-468.

[187] B. L. JENSEN, B. M. SUMER, J. FREDSE, *Turbulent oscillatory boundary layers at high Reynolds numbers*, J. Fluid Mech. **206** (1989), 265-297.

[188] R. S. JOHNSON, *On the development of a solitary wave moving over an uneven bottom*, Proc. Cambridge Philos. Soc. **73** (1973), 183-203.

[189] R. S. JOHNSON, *On the modulation of water waves in the neighbourhood of kh* $\sim$ 1.363, Proc. Roy. Soc. A **357** (1977), 131-141.

[190] R. S. JOHNSON, *A modern introduction to the mathematical theory of water waves*, Cambridge University Press, Cambridge, 1997.

[191] R. S. JOHNSON, *Camassa-Holm, Korteweg-de Vries and related models for water waves, J. Fluid Mech.* **457** (2002), 63-82.

[192] J.-L. JOLY, G. MÉTIVIER, J. RAUCH, *Diffractive nonlinear geometric optics with rectification*, Indiana Univ. Math. J. **47** (1998), 1167-1241.

[193] B. B. KADOMTSEV, V. I. PETVIASHVILI, *On the stability of solitary waves in weakly dispersive media*, Sov. Phys. Dokl. **15** (1971), 539-541.

[194] L. A. KALYAKIN, *Asymptotic decay of a one-dimensional wave packet in a nonlinear dispersive medium*, Math. USSR Sb. **60** (1988), 457-483.

[195] T. KAKUTANI, H. ONO, *Weak non-linear hydromagnetic waves in a cold collision-free plasma*, J. Phys. Soc. Japan **26** (1969), 1305-1318.

[196] T. KANO, *Une théorie trois-dimensionnelle des ondes de surface de l'eau et le développement de Friedrichs*, J. Math. Kyoto Univ. **26** (1986), 101-155 and 157-175.

[197] T. KANO, *L'équation de Kadomtsev-Petviashvili approchant les ondes longues de surface de l'eau en écoulement trois-dimensionnel.* Patterns and waves. Qualitative analysis of nonlinear differential equations. Stud. Math. Appl., **18**, 431-444. North-Holland, Amsterdam (1986).

[198] T. KANO, T. NISHIDA, *Sur les ondes de surface de leau avec une justification mathématique des équations des ondes en eau peu profonde*, J. Math. Kyoto Univ. **19** (1979), 335-370.

[199] T. KANO, T. NISHIDA, *A mathematical justification for Korteweg-de Vries equation and Boussinesq equation of water surface waves*, Osaka J. Math. **23** (1986), 389-413.

[200] T. KATO, *The Cauchy problem for quasi-linear symmetric hyperbolic systems*, Arch. Rational Mech. Anal **58** (1975), 181-205.

[201] T. KATO, *On the Korteweg-de Vries equation*, Manuscripta Math. **29** (1979), 89-99.

[202] T. KATO, G. PONCE, *Commutator estimates and the Euler and Navier-Stokes equations*, Comm. Pure Appl. Math. **41** (1988), 891-907.

[203] T. KAWAHARA, *Oscillatory solitary waves in dispersive media*, J. Phys. Soc. Japan **33** (1972), 260-264.

[204] C. E. KENIG, G. PONCE, L. VEGA, *Well-posedness and scattering results for the generalized Korteweg-de Vries equation via the contraction principle*, Comm. Pure Appl. Math. **46** (1993), 527-560.

[205] D. J. KAUP, *A higher-order water-wave equation and the method for solving it*, Progr. Theoret. Phys. **54** (1975), 396-408.

[206] N. KITA, J. SEGATA, *Well-posedness for the Boussinesq-type system related to the water wave*, Funkcial. Ekvac. **47** (2004), 329-350.

[207] R. KLEIN, A. MAJDA, *Systematic multi-scale models for the tropics*, J. Atmospheric Sci. **60** (2003), 173-196.

[208] C. KLEIN, J.-C. SAUT, *Numerical study of blow up and stability of solutions of generalized Kadomtsev-Petviashvili equations*, J. Nonlinear Science **22** (2012), 763-811.

[209] N. KOBAYASHI, E. A. KARJADI, B. D. JOHNSON, *Dispersion effects on longshore currents in surf zones*, J. Waterw. Port Coastal Ocean Eng. **123** (1997), 240-248.

[210] H. KOCH, J.-C. SAUT, *Local smoothing and local solvability for 3rd order dispersive equations*, SIAM Math. Anal. **38** (2007), 1528-1541.

[211] D. J. KORTEWEG, G. DE VRIES, London, Edinburgh Dublin Philos. Mag. J. Sci. **39** (1895), 422.

[212] J. L. LAGRANGE, *Mémoire sur la Théorie du Mouvement des Fluides*, Oeuvres de Lagrange, Gauthier-Villars, Paris, France (imprimé en 1867, J.A. Serret, Editeur), 1, 4, pp. 695-748, 1781.

[213] B. M. LAKE, H. C.YUEN, H. RUNGALDIER, W. E. FERGUSON, *Nonlinear deep-water waves: Theory and experiment. Part 2. Evolution of a continuous wave train*, J. Fluid Mech. **83** (1977) 49-74.

[214] H. LAMB, *Hydrodynamics*, Reprint of the 1932 sixth edition, Cambridge Univ. Press, 1993.

[215] D. LANNES, *Dispersive effects for nonlinear geometrical optics with rectification*, Asymptot. Anal. **18** (1998), 111-146.

[216] D. LANNES, *Secular growth estimates for hyperbolic systems, Journal of Differential Equations* **190** (2003), 466-503.

[217] D. LANNES, *Consistency of the KP Approximation*, Proceedings of the 4th International Conference on Dynamical Systems and Differential Equations, May 24-27, 2002, Wilmington, NC, USA, 517-525.

[218] D. LANNES, *Well-posedness of the water-waves equations*, J. Amer. Math. Soc. **18** (2005), 605-654.

[219] D. LANNES, *Sharp estimates for pseudo-differential operators with symbols of limited smoothness and commutators*, J. Funct. Anal. **232** (2006), 495-539.

[220] D. LANNES, *High-frequency nonlinear optics: From the nonlinear Schrödinger approximation to ultrashort-pulses equations*, Proc. Roy. Soc. Edinburgh Sect. A **141** (2011), 253-286.

[221] D. LANNES, *A stability criterion for two-fluid interfaces and applications*, Arch. Ration. Mech. Anal., to appear.

[222] D. LANNES, *Space time resonances*, Séminaire Bourbaki, 64ème année, 2011-2012, no 1053.

[223] D. LANNES, P. BONNETON, *Derivation of asymptotic two-dimensional time-dependent equations for surface water wave propagation*, Physics of Fluids **21** (2009).

[224] D. LANNES, J. RAUCH, *Validity of nonlinear geometric optics with times growing logarithmically*, Proc. Amer. Math. Soc. **129** (2001), 1087-1096.

[225] D. LANNES, J.-C. SAUT, *Weakly transverse Boussinesq systems and the KP approximation*, Nonlinearity **19** (2006), 2853-2875.

[226] P. S. LAPLACE, *Recherches sur plusieurs points du système du monde*, Mem. Acad. R. Sci. Paris (1978, 1979), reprinted in "Oeuvres," **9**, Gauthier-Villars, Paris, 1893.

[227] O. LE MÉTAYER, S. GAVRILYUK, S. HANK, *A numerical scheme for the Green-Naghdi model*, J. Comp. Phys. **229** (2010), 2034-2045.

[228] J. LENELLS, *Conservation laws of the Camassa-Holm equation, J. Phys. A* **38** (2005), 869-880.

[229] C. D. LEVERMORE, M. OLIVER, E. S. TITI, *Global Well-Posedness for the Lake Equations*, Physica D **98** (1996), 492-509.

[230] C. D. LEVERMORE, M. OLIVER, E. S. TITI, *Global well-posedness for models of shallow water in a basin with a varying bottom*, Indiana Univ. Math. J. **45** (1996), 479-510.

[231] T. D. LEVI-CIVITA, *Determination rigoureuse des ondes permanentes d'ampleur finie* (French), Math. Ann. **93** (1925), 264-314.

[232] Y. A. LI, *A shallow-water approximation to the full water wave problem*, Comm. Pure Appl. Math. **59** (2006), 1225-1285.

[233] Y. A. LI, *Linear stability of solitary waves of the Green-Naghdi equations*, Comm. Pure Appl. Math. **54** (2001), 501-536.

[234] Y. A. LI, *Hamiltonian structure and linear stability of solitary waves of the Green-Naghdi equations*, J. Nonlinear Math. Phys. **9** (2002), 99-105.

[235] M. J. LIGHTHILL, *Waves in Fluids*, Cambridge Univ. Press, London and New York (1978).

[236] H. LINDBLAD, *Well-posedness for the linearized motion of an incompressible liquid with free surface boundary*, Comm. Pure Appl. Math. **56** (2003), 153-197.

[237] H. LINDBLAD, *Well-posedness for the motion of an incompressible liquid with free surface boundary*, Ann. of Math. (2) **162** (2005), 109-194.

[238] J.-L. LIONS, R. TEMAM, S. WANG, *New formulations of the primitive equations of atmosphere and applications*, Nonlinearity **5** (1992), 237-288.

[239] E. LO, C. C. MEI, *A numerical study of water-wave modulation based on a higer-order nonlinear Schrödinger equation*, J. Fluid Mech. **150** (1985), 395-416.

[240] E. Y. Lo, C. C. Mei, *Slow evolution of nonlinear deep water waves in two horizontal directions: A numerical study*, Wave Motion **9** (1987), 245-259.

[241] M. S. Longuet-Higgins, *The instabilities of gravity waves of finite amplitude in deep water. I. Superharmonics*, Proc. Royal Society A **360** (1978), 471-488.

[242] M. S. Longuet-Higgins, *The instabilities of gravity waves of finite amplitude in deep water. II. Subharmonics*, Proc. Royal Society A **360** (1978), 489-505.

[243] J. C. Luke, *A variational principle for a fluid with a free surface*, J. Fluid Mech. **27** (1967), 395-397.

[244] J. H. Maddocks, R. L. Pego, *An unconstrained Hamiltonian formulation for incompressible fluid flow*, Comm. Math. Phys. **170** (1995), 207-217.

[245] P. A. Madsen, R. Murray, O. R. Sorensen, *A New Form of the Boussiensq Equations with Improved Linear Dispersion Characteristics*, Coastal Eng. **15** (1991), 371-388.

[246] P. A. Madsen, H. B. Bingham, H. Liu, *A new Boussinesq method for fully nonlinear waves from shallow to deep water*, J. Fluid Mech. **462** (2002), 1-30.

[247] A. Majda, *Introduction to PDEs and Waves for the Atmosphere and Ocean*, Courant Lecture Notes **9**, Amer. Math. Soc., 2003.

[248] N. Makarenko, A second long-wave approximation in the Cauchy Poisson problem, Dyn. Contin. Media **77** (1986), 5672 (in Russian).

[249] F. Marche, *Derivation of a new two-dimensional shallow water model with varying topography, bottom friction and capillary effects*, Eur. J. Mech. B/Fluids **26** (2007), 49-63.

[250] N. Masmoudi, *Examples of singular limits in hydrodynamics*, Chapter 3, 197269. Handbook of Differential Equations. Evolutionary Equations, volume 3 (C. M. Dafermos and E. Feireisl, eds.), North-Holland (2006), 652 pages.

[251] N. Masmoudi, F. Rousset, *Uniform regularity for the Navier-Stokes equation with Navier boundary condition*, preprint 2010, arXiv:1008.1678.

[252] N. Masmoudi, F. Rousset, *Uniform regularity and vanishing viscosity limit for the free surface Navier-Stokes equation*, preprint 2011.

[253] Y. Matsuno, *Nonlinear evolutions of surface gravity waves on fluid of finite depth, Phys. Rev. Lett.* **69** (1992), 609-611.

[254] Y. Matsuno, *Nonlinear evolution of surface gravity waves over an uneven bottom*, J. Fluid Mech. **249** (1993), 121-133.

[255] Y. Matsuno, *The N-soliton solution of the Degasperis-Procesi equation*, *Inverse Problems* **21** (2005), 2085-2101.

[256] C. C. Mei, *Resonant reflection of surface waves by bottom ripples*, J. Fluid Mech. **152** (1985), 315-335.

[257] C. C. Mei, B. Le Méhauté, *Note on the equations of long waves over an uneven bottom*, J. of Geophysical Research **71** (1966), 393-400.

[258] Y. Meyer, *Remarques sur un théorème de J.-M. Bony*, Rend. Circ. Mat. Palermo (2), 1981, suppl. 1-20.

[259] J. Miles, *On the Korteweg-de Vries equation for a gradually varying channel*, J. Fluid Mech. **91** (1979), 181-190.

[260] J. Miles, *The Korteweg-de Vries equation: A historical essay*, J. Fluid Mech. **106** (1981), 131-147.

[261] J. Miles, R. Salmon, *Weakly dispersive nonlinear gravity waves*, Journal of Fluid Mechanics **157** (1985), 519-531.

[262] M. Ming, Z. Zhang, *Well-posedness of the water-wave problem with surface tension*, J. Math. Pures Appliqués **92** (2009), 429-455.

[263] M. Ming, P. Zhang, Z. Zhang, *Large time well-posedness of the three dimensional capillary-gravity waves in the long wave regime*, Arch. Rational Mech. Anal. **204** (2012), 387-444.

[264] A. Nachbin, K. Sølna, *Apparent diffusion due to topographic microstructure in shallow waters*, Phys. Fluids, **15** (2002), 66-77.

[265] V. I. Nalimov, *The Cauchy-Poisson Problem* (in Russian). Dyn. Splosh. Sredy **18** (1974), 104-210.

[266] A. I. Nekrasov, *On steady waves*, Izv. Ivanovo-Voznesensk. Polit. Inst. **3** (1921), 52.

[267] A. C. Newell, J. V. Moloney, *Nonlinear optics* (Addison-Wesley, 1992).

[268] O. Nikodym, *Sur une classe de fonctions considérées dans l'étude du problème de Dirichlet*, Fundamenta Math. **21** (1933), 129-150.

[269] O. Nwogu, *Alternative form of Boussinesq equations for nearshore wave propagation*, J. Wtrwy., Port, Coast., and Oc. Engrg. **119** (1993), 616-638.

[270] L. V. Ovsjannikov, *To the shallow water theory foundation*, Arch. Mech. **26** (1974), 407-422.

[271] L. V. Ovsjannikov, *Cauchy problem in a scale of Banach spaces and its application to the shallow water theory justification*, Applications of methods of functional analysis to problems in mechanics, Lecture Notes in Math. **503**. Springer, Berlin, 1976.

[272] R. Pego, *Origin of the KdV Equation*, Notices Amer. Math. Soc. **45** (1997), p. 358.

[273] G. Ponce, J.-C. Saut, *Well-posedness for the Benney-Roskes/Zakharov-Rubenchik system*, Discrete and continuous dynamical systems **13** (2005), 1-16.

[274] L. Paumond, *A rigorous link between KP and a Benney-Luke equation*, Differ. Integral Equ. **16** (2003), 1039-1064.

[275] D. H. Peregrine, *Long waves on a beach*, Journal of Fluid Mechanics **27** (1967), 815-827.

[276] R. Rosales, G. Papanicolaou, *Gravity waves in a channel with a rough bottom*, Stud. Appl. Math. **68** (1983), 89-102.

[277] F. Rousset and N. Tzvetkov, *Transverse instability of the line solitary water-waves*, Invent.Math. **184** (2011), 257-388.

[278] F. Serre, *Contribution à l'étude des écoulements permanents et variables dans les canaux*, Houille Blanche **8** (1953), 374-388.

[279] J. S. Russell, *Experimental researches into the laws of certain hydrodynamical phenomena that accompany the motion of floating bodies, and have not perviously been reduced into conformity with the known laws of the resistance of fluids*, Royal Society of Edinburgh, Transactions **14** (1839), 47-109.

[280] J. S. Russell, *Report on Waves*, Report of the fourteenth meeting of the British Association for the Advancement of Science, York, September 1844. London: John Murray (1845), 311-390, Plates XLVII-LVII.

[281] A. B. de Saint-Venant, *Théorie du mouvement non permanent des eaux, avec application aux crues des rivières et à lintroduction des marées dans leur lit*, C. R. Acad. Sc. Paris **73** (1871), 147154.

[282] A. B. de Saint-Venant, *Sur la houle et le clapotis*, C. R. Acad. Sc. Paris **73** (1871), 521-528, 589-593.

[283] M. Sammartino, R. Caflisch, *Zero viscosity limit for analytic solutions, of the Navier-Stokes equation on a half-space. I. Existence for Euler and Prandtl equations*, Comm. Math. Phys. **192** (1998), 433-461.

[284] J.-C. Saut, *Remarks on the generalized Kadomtsev-Petviashvili equations*, Indiana Univ. Math. J. **42** (1993), 1011-1026.

[285] J.-C. Saut, N. Tzvetkov, *Global well-posedness for the KP-BBM equations*, AMRX Appl. Math. Res. Express 2004, 1-16.

[286] J.-C. Saut, N. Tzvetkov, *The Cauchy problem for the fifth order KP equations*, J. Math.Pures Appl. **79** (2000), 307-338.

[287] J.-C. Saut, L. Xu, *The Cauchy problem on large time for surface waves Boussinesq systems*, J. Math. Pures Appl. **97** (2012), 635-662.

[288] G. Schneider, *The long wave limit for a Boussinesq equation*, SIAM J. Appl. Math. **58** (1998), 1237-1245.

[289] G. Schneider, *Justification of modulation equations for hyperbolic systems via normal forms*, Nonlin. Diff. Eqns Applic. **5** (1998), 69-82.

[290] G. Schneider, *Approximation of the Korteweg-de Vries equation by the nonlinear Schrödinger equation*, J. Differential Equations **147** (1998), 333-354.

[291] G. Schneider, C. E. Wayne, *The long-wave limit for the water wave problem I. The case of zero surface tension*, Communication on Pure and Applied Mathematics **53** (2000), 1475-1535.

[292] G. Schneider, C. E. Wayne, *The rigorous approximation of long-wavelength capillary-gravity waves*, Arch. Ration. Mech. Anal. **162** (2002), 247-285.

[293] G. Schneider, C. E. Wayne *Justification of the NLS approximation for a quasilinear water wave model*, J. Differential Equations **251** (2011), 238-269.

[294] L. Schwartz, *Théorie des distributions (French)*, Publications de l'Institut de Mathématique de l'Université de Strasbourg, No. IX-X. Nouvelle édition, entièrement corrigée, refondue et augmentée. Hermann, Paris 1966 xiii+420 pp.

[295] B. SCHWEIZER, *On the three-dimensional Euler equations with a free boundary subject to surface tension*, Ann. Inst. H. Poincaré Anal. Non Linéaire **22** (2005), 753-781.

[296] F. J. SEABRA-SANTOS, D. P. RENOUARD, A. M. TEMPERVILLE, *Numerical and experimental study of the transformation of a solitary wave over a shelf or isolated obstacle*, J. Fluid Mech. **176** (1987), 117.

[297] J. SHATAH, C. ZENG, *Geometry and a priori estimates for free boundary problems of the Euler equation*, Comm. Pure Appl. Math. **61** (2008), 698-744.

[298] J. SHATAH, C. ZENG, *A priori estimates for fluid interface problems*, Comm. Pure Appl. Math. **61** (2008), 848-876.

[299] , J. SHATAH, C. ZENG, *Local well-posedness for fluid interface problems*, Arch. Rational Mech. Anal. **199** (2011), 653-705.

[300] M. SHINBROT, *The initial value problem for surface waves under gravity, I. The simplest case*, Indiana U. Math. J. **25** (1976), 281-300.

[301] A. SHNIRELMAN, *The geometry of the group of diffeomorphisms and the dynamics of an ideal incompressible fluid*, Mat. Sb. (N.S.) **128** (1985), 82-109, 144.

[302] G. G. STOKES, *On the theory of oscillatory waves*, Cambridge Trans. **8** (1847), 441-473.

[303] D. D. STRUIK, *Determination rigoureuse des ondes irrotationelles périodiques dans un canal à profondeur finie*, (French) Math. Ann. **95** (1926), 595-634.

[304] M.-Y. SU, M. BERGIN, P. MARLER, R. MYRICK, *Experiments on nonlinear instabilities and evolution of steep gravity-wave trains*, J. Fluid Mech. **124** (1982), 45-72.

[305] C. H. SU, C. S. GARDNER, *Korteweg-de Vries equation and generalizations. III. Derivation of the Korteweg-de Vries equation and Burgers equation*, J. Math. Phys. **10** (1969), 536.

[306] C. SULEM, P.-L. SULEM, *Finite time analyticity for the two- and three-dimensional Rayleigh-Taylor instability*, Trans. Amer. Math. Soc. **287** (1981), 127-160.

[307] C. SULEM, P.-L. SULEM, *The Nonlinear Schrödinger Equation: Self-Focusing and Wave Collapse*, Applied Mathematical Sciences, Volume 139. Springer, 1999.

[308] G. TAYLOR, *The instability of liquid surfaces when accelerated in a direction perpendicular to their planes. I.*, Proc. Roy. Soc. London. Ser. A. **201** (1950), 192-196.

[309] M. TAYLOR, *Commutator estimates*, Proc. Amer. Math. Soc. **131** (2003), 1501-1507.

[310] M. E. TAYLOR, *Partial Differential Equations II. Qualitative Studies of Linear Equations*, volume 116 of Applied Mathematical Sciences, Springer, 1997.

[311] M. E. TAYLOR, *Partial differential equations. III. Nonlinear Equations*, volume 117 of Applied Mathematical Sciences. Springer-Verlag, New York, 1997.

[312] R. TEMAM, M. ZIANE, *Some mathematical Problems in Geophysical Fluid Dynamics*, Handbook of Mathematical Fluid Dynamics, vol. III (S. Friedlander & D. Serre, eds., 2004).

[313] M. TISSIER, *Etude numérique de la transformation des vagues en zone littorale, de la zone de levée aux zones de surf et de jet de rive*, Thèse de l'Université Bordeaux 1, no. 4437 (2011).

[314] J. F. TOLAND, *On the existence of a wave of greatest height and Stokes's conjecture*, Proc. Roy. Soc. London Ser. A **363** (1978), 469-485.

[315] N. TOTZ, S. WU, *A Rigorous Justification of the Modulation Approximation to the 2D Full Water Wave Problem*, Comm. Math. Phys. **310** (2012), 817-883.

[316] F. TRÈVES, *Introduction to pseudodifferential and Fourier integral operators. Vol. 1. Pseudodifferential operators*, The University Series in Mathematics. Plenum Press, New York-London, 1980.

[317] S. UKAI, *Local solutions of the Kadomtsev-Petviashvili equation*, J. Fac. Sci. Univ. Tokyo IA (Math) **36** (1989), 193-209.

[318] J.-P. VILA, *A two moments closure of shallow water type for gravity driven flows*, preprint.

[319] E. WAHLEN, *A Hamiltonian formulation of water waves with constant vorticity*, Letters in Mathematical Physics **79** (2007), 303-315.

[320] C. E. WAYNE AND D. WRIGHT, *Higher order modulations equations for a Boussinesq equation. SIAM Journal of Applied Dynamical Systems* **1** (2002), 271-302.

[321] G. WEI, J. T. KIRBY, S. T. GRILLI, R. SUBRAMANYA, *A fully nonlinear Boussinesq model for surface waves. Part 1. Highly nonlinear unsteady waves*, Journal of Fluid Mechanics **294** (1995), 71-92.

[322] G. B. WHITHAM, *Variational methods and applications to water waves*, Proc. R. Soc. Lond., Ser. A **299** (1967), 6-25.

[323] G. B. WHITHAM, *Non-linear dispersion of water waves*, J. Fluid Mech. **27** (1967), 399-412.

[324] G. B. WHITHAM, *Linear and Nonlinear Waves*. Wiley, New York.

[325] J. D. WRIGHT, *Corrections to the KdV approximation for water waves*, SIAM J. Math. Anal. **37** (2005), 1161-1206.

[326] S. WU, *Well-posedness in Sobolev spaces of the full water wave problem in 2-D*, Invent. Math. **130** (1997), 39-72.

[327] S. WU, *Well-posedness in Sobolev spaces of the full water wave problem in 3-D*, J. Amer. Math. Soc., **12** (1999), 445-495.

[328] S. WU, *Almost global well-posedness of the 2-D full Water Wave Problem*, Invent. Math. **177** (2009), 45-135.

[329] S. WU, *Global well-posedness of the 3-D full water wave problem*, Invent. Math. **184** (2011), 125-220.

[330] H. YOSIHARA, *Gravity waves on the free surface of an incompressible perfect fluid of finite depth*, RIMS Kyoto **18** (1982), 49-96.

[331] H. YOSIHARA, *Capillary-gravity waves for an incompressible ideal fluid*, J. Math. Kyoto Univ. **23** (1983), 649-694.

[332] H. C. YUEN, B. M. LAKE, *Nonlinear deep water waves: Theory and experiment*, Phys. Fluids **18** (1975), 956-960.

[333] V. E. ZAKHAROV, *Stability of periodic waves of finite amplitude on the surface of a deep fluid*, J. Applied Mech. Tech. Phys. **9** (1968), 190-194.

[334] V. E. ZAKHAROV, A. M. RUBENCHIK, *Nonlinear interaction between high and low frequency waves*, Prikl. Mat. Techn. Fiz. **5** (1972), 84-98 (in Russian).

[335] V. E. ZAKHAROV, E. I. SCHULMAN, *Integrability of nonlinear systems and perturbation theory*. What is integrability?, (V. E. Zakharov, ed.), 185-250, Springer Series on Nonlinear Dynamics, Springer-Verlag, 1991.

[336] J. A. ZELT, *The run-up of nonbreaking and breaking solitary waves*, Coastal Eng. **15** (1991), 205-246.

[337] P. ZHANG, Z. ZHANG, *On the local wellposedness of 3-d water wave problem with vorticity*, Science in China Series A: Mathematics **50** (2007), 1065-1077.

[338] P. ZHANG, Z. ZHANG, *On the free boundary problem of three-dimensional incompressible Euler equations*, Comm. Pure Appl. Math. **61** (2008), 877-940.

Index

Selected Published Titles in This Series

188 **David Lannes,** The Water Waves Problem, 2013
187 **Nassif Ghoussoub and Amir Moradifam,** Functional Inequalities: New Perspectives and New Applications, 2013
186 **Gregory Berkolaiko and Peter Kuchment,** Introduction to Quantum Graphs, 2013
185 **Patrick Iglesias-Zemmour,** Diffeology, 2013
184 **Frederick W. Gehring and Kari Hag,** The Ubiquitous Quasidisk, 2012
183 **Gershon Kresin and Vladimir Maz'ya,** Maximum Principles and Sharp Constants for Solutions of Elliptic and Parabolic Systems, 2012
182 **Neil A. Watson,** Introduction to Heat Potential Theory, 2012
181 **Graham J. Leuschke and Roger Wiegand,** Cohen-Macaulay Representations, 2012
180 **Martin W. Liebeck and Gary M. Seitz,** Unipotent and Nilpotent Classes in Simple Algebraic Groups and Lie Algebras, 2012
179 **Stephen D. Smith,** Subgroup complexes, 2011
178 **Helmut Brass and Knut Petras,** Quadrature Theory, 2011
177 **Alexei Myasnikov, Vladimir Shpilrain, and Alexander Ushakov,** Non-commutative Cryptography and Complexity of Group-theoretic Problems, 2011
176 **Peter E. Kloeden and Martin Rasmussen,** Nonautonomous Dynamical Systems, 2011
175 **Warwick de Launey and Dane Flannery,** Algebraic Design Theory, 2011
174 **Lawrence S. Levy and J. Chris Robson,** Hereditary Noetherian Prime Rings and Idealizers, 2011
173 **Sariel Har-Peled,** Geometric Approximation Algorithms, 2011
172 **Michael Aschbacher, Richard Lyons, Stephen D. Smith, and Ronald Solomon,** The Classification of Finite Simple Groups, 2011
171 **Leonid Pastur and Mariya Shcherbina,** Eigenvalue Distribution of Large Random Matrices, 2011
170 **Kevin Costello,** Renormalization and Effective Field Theory, 2011
169 **Robert R. Bruner and J. P. C. Greenlees,** Connective Real K-Theory of Finite Groups, 2010
168 **Michiel Hazewinkel, Nadiya Gubareni, and V. V. Kirichenko,** Algebras, Rings and Modules, 2010
167 **Michael Gekhtman, Michael Shapiro, and Alek Vainshtein,** Cluster Algebras and Poisson Geometry, 2010
166 **Kyung Bai Lee and Frank Raymond,** Seifert Fiberings, 2010
165 **Fuensanta Andreu-Vaillo, José M. Mazón, Julio D. Rossi, and J. Julián Toledo-Melero,** Nonlocal Diffusion Problems, 2010
164 **Vladimir I. Bogachev,** Differentiable Measures and the Malliavin Calculus, 2010
163 **Bennett Chow, Sun-Chin Chu, David Glickenstein, Christine Guenther, James Isenberg, Tom Ivey, Dan Knopf, Peng Lu, Feng Luo, and Lei Ni,** The Ricci Flow: Techniques and Applications: Part III: Geometric-Analytic Aspects, 2010
162 **Vladimir Maz'ya and Jürgen Rossmann,** Elliptic Equations in Polyhedral Domains, 2010
161 **Kanishka Perera, Ravi P. Agarwal, and Donal O'Regan,** Morse Theoretic Aspects of p-Laplacian Type Operators, 2010
160 **Alexander S. Kechris,** Global Aspects of Ergodic Group Actions, 2010
159 **Matthew Baker and Robert Rumely,** Potential Theory and Dynamics on the Berkovich Projective Line, 2010
158 **D. R. Yafaev,** Mathematical Scattering Theory, 2010

For a complete list of titles in this series, visit the AMS Bookstore at **www.ams.org/bookstore/survseries/**.